Lecture Notes in Mathematics

1931

Editors:
J.-M. Morel, Cachan
F. Takens, Groningen
B. Teissier, Paris

FONDAZIONE CIME

ROBERTO CONTI

CENTRO INTERNAZIONALE MATEMATICO ESTIVO
INTERNATIONAL MATHEMATICAL SUMMER CENTER

C.I.M.E. means Centro Internazionale Matematico Estivo, that is, International Mathematical Summer Center. Conceived in the early fifties, it was born in 1954 and made welcome by the world mathematical community where it remains in good health and spirit. Many mathematicians from all over the world have been involved in a way or another in C.I.M.E.'s activities during the past years.

So they already know what the C.I.M.E. is all about. For the benefit of future potential users and co-operators the main purposes and the functioning of the Centre may be summarized as follows: every year, during the summer, Sessions (three or four as a rule) on different themes from pure and applied mathematics are offered by application to mathematicians from all countries. Each session is generally based on three or four main courses (24−30 hours over a period of 6−8 working days) held from specialists of international renown, plus a certain number of seminars.

A C.I.M.E. Session, therefore, is neither a Symposium, nor just a School, but maybe a blend of both. The aim is that of bringing to the attention of younger researchers the origins, later developments, and perspectives of some branch of live mathematics.

The topics of the courses are generally of international resonance and the participation of the courses cover the expertise of different countries and continents. Such combination, gave an excellent opportunity to young participants to be acquainted with the most advance research in the topics of the courses and the possibility of an interchange with the world famous specialists. The full immersion atmosphere of the courses and the daily exchange among participants are a first building brick in the edifice of international collaboration in mathematical research.

C.I.M.E. Director
Pietro ZECCA
Dipartimento di Energetica "S. Stecco"
Università di Firenze
Via S. Marta, 3
50139 Florence
Italy
e-mail: zecca@unifi.it

C.I.M.E. Secretary
Elvira MASCOLO
Dipartimento di Matematica
Università di Firenze
viale G.B. Morgagni 67/A
50134 Florence
Italy
e-mail: mascolo@math.unifi.it

For more information see CIME's homepage: http://www.cime.unifi.it

CIME's activity is supported by:

− Istituto Nazionale di Alta Mathematica "F. Severi"
− Ministero dell'Istruzione, dell'Università e delle Ricerca
− Ministero degli Affari Esteri, Direzione Generale per la Promozione e la Cooperazione, Ufficio V

Michael Cowling · Edward Frenkel
Masaki Kashiwara · Alain Valette
David A. Vogan, Jr. · Nolan R. Wallach

Representation Theory and Complex Analysis

Lectures given at the
C.I.M.E. Summer School
held in Venice, Italy
June 10–17, 2004

Editors: Enrico Casadio Tarabusi
        Andrea D'Agnolo
        Massimo Picardello

 Springer

FONDAZIONE
CIME
ROBERTO CONTI

Enrico Casadio Tarabusi

Dipartimento di Matematica "G. Castelnuovo"
Sapienza Università di Roma
Piazzale A. Moro 2
00185 Roma, Italy
casadio@mat.uniroma1.it

Michael Cowling

School of Mathematics
University of New South Wales
UNSW Sydney 2052, Australia
michaelc@maths.unsw.edu.au

Andrea D'Agnolo

Dipartimento di Matematica Pura ed Applicata
Università degli Studi di Padova
Via Trieste, 63
35121 Padova, Italy
dagnolo@math.unipd.it

Edward Frenkel

Department of Mathematics
University of California
Berkeley, CA 94720, USA
frenkel@math.berkeley.edu

Masaki Kashiwara

Research Institute for Mathematical Sciences
Kyoto University
Kyoto 606-8502, Japan
masaki@kurims.kyoto-u.ac.jp

Massimo Picardello

Dipartimento di Matematica
Università di Roma "Tor Vergata"
Via della Ricerca Scientifica
00133 Roma, Italy
picard@mat.uniroma2.it

Alain Valette

Institut de Mathématiques
University of Neuchâtel
11, Rue Emile Argand, BP 158
2009 Neuchâtel, Switzerland
alain.valette@unine.ch

David A. Vogan, Jr.

Department of Mathematics
Massachusetts Institute of Technology
Cambridge, Massachusetts 02139, USA
dav@math.mit.edu

Nolan R. Wallach

Department of Mathematics
University of California
San Diego, La Jolla
CA 92093-0112, USA
nwallach@ucsd.edu

ISBN: 978-3-540-76891-3   e-ISBN: 978-3-540-76892-0
DOI: 10.1007/978-3-540-76892-0

Lecture Notes in Mathematics ISSN print edition: 0075-8434
ISSN electronic edition: 1617-9692

Library of Congress Control Number: 2007942546

Mathematics Subject Classification (2000): 22EXX, 43AXX, 17BXX, 58GXX, 81P68

Cover design: *design & production* GmbH, Heidelberg

Printed on acid-free paper

9 8 7 6 5 4 3 2 1

springer.com

Preface

This volume collects the notes of six series of lectures given on the occasion of the CIME session *Representation Theory and Complex Analysis* held in Venice on July 10–17, 2004. We thank Venice International University for its hospitality at the beautiful venue of San Servolo island.

Our aim in organizing this meeting was to present the audience with a wide spectrum of recent results on the subject of the title, ranging from topics with an analytical flavor, to more algebraic or geometric oriented ones, without neglecting interactions with other domains, such as quantum computing.

Two papers present a general introduction to ideas and properties of analysis on semi-simple Lie groups and their unitary representations. MICHAEL COWL-ING presents a panorama of various interactions between representation theory and harmonic analysis on semisimple groups and symmetric spaces. Unexpected phenomena occur in this context, as for instance the Kunze–Stein property, that reveal a dramatic difference between these groups and group actions and the classical amenable group (an extension of abelian groups). Results of this type are strongly related to the vanishing of coefficients of unitary representations. Complementarily, ALAIN VALETTE recalls the notion of amenability and investigates its relations with vanishing of coefficients of unitary representations of semi-simple groups and with ergodic actions. He applies these ideas to show another surprising property of representations of semi-simple groups and their lattices, namely Margulis' super-rigidity.

Three papers deal in full detail with the hard analysis of semisimple group representations. Ideally, this analysis could be split into representations of real groups or complex groups, or of algebraic groups over local fields. A deep account of the interaction between the real and complex world is given by MASAKI KASHIWARA, whose paper studies the relations between the representation theory of real semisimple Lie groups and the (microlocal) geometry of the flag manifolds associated with the corresponding complex algebraic groups. These results, a considerable part of which are joint work with W. Schmid, were announced some years ago, and are published here in complete form for the first time. DAVID VOGAN expresses unitary representations of real or complex semi-simple groups using tools of complex analysis, such as minimal globalizations realized on Dolbeault cohomology with compact supports. EDWARD FRENKEL describes the geometric Langlands correspondence for complex algebraic curves, concentrating on the ramified case where a finite number of regular singular points is allowed.

Finally, NOLAN WALLACH illustrates briefly a surprising application that could be relevant for the future of computing and its complexity: his paper studies how representation theory is related to quantum computing, focusing attention in particular on the study of qubit entanglement.

We wish to thank all the lecturers for the excellence of their live and written contributions, as well as the many participants from all age ranges and parts of the world, who created a very pleasant working atmosphere.

Roma and Venezia, November 2006   *Enrico Casadio Tarabusi*
Andrea D'Agnolo
Massimo Picardello

Contents

Ramifications of the Geometric Langlands Program

**Equivariant Derived Category and Representation of Real
Semisimple Lie Groups**
Masaki Kashiwara 137

Applications of Representation Theory to Harmonic Analysis of Lie Groups (and Vice Versa)

Michael Cowling

School of Mathematics, University of New South Wales, UNSW Sydney 2052, Australia
michaelc@maths.unsw.edu.au

These notes began as lectures that I intended to deliver in Edinburgh in April, 1999. Unfortunately I was not able to leave Australia at the time. Since then there has been progress on many of the topics, some of which is reported here, and I have added another lecture, on uniformly bounded representations, so that these notes are expanded on the original version in several ways.

I have tried to make these notes an understandable introduction to the subject for mathematicians with little experience of analysis on Lie groups or Lie theory. I aimed to present a wide panorama of different aspects of harmonic analysis on semisimple groups and symmetric spaces, and to try to illuminate some of the links between these aspects; I may well not have succeeded in this aim. Many readers will find much of what is written here to be elementary, and others may well disagree with my perspective. I apologise in advance to both the neophytes for whom my outline is too sketchy and to the experts for whom these notes are worthless.

I had hoped to produce an extensive bibliography, but I have not found the time to do so. Consequently I must bear the responsibility for the many omissions of important references in the subject.

Whoever wishes to delve into this subject more deeply will need a more complete introduction. There are many possibilities; the books of S. Helgason [59, 60, 62] and of A.W. Knapp [71] come to mind immediately as essential reading.

1 Basic Facts of Harmonic Analysis on Semisimple Groups and Symmetric Spaces

I will deal with noncompact classical algebraic semisimple Lie groups, such as $SO(p, q)$, $SU(p, q)$, $Sp(p, q)$, $SL(n, \mathbb{R})$, $SL(n, \mathbb{C})$, and $SL(n, \mathbb{H})$. The definitions of these may be found in [59, pp. 444–447] or [71, pp. 3–6].

All noncompact algebraic semisimple Lie groups have various standard subgroups and decompositions. I begin by describing these, then describe families of unitary representations parametrised by representations of some of these subgroups. Finally, I discuss the Plancherel formula. The fact that most of the important representations are parametrised by representations of subgroups allows arguments involving induction on the rank of the group.

1.1 Structure of Semisimple Lie Algebras

First, fix a *Cartan involution* θ of the Lie algebra $\mathfrak{g}$ of the group G, and write $\mathfrak{k}$ and $\mathfrak{p}$ for the $+1$ and -1 eigenspaces of θ. Then $\mathfrak{k}$ is a maximal compact subalgebra of $\mathfrak{g}$, and $\mathfrak{p}$ is a subspace; $[X, Y] \in \mathfrak{k}$ for all $X, Y \in \mathfrak{p}$. Since θ is an involution, we have the Cartan decomposition of the Lie algebra:

$$\mathfrak{g} = \mathfrak{k} \oplus \mathfrak{p}.$$

In this and future formulae about the Lie algebra, $\oplus$ means "vector space direct sum". All Cartan involutions are conjugate in the group of Lie algebra automorphisms of $\mathfrak{g}$, which is a finite extension of the group generated by $\{\exp \operatorname{ad} X : X \in \mathfrak{g}\}$. The Cartan involution θ extends to an automorphism Θ of the group G, whose fixed point set is a maximal compact subgroup K of G.

Next choose a maximal subalgebra of $\mathfrak{p}$; this is abelian, and is denoted by $\mathfrak{a}$. All such subalgebras are conjugate under K. Let $\operatorname{ad}(X)$ denote the derivation $Y \mapsto [X, Y]$ of $\mathfrak{g}$. Then the Killing form B, given by

$$B(X, Y) = \operatorname{tr}(\operatorname{ad}(X) \operatorname{ad}(Y)) \qquad \forall X, Y \in \mathfrak{g},$$

gives rise to an inner product on $\mathfrak{a}$:

$$(X, Y)_B = -B(X, \theta Y) \qquad \forall X, Y \in \mathfrak{g},$$

which gives rise to a dual inner product, denoted in the same way, on $\mathfrak{a}^*$, which in turn extends to a bilinear form on $\mathfrak{a}_\mathbb{C}$, also denoted in the same way.

The third step in the description and construction of the various special subalgebras of $\mathfrak{g}$ and corresponding subgroups of G is to decompose $\mathfrak{g}$ as a direct sum of *root spaces* $\mathfrak{g}_\alpha$ and a subalgebra $\mathfrak{g}_0$. Simultaneously diagonalise the operators $\operatorname{ad}(H)$, for H in $\mathfrak{a}$. For α in the real dual $\mathfrak{a}^*$ of $\mathfrak{a}$ (that is, $\mathfrak{a}^* = \operatorname{Hom}_\mathbb{R}(\mathfrak{a}, \mathbb{R})$), define

$$\mathfrak{g}_\alpha = \{X \in \mathfrak{g} : [H, X] = \alpha(H)X \quad \forall H \in \mathfrak{a}\}.$$

For most α in $\mathfrak{a}^*$, $\mathfrak{g}_\alpha = \{0\}$, but when $\alpha = 0$, then $\mathfrak{a} \subseteq \mathfrak{g}_0$, so $\mathfrak{g}_0 \neq \{0\}$. There are finitely many nonzero α in $\mathfrak{a}^*$ for which $\mathfrak{g}_\alpha \neq \{0\}$; these α are called the *real roots* of $(\mathfrak{g}, \mathfrak{a})$, and the set thereof is written Σ. This set is a *root system*, a highly symmetric subset of $\mathfrak{a}^*$. Because $\mathfrak{g}_0$ is θ-stable,

$$\mathfrak{g}_0 = (\mathfrak{g}_0 \cap \mathfrak{k}) \oplus (\mathfrak{g}_0 \cap \mathfrak{p}) = \mathfrak{m} \oplus \mathfrak{a},$$

say, where $\mathfrak{m}$ is the subalgebra of $\mathfrak{k}$ of elements which commute with $\mathfrak{a}$. Using the fact that $\mathrm{ad}(H)$ is a derivation of $\mathfrak{g}$ for each H in $\mathfrak{a}$, it is easy to check that

(1.1) $$[\mathfrak{g}_\alpha, \mathfrak{g}_\beta] \subseteq \mathfrak{g}_{\alpha+\beta}.$$

In particular, $\mathfrak{g}_0$ is a subalgebra, and $\mathfrak{g}_\alpha$ and $\mathfrak{g}_\beta$ commute when $\mathfrak{g}_{\alpha+\beta} = \{0\}$. Clearly

$$\mathfrak{g} = \mathfrak{g}_0 \oplus \sum_{\alpha \in \Sigma}^{\oplus} \mathfrak{g}_\alpha.$$

Now order the roots. The hyperplanes $\{H \in \mathfrak{a} : \alpha(H) = 0\}$, for α in Σ, divide $\mathfrak{a}$ into finitely many connected open cones, known as Weyl chambers. Pick one of these (arbitrarily) and fix it; it is called the positive Weyl chamber, and written $\mathfrak{a}^+$. A root α is now said to be *positive* or *negative* as $\alpha(H) > 0$ or $\alpha(H) < 0$ for all H in $\mathfrak{a}^+$. Write Σ^+ for the set of positive roots; then $\Sigma = \Sigma^+ \cup -(\Sigma^+)$. For some roots α and real numbers t, $t\alpha$ is also a root; the possibilities are that $t = \pm 1$ (this always happens), $t = \pm 1/2$ or $t = \pm 2$ (these last four possibilities may or may not occur). If $(1/2)\alpha$ is not a root, then α is said to be *indivisible*; denote by Σ_0^+ the set of indivisible positive roots. We can now define some more important subalgebras: let

$$\mathfrak{n} = \sum_{\alpha \in \Sigma^+} \mathfrak{g}_\alpha \qquad \text{and} \qquad \bar{\mathfrak{n}} = \sum_{\alpha \in \Sigma^+} \mathfrak{g}_{-\alpha};$$

it is easy to deduce from formula (1.1) that $\mathfrak{n}$ and $\bar{\mathfrak{n}}$ are *nilpotent* subalgebras of $\mathfrak{g}$. Define ρ by the formula

$$\rho(H) = \frac{1}{2} \mathrm{tr}(\mathrm{ad}(H)|_{\mathfrak{n}}) \qquad \forall H \in \mathfrak{a};$$

then $\rho = (1/2) \sum_{\alpha \in \Sigma^+} \dim(\mathfrak{g}_\alpha)\, \alpha$. We now have the ingredients for two more decompositions of $\mathfrak{g}$: the Iwasawa decomposition and the Bruhat decomposition, written

$$\mathfrak{g} = \mathfrak{k} \oplus \mathfrak{a} \oplus \mathfrak{n} \qquad \text{and} \qquad \mathfrak{g} = \bar{\mathfrak{n}} \oplus \mathfrak{m} \oplus \mathfrak{a} \oplus \mathfrak{n}.$$

The proof of the second (Bruhat) decomposition is immediate. For the first (Iwasawa) decomposition, note that if $X \in \mathfrak{g}_\alpha$, then $\theta X \in \mathfrak{g}_{-\alpha}$, so that, if $X \in \bar{\mathfrak{n}}$, then

$$X = (X + \theta X) - \theta X \in \mathfrak{k} \oplus \mathfrak{n}.$$

1.2 Decompositions of Semisimple Lie Groups

At the group level, there are similar decompositions (usually known as factorisations in undergraduate linear algebra courses). Let K, A, N and $\overline{N}$ denote the connected subgroups of G with Lie algebras $\mathfrak{k}$, $\mathfrak{a}$, $\mathfrak{n}$ and $\overline{\mathfrak{n}}$, and let A^+ and $\overline{A}^+$ be the subsemigroup $\exp(\mathfrak{a}^+)$ of A and its closure. Let M and M' be the centraliser and normaliser of $\mathfrak{a}$ in K. Then both M and M' have $\mathfrak{m}$ as their Lie algebra. The group M' is never connected, while M is connected in some examples and is not in others. However, M'/M is always finite. In fact, the adjoint action Ad of M' on $\mathfrak{a}$ induces an isomorphism of M'/M with a finite group of orthogonal transformations of $\mathfrak{a}$, generated by reflections. This is the Weyl group, $W(\mathfrak{g},\mathfrak{a})$. It acts simply transitively on the space of Weyl chambers, that is, every Weyl chamber is the image of $\mathfrak{a}^+$ under a unique element of the Weyl group. By duality, this group also acts on $\mathfrak{a}^*$, and permutes the roots amongst themselves. Take a representative s_w in M' of each w in the Weyl group.

At the group level, there are three important decompositions:

$$(1.2) \qquad\qquad G = K\overline{A}^+K,$$

$$(1.3) \qquad\qquad G = KAN,$$

$$(1.4) \qquad\qquad G = \bigsqcup_{w\in W} MANs_wMAN$$

(this last formula involves a disjoint union). The Cartan decomposition (1.2) arises from the "polar decomposition" $G = K\exp(\mathfrak{p})$, in which the map $(k,X) \mapsto k\exp(X)$ is a diffeomorphism from $K \times \mathfrak{p}$ onto G; every element of $\mathfrak{p}$ is conjugate to an element of $\overline{\mathfrak{a}}^+$ by an element of K. In the Iwasawa decomposition (1.3), the map $(k,a,n) \mapsto kan$ is a diffeomorphism from $K\times A\times N$ onto G. In the Bruhat decomposition (1.4), each of the sets $MANs_wMAN$ is a submanifold of G, and the $|W|$ submanifolds are pairwise disjoint. There is a unique *longest element* $\overline{w}$ of the Weyl group, which maps $\mathfrak{a}^+$ to $-\mathfrak{a}^+$; the corresponding submanifold of G is open and its complement is a union of submanifolds of lower dimension. More precisely,

$$\begin{aligned} G &= \bigsqcup_{w\in W} s_{\overline{w}}MANs_wMAN \\ &= \bigsqcup_{w\in W} s_{\overline{w}}s_ws_w^{-1}NAMs_wMAN \\ &= \bigsqcup_{w\in W} s_{\overline{w}w}\overline{N}_wMAN, \end{aligned}$$

where $\overline{N}_w = s_w^{-1}NS_w \cap \overline{N}$; each $\overline{N}_w$ is a Lie subgroup of $\overline{N}$, of lower dimension unless $w = \overline{w}$, and the map $(\overline{n},m,a,n) \mapsto \overline{n}man$ is a diffeomorphism from $\overline{N}_w \times M \times A \times N$ onto $\overline{N}_wMAN$.

For many purposes it is sufficient to think of the Bruhat decomposition in the following way: the map $(\overline{n}, m, a, n) \mapsto \overline{n}MAN$ of $\overline{N} \times M \times A \times N$ to G is a diffeomorphism of $\overline{N}MAN$ onto an open dense subset of G whose complement is a finite union of lower dimensional submanifolds. In particular, $\overline{N}MAN$ is of full measure in G/MAN, equipped with any of the natural measures. I will use the abusive notation $G \simeq \overline{N}MAN$ to indicate this sort of "quasi-decomposition".

There are integral formulae associated with these group decompositions. In particular, we will use the formula

(1.5)
$$\int_G u(x)\, dx = C \int_K \int_{\mathfrak{a}^+} \int_K u(k_1 \exp(H)k_2) \\ \prod_{\alpha \in \Sigma} \sinh(\alpha(H))^{\dim(\mathfrak{g}_\alpha)}\, dk_1\, dH\, dk_2,$$

which relates the Haar measure on G with the Haar measure dk on K and a weighted variant of Lebesgue measure dH on $\mathfrak{a}^+$. For the formulae for the Iwasawa and Bruhat decompositions, see [60, Propositions I.5.1 and I.5.21].

1.3 Parabolic Subgroups

The subgroup MAN, often written P, is known as a *minimal parabolic subgroup*. Any subgroup P_1 of G containing MAN is known as a parabolic subgroup; such a group may be decomposed in the form

$$P_1 = M_1 A_1 N_1,$$

where $M_1 \supseteq M$, $A_1 \subseteq A$, and $N_1 \subseteq N$. The group M_1 is a semisimple subgroup of G, and has its own Iwasawa and Bruhat decompositions:

$$M_1 = K^1 A^1 N^1 \qquad \text{and} \qquad M_1 \simeq \overline{N}^1 M^1 A^1 N^1.$$

In these formulae, $K^1 \subseteq K$, $A^1 \subseteq A$, $N^1 \subseteq N$, $M^1 \supseteq M$, and $\overline{N}^1 \subseteq \overline{N}$; moreover, $\overline{N}^1 = \Theta N^1$. If $\mathfrak{a}_1$, $\mathfrak{a}^1$, $\mathfrak{n}_1$ and $\mathfrak{n}^1$ denote the subalgebras of $\mathfrak{a}$ and $\mathfrak{n}$ corresponding to A_1, A^1, N_1 and N^1, then $\mathfrak{a} = \mathfrak{a}_1 \oplus \mathfrak{a}^1$ and $\mathfrak{n} = \mathfrak{n}_1 \oplus \mathfrak{n}^1$. To each parabolic subgroup P_1, we associate ρ_1 on $\mathfrak{a}_1$, defined similarly to ρ; more precisely,

$$\rho_1(H) = \frac{1}{2} \operatorname{tr}(\operatorname{ad}(H)|_{\mathfrak{n}_1}) \qquad \forall H \in \mathfrak{a}_1$$

The point of this is mainly that the set of all subgroups P_1 of G containing P is well understood: it is a finite lattice with a well determined structure.

We conclude our discussion of the structure of G with one more definition. A parabolic subgroup P_1 of G is called *cuspidal* if M_1 has a compact Cartan subgroup, that is, if there is a compact abelian subgroup of K_1 which cannot be extended to a larger abelian subgroup of M_1. Since M is compact,

P is automatically cuspidal. It is a deep theorem of Harish-Chandra that the semisimple groups which have discrete series representations, that is, irreducible unitary representations which are subrepresentations of the regular representation, are precisely those with compact Cartan subgroups.

1.4 Spaces of Homogeneous Functions on G

For this section, fix a parabolic subgroup $M_1A_1N_1$ of G. Take an irreducible unitary representation μ of M_1 and λ in the complexification $\mathfrak{a}_{1\mathbb{C}}^*$ of $\mathfrak{a}_1^*$ (that is, $\mathfrak{a}_{1\mathbb{C}}^* = \mathrm{Hom}_{\mathbb{R}}(\mathfrak{a}_1, \mathbb{C})$). Let $\mathcal{H}_\mu$ denote the Hilbert space on which the representation μ acts. Consider the vector space $\mathcal{V}^{\mu,\lambda}$ of all smooth (infinitely differentiable) $\mathcal{H}_\mu$-valued functions ξ on G with the property that

$$\xi(xman) = e^{(i\lambda - \rho_1)(\log a)} \mu(m)^{-1} \xi(x),$$

for all x in G, all m in M_1, all a in A_1 and all n in N_1. These functions may also be viewed as functions on G/N_1, since $\xi(xn) = \xi(x)$ for all x in G and n in N_1, or as sections of a vector bundle over G/P_1. I shall take the naive viewpoint that they are functions on G, even though there are often good geometric reasons for using vector bundle terminology. Write $\pi^{\mu,\lambda}$ for the left translation representation on $\mathcal{V}^{\mu,\lambda}$:

$$[\pi^{\mu,\lambda}(y)\xi](x) = \xi(y^{-1}x) \qquad \forall x, y \in G.$$

The inner product on $\mathcal{H}_\mu$ induces a pairing $\mathcal{V}^{\mu,\lambda'} \times \mathcal{V}^{\mu,\lambda} \to \mathcal{V}^{1,\lambda'-\bar{\lambda}+i\rho_1}$: indeed,

$$\langle \xi(xman), \eta(xman) \rangle$$
$$= \langle e^{(i\lambda'-\rho_1)(\log a)} \mu(m)^{-1}\xi(x), \, e^{(i\lambda-\rho_1)(\log a)} \mu(m)^{-1}\eta(x) \rangle$$
$$= e^{(i\lambda'-i\bar{\lambda}-2\rho_1)(\log a)} \langle \xi(x), \eta(x) \rangle,$$

so the complex-valued function $x \mapsto \langle \xi(x), \eta(x) \rangle$ indeed satisfies the covariance condition characterising $\mathcal{V}^{1,\lambda'-\bar{\lambda}+i\rho_1}$.

Lemma 1.1. *There is a G-invariant positive linear functional I_{P_1} on $\mathcal{V}^{1,i\rho_1}$, which is unique up to a constant. It may be defined as (a constant multiple of) the Haar measure on K,*

$$\xi \mapsto \int_K \xi(k)\, dk,$$

or as (a constant multiple of) the Haar measure on $\overline{N}_1$,

$$\xi \mapsto \int_{\overline{N}_1} \xi(\overline{n})\, d\overline{n}.$$

Proof. This can be proved by fairly explicit means, involving calculation of Jacobians, to show that there is a constant c such that

$$\int_K \xi(k)\, dk = c \int_{\overline{N}} \xi(\overline{n})\, d\overline{n},$$

and then deducing that this expression is K-invariant and $\overline{N}$-invariant, and so invariant under the group generated by K and $\overline{N}$, which is G itself.

Alternatively, this may be proved by using the fact that the "modular function" of $M_1 A_1 N_1$ is given by $man \mapsto e^{-2\rho_1(\log a)}$. See, for instance, [71, pp. 137–141].

Normalise I_{P_1} so that

$$I_{P_1}(\xi) = \int_K \xi(k)\, dk.$$

An immediate corollary of this lemma is that the spaces $\mathcal{V}^{\mu,\lambda}$ and $\mathcal{V}^{\mu,\overline{\lambda}}$ are in duality: the map

$$\langle \xi, \eta \rangle \mapsto I_{P_1}\langle \xi(\cdot), \eta(\cdot) \rangle$$

is well defined. Further, it is easy to check that

$$\langle \pi^{\mu,\lambda}(y)\xi, \pi^{\mu,\overline{\lambda}}(y)\eta \rangle = \langle \xi, \eta \rangle \quad \forall y \in G.$$

In particular, if λ is real, then the duality on $\mathcal{V}^{\mu,\lambda} \times \mathcal{V}^{\mu,\lambda}$ gives an inner product on $\mathcal{V}^{\mu,\lambda}$ relative to which $\pi^{\mu,\lambda}$ acts unitarily. In this case, $\mathcal{V}^{\mu,\lambda}$ may be completed to obtain a Hilbert space $\mathcal{H}^{\mu,\lambda}$ on which $\pi^{\mu,\lambda}$ acts unitarily.

In some cases, when λ is not real, it is possible to find an inner product on $\mathcal{V}^{\mu,\lambda}$ relative to which $\pi^{\mu,\lambda}$ acts unitarily. The unitary representations which arise by completing $\mathcal{V}^{\mu,\lambda}$ relative to this inner product are known as *complementary series* representations. It is also possible to work with other completions of $\mathcal{V}^{\mu,\lambda}$. For example, if $1 \leq p \leq \infty$, and $\mathrm{Im}(\lambda) = (2/p - 1)\rho_1$, then

$$\|\xi(xman)\|_{\mathcal{H}_\mu}^p = \left\| \mu(m) e^{(i\lambda - \rho_1)(\log a)} \xi(x) \right\|_{\mathcal{H}_\mu}^p$$

$$= e^{-p(\mathrm{Im}\,\lambda + \rho_1)(\log a)} \|\xi(x)\|_{\mathcal{H}_\mu}^p$$

$$= e^{-2\rho_1(\log a)} \|\xi(x)\|_{\mathcal{H}_\mu}^p$$

for all x in G, all m in M, all a in A and all n in N, so $\|\xi(\cdot)\|_{\mathcal{H}_\mu}^p \in \mathcal{V}^{1,i\rho_1}$. In this case, $\pi^{\mu,\lambda}$ acts isometrically on the completion of $\mathcal{V}^{\mu,\lambda}$ in the L^p-norm $\left(I_{P_1}(\|\cdot\|_{\mathcal{H}_\mu}^p) \right)^{1/p}$, and to all intents and purposes we are dealing with a representation on a $\mathcal{H}_\mu$-valued L^p-space.

It is a notable fact that for the case where $G = \mathrm{SO}(1,n)$, the representations $\pi^{\mu,\lambda}$ may be completed to obtain unitary representations in Sobolev spaces, as well as isometric representations on L^p-spaces, and the Sobolev spaces and the L^p-spaces are linked as in the Sobolev embedding theorem: the degree of differentiation involved is such that the Sobolev space is either included in the L^p-space (when $p > 2$) or the L^p-space is included in the Sobolev space (when $p < 2$). To understand the corresponding result for other families of semisimple Lie groups, such as $\mathrm{SU}(1, n)$, is an important problem, to be discussed in the last lecture.

It is known that the representations $\pi^{\mu,\lambda}$ on $\mathcal{V}^{\mu,\lambda}$ are mostly irreducible—for a given μ, the set of λ in $\mathfrak{a}_{1\mathbb{C}}^*$ for which $\pi^{\mu,\lambda}$ is reducible is a countable union of hyperplanes in $\mathfrak{a}_{1\mathbb{C}}^*$. Here, reducible means that there are nontrivial closed (in the C^∞-topology) G-invariant subspaces of $\mathcal{V}^{\mu,\lambda}$.

1.5 The Plancherel Formula

The Plancherel formula for semisimple Lie groups was proved by Harish-Chandra [53, 54, 55], following previous work by various people for various special cases. The representations involved are the representations $\pi^{\mu,\lambda}$, where μ is a discrete series representation of M_1 (written $\mu \in \widehat{M}_{1d}$), and λ in $\mathfrak{a}_{1\mathbb{C}}^*$ is real. Such representations are sometimes called *unitary principal series* representations—in this nomenclature, the *principal series* is the collection of all the $\pi^{\mu,\lambda}$ without the restriction on λ. Other authors call the smaller collection of representations the unitary principal series, and the larger collections of representations is then known as the *analytic continuation* of the principal series. All the cuspidal parabolic subgroups are involved.

A bit more notation is needed to state the Plancherel theorem: for u in $C_c^\infty(G)$, the operator $\pi^{\mu,\lambda}(u)$ is given by the formula

$$\pi^{\mu,\lambda}(u) = \int_G u(y)\, \pi^{\mu,\lambda}(y)\, dy,$$

which is to be interpreted as an operator-valued integral. Let $\mathcal{P}$ be a set of nonconjugate cuspidal parabolic subgroups of G, and $\mathbf{c}$ be the more or less explicitly determined function known as the Harish-Chandra $\mathbf{c}$-function.

Theorem 1.1. *Suppose that* $u \in C_c^\infty(G)$. *Then the operators* $\pi^{\mu,\lambda}(u)$ *are trace-class for all* μ *in* $\widehat{M}_{1d}$ *and* λ *in* $\mathfrak{a}_{1\mathbb{C}}^*$, *and the* $L^2(G)$-*norm of* u *is given by*

$$\|u\|_2^2 = \sum_{P_1 \in \mathcal{P}} \sum_{\mu \in \widehat{M}_{1d}} c_{P_1} \int_{\mathfrak{a}_1^*} \operatorname{tr}\left(\pi^{\mu,\lambda}(u)^* \pi^{\mu,\lambda}(u)\right) |\mathbf{c}(P_1, \mu, \lambda)|^{-2}\, d\lambda.$$

Fortunately, a simpler formula is available for most of the analysis in the following lectures. If the function u is K-invariant, on the left or the right or both, then $\pi^{\mu,\lambda}(u) = 0$ unless P_1 is the minimal parabolic and μ is the trivial representation 1 of M. This reduces substantially the number of terms in the Plancherel formula. Further, the operators $\pi^{1,\lambda}(u)$ are rank one operators, so that the Hilbert–Schmidt norm and the operator norm $\| \cdot \|$ (and all the other Schatten p-norms) coincide. Thus

$$\|u\|_2^2 = c_G \int_{\mathfrak{a}^*} \left\| \pi^{\mu, \lambda}(u) \right\|^2 |\mathbf{c}(\lambda)|^{-2} \, d\lambda.$$

Let 1^λ denote the function in $V^{1,\lambda}$ which is identically equal to 1 on K. When u is K-bi-invariant, the formula simplifies further: $\pi^{1,\lambda}(u)$ is a multiple of the projection onto $\mathbb{C}1^\lambda$. This multiple, denoted $\widetilde{u}(\lambda)$, is given by

$$\widetilde{u}(\lambda) = \int_G u(x)\, \varphi_\lambda(x)\, dx$$

where

$$\varphi_\lambda(x) = \langle \pi^{1,\lambda}(x)1^\lambda, 1^{\overline{\lambda}} \rangle \qquad \forall x \in G.$$

For K-bi-invariant functions u, the Plancherel formula becomes

(1.6) $$\|u\|_2^2 = c_G \int_{\mathfrak{a}^*} |\widetilde{u}(\lambda)|^2 |\mathbf{c}(\lambda)|^{-2} \, d\lambda;$$

the corresponding inversion formula is

(1.7) $$u(x) = c_G \int_{\mathfrak{a}^*} \widetilde{u}(\lambda) \overline{\varphi}_\lambda(x) \, |\mathbf{c}(\lambda)|^{-2} \, d\lambda \qquad \forall x \in G.$$

There are a number of integral formulae for φ_λ, which may be found in [60, Chapter IV]. One of these is the following: for any x in G, denote by $A(x)$ the unique element of $\mathfrak{a}$ such that $x \in N \exp A(x)\, K$. For any λ in $\mathfrak{a}_{\mathbb{C}}^*$, the spherical function φ_λ is given by

$$\varphi_\lambda(x) = \int_K \exp\big((i\lambda + \rho)(A(kx))\big) \, dk \qquad \forall x \in G.$$

It is worth pointing out that when G is complex, then there are explicit formulae for φ_λ, in terms of elementary functions, and for several other cases where $\dim(\mathfrak{a})$ is small, there are formulae in terms of hypergeometric functions or other less elementary functions (see, for instance, [64]). Perhaps the most important technique for understanding spherical functions in a fairly general context is M. Flensted-Jensen's method [48] of reducing the case of a normal real form to the complex case. It is possible to work with spherical functions fairly effectively: one can formulate conjectures using the complex case as a guide, and prove many of these for some or all general semisimple groups.

2 The Equations of Mathematical Physics on Symmetric Spaces

Let G be a semisimple Lie group with a maximal compact subgroup K; then the quotient space $X = G/K$ is, in a natural way, a negatively curved Riemannian manifold. In particular, when $G = SO(1,n)$, $SU(1,n)$, or $Sp(1,n)$, the manifold X is a real, complex, or quaternionic hyperbolic space. The Laplace–Beltrami operator on X is a natural second order elliptic differential operator. I prefer to deal with the positive operator Δ, equal to minus the Laplace–Beltrami operator. The L^2 spectrum of Δ is the interval $[b, \infty)$, where $b = (\rho, \rho)_B$. It is natural to study not only Δ, but also $\Delta - b$, which is still a positive operator and from some geometric points of view is more canonical than Δ. We shall consider $\Delta - \theta b$, where $\theta \in [0, 1]$.

This lecture deals with the equations of mathematical physics on X, that is, with the solutions of the equations

$$\frac{\partial}{\partial t} u_1(x,t) + (\Delta - \theta b) u_1(x,t) = 0$$

$$\frac{\partial^2}{\partial t^2} u_2(x,t) - (\Delta - \theta b) u_2(x,t) = 0$$

$$\frac{\partial^2}{\partial t^2} u_3(x,t) + (\Delta - \theta b) u_3(x,t) = 0$$

$$\frac{\partial}{\partial t} u_4(x,t) + i(\Delta - \theta b) u_4(x,t) = 0$$

for all (x,t) in $X \times \mathbb{R}^+$, with boundary conditions $u_k(\cdot, 0) = f$ for all k in $\{1, 2, 3, 4\}$, $u_2(\cdot, t) \to 0$ (in some sense, depending on f) as $t \to \infty$, and $\partial u_3/\partial t(\cdot, 0) = i(\Delta - \theta b)^{1/2} f$. These equations are the heat equation, Laplace's equation, the wave equation, and the Schrödinger equation. In the Euclidean case, these equations can be solved using the Fourier transform. The same is true in this case.

2.1 Spherical Analysis on Symmetric Spaces

Any function f on X gives rise canonically to a K-right-invariant function on G, also denoted by f. The key to the Fourier transform approach to these equations is that, for any f in $C_c^\infty(X)$,

$$\pi^{1,\lambda}(\Delta f) = ((\lambda, \lambda)_B + (\rho, \rho)_B)\pi^{1,\lambda}(f) \qquad \forall \lambda \in \mathfrak{a}_\mathbb{C}^*,$$

that is, Δ is a *Fourier multiplier*. Because Δ corresponds to a positive operator on $L^2(X)$, it is possible to use spectral theory to define $m(\Delta)$ for Borel measurable functions on $[b, \infty)$. For bounded m, the operator $m(\Delta)$ is defined by

$$m(\Delta)f = \int_b^\infty m(\zeta) \, dP_\zeta f \qquad \forall f \in L^2(X),$$

where $\{P_\zeta\}$ is the spectral resolution of the identity such that

$$\Delta f = \int_b^\infty \zeta \, dP_\zeta f \qquad \forall f \in \mathrm{Dom}(\Delta).$$

Define the quadratic function Q_θ by the formula

$$Q_\theta(\lambda) = (\lambda, \lambda)_B + (1 - \theta)(\rho, \rho)_B \qquad \forall \lambda \in \mathfrak{a}_\mathbb{C}^*.$$

By spectral theory,

$$\pi^{1,\lambda}((\Delta - \theta b)f) = Q_\theta(\lambda)\,\pi^{1,\lambda}(f)$$

and

$$\pi^{1,\lambda}(m(\Delta - \theta b)f) = m(Q_\theta(\lambda))\,\pi^{1,\lambda}(f).$$

At least formally, the solutions of the equations of mathematical physics on $L^2(X)$ are given by

$$u_1(\cdot, t) = e^{-t(\Delta - \theta b)} f$$

$$u_2(\cdot, t) = e^{-t(\Delta - \theta b)^{1/2}} f$$

$$u_3(\cdot, t) = e^{it(\Delta - \theta b)^{1/2}} f$$

$$u_4(\cdot, t) = e^{it(\Delta - \theta b)} f$$

for all t in $\mathbb{R}^+$. These solutions may also be expressed in terms of convolutions with kernels:

$$u_1(xK, t) = \mathcal{H}_{t,\theta} f(xK) = f \star h_{t,\theta}(x)$$

$$u_2(xK, t) = \mathcal{L}_{t,\theta} f(xK) = f \star l_{t,\theta}(x)$$

$$u_3(xK, t) = \mathcal{W}_{t,\theta} f(xK) = f \star w_{t,\theta}(x)$$

$$u_4(xK, t) = \mathcal{S}_{t,\theta} f(xK) = f \star s_{t,\theta}(x),$$

for all x in G and t in $\mathbb{R}^+$; here, at least formally, the kernels are the K-bi-invariant objects on G such that

$$\widetilde{h}_{t,\theta} = e^{-tQ_\theta}$$

$$\widetilde{l}_{t,\theta} = e^{-tQ_\theta^{1/2}}$$

$$\widetilde{w}_{t,\theta} = e^{itQ_\theta^{1/2}}$$

$$\widetilde{s}_{t,\theta} = e^{itQ_\theta}.$$

These formulae should be compared with the results in the Euclidean case obtained by classical Fourier analysis. For example, for the heat equation in $\mathbb{R}^n$,

$$u(x,t) = f \star g_t$$

$$\mathcal{F}u(\xi,t) = (\mathcal{F}f)(\xi)\,e^{-t|\xi|^2}$$

for all x and ξ in $\mathbb{R}^n$ and t in $\mathbb{R}^+$, where g_t is the appropriately normalised Gaussian kernel on $\mathbb{R}^n$ and $\mathcal{F}$ denotes the spatial Fourier transform.

In order to obtain useful information from these formulae for the solutions to these equations, we need to have information about the kernels $h_{t,\theta}$, $l_{t,\theta}$, $w_{t,\theta}$, and $s_{t,\theta}$. This information can be of several types, for instance, pointwise estimates, L^p estimates, or parametrix expressions (the latter means an expression of the kernel as a sum of distributions). It is also important to understand the regularity properties of Δ itself.

2.2 Harmonic Analysis on Semisimple Groups and Symmetric Spaces

This section outlines some of the features of harmonic analysis on noncompact semisimple Lie groups and symmetric spaces. In particular, we describe the spherical Fourier transformation, and the Plancherel measure and the c-function. We also prove a Hausdorff–Young type theorem about the Fourier transform of an $L^p(G)$-function for p in $(1,2)$, and a partial converse.

We first discuss the spherical Fourier transform of an $L^1(G)$-function. Let $\mathbf{W}_1$ be the interior of the convex hull in $\mathfrak{a}^*$ of the images of ρ under the Weyl group W of $(\mathfrak{g}, \mathfrak{a})$. For δ in $(0,1)$, denote by $\mathbf{W}_\delta$ and $\mathbf{T}_\delta$ the dilate of $\mathbf{W}_1$ by δ and the tube over the polygon $\mathbf{W}_\delta$, that is, $\mathbf{T}_\delta = \mathfrak{a}^* + i\delta\,\mathbf{W}_1$; $\overline{\mathbf{W}}_\delta$ and $\overline{\mathbf{T}}_\delta$ denote the closures of these sets in $\mathfrak{a}^*$ and $\mathfrak{a}^*_{\mathbb{C}}$ respectively.

If $\lambda = \lambda_0 - i\rho$, where λ_0 lies in $\mathfrak{a}^*$, then the formula (1.2) defining the spherical function φ_λ becomes

$$\varphi_\lambda(x) = \int_K \exp\bigl(i\lambda_0(A(kx))\bigr)\,dk \qquad \forall x \in G,$$

which implies immediately that φ_λ is bounded. The spherical functions are invariant under the Weyl group action on $\mathfrak{a}^*_{\mathbb{C}}$, that is, $\varphi_\lambda = \varphi_{w\lambda}$ for all w in W. Hence φ_λ is bounded whenever λ lies in $\mathfrak{a}^* - iw\rho$, for any w in W, and now a straightforward interpolation argument implies that φ_λ is bounded whenever λ lies in $\mathbf{T}_1$. A full proof of this is in Helgason [60, IV.8].

Further, the map $\lambda \mapsto \varphi_\lambda$ from $\mathfrak{a}^*_{\mathbb{C}}$ to $C(G)$, endowed with the topology of uniform convergence on compact sets, is holomorphic and so, in particular, continuous. It follows that if f is in $L^1(G)$, then $\tilde{f}$ extends to a continuous function in $\overline{\mathbf{T}}_1$, holomorphic in $\mathbf{T}_1$. If f is a distribution on G which convolves $L^1(G)$ into itself, then f is a bounded measure on G, and similarly, $\tilde{f}$ also extends continuously to $\overline{\mathbf{T}}_1$, and holomorphically to $\mathbf{T}_1$.

We now discuss the Plancherel formula. Recall (1.6): for K-bi-invariant functions u,

$$\|u\|_2^2 = c_G \int_{\mathfrak{a}^*} \bigl|\tilde{u}(\lambda)\bigr|^2 \,|\mathbf{c}(\lambda)|^{-2}\,d\lambda.$$

The Gindikin–Karpelevič formula for $\mathbf{c}$ states that

$$|\mathbf{c}(\lambda)|^{-2} = \prod_{\alpha \in \Sigma_0^+} |\mathbf{c}_\alpha((\alpha, \lambda)_B)|^{-2} \qquad \forall \lambda \in \mathfrak{a}^*,$$

where each "Plancherel factor" $|\mathbf{c}_\alpha(\cdot)|^{-2}$, which is given by an explicit formula involving several Γ-functions, extends to an analytic function in a neighbourhood of the real axis and satisfies

$$(2.1) \qquad |\mathbf{c}_\alpha(z)|^{-2} \sim |z|^2 \, (1 + |z|)^{d_\alpha - 2} \qquad \forall z \in \mathbb{R},$$

where $d_\alpha = \dim(\mathfrak{g}_\alpha) + \dim(\mathfrak{g}_{2\alpha})$. This and other useful results about the c-function may be found in Helgason's book [60, IV.6]. It follows easily that there exists a positive constant C such that

$$|\mathbf{c}(\lambda)|^{-2} \le C \, |\lambda|^{\nu - \ell} \, (1 + |\lambda|)^{n - \nu} \qquad \forall \lambda \in \mathfrak{a}^*.$$

We shall use a modified version μ of the Plancherel measure as well as an auxiliary function Υ on $\mathfrak{a}^*$ defined by the rule

$$d\mu(\lambda)/d\lambda = \Upsilon(\lambda) = \prod_{\alpha \in \Sigma_0^+} (1 + |(\alpha, \lambda)_B|)^{d_\alpha}.$$

The estimate (2.1) implies that

$$\|f\|_2 \le C \left(\int_{\mathfrak{a}^*} |\widetilde{f}(\lambda)|^2 \, d\mu(\lambda) \right)^{1/2} \qquad \forall f \in L^2(K \backslash X).$$

The modified Plancherel measure is invariant under the action of the Weyl group W (like the Plancherel measure), and moreover it is quasi-invariant under translations, in the sense that for any measurable subset S of $\mathfrak{a}^*$,

$$\mu(S + \lambda) \le \Upsilon(\lambda) \, \mu(S) \qquad \forall \lambda \in \mathfrak{a}^*.$$

We now describe a version of the Hausdorff–Young inequality valid for semisimple Lie groups. Write $\delta(p)$ for $2/p - 1$.

Theorem 2.1. *Equip $\mathfrak{a}^*$ with the modified Plancherel measure μ. Suppose that $1 < p < 2$, and that f lies in $L^p(G)$. Then $\widetilde{f}$ may be extended to a measurable function on the tube $\overline{\mathbf{T}}_{\delta(p)}$, holomorphic in $\mathbf{T}_{\delta(p)}$, such that $\lambda_0 \mapsto \widetilde{f}(\cdot + i\lambda_0)$ is continuous from $\overline{\mathbf{W}}_{\delta(p)}$ to $L^{p'}(\mathfrak{a}^*)$, and such that*

$$\left(\int_{\mathfrak{a}^*} |\widetilde{f}(\lambda + i\lambda_0)|^{p'} d\mu(\lambda) \right)^{1/p'} \le C \, \|f\|_p \qquad \forall f \in L^p(G) \quad \forall \lambda_0 \in \overline{\mathbf{W}}_{\delta(p)}.$$

Further, for any closed subtube $\mathbf{T}$ of $\mathbf{T}_{\delta(p)}$, there exists a constant C such that

$$|\widetilde{f}(\lambda)| \le C \, \Upsilon(\lambda)^{-1/p'} \, \|f\|_p \qquad \forall f \in L^p(G) \quad \forall \lambda \in \mathbf{T}.$$

Proof. See [36].

It follows from this that if $1 < p < 2$ and λ is in $\mathbf{T}_{\delta(p)}$, then φ_λ is in $L^{p'}(G)$, and for every closed subtube $\mathbf{T}$ of $\mathbf{T}_{\delta(p)}$, there exists a constant C such that $\|\varphi_\lambda\|_{p'} \leq C\Upsilon(\lambda)^{-1/p'}$. This is a little sharper than the standard result, that $\|\varphi_\lambda\|_{p'} \leq C$, which is based on the pointwise inequality $|\varphi_{\lambda_1+i\lambda_2}| \leq \varphi_{i\lambda_2}$, trivially true when λ_1 and λ_2 lie in $\mathfrak{a}^*$.

Using only the fact that the spherical functions φ_λ are in $L^{p'}(G)$ when λ is in $\mathbf{T}_{\delta(p)}$ and $1 \leq p < 2$ (which may be proved by interpolation, as above, or by careful estimates on the spherical functions), J.-L. Clerc and E.M. Stein [23, Theorem 1] showed that if $1 \leq p < 2$, then the spherical Fourier transform of an L^p-function extends to a holomorphic function in $\mathbf{T}_{\delta(p)}$, bounded in closed subtubes thereof, and that if f convolves $L^p(G)$ into itself, then its spherical Fourier transform extends to a bounded holomorphic function in $\mathbf{T}_{\delta(p)}$.

Another consequence of Theorem 2.1 is the K-bi-invariant version of the Kunze–Stein phenomenon: if $1 \leq p < 2$ and k is in $L^p(K\backslash X)$, then the maps $f \mapsto f * k$ and $f \mapsto k * f$ are bounded on $L^2(G)$. Indeed, without loss of generality we may assume that $k \geq 0$, and then Herz' *principe de majoration* [63] shows that it suffices to have $\widetilde{k}(0)$ bounded. Some of our computations require this result, and others the stronger result that the maps $f \mapsto f*k$ and $f \mapsto k*f$ are bounded on $L^2(G)$, if k is in $L^p(G)$. This stronger result is known as the Kunze–Stein phenomenon. The Kunze–Stein phenomenon (which is discussed in more detail in Section 3.3) and the generalisation of Young's inequality to locally compact groups have the following consequences.

Theorem 2.2. *Suppose that $1 \leq r \leq \infty$. For a function k in $L^r(G)$, denote by $\mathcal{K}$ and $\mathcal{K}'$ the operators $f \mapsto k * f$ and $f \mapsto f * k$ from $S(G)$ to $C(G)$. Then $\mathcal{K}$ and $\mathcal{K}'$ are bounded from $L^p(G)$ to $L^q(G)$, with a corresponding operator norm inequality, provided that one of the following conditions holds:*

(i) if $r = 1$, and $1 \leq p = q \leq \infty$;
(ii) if $1 < r \leq 2$, $q \geq r$, $p \leq r'$, $0 \leq 1/p - 1/q \leq 1/r'$, and $(p,q) \neq (r,r)$ or (r',r');
(iii) if $2 < r < \infty$, $q \geq r$, $p \leq r'$, $0 \leq 1/p - 1/q \leq 1/r'$, and $(p,q) \neq (r,r')$;
(iv) if $r = \infty$, $p = 1$, and $q = \infty$.

Consequently, if k is in $L^r(G)$ for all r in $(2,\infty]$, $\mathcal{K}$ and $\mathcal{K}'$ are bounded from $L^p(G)$ to $L^q(G)$ when $1 \leq p < 2 < q \leq \infty$.

Proof. We give the details of this proof because it is short and indicative of the differences between harmonic analysis on symmetric spaces and on Euclidean spaces. Without loss of generality, we may restrict our attention to the operator $\mathcal{K}$, because G is unimodular, so the mapping $f \mapsto \check{f}$, where $\check{f}(x) = f(x^{-1})$ for all x in G, which has the property that $(f*k)\check{} = \check{k}*\check{f}$, acts isometrically on each of the spaces $L^s(G)$. Thus $\mathcal{K}$ is bounded from $L^p(G)$ to $L^q(G)$ if and only if $\mathcal{K}'$ is.

For the cases where $r = 1$ and $r = \infty$, the result is standard. If $1 < r < 2$, then it suffices to prove that $L^r(G) * L^p(G) \subseteq L^r(G)$ when $1 \leq p < r$ and $L^r(G) * L^p(G) \subseteq L^p(G)$ when $r < p \leq 2$. For then duality arguments establish that $L^r(G) * L^{r'}(G) \subseteq L^p(G)$ when $r' < p \leq \infty$ and $L^r(G) * L^p(G) \subseteq L^p(G)$ when $2 \leq p < r'$, and interpolation arguments establish the boundedness in the set claimed. The first inclusion follows by multilinear interpolation between the inclusions $L^1(G) * L^1(G) \subseteq L^1(G)$ and $L^2(G) * L^s(G) \subseteq L^2(G)$, for any s in $[1, 2)$. The second inclusion follows by multilinear interpolation between the inclusions $L^1(G) * L^s(G) \subseteq L^s(G)$ and $L^t(G) * L^2(G) \subseteq L^2(G)$, for any s and t in $[1, 2)$.

When $r = 2$, the result is easy: one first reformulates the Kunze–Stein phenomenon to show that $L^2(G) * L^2(G) \subseteq L^s(G)$ when $2 < s \leq \infty$, then applies duality and interpolation.

When $2 < r < \infty$, the result follows by multilinear interpolation between the results when $r = 2$ and when $r = \infty$.

The final consequence is proved by combining the results of (iii) and (iv).

We shall deal with functions which belong to $L^r(G)$ for all r in $(2, \infty]$; the convolution properties thereof are described in the theorem just proved. As a corollary of Theorem 2.1, we may establish a criterion for a function on $\mathfrak{a}^*$ to be the spherical Fourier transform of such a function. A technical definition is necessary: for any small positive ϵ, let $H^\infty(\mathbf{T}_\epsilon)$ denote the space of all bounded holomorphic functions in $\mathbf{T}_\epsilon$, with the supremum norm; clearly when $\delta < \epsilon$, $H^\infty(\mathbf{T}_\epsilon)$ may be injected into $H^\infty(\mathbf{T}_\delta)$ by restricting $H^\infty(\mathbf{T}_\epsilon)$ functions to $\mathbf{T}_\delta$. The inductive limit space $\bigcup_{\epsilon > 0} H^\infty(\mathbf{T}_\epsilon)$ is denoted by $A(\mathfrak{a}^*)$. An element of the dual space of $A(\mathfrak{a}^*)$ will be called, somewhat abusively, an analytic functional on $\mathfrak{a}^*$.

Corollary 2.1. *Suppose that T is an analytic functional on $\mathfrak{a}^*$. Then there exists a K-bi-invariant function on G, k say, which belongs to $L^r(G)$ for all r in $(2, \infty]$, such that*

$$\int_G k(x)\, f(x)\, dx = T(\widetilde{f}) \qquad \forall f \in S(G).$$

If $\mathfrak{a}^*$ is endowed with the Plancherel measure and if

$$T(\widetilde{f}) = \int_{\mathfrak{a}^*} \widetilde{f}(\lambda)\, t(\lambda)\, |\mathbf{c}(\lambda)|^{-2}\, d\lambda \qquad \forall f \in S(G),$$

where t is Weyl group invariant and lies in $L^1(\mathfrak{a}^*) \cap L^2(\mathfrak{a}^*)$, then $\widetilde{k} = t$.

Proof. This is an immediate consequence of Theorem 2.1. Fix s in $[1, 2)$. If f is in $L^s(G)$ then $\widetilde{f}$ is in $A(\mathfrak{a}^*)$, and so composition with T provides a linear functional on $L^s(G)$. Thus there exists k in $L^{s'}(G)$ such that

$$\int_G k(x)\, f(x)\, dx = T(\widetilde{f}) \qquad \forall f \in L^s(G).$$

Since this works for arbitrary s, k has the properties claimed. The second part of the corollary follows from the Plancherel formula.

It is clear that if T is an analytic functional on $\mathfrak{a}^*$, then for any h belonging to $H^\infty(\mathbf{T}_\epsilon)$ for some positive ϵ, hT, defined by the rule $hT(g) = T(hg)$ for all g in $A(\mathfrak{a}^*)$, is also an analytic functional on $\mathfrak{a}^*$.

Now we consider the question whether Theorem 2.1 has a converse: if $\widetilde{f}$ extends to a holomorphic function in $\mathbf{T}_{\delta(p)}$, is it necessarily true that f must lie in $L^p(G)$? It is too much to expect that we will be able to prove the converse in the semisimple case, when even in Euclidean Fourier analysis this is impossible. However, we can prove a result which is useful.

To find an estimate for $\|f\|_p$, it is tempting to try to interpolate between estimates for $\|f\|_1$ and $\|f\|_2$. Unfortunately, the straightforward interpolation argument gives

$$\|f\|_p \leq \|f\|_1^{\delta(p)} \|f\|_2^{1-\delta(p)},$$

which is useless unless $\widetilde{f}$ is holomorphic in $\mathbf{T}_1$, for otherwise f could not be in $L^1(G)$. To obviate this problem, the obvious estimate is replaced by an estimate

$$\|f\|_p \leq \|f\varphi_{ic\rho}\|_1^{\delta(p)} \|f\varphi_{ic'\rho}\|_2^{1-\delta(p)},$$

where the spherical functions $\varphi_{ic\rho}$ and $\varphi_{ic'\rho}$ are chosen so that the first factor lies in $L^1(G)$. A technique of L. Vretare [101] enables us to compute the second factor in terms of an L^2-norm of $\widetilde{f}$.

Theorem 2.3. *Suppose* $1 < p < 2$, *that* f *is a measurable* K-*bi-invariant function on* G, *and that* $f\varphi_{i\delta(p)\rho}$ *lies in* $L^1(G)$. *Then the spherical Fourier transform of* f *extends holomorphically into the tube* $\mathbf{T}_{\delta(p)}$, *and continuously to* $\overline{\mathbf{T}}_\delta(p)$. *If moreover* $N < \infty$, *where*

$$N = \left(\int_{\mathfrak{a}^*} \left| \widetilde{f}(\lambda + i\delta(p)\rho) \right|^2 d\mu(\lambda) \right)^{1/2},$$

then f *lies in* $L^p(G)$, *and*

$$\|f\|_p \leq C \left\| f\varphi_{i\delta(p)\rho} \right\|_1^{\delta(p)} N^{1-\delta(p)}.$$

Vretare [102] also proved a form of inverse Hausdorff–Young theorem for semisimple groups.

2.3 Regularity of the Laplace–Beltrami Operator

The following result, taken from [36], but based on much previous work, encapsulates the various Sobolev-type regularity theorems for the Laplace–Beltrami operator. In the following, we denote by n the dimension of X, by ℓ its real rank, that is, the (real) dimension of A, and by ν the *pseudo-dimension* or *dimension at infinity* $2\left|\Sigma_0^+\right| + \ell$, where $\left|\Sigma_0^+\right|$ is the cardinality of the set of the indivisible positive roots. The pseudo-dimension ν may very well be strictly larger than the dimension n, as, for instance, in the case of $\mathrm{SL}(p, \mathbb{R})$.

If $1 \leq p, q \leq \infty$, we denote by $\|T\|_{p;q}$ the norm of the linear operator T from $L^p(X)$ to $L^q(X)$. In the case where $p = q$ we shall simply write $\|T\|_p$. By C we denote a constant which may not be the same at different occurrences. The expression

$$A(t) \sim B(t) \qquad \forall t \in \mathbf{D},$$

where $\mathbf{D}$ is some subset of the domains of A and of B, means that there exist (positive) constants C and C' such that

$$CA(t) \leq B(t) \leq C'A(t) \qquad \forall t \in \mathbf{D};$$

C and C' may depend on any quantifiers written *before* the displayed formula.

Finally, p_θ denotes $2/\left[1 + (1 - \theta)^{1/2}\right]$, and for α in $\mathbb{C}$ with positive real part, $\mathbf{I}_\alpha$ denotes the interval $[2, 2n/(n-2\operatorname{Re}\alpha)]$ if $0 \leq \operatorname{Re}\alpha < n/2$, the interval $[2, \infty)$ if $\operatorname{Re}\alpha = n/2$, and the interval $[2, \infty]$ if $\operatorname{Re}\alpha > n/2$, while $\mathbf{P}_\alpha$ denotes the set of all (p, q) satisfying the following conditions:

(i) if $\alpha = 0$ then $1 \leq p = q \leq \infty$;
(ii) if $\operatorname{Re}\alpha = 0$ and $\operatorname{Im}\alpha \neq 0$, then $1 < p = q < \infty$;
(iii) if $0 < \operatorname{Re}\alpha < n$, then $1 \leq p \leq q \leq \infty$ and either $1/p - 1/q < \operatorname{Re}\alpha/n$ or $1/p - 1/q = \operatorname{Re}\alpha/n$, $p > 1$, and $q < \infty$;
(iv) if $\operatorname{Re}\alpha = n$, then $1 \leq p \leq q \leq \infty$ and either $1/p - 1/q < \operatorname{Re}\alpha/n$ or $\operatorname{Im}\alpha \neq 0$, $p = 1$, and $q = \infty$;
(v) if $\operatorname{Re}\alpha > n$, then $1 \leq p \leq q \leq \infty$.

Theorem 2.4. *Suppose that* $0 \leq \theta < 1$, $1 \leq p \leq \infty$, *and* $\operatorname{Re}\alpha \geq 0$. *The operator* $(\Delta - \theta b)^{-\alpha/2}$ *is bounded on* $L^p(X)$ *if and only if one of the following conditions holds:*

(i) $\alpha = 0$;
(ii) $\operatorname{Re}\alpha = 0$, $\alpha \neq 0$, $1 < p < \infty$, *and* $p_\theta \leq p \leq p'_\theta$;
(iii) $\operatorname{Re}\alpha > 0$ *and* $p_\theta < p < p'_\theta$.

Suppose that $0 \leq \theta < 1$, $\operatorname{Re}\alpha > 0$, *and* $1 \leq p < q \leq \infty$. *Then the operator* $(\Delta - \theta b)^{-\alpha/2}$ *is bounded from* $L^p(X)$ *to* $L^q(X)$ *if and only if the following conditions both hold:*

(iv) *either* $0 < \operatorname{Re}\alpha < (\ell + 1)/p'_\theta$, $p \leq p'_\theta$, *and* $q \geq p_\theta$ *or* $\operatorname{Re}\alpha \geq (\ell + 1)/p'_\theta$, $p < p'_\theta$, *and* $q > p_\theta$;
(v) (p, q) *is in* $\mathbf{P}_\alpha$.

Suppose that $\operatorname{Re}\alpha \geq 0$ *and* $1 \leq p \leq q \leq \infty$. *Then* $(\Delta - b)^{-\alpha/2}$ *is bounded from* $L^p(X)$ *to* $L^q(X)$ *if and only if one of the following conditions holds:*

(vi) $p = q = 2$ *and* $\operatorname{Re}\alpha = 0$;
(vii) $p = 2 < q$, $0 < \operatorname{Re}\alpha < \nu/2$, *and* q *is in* $\mathbf{I}_\alpha$;

(viii) $p < 2 = q$, $0 < \operatorname{Re}\alpha < \nu/2$, and p' is in $\mathbf{I}_\alpha$;

(ix) $p < 2 < q$, $\operatorname{Re}\alpha - \nu$ is not in $2\mathbb{N}$ and $1/p - 1/q < \operatorname{Re}\alpha/n$.

(x) $p < 2 < q$, $\operatorname{Re}\alpha - \nu$ is not in $2\mathbb{N}$, $1/p - 1/q = \operatorname{Re}\alpha/n$, and if $p = 1$ or $q = \infty$ then both

(xi) $\operatorname{Re}\alpha = n$ and $\operatorname{Im}\alpha \neq 0$.

2.4 Approaches to the Heat Equation

Quite a bit is known about the heat kernel $h_{t,\theta}$. For complex Lie groups, an explicit expression is available [57]. For other Lie groups, less explicit but nevertheless useful pointwise formulae are known. In particular, P. Sawyer [91, 92] estimated the heat kernels in a number of special cases. In the general case, the best estimates are due to J.-Ph. Anker [3], Anker and L. Ji [4, 5], and Anker and P. Ostellari [6]. Those who study these questions often speak of having problems "at the walls"; this is, for example, the major problem with the pointwise estimates for the spherical functions. To a large extent, these problems are unimportant, since, for instance, it is possible to show that the heat kernel is very small near the walls so that the precise behaviour there is irrelevant. By the inversion theorem for the spherical Fourier transform (1.7),

$$h_{t,\theta}(x) = c_G \int_{\mathfrak{a}^*} e^{-tQ_\theta(\lambda)}\varphi_\lambda(x)\,|\mathbf{c}(\lambda)|^{-2}\,d\lambda.$$

To estimate $h_{t,\theta}$, one needs to know about the functions φ_λ. By the Cartan decomposition (1.2), it suffices to consider x in A. The "difficulties at the walls" are two-fold: for H in $\mathfrak{a}$, $\varphi_\lambda(\exp H)$ is hard to handle when λ is close to a wall of a Weyl chamber in $\mathfrak{a}^*$ or when H is close to a wall of a Weyl chamber in $\mathfrak{a}$. Even in the complex case, obtaining estimates "close to the walls" is tricky, especially if one wants estimates which are uniform in both λ and H. Some progress on this problem has been made recently by Cowling and A. Nevo [42], based on an idea of H. Gunawan [52].

The other approach to the problem is to try to obtain other sorts of estimates for the kernels. For the heat equation, this has been quite effective. In the next section, we summarise some of the results of [36, 37], and extend one of these a little.

2.5 Estimates for the Heat and Laplace Equations

Putting together the facts of the previous section, it is relatively easy to obtain estimates for the operators arising in the heat and Laplace equations.

Theorem 2.5. *Let $(\mathcal{H}_t)_{t>0}$ be the heat semigroup. Then the following hold:*

(i) for all p in $[1,\infty]$,

$$\|\mathcal{H}_t\|_p = \exp\left(-(1 - \delta(p)^2)bt\right) \qquad \forall t \in \mathbb{R}^+;$$

(ii) for all p, q such that $1 \leq p \leq q \leq \infty$,

$$\|\mathcal{H}_t\|_{p;q} \sim t^{-n(1/p-1/q)/2} \qquad \forall t \in (0, 1];$$

(iii) for all p, q such that either $1 \leq p < q = 2$ or $2 = p < q \leq \infty$,

$$\|\mathcal{H}_t\|_{p;q} \sim t^{-\nu/4} \exp(-bt) \qquad \forall t \in [1, \infty);$$

(iv) for all p, q such that $1 \leq p < 2 < q \leq \infty$,

$$\|\mathcal{H}_t\|_{p;q} \sim t^{-\nu/2} \exp(-bt) \qquad \forall t \in [1, \infty);$$

(v) for all p, q such that $1 \leq p < q < 2$,

$$\|\mathcal{H}_t\|_{p;q} \sim t^{-\ell/2q'} \exp\left(-(1 - \delta(q)^2)bt\right) \qquad \forall t \in [1, \infty);$$

(vi) for all p, q such that $2 < p < q \leq \infty$,

$$\|\mathcal{H}_t\|_{p;q} \sim t^{-\ell/2p} \exp\left(-(1 - \delta(p)^2)bt\right) \qquad \forall t \in [1, \infty).$$

Theorem 2.6. *Suppose that* $0 \leq \theta \leq 1$ *and* $1 \leq p, q \leq \infty$. *The following hold:*

(i) *if* $t > 0$, *then* $\mathcal{L}_{t,\theta}$ *is bounded from* $L^p(X)$ *to* $L^q(X)$ *only if* $p \leq q$, $p \leq p'_\theta$, *and* $q \geq p_\theta$;

(ii) *if* $p_\theta \leq p \leq p'_\theta$, *then*

$$\|\mathcal{L}_{t,\theta}\|_{p;p} = \exp\left(-[(\frac{4}{pp'} - \theta)b]^{1/2}t\right) \qquad \forall t \in \mathbb{R}^+;$$

(iii) *if* $p \leq q$, $p \leq p'_\theta$ *and* $q \geq p_\theta$, *then*

$$\|\mathcal{L}_{t,\theta}\|_{p;q} \sim t^{-n(1/p-1/q)} \qquad \forall t \in (0, 1];$$

(iv) *if* $p < q = 2$ *or* $2 = p < q$, *then*

$$\|\mathcal{L}_{t,\theta}\|_{p;q} \sim t^{-\nu/4} \exp\left(-[(1 - \theta)b]^{1/2}t\right) \qquad \forall t \in [1, \infty);$$

(v) *if* $p < 2 < q$, *then*

$$\|\mathcal{L}_{t,\theta}\|_{p;q} \sim t^{-\nu/2} \exp\left(-[(1 - \theta)b]^{1/2}t\right) \qquad \forall t \in [1, \infty);$$

(vi) *if* $p < q < 2$ *and* $q > p_\theta$, *then*

$$\|\mathcal{L}_{t,\theta}\|_{p;q} \sim t^{-\ell/2q'} \exp\left(-[(\frac{4}{qq'} - \theta)b]^{1/2}t\right) \qquad \forall t \in [1, \infty);$$

(vii) *if* $p < q < 2$ *and* $q > p_\theta$, *then*

$$\|\mathcal{L}_{t,\theta}\|_{p;q} \sim t^{-\ell/2q'} \exp\left(-[(\frac{4}{qq'} - \theta)b]^{1/2}t\right) \qquad \forall t \in [1, \infty);$$

(viii) if $p < q = p_\theta$, then

$$\|\mathcal{L}_{t,\theta}\|_{p;q} \sim t^{-(\ell+1)/q'} \qquad \forall t \in [1, \infty);$$

(ix) if $2 < p < q$ and $p < p'_\theta$, then

$$\|\mathcal{L}_{t,\theta}\|_{p;q} \sim t^{-\ell/2p} \exp\left(-[(\frac{4}{pp'} - \theta)b]^{1/2}t\right) \qquad \forall t \in [1, \infty);$$

(x) if $p'_\theta = p < q$, then

$$\|\mathcal{L}_{t,\theta}\|_{p;q} \sim t^{-(\ell+1)/p} \qquad \forall t \in [1, \infty).$$

It is worth pointing out that the significant difference in the behaviour of the solutions of the heat and Laplace equations is due to the fact that the function $\exp(-tQ_\theta(\cdot))$ extends to an entire function in $\mathfrak{a}_{\mathbb{C}}^*$ while the function $\exp(-tQ_\theta(\cdot)^{1/2})$ extends into a tube $\mathbf{T}_\delta$, where $\delta = (1 - \theta)^{1/2}$, but into no larger tube. This shows clearly that harmonic analysis on a noncompact symmetric space involves phenomena with no Euclidean analogue.

2.6 Approaches to the Wave and Schrödinger Equations

Dealing with the wave equation is tricky. There is a method, due originally to Hadamard, for obtaining a parametrix for the fundamental solutions, but for large values of the time parameter this is not very easy to deal with except in the complex case and a few other relatively simple special cases.

Despite this, considerable progress has been made, and there are many important papers on this topic, starting, perhaps, with work of Helgason [61]. T. Branson, G. Ólafsson, and H. Schlichtkrull, in various combinations (see [13] and the references cited there), have studied the heat equation by analytic methods, while O.A. Chalykh and A.P. Veselov [19] used algebraic methods.

S. Giulini, S. Meda and I [38] have given a parametrix expression for the wave operator and used this to obtain L^p-L^q mapping estimates for the complexified Poisson semigroup, and we have also [39] obtained L^p-L^q estimates for the operator with convolution kernel w_θ^α, defined by the condition

$$\widetilde{w}_\theta^\alpha(\lambda) = Q_\theta(\lambda)^{-\alpha/2} \exp(iQ_\theta(\lambda)^{1/2}),$$

using a representation of this operator originating in T.P. Schonbek [94] in the Euclidean case. There is interest in obtaining similar inequalities where the exponential is replaced by $\exp(itQ_\theta(\lambda)^{1/2})$, and t is allowed to vary. Such inequalities, which belong to the family known as Strichartz estimates, are now a standard tool in hyperbolic partial differential equations.

Last but not least, let us consider the Schrödinger operator. Here, not very much is known except in the usual special cases (complex groups and real rank one groups). Note, however, that the spherical Fourier transform $\widetilde{s}_{t,\theta}$ extends to an entire function, but this grows exponentially in any tube T_δ when $\delta > 0$; this suggests that, as far as restrictions on the indices are concerned, the results should resemble those for the heat equation case more than those for Laplace's equation. Note also that the rapid oscillation of $\widetilde{s}_{t,\theta}(\lambda)$ as $\lambda \to \infty$ in $\mathfrak{a}^*$ implies that $\widetilde{s}_{t,\theta}(\lambda)$ defines an analytic functional, so that the kernel is in $L^{2+\epsilon}(G)$ for all positive ϵ. In a "discrete symmetric space" (see later for an indication of what this might mean), some results have been obtained by A.G. Setti [95].

2.7 Further Results

Much more is known about harmonic analysis on semisimple groups than is outlined here. In the area of spherical Fourier analysis, it is more than appropriate to mention the book of R. Gangolli and V.S. Varadarajan [50], which presents a complete picture of the Harish-Chandra viewpoint. Much of the theory of spherical functions may be viewed as statements about certain special functions, and generalised. T. Koornwinder [72] presents a pleasant account of this interface between group theory and special functions. As mentioned above, in the complex case, there are explicit formulae for the spherical functions. For some other groups, there are ways of getting some control of the spherical functions.

For the purposes of harmonic analysis, there are a number of important characterisations of the image under the spherical Fourier transformation of spaces on the semisimple group G. In particular, there is a family of "Schwartz spaces" on G, whose images were characterised by various authors, as well as a Paley–Wiener theorem characterising the compactly supported functions. The original Paley–Wiener results are due to Helgason [58] and Gangolli [49], and the Schwartz space results are due to P.C. Trombi and V.S. Varadarajan [100]. The proofs have been simplified since then. See [2, 24, 60] for more in this direction.

A lot of effort has been put into determining conditions on a distribution or on its Fourier transform which imply that it convolves $L^p(G)$ into itself. The major contributions here include [1, 77, 78, 96]. In the more general setting of a Riemannian manifold, there are many results on the functional calculus for the Laplace–Beltrami operator. Arguably, the key technique here has been the use of the finite propagation speed of the solutions to the heat equation, pioneered by J. Cheeger, M. Gromov, and especially M.E. Taylor [20, 99].

3 The Vanishing of Matrix Coefficients

Semisimple groups differ from other locally compact groups in the sense that their unitary representations may be characterised by the rate of decay of their matrix coefficients. In this lecture, I make this statement more precise, describing different ways in which this decay can be quantified.

3.1 Some Examples in Representation Theory

Suppose that G is a Lie group. A *unitary representation* $(\pi, \mathcal{H}_\pi)$ of G is a Hilbert space $\mathcal{H}_\pi$ and a homomorphism π from G into $\mathcal{U}(\mathcal{H}_\pi)$, the group of unitary operators on $\mathcal{H}_\pi$. We always suppose that π is continuous when $\mathcal{U}(\mathcal{H}_\pi)$ is equipped with the strong operator topology. We usually abuse notation a little and talk of "the representation π", the Hilbert space being implicit.

Recall that the representation π is said to be *reducible* if there are non-trivial G-invariant closed subspaces of $\mathcal{H}_\pi$, and *irreducible* otherwise. A vector ξ in $\mathcal{H}_\pi$ is said to be *smooth* if the $\mathcal{H}_\pi$-valued function $x \mapsto \pi(x)\xi$ on G is smooth, or equivalently if all the $\mathbb{C}$-valued functions $x \mapsto \langle \pi(x)\xi, \eta \rangle$ (where η varies over $\mathcal{H}_\pi$) are smooth. Similar definitions may be made when the Hilbert space $\mathcal{H}_\pi$ is replaced by a Banach space.

For unitary representations, as distinct from Banach space representation, there is a reasonably satisfactory theory of decompositions into irreducible representations. An arbitrary unitary representation can be written as a *direct integral* (a generalisation of a direct sum) of irreducible representations. Many Lie groups, including semisimple Lie groups and real algebraic groups (groups of matrices defined by algebraic equation in the entries), have the property that this direct integral decomposition is essentially unique. On the other hand, many groups, such as noncommutative free groups, do not have unique direct integral decompositions; this makes analysis on these groups harder. To give an indication of the sorts of decomposition which appear for "good" groups, we give two examples.

For the group $\mathbb{R}^n$, the irreducible representations are the characters $\chi_y \colon x \mapsto \exp(-2\pi i y \cdot x)$, where y varies over $\mathbb{R}^n$. Given a positive Borel measure ν on $\mathbb{R}^n$ with support S_ν, form the usual Hilbert space $L^2(S_\nu, \nu)$ of complex-valued functions on S_ν, and define the representation π_ν on $L^2(S_\nu, \nu)$ by the formula

$$[\pi_\nu(x)\xi](y) = \chi_y(x)\,\xi(y) \qquad \forall y \in S_\nu$$

for all ξ in $L^2(S_\nu, \nu)$ and all x in $\mathbb{R}^n$. Any unitary representation of $\mathbb{R}^n$ is unitarily equivalent to a direct sum of representations π_ν, with possibly different ν's.

The *ax + b group* Q is defined to be the group of all matrices $M_{a,b}$ of the form

$$M_{a,b} = \begin{bmatrix} a & b \\ 0 & a^{-1} \end{bmatrix},$$

where $a \in \mathbb{R}^+$ and $b \in \mathbb{R}$, equipped with the obvious topology.

The set N of matrices $M_{1,b}$, with b in $\mathbb{R}$, is a closed normal subgroup of Q, and the quotient group Q/N is isomorphic to the multiplicative group $\mathbb{R}^+$. By identifying x in $\mathbb{R}$ with the vector $\begin{pmatrix} x \\ 1 \end{pmatrix}$ in $\mathbb{R}^2$, we obtain an action of Q on $\mathbb{R}$, given by

$$M_{a,b} \circ x = a^2 x + ab \qquad \forall x \in \mathbb{R} \quad \forall M_{a,b} \in Q.$$

Using this action, we define the representation σ of Q on $L^2(\mathbb{R})$ by the formula

$$\sigma(M_{a,b})\xi(x) = a^{-1}\xi(M_{a,b}^{-1} \circ x) \qquad \forall x \in \mathbb{R}.$$

Under the representation σ of Q, the Hilbert space $L^2(\mathbb{R})$ breaks up into two irreducible subspaces, $L^2(\mathbb{R})_+$ and $L^2(\mathbb{R})_-$, containing those functions whose Fourier transforms are supported in $[0, +\infty)$ and in $(-\infty, 0]$ respectively. The restrictions of σ to these two subspaces are denoted σ_+ and σ_-. Any unitary representation π of the group Q decomposes as a sum $\pi_1 \oplus \pi_0$, where π_1 is trivial on N and hence is essentially a representation of the quotient group Q/N, and π_0 is a direct sum of copies of the representations σ_+ and σ_-.

Unitary representations of semisimple groups have been described, albeit incompletely and briefly, in the first lecture. Much more is known—see the references there.

Associated to a unitary representation π of G, there are *matrix coefficients*. For ξ and η in $\mathcal{H}$, $x \mapsto \langle \pi(x)\xi, \eta \rangle$ is a bounded continuous function on G, written $\langle \pi(\cdot)\xi, \eta \rangle$. If ξ and η run over an orthogonal basis of $\mathcal{H}_\pi$, then we obtain a matrix of functions corresponding to the representation of $\pi(\cdot)$ as a matrix in this basis. Thus the collection of all matrix coefficients contains complete information about π; for many purposes, however, it is easier to deal with spaces of functions on G rather than representations. As we may decompose a unitary representation π of G, so we may decompose the matrix coefficients of π into sums of matrix coefficients of "smaller" representations. In the case of $\mathbb{R}^n$, this decomposition writes a function on $\mathbb{R}^n$ as a sum or an integral of characters—this is just Fourier analysis under a different guise.

When $G = \mathbb{R}^n$, the matrix coefficients of the irreducible representations of G are (multiples of) characters. These are constant in absolute value, and in particular do not vanish at infinity or belong to any L^p space with finite p. However, the regular representation λ of G on $L^2(G)$ has matrix coefficients which decay at infinity: if $\xi, \eta \in L^2(G)$, then

$$\langle \lambda(x)\xi, \eta \rangle = \int_G \lambda(x)\xi(y)\,\overline{\eta}(y)\,dy$$

$$= \int_{\mathbb{R}^n} \xi(y - x)\,\overline{\eta}(y)\,dy \qquad \forall x \in \mathbb{R}^n,$$

and $\langle \lambda(\cdot)\xi, \eta \rangle$ has compact support if ξ and η do, and is in $C_0(G)$ in general. It is easy to show that, by choosing ξ and η appropriately, it is possible to make $\langle \lambda(\cdot)\xi, \eta \rangle$ decay arbitrarily slowly.

In the $ax + b$ case, all the matrix coefficients of a representation π_1 which is trivial on N are constant on cosets of N in Q, and do not vanish at infinity, no matter how π_1 behaves on the quotient group Q/N. On the other hand, the representations σ_+ and σ_- have the property that all their matrix coefficients vanish at infinity; as in the $\mathbb{R}^n$ case, this decay may be arbitrarily slow.

3.2 Matrix Coefficients of Representations of Semisimple Groups

All semisimple Lie groups are "almost direct products" of "simple factors"; unitary representations break up as "outer tensor products" of representations of the various factors, and the matrix coefficients decompose similarly. There is therefore little loss of generality in restricting attention to simple Lie groups, that is, those whose Lie algebra is simple, for the rest of this lecture.

An important notion in the study of representations of a semisimple Lie group is K-finiteness. A vector ξ in $\mathcal{H}_\pi$ is said to be K-finite if $\{\pi(k)\xi : k \in K\}$ spans a finite dimensional subspace. The set of K-finite vectors is a dense subspace of $\mathcal{H}_\pi$.

The first key fact about a unitary representation π of a simple Lie group G is that it decomposes into two pieces, π_1 and π_0. The representation π_1 is a multiple of the trivial representation, and the associated matrix coefficients are constants, while all the matrix coefficients of π_0 vanish at infinity in G. Several of the proofs of this involve looking at subgroups R of G similar to the group Q described above, and "lifting" to G the decomposition from R. The difficulty of the proof is in showing that G acts trivially on the vectors where the normal subgroup N of R acts trivially.

The remarkable fact is that we can often say more than this: for most representations of interest, there is control on the rate of decay. There are two ways to quantify the rate of decay of matrix coefficients: uniform estimates and L^{p+} estimates.

Recall, from Lecture 1, the Cartan decomposition: every x in G may be written in the form

$$x = k_1 a k_2,$$

where $k_1, k_2 \in K$ and $a \in \overline{A}^+$, the closure of $\exp(\mathfrak{a}^+)$ in A. A uniform estimate for a matrix coefficient u is an estimate of the form

$$|u(k_1 a k_2)| \leq C\,\phi(a) \qquad \forall k_1, k_2 \in K \quad \forall a \in \overline{A}^+,$$

for some function ϕ in $C_0(\overline{A}^+)$. An L^{p+} estimate for u is the statement that, for any positive ϵ, the function u is in $L^{p+\epsilon}(G)$, and $\|u\|_{p+\epsilon}$, the $L^{p+\epsilon}(G)$ norm of u, may be estimated. Matrix coefficients of unitary representations are always bounded, so that if $u \in L^p(G)$, then $u \in L^q(G)$ for all q in $[p, \infty]$. Thus the set of q such that $u \in L^q(G)$ is an interval containing ∞.

For irreducible representations of G, the K-finite matrix coefficients are solutions of differential equations on A. These are ℓ-dimensional generalisations of the hypergeometric differential equation, and an extension of the analysis of differential equations with regular singular points to A leads to proofs of the existence of asymptotic expressions for matrix coefficients (first carried out by Harish-Chandra, but published in an improved and simplified version by W. Casselman and D. Miličić [18]). In particular, it can be shown that, for all K-finite matrix coefficients of an irreducible unitary representation,

$$\langle \pi(\exp(H))\xi, \eta \rangle \sim \sum_{\gamma \in I} C(\xi, \eta, \gamma) \wp_\gamma(H) e^{-(\rho+\gamma)(H)}$$

as $H \to \infty$ in $\mathfrak{a}^+$, keeping away from the walls of $\mathfrak{a}^+$, for some finite subset I of $\mathfrak{a}_{\mathbb{C}}^*$ with the property that $\operatorname{Re}\gamma(H) \geq 0$ for all H in $\mathfrak{a}^+$ and γ in I, and some polynomials $\wp_\gamma$ of bounded degree; both the set I of "leading terms" and the polynomials $\wp_\gamma$ are independent of ξ and η.

From this fact, it appears that the best sort of uniform estimate to consider is one of the form

$$|u(k_1 \exp(H)k_2)| \leq C \, \wp(H) e^{-\gamma(H)} \qquad \forall k_1, k_2 \in A \quad \forall H \in \overline{\mathfrak{a}}^+,$$

where $\gamma \in \mathfrak{a}^+$ and $\wp$ is a polynomial. It can be shown that such estimates hold for K-finite matrix coefficients of irreducible representations.

Let us now formulate a conjecture.

Conjecture 3.1. *Suppose that π is a unitary representation of a simple Lie group G, that $\gamma \in \mathfrak{a}^*$, and that $0 < \gamma(H) \leq \rho(H)$ for all H in $\mathfrak{a}^+$. Then*

$$|\langle \pi(k_1 \exp(H)k_2)\xi, \eta \rangle| \leq C(\xi, \eta) \, \wp_\pi(H) e^{-\gamma(H)} \qquad \forall k_1, k_2 \in A \quad \forall H \in \overline{\mathfrak{a}}^+$$

if and only if a similar inequality, with the same γ, holds for each of the irreducible representations involved in the decomposition of π.

Note that some of the representations involved in the decomposition of π may admit uniform estimates with much more rapid decay rates. I know of no proof of this conjecture in general, but it is certainly true in a few simple cases.

By using L^{p+} estimates, we may prove a version of Conjecture 3.1 for the case where $\alpha = (1/m)\rho$, for some positive integer m. To do this, we have to find a connection between uniform estimates and L^{p+} estimates.

Suppose that $0 < t \leq 1$, that $\wp$ is a positive polynomial on $\overline{\mathfrak{a}}^+$, and that

$$\phi(\exp(H)) = \wp(H) e^{-t\rho(H)} \qquad \forall H \in \overline{\mathfrak{a}}^+.$$

If $pt \geq 2$, and

$$|u(k_1 a k_2)| \leq \phi(a) \qquad \forall k_1, k_2 \in K \quad \forall a \in \overline{A}^+,$$

then u satisfies L^{p+} estimates. Indeed, from (1.5),

$$\|u\|_{p+\epsilon} = \left(C \int_K \int_{\overline{\mathfrak{a}}^+} \int_K |u(k_1 \exp(H)k_2)|^{p+\epsilon}\right.$$

$$\left.\prod_{\alpha \in \Sigma} \sinh(\alpha(H))^{\dim(\mathfrak{g}_\alpha)} \, dk_1 \, dH \, dk_2\right)^{1/(p+\epsilon)}$$

$$\leq \left(C \int_{\overline{\mathfrak{a}}^+} \left|\wp(H)e^{-t\rho(H)}\right|^{p+\epsilon} e^{2\rho(H)} \, dH\right)^{1/(p+\epsilon)}$$

$$= \left(C \int_{\overline{\mathfrak{a}}^+} |\wp(H)|^{p+\epsilon} \exp(-t(p+\epsilon)\rho(H) + 2\rho(H)) \, dH\right)^{1/(p+\epsilon)}$$

$$\leq \left(C \int_{\overline{\mathfrak{a}}^+} |\wp(H)|^{p+\epsilon} \exp(-t\epsilon\rho(H)) \, dH\right)^{1/(p+\epsilon)}$$

$$= C_{G,p,\epsilon} < \infty.$$

More generally, if $\alpha \in \mathbf{T}_t$, and

$$|u(k_1 \exp(H)k_2)| \leq \wp(H)e^{-\alpha(H)} \qquad \forall H \in \overline{\mathfrak{a}}^+,$$

for some polynomial $\wp$, then u satisfies $L^{2/t+}$ estimates. This shows that uniform estimates imply L^{p+} estimates.

Conversely, good L^{p+} estimates imply uniform estimates. More precisely, if we know that $\langle \pi(\cdot)\xi, \eta \rangle \in L^{p+\epsilon}(G)$ for all positive ϵ and all K-finite smooth vectors ξ and η in $\mathcal{H}_\pi$, then we argue that the function $\langle \pi(\cdot)\xi, \eta \rangle$ and lots of its derivatives lie in $L^{p+\epsilon}(G)$ for all positive ϵ; using an argument involving the Sobolev embedding theorem and the exponential growth of G, we may then show that

$$|\langle \pi(k_1 \exp(H)k_2)\xi, \eta \rangle| \leq C(\epsilon, \xi, \eta)e^{-(2/p+\epsilon)\rho(H)} \qquad \forall k_1, k_2 \in K \quad \forall H \in \overline{\mathfrak{a}}^+.$$

The details of this may be found in [30].

Observe that the uniform estimates are "nicer" than the L^{p+} estimates because they contain more information: decay can be faster in some directions than others. However, they have the weakness that they are not translation-invariant: if

$$|u(k_1 a k_2)| \leq \phi(a) \qquad \forall k_1, k_2 \in K \quad \forall a \in \overline{A}^+,$$

and $v(x) = u(xy)$ for some y and all x in G, it need not follow that

$$|v(k_1 a k_2)| \leq \phi(a) \qquad \forall k_1, k_2 \in K \quad \forall a \in \overline{A}^+.$$

On the other hand, spaces of matrix coefficients are translation-invariant: indeed

$$\langle \pi(\cdot y)\xi, \eta \rangle = \langle \pi(\cdot)(\pi(y)\xi), \eta \rangle.$$

3.3 The Kunze–Stein Phenomenon

Possibly the most important decay estimate for matrix coefficients of simple (or semisimple) Lie groups is the Kunze–Stein phenomenon. This says that L^{2+} estimates hold for the matrix coefficients of the regular representation λ of G on $L^2(G)$. More precisely, for all positive ϵ, there exists a constant C_ϵ such that, if ξ and η are in $L^2(G)$, then $\langle\lambda(\cdot)\xi,\eta\rangle \in L^{2+\epsilon}(G)$ and

$$\|\langle\lambda(\cdot)\xi,\eta\rangle\|_{2+\epsilon} \le C_\epsilon \,\|\xi\|_2 \,\|\eta\|_2 \,.$$

This result was first observed by R.A. Kunze and Stein for the case where $G = \mathrm{SL}(2,\mathbb{R})$, then extended to a number of other simple Lie groups by Kunze and Stein, and by others. Inspired by Kunze and Stein, C.S. Herz [63] and then P. Eymard and N. Lohoué [46] made inroads into the general case. The first general proof, which uses a simplified version of the argument of Kunze and Stein, may be found in [29]. An important corollary of a little functional analysis combined with the Kunze–Stein phenomenon is that, if π is any representation of a simple Lie group G and, for some positive integer m, $\langle\pi(\cdot)\xi,\eta\rangle \in L^{2m+\epsilon}(G)$ for all positive ϵ and all ξ and η in a dense subspace of $\mathcal{H}_\pi$, then $\langle\pi(\cdot)\xi,\eta\rangle \in L^{2m+\epsilon}(G)$ for all positive ϵ and all ξ and η in $\mathcal{H}_\pi$; further there exists a constant $C(G,\epsilon,m)$ such that

$$\|\langle\pi(\cdot)\xi,\eta\rangle\|_{2m+\epsilon} \le C(G,\epsilon,m) \,\|\xi\|\,\|\eta\| \qquad \forall \xi,\eta \in \mathcal{H}_\pi.$$

The functional analysis serves to show that the m-fold tensor product $\pi\otimes^m$ of π is "weakly contained in the regular representation"; the Kunze–Stein phenomenon then gives $L^{2+\epsilon/m}$ estimates for $\langle\pi(\cdot)\xi,\eta\rangle^m$. See [35] for more details.

It may be conjectured that, if $q \ge 2$ and $\langle\pi(\cdot)\xi,\eta\rangle \in L^{q+\epsilon}(G)$ for all positive ϵ and all ξ and η in a dense subspace of $\mathcal{H}_\pi$, then $\langle\pi(\cdot)\xi,\eta\rangle \in L^{q+\epsilon}$ for all ξ and η in $\mathcal{H}$, and L^{q+} estimates hold. This is certainly correct for the cases when $G = \mathrm{SO}(1,n)$ or $\mathrm{SU}(1,n)$, but the proof for these groups does not generalise. No such result can hold if $q < 2$: see [30] for the argument.

In any case, these ideas, together with the links between uniform estimates and L^{p+} estimates, establish the claim earlier that Conjecture 3.1 holds when $\alpha = (1/m)\rho$, for some positive integer m.

Recently, sharper versions of the Kunze–Stein phenomenon have been discovered, at least for groups of real rank one. In particular, the Kunze–Stein estimates are dual to the convolution estimate $L^p(G) * L^2(G) \subseteq L^2(G)$; by interpolation with the obvious result $L^1(G) * L^1(G) \subseteq L^1(G)$, this implies that $L^r(G) * L^s(G) \subseteq L^s(G)$ when $1 \le r < s \le 2$. By using Lorentz spaces $L^{p,q}(G)$, it is possible to formulate versions of the Kunze–Stein convolution theorem such as $L^{p,1}(G) * L^p(G) \subseteq L^p(G)$ for the case where $p < 2$ (see [32]) and $L^{2,1}(G) * L^{2,1}(G) \subseteq L^{2,\infty}(G)$ (see [66]). The study of related operators, such as maximal operators, has also begun; the major result here is that of J.-O. Strömberg [97].

3.4 Property T

Certain Lie groups (and many more locally compact groups) have property T. This is a property with several equivalent formulations, one of which is that the trivial representation is isolated in the unitary dual $\hat{G}$ of G, that is, the set of all equivalence classes of irreducible unitary representations of G, equipped with a natural topology.

This property was introduced by D.A. Kazhdan [70] who proved that $SL(3, \mathbb{R})$, and many other simple Lie groups with $\dim(A) \geq 2$, have it. Shortly after, S.P. Wang [103] observed that Kazhdan's argument could be developed to prove that all simple Lie groups with $\dim(A) \geq 2$ have property T. At about the same times, B. Kostant [73, 74] established that $Sp(1, n)$ and $F_{4,-20}$ have property T, while $SO(1, n)$ and $SU(1, n)$ do not. Since then property T has appeared in a number of different applications of representation theory, including the proof (of J.M. Rosenblatt, D. Sullivan and G.A. Margulis [82, 88, 98]) that Lebesgue measure is the only finitely additive rotation-invariant position additive set function on the sphere S^k, for $k \geq 5$, and the construction by Margulis [81, 83] of "expanders", graphs with a very high degree of connectivity. The monograph of P. de la Harpe and A. Valette [56] presents a detailed account of these applications, and much more; for more recent applications, see also the monographs of P. Sarnak [89] and A. Lubotzky [79]. If G is a simple Lie group with property T, then there exists p_G in $(2, \infty)$ such that

$$\langle \pi(\cdot)\xi, \eta \rangle \in L^{p_G + \epsilon}(G) \qquad \forall \xi, \eta \in \mathcal{H}_\pi$$

for all unitary representations π of G with no trivial subrepresentations, and further

$$\|\langle \pi(\cdot)\xi, \eta \rangle\|_{p_G + \epsilon} \leq C_{G,\epsilon} \|\xi\| \|\eta\| \qquad \forall \xi, \eta \in \mathcal{H}_\pi.$$

In other words, there is uniform vanishing at infinity of all matrix coefficients which vanish at infinity. This can also be expressed with uniform estimates for smooth K-finite matrix coefficients.

Kazhdan's proof that $SL(3, \mathbb{R})$ has property T uses an argument like the argument already given to show that matrix coefficients decay at infinity. Indeed, $SL(3, \mathbb{R})$ contains the subgroup Q_1, of all elements of the form

$$\begin{bmatrix} a & b & x \\ c & d & y \\ 0 & 0 & 1 \end{bmatrix},$$

where $a, b, c, d, x, y \in \mathbb{R}$, and $ad - bc = 1$. The subgroups M_1 and N_1 of Q_1 are defined by the conditions that $x = y = 0$ (for M_1) and $a = d = 1$ and $b = c = 0$ (for N_1).

Representation theory ("the Mackey machine") shows that any unitary representation π of Q_1 splits into two: $\pi = \pi_1 \oplus \pi_0$, where π_1 is trivial on N_1 and $\pi_0|_{M_1}$ is a subrepresentation of the regular representation of M_1. A representation whose matrix coefficients vanish at infinity cannot have a π_1 component. Kazhdan used this analysis to deduce that π cannot approach 1. In [30], it is shown that the matrix coefficients of π, restricted to M_1, satisfy $L^{2+\epsilon}$ estimates; this is then used to show that the matrix coefficients satisfy L^{p_G+} estimates on G. Later, R.E. Howe [65], R. Scaramuzzi [93], J.-S. Li [75, 76] and H. Oh [85, 86] analysed the various possibilities more carefully, and found optimal values for p_G.

3.5 The Generalised Ramanujan–Selberg Property

Suppose that the simple Lie group G acts on a probability space Ω, preserving the measure. Then there is a unitary representation π of G on $L^2(\Omega)$ given by the formula

$$[\pi(x)\xi](\omega) = \xi(x^{-1}\omega) \qquad \forall \omega \in \Omega \quad \forall \xi \in L^2(\Omega).$$

The constant functions form a 1-dimensional G-invariant subspace of $L^2(\Omega)$; denote by $L^2(\Omega)_0$ its orthogonal complement. When G acts ergodically on Ω, there is no invariant vector in $L^2(\Omega)_0$ (this may be taken as the definition of ergodicity). It follows that all the matrix coefficients of the restriction π_0 of π to $L^2(\Omega)_0$ vanish at infinity. If G has property T, then these matrix coefficients satisfy a L^{p_G+} estimate. We define an action of G on a probability space Ω to be a T-action if there is some finite q such that the restricted representation π_0 has matrix coefficients which satisfy a L^{q+} estimate. Then every ergodic action of a group with property T is a T-action.

If the action of G on a probability space Ω is an T-action, then information about the representation of G on $L^p(\Omega)$ comes from complex interpolation. Indeed, suppose that $p < 2$, and that $\xi \in L^p(\Omega)_0$ and $\eta \in L^{p'}(\Omega)_0$, that is, $\xi \in L^p(\Omega)$, $\eta \in L^{p'}(\Omega)$, and both have zero mean on Ω. For a complex number z with $\mathrm{Re}(z)$ in $[0,1]$, define ξ_z and η_z:

$$\xi_z = |\xi|^{p(1-z/2)-1}\xi - \int_\Omega |\xi|^{p(1-z/2)-1}\xi$$

and

$$\overline{\eta}_z = |\eta|^{p'z/2-1}\overline{\eta} - \int_\Omega |\eta|^{p'z/2-1}\overline{\eta}.$$

If $\mathrm{Re}(z) = 0$, then $\left||\xi|^{p(1-z/2)-1}\xi\right| = |\xi|^p$ and $\left||\eta|^{p'z/2-1}\overline{\eta}\right| = 1$, whence

$$\|\xi_z\|_1 \leq 2\|\xi\|_p^p \qquad \text{and} \qquad \|\eta_z\|_\infty \leq 2;$$

similarly if $\mathrm{Re}(z) = 1$, then

$$\|\xi_z\|_2 \leq 2\|\xi\|_p^{p/2} \qquad \text{and} \qquad \|\eta_z\|_2 \leq 2\|\eta\|_{p'}^{p'/2}.$$

Consider the analytic family of functions on $\{z \in \mathbb{C} : \mathrm{Re}(z) \in [0,1]\}$ given by

$$z \mapsto \langle \pi(\cdot)\xi_z, \eta_z \rangle.$$

When $\operatorname{Re}(z) = 0$, these functions are bounded on G, while when $\operatorname{Re}(z) = 1$, these functions are coefficients of π_0, so satisfy L^{q+} estimates. When $z = 2/p'$, we get $\langle \pi(\cdot)\xi, \eta \rangle$. By a standard complex interpolation argument, this function on G satisfies $L^{2q/p'+}$ estimates.

There are some important examples of T-actions of groups which do not have property T. In particular, if $G = \operatorname{SL}(2, \mathbb{R})$ and $X = G/\Gamma$, where Γ is a congruence subgroup, that is, for some N in $\mathbb{Z}^+$,

$$\Gamma = \left\{ \begin{bmatrix} a & b \\ c & d \end{bmatrix} \in \operatorname{SL}(2, \mathbb{Z}) : a - 1 \equiv d - 1 \equiv b \equiv c \equiv 0 \mod N \right\},$$

then the action of G on X is a T-action. Indeed, π_0 satisfies a L^{4+} estimate. This is a reformulation of a celebrated result of A. Selberg (generalising a result of G. Roelcke for $\operatorname{SL}(2, \mathbb{R})/\operatorname{SL}(2, \mathbb{Z})$ which states, in our language, that for this choice of Γ, the representation π_0 satisfies a L^{2+} estimate). Selberg also conjectured that π_0 satisfies an L^{2+} estimate. This result is usually phrased in terms of the first nonzero eigenvalue of the Laplace–Beltrami operator on the space $K \backslash G/\Gamma$, a quotient of the hyperbolic upper half plane; the representation theoretic version is due to I. Satake [90].

Similar results were discovered by M. Burger, J.-S. Li and Sarnak [16, 17], and formulated in terms of the "Ramanujan dual". It is now known that every action of a real simple algebraic group G on the quotient space G/Γ is a T-action, for any lattice Γ in G (arithmetic or not). As pointed out by M.E.B. Bekka, this follows from the Burger–Sarnak argument and work of A. Borel and H. Garland [12] (see [9] for more details). The Burger–Sarnak argument has been reworked by a number of people, including L. Clozel, Oh and E. Ullmo [26] and Cowling [33].

An observation by C.C. Moore [84] is relevant here; representations of simple Lie groups with finite centres which do not weakly contain the trivial representation automatically satisfy L^{p+} estimates.

4 More General Semisimple Groups

In this lecture, I look at questions in graph theory and number theory. The initial motivation is to shed light on analytical problems, such as finding the behaviour of the eigenvalues of the Laplace–Beltrami operator on Riemannian manifolds, by studying this problem for the eigenvalues of graph Laplacians. It turns out that certain problems in discrete mathematics can also be attacked effectively using approaches and results from analysis.

4.1 Graph Theory and its Riemannian Connection

A *graph* $\mathcal{G}(V, E)$, usually written $\mathcal{G}$, is a set V of *vertices* and a set E of *edges*, that is, a symmetric subset of $V \times V$. A *path* in $\mathcal{G}$ from v_0 to v_n of length n is a list of vertices $[v_0, v_1, \ldots, v_n]$ with the property that $(v_{i-1}, v_i) \in E$ when $i = 1, 2, \ldots, n$; we consider $[v_0]$ to be a path of length 0. The graph is said to be *connected* if there is a path between any two vertices. The *distance* $d(v, w)$ between vertices v and w in a connected graph is the length of a shortest path between them. The *diameter* of a finite connected graph is the greatest distance between any pair of vertices. The *degree* of a vertex v, written $\deg v$, is the cardinality of the set of vertices at distance 1 from v; the degree of the graph $\mathcal{G}$ is the supremum of the degrees of the vertices. We shall deal with connected graphs of finite degree.

For a connected graph $\mathcal{G}$ of finite degree, the graph Laplacian $\Delta_\mathcal{G}$ is defined as a map on functions on V:

$$\Delta_\mathcal{G} f(v) = f(v) - \frac{1}{\deg v} \sum_{\substack{w \in V \\ d(v, w) = 1}} f(w).$$

This operator is bounded on $L^2(V)$ and self-adjoint. It is a natural analogue of the Laplace–Beltrami operator Δ_M on a Riemannian manifold M but, being bounded, is easier to analyse.

There is already an extensive theory of "approximation" of a Riemannian manifold M and Δ_M by a graph $\mathcal{G}$ and $\Delta_\mathcal{G}$. The underlying philosophy is that properties of Δ_M which are "local" in the manifold and are reflected spectrally at infinity are lost when the manifold is discretised, but that properties of Δ_M which are "global" in the manifold and are reflected in the spectrum "near 0" will be seen in the properties of $\Delta_{\mathcal{G}_0}$. For some examples in this direction, see [27, 28, 69].

Interesting examples of graphs arise in the study of semisimple groups in two ways: as Cayley graphs of discrete groups and as "discrete symmetric spaces". We next consider Cayley graphs and then describe the p-adic numbers and discrete symmetric spaces.

4.2 Cayley Graphs

Suppose X is a set of generators for a group G, closed under the taking of inverses. The Cayley graph of (G, X) is the graph $\mathcal{G}(G, E)$, where E is the subset of $G \times G$ defined by the condition that $(x, y) \in E$ if and only if $xy^{-1} \in X$ (or equivalently $yx^{-1} \in X$). The group G acts simply transitively and isometrically on $\mathcal{G}(G, E)$ by left multiplication, so that Cayley graphs are homogeneous: all points "look alike". Cayley graphs are good for obtaining examples of graphs of small degree and small diameter but high cardinality (these are "expanders", which are important in discrete mathematics).

Suppose that $(u_n : n \in \mathbb{N})$ is a sequence of positive definite functions on G, normalised in the sense that $u_n(e) = 1$ for all n (where e denotes the identity of G). Then $u_n = \langle \pi_n(\cdot)\xi_n, \xi_n \rangle$, where $\|\xi_n\| = 1$. Suppose that $|u_n(x) - 1| < \epsilon_n$ for all x in X. Then

$$\|\pi_n(x)\xi_n - \xi_n\|^2 = \langle \pi_n(x)\xi_n - \xi_n, \pi_n(x)\xi_n - \xi_n \rangle$$
$$= 2 - 2\operatorname{Re}\langle \pi_n(x)\xi_n, \xi_n \rangle,$$

and so

$$\|\pi_n(x)\xi_n - \xi_n\| \leq (2\epsilon)^{1/2}.$$

It follows, by induction, that

$$\|\pi_n(x_1 \ldots x_m)\xi_n - \xi_n\| \leq m(2\epsilon_n)^{1/2} \qquad \forall x_1, \ldots, x_m \in X;$$

indeed

$$\|\pi_n(x_1 \ldots x_m)\xi_n - \xi_n\| \leq \|\pi_n(x_1 \ldots x_{m-1})(\pi_n(x_m)\xi_n - \xi_n)\|$$
$$+ \|\pi_n(x_1 \ldots x_{m-1})\xi_n - \xi_n\|.$$

Thus if $u_n(x) \to 1$ as $n \to \infty$ for all x in X, $u_n \to 1$ as $n \to \infty$ locally uniformly on G.

Property T may be expressed in the following form: if none of the unitary representations π_n has a trivial subrepresentation, then the corresponding matrix coefficients u_n cannot tend to 1 locally uniformly. It becomes possible to quantify property T, by finding numbers τ_G such that

$$\sup_{x \in X} |u(x) - 1| > \tau_G,$$

or perhaps (if G is finitely generated)

$$\sum_{x \in X} |u(x) - 1| > \tau_G \qquad \text{or} \qquad \left(\sum_{x \in X} |u(x) - 1|^2 \right)^{1/2} > \tau_G,$$

for all normalised positive definite functions u which are associated to unitary representations without trivial subrepresentations. (It might be argued, on the basis of results about the free group, and of its utility in formulae like inequality (4.1) that this final definition is the best). This quantification of property T leads to estimates for a spectral gap for $\Delta_{\mathcal{G}}$, acting on matrix coefficients of unitary representations. For the normalised positive definite function u, equal to $\langle \pi(\cdot)\xi, \xi \rangle$,

$$|u(x)| = |\langle \pi(x)\xi, \xi \rangle| \le \|\pi(x)\xi\| \, \|\xi\| \le 1,$$

and for any complex number z in the closed unit disc,

$$\mathrm{Re}(1 - z) \ge \frac{|1 - z|^2}{2}.$$

Thus

$$(4.1) \qquad \Delta_{\mathcal{G}} u(e) = \frac{1}{\deg e} \, \mathrm{Re}\left(\sum_{x \in X} (1 - u(x)) \right) \ge \frac{1}{2|X|} \sum_{x \in X} |1 - u(x)|^2,$$

which is bounded away from 0 if τ_G is bounded away from 0 (using any of the above definitions).

4.3 An Example Involving Cayley Graphs

Let G denote the group $\mathrm{SL}(3, \mathbb{Z})$ of all 3×3 integer matrices with determinant 1, and let X denote the symmetric generating subset

$$\left\{ \begin{bmatrix} 1 & x_{12} & x_{13} \\ x_{21} & 1 & x_{23} \\ x_{31} & x_{32} & 1 \end{bmatrix} : x_{ij} \in \{0, \pm 1\}, \quad |x_{12}| + \cdots + |x_{32}| = 1 \right\}$$

(that is, all x_{ij} but one are equal to 0). Burger [15] estimated τ_G for this group (using the first definition). Let π be a unitary representation of $\mathrm{SL}(3, \mathbb{Z})$ in a Hilbert space $\mathcal{H}_\pi$, and S a finite set of generators of $\mathrm{SL}(3, \mathbb{Z})$. For various examples of S and π, he obtains an explicit positive ε such that $\max_{\gamma \in S} \|\pi(\gamma)\xi - \xi\| / \|\xi\| \ge \varepsilon$, for all ξ in the space of π. As he observes, these results give a partial solution to the problem of giving a quantitative version of Kazhdan's property (T) for $\mathrm{SL}(3, \mathbb{Z})$.

For any prime number p, let G_p denote the finite group $\mathrm{SL}(3, \mathbb{F}_p)$, that is, the group of matrices of determinant 1 with entries in the finite field $\mathbb{F}_p$ with p elements. This is a quotient group of G. Indeed, define the normal subgroup Γ_p of G by

$$\Gamma_p = \left\{ \begin{bmatrix} x_{11} & x_{12} & x_{13} \\ x_{21} & x_{22} & x_{23} \\ x_{31} & x_{32} & x_{33} \end{bmatrix} \in \mathrm{SL}(3, \mathbb{Z}) : x_{ij} \equiv \delta_{ij} \mod p \right\},$$

where δ_{ij} is the Kronecker delta; then G_p is isomorphic to G/Γ_p. A unitary representation of G/Γ_p with no trivial subrepresentation lifts canonically to a unitary representation of G with no trivial subrepresentation, and so the estimates on the matrix coefficients of all unitary representations of G imply, in particular, estimates for the matrix coefficients of these lifted representations. Thus we obtain estimates on the degree of isolation of the trivial representation of G_p which are uniform in p. A number of estimates of this type have recently been summarised in the survey of P. Diaconis and L. Saloff-Coste [44].

4.4 The Field of p-adic Numbers

For a prime number p, the p-adic norm on the set of rational numbers $\mathbb{Q}$ is defined by

$$|0|_p = 0 \quad \text{and} \quad |x|_p = p^{-\alpha},$$

where $x = mp^\alpha/n$, m and n being integers with no factors of p. It is easy to check that $|x|_p = 0$ only if $x = 0$, that $|xy|_p = |x|_p|y|_p$, and that

$$|x + y|_p \leq \max\{|x|_p, |y|_p\} \quad \forall x, y \in \mathbb{Q}.$$

The completion of $\mathbb{Q}$ in the associated distance d_p, that is, $d_p(x, y) = |x-y|_p$, is a totally disconnected locally compact field, called the field of p-adic numbers, and written $\mathbb{Q}_p$. The algebraic operations of $\mathbb{Q}_p$ are those of formal series of the form

$$\sum_{n=N}^{\infty} a_n p^n,$$

where $N \in \mathbb{Z}$ and $a_n \in \{0, 1, 2, \ldots, p-1\}$, with "carrying", for instance,

$$
\begin{aligned}
(-1)p^k &= (p-1)p^k + (-1)p^{k+1} \\
&= (p-1)p^k + (p-1)p^{k+1} + (-1)p^{k+2} \\
&= (p-1)p^k + (p-1)p^{k+1} + (p-1)p^{k+2} + \ldots.
\end{aligned}
$$

The subset of $\mathbb{Q}_p$ of all series where $N \geq 0$ is an open and closed subring of $\mathbb{Q}_p$, known as the ring of p-adic integers, and written $\mathcal{O}_p$; this is the completion of $\mathbb{Z}$ in $\mathbb{Q}_p$. The field $\mathbb{Q}_p$ presents a few surprises to the uninitiated: for example, if $p \equiv 1 \mod 4$, then $\mathbb{Q}_p$ contains a square root of -1. However, $\mathbb{Q}_p$ does not contain very many new roots, and the algebraic completion of $\mathbb{Q}_p$ is of infinite degree over $\mathbb{Q}_p$.

Apart from $\mathbb{R}$ and $\mathbb{C}$, the real and complex numbers, the locally compact complete normed fields are "local fields", that is, they are totally disconnected. Every local field is either a finite algebraic extension of $\mathbb{Q}_p$ or a field of Laurent series in one variable over a finite field. Like $\mathbb{Q}_p$, these all have a compact open "ring of integers" $\mathcal{O}$. There is a unique translation-invariant measure on any local field which assigns measure 1 to $\mathcal{O}$.

4.5 Lattices in Vector Spaces over Local Fields

Let V be the vector space $\mathbb{Q}_p^n$ over the local field $\mathbb{Q}_p$, with the standard basis $\{e_1, \ldots, e_n\}$. A *lattice* L in V is a subset of V of the form

$$L = \{m_1 v_1 + m_2 v_2 + \cdots + m_n v_n : m_1, m_2, \ldots, m_n \in \mathcal{O}_p\},$$

where $\{v_1, \ldots, v_n\}$ is a basis for V. The set $\{v_1, \ldots, v_n\}$ is also called a basis for L over $\mathcal{O}_p$, for obvious reasons. The standard lattice L_0 is the lattice with basis $\{e_1, \ldots, e_n\}$. All lattices are compact open subsets of V.

Given two lattices L_1 and L_2, it is possible to find a basis $\{v_1, \ldots, v_n\}$ of L_1 over $\mathcal{O}_p$ such that, for suitable integers $a_1, \ldots, a_n$,

$$(4.2) \quad L_2 = \{m_1 p^{a_1} v_1 + m_2 p^{a_2} v_2 + \cdots + m_n p^{a_n} v_n : m_1, m_2, \ldots, m_n \in \mathcal{O}_p\}.$$

The order of the numbers a_i may depend on the basis chosen, but the numbers themselves do not. This result, known as the invariant factor theorem, may be found in many texts on algebra, such as C.W. Curtis and I. Reiner [43, pp. 150–153].

The group $\mathrm{GL}(n, \mathbb{Q}_p)$ acts on the vector space V and hence on the space $\mathcal{L}$ of lattices in V. The stabiliser of the standard lattice L_0 is the compact subgroup $\mathrm{GL}(n, \mathcal{O}_p)$ of invertible $\mathcal{O}_p$-valued $n \times n$ matrices whose inverses are also $\mathcal{O}_p$-valued (equivalently, whose determinant has norm 1). Thus the space $\mathcal{L}$ may be identified with the coset space $\mathrm{GL}(n, \mathbb{Q}_p)/\mathrm{GL}(n, \mathcal{O}_p)$. The group $\mathrm{GL}(n, \mathbb{Q}_p)$ is not semisimple, and this is not quite the analogue of a Riemannian symmetric space.

One example of a discrete symmetric space may be obtained by restricting attention to the space $\mathcal{L}_1$ of lattices whose volume is equal to that of the standard lattice. It follows from the invariant factor theorem that $\mathcal{L}_1$ may be identified with the coset space $\mathrm{SL}(n, \mathbb{Q}_p)/\mathrm{SL}(n, \mathcal{O}_p)$.

The more standard example of a discrete symmetric space is a quotient space of $\mathcal{L}$. Define an equivalence relation $\sim$ on $\mathcal{L}$ by the stipulation that $L_1 \sim L_2$ if $L_1 = \lambda L_2$ for some λ in $\mathbb{Q}_p$; the equivalence class of L is written $[L]$. Define $d \colon \mathcal{L} \times \mathcal{L} \to \mathbb{N}$ by

$$d(L_1, L_2) = \max\{a_i : 1 \leq i \leq n\} - \min\{a_i : 1 \leq i \leq n\},$$

where $\{a_i : 1 \leq i \leq n\}$ is as in formula (4.2) above. It is simple to check that d factors to a distance function on the space $[\mathcal{L}]$ of equivalence classes of lattices. We may identify $[\mathcal{L}]$ with the coset space $\mathrm{PGL}(n, \mathbb{Q}_p)/\mathrm{PGL}(n, \mathcal{O}_p)$, where $\mathrm{PGL}(n, \mathbb{Q}_p)$ is the quotient group $\mathrm{GL}(n, \mathbb{Q}_p)/Z$, Z being its centre (that is, the group of nonzero diagonal matrices), and $\mathrm{PGL}(n, \mathcal{O}_p)$ is the image of $\mathrm{GL}(n, \mathcal{O}_p)$ in $\mathrm{PGL}(n, \mathbb{Q}_p)$. The key to this identification is the observation that the scalar matrix λI moves L to λL, and preserves the equivalence classes. The space $[\mathcal{L}]$ has the structure of a simplicial complex, in which the vertices are the equivalence classes $[L]$, and the edges are pairs $([L_1], [L_2])$ where $d(L_1, L_2) = 1$.

Similar constructions apply when $\mathbb{Q}_p$ is replaced by another local field.

Perhaps the moral of this is just that a discrete symmetric space is a well defined combinatorial object; it has "invariant difference operators" analogous to the "invariant differential operators" on a symmetric space, and various functions and function spaces on a symmetric space have discrete analogues which are easier to work with. In particular, the formulae for the spherical functions are easier to deal with, so that, for example, it should be easier to analyse the heat equation on a discrete symmetric space than on a normal symmetric space. For the rank one case, compare [36] and [40, 41].

Useful bibliography on discrete symmetric spaces includes [14, 87] (for the geometric and combinatorial structure), and [80] (for the spherical functions, Plancherel theorem, ...). For the rank one case, these are "trees"(that is, simply connected graphs), and analysis on these structures was developed by A. Figà-Talamanca and C. Nebbia [47]. More detailed analysis on trees may be found in, for instance, [40, 41].

4.6 Adèles

We conclude this outline of some of the generalisations of semisimple Lie groups with a brief discussion of the adèles and adèle groups.

The ring of adèles, $\mathbb{A}$, is the "restricted direct product" $\mathbb{R} \times \Pi_{p \in P} \mathbb{Q}_p$, where P is the set of prime numbers. An adèle is a "vector" $(x_\infty, x_2, x_3, \ldots, x_p, \ldots)$, where $x_\infty \in \mathbb{R}$, and $x_p \in \mathbb{Q}_p$; further, $|x_p|_p > 1$ for only finitely many p in P. The operations in the ring are componentwise addition, subtraction, and multiplication. For an adèle to be invertible, it is necessary and sufficient that no component be zero, and that $|x_p|_p \neq 1$ for only finitely many components.

The ring of adèles may be topologised by defining a basis of open sets at 0 to be all sets of the form $U_\infty \times U_2 \times U_3 \times \cdots \times U_p \times \cdots$, where U_∞ is an open set containing 0, as is each U_p, and all but finitely many U_p are equal to $\mathcal{O}_p$. The translates of these sets by x then form a basis for the topology at x. Similarly $\mathbb{A}$ may be equipped with product measure.

It is possible to form groups such as $\mathrm{SL}(2, \mathbb{A})$. One may think either of matrices with "vector" entries, or equivalently as a "vector" of matrices:

$$\begin{bmatrix} (a_\infty, a_2, \cdots) & (b_\infty, b_2, \cdots) \\ (c_\infty, c_2, \cdots) & (d_\infty, d_2, \cdots) \end{bmatrix} \sim \left(\begin{bmatrix} a_\infty & b_\infty \\ c_\infty & d_\infty \end{bmatrix}, \begin{bmatrix} a_2 & b_2 \\ c_2 & d_2 \end{bmatrix}, \cdots \right).$$

Because the operations are component-by-component, these are equivalent formulations.

The rational numbers may be injected diagonally into the adèles, that is, the rational number r corresponds to the adèle $(r, r, r, \ldots)$. Then $\mathbb{Q}$ "is" a discrete subring of $\mathbb{A}$, and $\mathbb{A}/\mathbb{Q}$ is compact.

One of the major goals of number theorists is to understand the unitary representation λ of $\mathrm{SL}(2, \mathbb{A})$ on the space $L^2(\mathrm{SL}(2, \mathbb{A})/\mathrm{SL}(2, \mathbb{Q}))$, and similar representations involving other groups, such as $\mathrm{SL}(n, \mathbb{A})$. For the $\mathrm{SL}(2)$ case, much information is contained in Gel'fand, Graev, and Pyateckii-Shapiro [51].

The group $SL(2, \mathbb{A})$ has unitary representations, which are "restricted tensor products" of unitary representations of the factors, and quite a bit is known about how λ decomposes into irreducible components. A number of important conjectures in number theory may be reformulated in terms of the harmonic analysis of $SL(2, \mathbb{A})$.

An important recent result is concerned with the representation of $G(\mathbb{A})$ on $L^2(G(\mathbb{A})/G(\mathbb{Q}))$. The space $G(\mathbb{A})/G(\mathbb{Q})$ has finite volume, so the constant functions lie in the Hilbert space, and the representation contains a trivial subrepresentation. Clozel [25] has recently proved a conjecture of A. Lubotzky and R.J. Zimmer that all the other components of this representation are isolated away from the trivial representation. A consequence of this is that there exists p such that each of the other components satisfies an L^{p+} estimate, which in turn implies that the restriction of many unitary representations of $G(\mathbb{A})$ to $G(\mathbb{Q})$ are irreducible (see [10, 11]).

4.7 Further Results

The work of I. Cherednik [21, 22] offers another unification of the real and p-adic settings.

5 Carnot–Carathéodory Geometry and Group Representations

In this lecture, following [7], I construct some unitary and uniformly bounded representations of simple Lie groups of real rank one, using geometric methods.

5.1 A Decomposition for Real Rank One Groups

Theorem 5.1. *Suppose that G is a real rank one simple Lie group, with an Iwasawa decomposition KAN. Then $G = KNK$, in the sense that every element g of G may be written (not uniquely) in the form $k_1 n k_2$, where $n \in N$ and $k_1, k_2 \in K$.*

Proof. Consider the action of the group G on the associated symmetric space X (which may be identified with G/K). It will suffice to show that any point x in X may be written in the form kno, where o is the base point of X (that is, the point stabilised by K). Suppose that the distance of x from o is d. As n varies over the connected group N, the point no varies over a subset No of X which contains o. Since this subset is connected and unbounded, there exists a point no in No whose distance from o is d. Now K acts transitively on all the spheres with centre o, so there exists k in K such that $kno = x$, as required.

5.2 The Conformal Group of the Sphere in $\mathbb{R}^n$

Stereographic projection from $\mathbb{R}^n$ to S^n may be defined by the formula

$$\sigma(x) = \left(1 + \frac{|x|^2}{4}\right)^{-1}\left(x, 1 - \frac{|x|^2}{4}\right) \qquad \forall x \in \mathbb{R}^n,$$

where (x, t) is shorthand for $(x_1, \ldots, x_n, t)$. It is a conformal map, that is, its differential is a multiple D_σ of an orthogonal map. Its Jacobian J_σ is the nth power of this multiple, that is,

$$J_\sigma(x) = \left(1 + \frac{|x|^2}{4}\right)^{-n} \qquad \forall x \in \mathbb{R}^n.$$

It is well known that G, the conformal group of the sphere, that is, the group of all orientation-preserving conformal diffeomorphisms of S^n, may be identified with $\mathrm{SO}_0(1, n+1)$. Let P denote the subgroup of G of conformal maps which fix the north pole b. By conjugation with σ, we may identify P with the group of all conformal diffeomorphisms of $\mathbb{R}^n$. This is the Euclidean motion group, which is the semidirect product of the group $\mathrm{SO}(n) \times \mathbb{R}^+$ of conformal linear maps of $\mathbb{R}^n$ (which are all products of rotations and dilations) and the group of translations of $\mathbb{R}^n$ (isomorphic to $\mathbb{R}^n$ itself). Then P may be decomposed as MAN, where MA is the subgroup of P giving linear conformal maps of

$\mathbb{R}^n$ (that is, $\sigma(\mathrm{SO}(n) \times \mathbb{R}^n)\sigma^{-1}$ and $N = \sigma\{\tau_x : x \in \mathbb{R}^n\}\sigma^{-1}$, where τ_x denotes translation by x on $\mathbb{R}^n$ (that is, $\tau_x y = x + y$). Write K for $\mathrm{SO}(n)$. Then the groups K, M, A and N are those which arise in the Iwasawa and Bruhat decompositions of G, described in Lecture 1. We can also establish these decompositions geometrically: for instance, given g in G, there exists a rotation k of S^n such that $gb = kb$. Then $k^{-1}gb = b$, so $k^{-1}g \in P$, and $k^{-1}g$ may be written in the form man, where $m \in M$, $a \in A$, and $n \in N$. Consequently, $g = (km)an$, and $km \in K$; thus we have shown that $G = KAN$.

Theorem 5.2. *Suppose that $\mathfrak{F}_z(S^n)$ and $\mathfrak{F}_z(\mathbb{R}^n)$ are function spaces on S^n and $\mathbb{R}^n$, such that $T_z \colon \mathfrak{F}_z(S^n) \to \mathfrak{F}_z(\mathbb{R}^n)$ is an isomorphism, where*

$$T_z f = J_\sigma^{z/n+1/2} f \circ \sigma,$$

and that translations and rotations act isometrically on $\mathfrak{F}(\mathbb{R}^n)$ and $\mathfrak{F}_z(S^n)$ respectively. Then $\pi_z \colon G \to \mathrm{End}(\mathfrak{F}_z(S^n))$, given by

$$\pi_z(g)f(x) = J_{g^{-1}}^{z/n+1/2}(x)f(g^{-1}x)$$

(where J_g is the Jacobian of the conformal map g on S^n), is a representation of G on $\mathfrak{F}_z(S^n)$ by isomorphisms. If the maps T_z are isometric, then so is π_z.

Proof. Since $G = KNK$, it suffices to show that N acts by isomorphisms (which are isometric if the map T_z is isometric), since K acts isometrically. Now

$$\pi_z(\sigma\tau_x^{-1}\sigma^{-1})f(y) = \left(\frac{d\sigma\tau_x\sigma^{-1}y}{dy}\right)^{z/n+1/2} f(\sigma\tau_x\sigma^{-1}y)$$

(5.1)
$$= \left(\frac{d\sigma\tau_x\sigma^{-1}y}{d\tau_x\sigma^{-1}y}\frac{d\tau_x\sigma^{-1}y}{d\sigma^{-1}y}\frac{d\sigma^{-1}y}{dy}\right)^{z/n+1/2} f(\sigma\tau_x\sigma^{-1}y)$$

$$= (T_z^{-1}\tau_x^{-1}T_z f)(y),$$

by the chain rule and the definitions, so

$$\|\pi_z(n)f\| \le \|T_z^{-1}\| \, \|T_z\| \, \|f\|.$$

Clearly, if T_z is isometric, then π_z is isometric.

Here are some examples. First, if $\mathfrak{F}_z = L^p$, where $\mathrm{Re}(z) = n(1/p - 1/2)$, then T_z is an isometry, by definition. In particular, if $\mathrm{Re}(z) = 0$, then π_z acts unitarily on $L^2(S)$, giving us the "unitary class-one principal series" of representations, indicated by the heavy vertical line in the diagram at the end of this section.

Next, if $\mathfrak{F} = H^s$, where

$$s = -\mathrm{Re}(z) \in \left(-\frac{n}{2}, \frac{n}{2}\right),$$

then T_z is an isomorphism. Here

$$H^s(\mathbb{R}^n) = \{f \in \mathcal{S}'(\mathbb{R}^n) : |\cdot|^s \hat{f} \in L^2(\mathbb{R}^n)\}$$
$$H^s(S^n) = \{f \in L^1(S^n) : \sum_k (1+k)^s f_k \in L^2(S^n)\},$$

where $\sum_k f_k$ is the decomposition of f as a sum of spherical harmonics f_k of degree k.

One proof of this uses the remarkable formula

(5.2) $$|1 - \sigma(x) \cdot \sigma(y)| = J_\sigma(x)^{1/2n} J_\sigma(y)^{1/2n} |x - y| \qquad \forall x, y \in N.$$

Note that $|x - y|$ is the Euclidean distance between x and y in $\mathbb{R}^n$, and that $|1 - p \cdot q|$, henceforth written $d_{S^n}(p, q)$, is the distance between p and q in S^n (the length of the chord joining them, not the geodesic distance in the sphere). This implies that, if $s \in (-n/2, n/2)$, then

(5.3)
$$\begin{aligned}
&\int_{S^n} \int_{S^n} f(p) g(q) d_{S^n}(p, q)^{-2s-n} \, dp \, dq \\
&= \int_{\mathbb{R}^n} \int_{\mathbb{R}^n} f(x) g(y) J_\sigma(x)^{-s/n+1/2} J_\sigma(y)^{-s/n+1/2} |x - y|^{-2s-n} \, dx \, dy \\
&= \int_{\mathbb{R}^n} \int_{\mathbb{R}^n} T_{-s} f(x) T_{-s} g(y) |x - y|^{-2s-n} \, dx \, dy
\end{aligned}$$

where we integrate using the Riemannian volume element on the sphere. By putting $f = \bar{g}$ and taking square roots, we deduce that

$$\|T_{-s} f\|_{H^s(\mathbb{R}^n)} = \|f\|_{\tilde{H}^s(S^n)},$$

where the Hilbert space $\tilde{H}^s(S^n)$ is defined like $H^s(S^n)$, but with $(1+k)^s$ replaced by a quotient of Γ functions determined by the spherical harmonic decomposition of $d_{S^n}(\cdot, \cdot)$ (see, for instance, [67]). More precisely,

$$\|f\|_{H^s(S^n)} = \left\| \sum_k (1+k)^s f_k \right\|_2 = \left(\sum_k (1+k)^{2s} \|f_k\|_2^2 \right)^{1/2}$$

$$\|f\|_{\tilde{H}^s(S^n)} = C_s \left(\sum_k \frac{\Gamma(k+s)}{\Gamma(k-s)} \|f_k\|_2^2 \right)^{1/2},$$

where C_s depends only on n and s. Hence π_{-s} acts unitarily on $\tilde{H}^s(S^n)$ when $s \in (-n/2, n/2)$, indicated by the heavy horizontal line in the diagram at the end of this section.

To show that π_z acts uniformly boundedly on the Hilbert space $\tilde{H}^s(S^n)$ when $\text{Re}(z) = -s$ and $s \in (-n/2, n/2)$, we use the fact that $T_{\text{Re} z}$ is a unitary map from $\tilde{H}^s(S^n)$ to $H^s(\mathbb{R}^n)$. Now $T_z f = m_{i \, \text{Im} \, z/n} T_{\text{Re} z}$, where m_{iy} denotes pointwise multiplication by the function J_σ^{iy}, so it suffices to show that the functions m_{iy} multiply the spaces $H^s(\mathbb{R}^n)$ pointwise. This is a little tedious, but not hard (see [7]). Thus the region where the representations can be made uniformly bounded is between the dashed lines in the diagram at the end of this section.

It is easy to show that π_z, as defined in this section, is the same as $\pi^{1,\lambda}$, as defined in Lecture 1, where $\lambda = (2iz/n)\rho$.

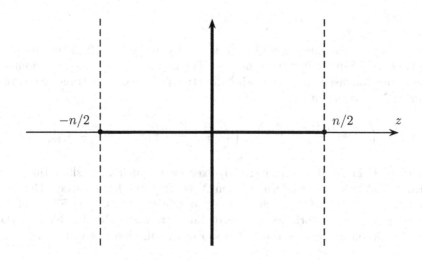

Class-one Unitary Representations of $SO_0(1, n+1)$

5.3 The Groups $SU(1, n+1)$ and $Sp(1, n+1)$

Denote by $\mathbb{F}$ either the complex numbers $\mathbb{C}$ or the quaternions $\mathbb{H}$; define the "$\mathbb{F}$-valued inner product" on the right vector space $\mathbb{F}^n$ by

$$x \cdot x' = \sum_{j=1}^{n} x'_j \bar{x}_j \qquad \forall x, x' \in \mathbb{F}^n.$$

We denote the projection of $\mathbb{F}$ onto the subspace of purely imaginary elements by $\Im$. The spheres S^{2n+1} in $\mathbb{C}^n$ and S^{4n+3} in $\mathbb{H}^{n+1}$ are Carnot–Carathéodory (generalised CR) manifolds. For p in S^{2n+1} or S^{4n+3}, we denote the subspace of the tangent space T_p to the sphere at p of vectors orthogonal to $p\mathbb{C}$ or $p\mathbb{H}$ by U_p, and endow U_p with the restriction of the standard Riemannian metric. We denote either of these spheres, with its Carnot–Carathéodory structure (that is, a privileged nonintegrable subbundle of the tangent bundle, equipped with an inner product), by S.

A diffeomorphism $f: S \to S$ is said to be Carnot–Carathéodory contact if f_* maps U_p into $U_{f(p)}$ for all p, and Carnot–Carathéodory conformal if it is contact and in addition $f_*\big|_{U_p}$ is a multiple of an orthogonal map for all p. The groups $SU(1, n+1)$ and $Sp(1, n+1)$ may be identified with the conformal groups of S^{2n+1} (in the complex case) and S^{4n+3} (in the quaternionic case).

The analogue of the stereographic projection is the Cayley transform from a Heisenberg-type group N to the sphere S. We define N to be the set $\mathbb{F}^n \times \Im(\mathbb{F})$, equipped with the multiplication

$$(x, t)(x', t') = (x + x', t + t' + \frac{1}{2}\Im(x' \cdot x)) \qquad \forall (x, t), (x', t') \in \mathbb{F}^n \times \Im(\mathbb{F}).$$

The homogeneous dimension Q of N is defined to be $2n + 2$ in the complex case and $4n + 6$ in the quaternionic case. The homogeneous dimension double-counts the dimension of the "missing directions". The Cayley transform is the map $\sigma : N \to S$, given by

$$\sigma(x, t) = \left(\left(\left(1 + \frac{|x|^2}{4}\right)^2 + |z|^2\right)^{-1}\left(\left(1 + \frac{|x|^4}{4} - z\right)x, -1 + \frac{|x|^4}{16} + |z|^2 + 2z\right)\right)$$

for all (x, t) in N. It is a (nontrivial) exercise in calculus to show that σ is Carnot–Carathéodory conformal when N is given the left-invariant Carnot–Carathéodory structure which at the group identity $(0, 0)$ is $\mathbb{F}^n$ (that is, $\{(x, 0) : x \in \mathbb{F}^n\}$) with its standard inner product. The Jacobian of the transformation is given by the Qth power of the dilation factor, that is,

$$J_\sigma(x, t) = \left(\left(1 + \frac{|x|^2}{4}\right)^2 + |z|^2\right)^{-Q/2}.$$

The representations that we wish to investigate are given by the formula

$$\pi_z(g)f(x) = J_{g^{-1}}^{z/Q+1/2}(x)f(g^{-1}x).$$

We define the nonhomogeneous distance d_S on S by

$$d_S(p, q) = |1 - p \cdot q|,$$

and the nonhomogeneous distance d_N on N to be the left-invariant distance such that

$$d_N((0, 0), (x, z)) = \left(\frac{|x|^4}{16} + |z|^2\right)^{1/4}.$$

Then, as proved geometrically in [7], the analogue of the remarkable formula (5.2) is

$$d_S(\sigma(x), \sigma(y)) = J_\sigma(x)^{1/2Q} J_\sigma(y)^{1/2Q} d_N(x, y).$$

We now define the map T_z taking functions on S to functions on N much as before:

$$T_z f(x, t) = J_\sigma(x, t)^{z/Q+1/2} f(\sigma(x, t))$$

(compare with (5.1)), and a very similar calculation to (5.3) shows that

$$\|T_{-s}f\|_{\tilde{H}^2(N)} = \|f\|_{\tilde{H}^s(S)},$$

where $\tilde{H}^s(N)$ is the "Sobolev space" of functions f such that

$$\left(\int_N \int_N f(x)\,\bar{f}(y)\,d_N(x,y)^{-2s-Q}\,dx\,dy\right)^{1/2} < \infty,$$

and $\tilde{H}^s(S)$ is defined similarly on S.

If $G = SU(1, n+1)$, then $S = S^{2n+1}$ and N is $\mathbb{C} \times (i\mathbb{R})$ (setwise). In this case, π_z acts unitarily on $L^2(S)$ where $\operatorname{Re}(z) = 0$, giving us the unitary class-one principal series of representations (indicated by the heavy vertical line in the diagram below). Further, the two kernels d_S^{-2s-Q} and d_N^{-2s-Q} are positive definite and the "Sobolev spaces" are defined for all s in $(-Q/2, Q/2)$ (the calculations may be found in [67] for d_S, and in [31] for d_N). This gives a construction of the class-one complementary series (indicated by the heavy horizontal line in the diagram below). Further, it can be shown that pointwise multiplication by purely imaginary powers of J_σ is a bounded map on $\tilde{H}^s(N)$ for all s in this range, and so when $\operatorname{Re}(z) \in (-Q/2, Q/2)$ the representations π_z are uniformly bounded on $\tilde{H}^s(S)$, where $s = -\operatorname{Re} z$ (see [7]).

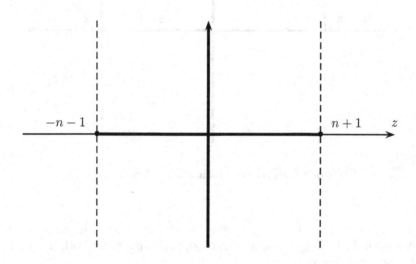

Class-one Unitary Representations of $SU_0(1, n+1)$

For $Sp(1, n+1)$, the picture is different. The representations π_z act by isometries on L^p spaces when $-2n - 3 \leq \operatorname{Re}(z) \leq 2n + 3$, and we might hope that all the representations π_x for x in $(-2n - 3, 2n + 3)$ might be unitary on some Hilbert space, and all the representations π_z for z inside this strip might be made to act uniformly boundedly on some Hilbert space. When

Re(z) = 0, we again obtain unitary representations on $L^2(S)$, the unitary class-one principal series, represented by the heavy vertical line in the diagram below.

However, $\tilde{H}^s(S)$ and $\tilde{H}^s(N)$ are only Hilbert spaces when the kernels $d_S(\cdot,\cdot)^{-2s-Q}$ and $d_N(\cdot,\cdot)^{-2s-Q}$ are positive (semi) definite, and this is only while

$$-2n - 1 \le s \le 2n + 1.$$

Nevertheless, at the expense of loosing the isometry of the representation it is possible to modify the spaces $\tilde{H}^s$, taking more standard Sobolev spaces $H^s(S)$ (spaces of functions with s derivatives in $L^2(S)$—but only derivatives in the Carnot–Carathéodory directions) and show that the representations may still be made uniformly bounded. This is achieved in [7].

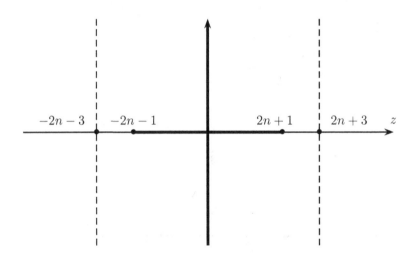

Class-one Unitary Representations of $Sp_0(1, n + 1)$

It is easy to show that π_z, as defined in this section, is the same as $\pi^{1,\lambda}$, as defined in Lecture 1, where $\lambda = (2iz/Q)\rho$; this applies for both the unitary and symplectic groups.

Cowling [30] showed that it is possible to make the representations inside the strip uniformly bounded, but this paper provides no control on the norms of the representations. Then Cowling and Haagerup [34] showed that it is possible to control the spherical functions associated to the representations π_x, where $-Q/2 < x < Q/2$. A.H. Dooley [45] has recently shown that, for the group $Sp(1, n + 1)$, it is possible to choose a "Sobolev type" norm on N (this is called the "noncompact picture") so that $\|\pi_x\|^2 \le 2n - 1$ whenever $-Q/2 < x < Q/2$. This is the best possible constant. It has also been shown, working on the sphere (this is the "compact picture"), that the representations of $\mathrm{Sp}(1, n+1)$ have special properties which show that this group satisfies the Baum–Connes conjecture "with coefficients" (see [7, 68]). It would be nice to obtain a similar best possible result for the "picture changing" version, that is, to show that $\|T_s\|\|T_s^{-1}\| \leq (2n-1)^{1/2}$, as this would unify all the known results in this direction, and be best possible. This result should presumably be approached from a geometric point of view; more relevant information on the geometry of the spaces N and K/M is in the papers [7, 8]. There has also been much work on understanding these representations from a differential geometric point of view; see, for instance, [13], and other papers by these authors.

It may also be possible to extend some of the ideas here to more general semisimple groups: the Cayley transform treated here arises by composition of various natural mappings which appear when one considers the Iwasawa and Bruhat decompositions:

$$N \to NM A\bar{N}/M A\bar{N} \to G/M A\bar{N} = K A\bar{N}/M A\bar{N} \cong K/M.$$

References

1. J.-Ph. Anker, L_p Fourier multipliers on Riemannian symmetric spaces of the noncompact type, *Ann. of Math.* **132** (1990), 597–628.
2. J.-Ph. Anker, The spherical Fourier transform of rapidly decreasing functions. A simple proof of a characterization due to Harish-Chandra, Helgason, Trombi, and Varadarajan, *J. Funct. Anal.* **96** (1991), 331–349.
3. J.-Ph. Anker, Sharp estimates for some functions of the Laplacian on noncompact symmetric spaces, *Duke Math. J.* **65** (1992), 257–297.
4. J.-Ph. Anker and L. Ji, Heat kernel and Green function estimates on noncompact symmetric spaces. I, *Geom. Funct. Anal.* **9** (1999), 1035–1091.
5. J.-Ph. Anker and L. Ji, Heat kernel and Green function estimates on noncompact symmetric spaces. II, pages 1–9 in: *Topics in probability and Lie groups: boundary theory.* CMS Conf. Proc. 28. Amer. Math. Soc., Providence, RI, 2001.
6. J.-Ph. Anker and P. Ostellari, The heat kernel on noncompact symmetric spaces, pages 27–46 in: *Lie groups and symmetric spaces.* Volume in memory of F. I. Karpelevitch; S.G. Gindikin ed. Amer. Math. Soc. Transl. Ser. 2, 210. Amer. Math. Soc., Providence, RI, 2003.
7. F. Astengo, M. Cowling and B. Di Blasio, The Cayley transform and uniformly bounded representations, *J. Funct. Anal.* **213** (2004), 241–269.
8. A.D. Banner, Some properties of boundaries of symmetric spaces of rank one, *Geom. Dedicata* **88** (2001), 113–133.
9. M.E.B. Bekka, On uniqueness of invariant means, *Proc. Amer. Math. Soc.* **126** (1998), 507–514.
10. M.E.B. Bekka and M. Cowling, Some unitary representations of $G(K)$ for a simple algebraic group over a field K, *Math. Z.* **241** (2002), 731–741.
11. M.E.B. Bekka and M. Cowling, Addendum to "Some unitary representations of $G(K)$ for a simple algebraic group over a field K", in preparation.
12. A. Borel and H. Garland, Laplacian and the discrete spectrum of an arithmetic group, *Amer. J. Math.* **105** (1983), 309–335.
13. T. Branson, G. Ólafsson and H. Schlichtkrull, Huyghens' principle in Riemannian symmetric spaces, *Math. Ann.* **301** (1995), 445–462.
14. K.S. Brown, *Buildings.* Springer-Verlag, Berlin New York, 1989.
15. M. Burger, Kazhdan constants for SL(3, $\mathbb{Z}$), *J. reine angew. Math.* **413** (1991), 36–67.
16. M. Burger, J.-S. Li and P. Sarnak, Ramanujan duals and automorphic spectrum, *Bull. Amer. Math. Soc. (N.S.)* **26** (1992), 253–257.
17. M. Burger and P. Sarnak, Ramanujan duals. II, *Invent. Math.* **106** (1991), 1–11.
18. W. Casselman and D. Miličić, Asymptotic behavior of matrix coefficients of admissible representations, *Duke Math. J.* **49** (1982), 869–930.
19. O.A. Chalykh and A.P. Veselov, Integrability and Huygens' principle on symmetric spaces, *Comm. Math. Phys.* **178** (1996), 311–338.
20. J. Cheeger, M. Gromov, and M.E. Taylor, Finite propagation speed, kernel estimates for functions of the Laplace operator, and the geometry of complete Riemannian manifolds, *J. Differential Geom.* **17** (1982), 15–53.
21. I. Cherednik, Macdonald's evaluation conjectures and difference Fourier transform, *Invent. Math.* **122** (1995), 119–145.
22. I. Cherednik, Double affine Hecke algebras and Macdonald's conjectures, *Ann. of Math.* **141** (1995), 191–216.

23. J.-L. Clerc and E.M. Stein, L^p multipliers for noncompact symmetric spaces, *Proc. Nat. Acad. Sci. U.S.A.* **71** (1974), 3911–3912.

24. J.-L. Clerc, Transformation de Fourier sphérique des espaces de Schwartz, *J. Funct. Anal.* **37** (1980), 182–202.

25. L. Clozel, Démonstration de la conjecture τ, *Invent. Math.* **151** (2003), 297–328.

26. L. Clozel, H. Oh and E. Ullmo, Hecke operators and equidistribution of Hecke points, *Invent. Math.* **144** (2001), 327–351.

27. T. Coulhon, Noyau de la chaleur et discrétisation d'une variété riemannienne, *Israel J. Math.* **80** (1992), 289–300.

28. T. Coulhon and L. Saloff-Coste, Variétés riemanniennes isométriques à l'infini, *Rev. Mat. Iberoamericana* **11** (1995), 687–726.

29. M. Cowling, The Kunze–Stein phenomenon, *Ann. of Math. (2)* **107** (1978), 209–234.

30. M. Cowling, Sur les coefficients des représentations unitaires des groupes de Lie simples, pages 132–178 in: *Analyse harmonique sur les groupes de Lie II (Sém. Nancy-Strasbourg 1976–1978)*. Lecture Notes in Math. **739**. Springer, Berlin, 1979.

31. M. Cowling, Unitary and uniformly bounded representations of some simple Lie groups, pages 49–128 in: *Harmonic Analysis and Group Representations (C.I.M.E. II ciclo 1980)*. Liguori, Naples, 1982.

32. M. Cowling, Herz's "principe de majoration" and the Kunze–Stein phenomenon, pages 73–88 in: *Harmonic analysis and number theory, Montreal 1996*. S.W. Drury and M. Ram Murty (eds), CMS Conf. Proc. 21. Amer. Math. Soc., 1997.

33. M. Cowling, Measure theory and automorphic representations,, to appear. *Bull. Kerala Math. Assoc.* **3 (Special issue)** (2006), 139–153.

34. M. Cowling and U. Haagerup, Completely bounded multipliers of the Fourier algebra of a simple Lie group of real rank one, *Invent. Math.* **96** (1989), 507–549.

35. M. Cowling, U. Haagerup and R.E. Howe, Almost L^2 matrix coefficients, *J. reine angew. Math.* **387** (1988), 97–100.

36. M. Cowling, S. Giulini and S. Meda, L^p-L^q estimates for functions of the Laplace–Beltrami operator on noncompact symmetric spaces I, *Duke Math. J.* **72** (1993), 109–150.

37. M. Cowling, S. Giulini and S. Meda, L^p-L^q-estimates for functions of the Laplace–Beltrami operator on noncompact symmetric spaces II, *J. Lie Theory* **5** (1995), 1–14.

38. M. Cowling, S. Giulini and S. Meda, L^p-L^q-estimates for functions of the Laplace–Beltrami operator on noncompact symmetric spaces III, *Annales Inst. Fourier (Grenoble)* **51** (2001), 1047–1069.

39. M. Cowling, S. Giulini and S. Meda, Oscillatory multipliers related to the wave equation on noncompact symmetric spaces, *J. London Math. Soc.* **66** (2002), 691–709.

40. M. Cowling, S. Meda and A.G. Setti, Estimates for functions of the Laplace operator on homogeneous trees, *Trans. Amer. Math. Soc.* **352** (2000), 4271–4298.

41. M. Cowling, S. Meda and A.G. Setti, An overview of harmonic analysis on the group of isometries of a homogeneous tree, *Expositiones Math.* **16** (1998), 385–423.

42. M. Cowling and A. Nevo, Uniform estimates for spherical functions on complex semisimple Lie groups, *Geom. Funct. Anal.* **11** (2001), 900–932.
43. C.W. Curtis and I. Reiner, *Representation Theory of Finite Groups and Associative Algebras*. Pure and Applied Mathematics **11**. Interscience, New York, 1962.
44. P. Diaconis and L. Saloff-Coste, Random walks on finite groups: a survey of analytic techniques, pages 44–75 in: *Probability Measures on Groups and Related Structures, XI (Oberwolfach, 1994)*. World Sci. Publishing, River Edge, NJ, 1995.
45. A.H. Dooley, Heisenberg-type groups and intertwining operators, *J. Funct. Anal.* **212** (2004), 261–286.
46. P. Eymard and N. Lohoué, Sur la racine carrée du noyau de Poisson dans les espaces symétriques et une conjecture de E. M. Stein, *Ann. Sci. École Norm. Sup. (4)* **8** (1975), 179–188.
47. A. Figà-Talamanca and C. Nebbia, *Harmonic Analysis and Representation Theory for Groups Acting on Homogeneous Trees*. London Mathematical Society Lecture Notes **162**. Cambridge University Press, Cambridge, 1991.
48. M. Flensted-Jensen, Spherical functions of a real semisimple Lie group. A method of reduction to the complex case, *J. Funct. Anal.* **30** (1978), 106–146.
49. R. Gangolli, On the Plancherel formula and the Paley–Wiener theorem for spherical functions on semisimple Lie groups, *Ann. of Math.* **93** (1971), 150–165.
50. R. Gangolli and V.S. Varadarajan, *Harmonic Analysis of Spherical Functions on Real Reductive Groups*. Ergeb. Math. Grenzgeb. **101**. Springer-Verlag, 1988.
51. I.M. Gel'fand, M.I. Graev and I.I. Pyatetskii-Shapiro, *Representation Theory and Automorphic Functions*. Translated from the Russian by K.A. Hirsch. W.B. Saunders Co., Philadelphia, PA, 1969.
52. H. Gunawan, A generalization of maximal functions on compact semisimple Lie groups, *Pacific J. Math.* **156** (1992), 119–134.
53. Harish-Chandra, Harmonic analysis on real reductive groups. I. The theory of the constant term, *J. Funct. Anal.* **19** (1975), 104–204.
54. Harish-Chandra, Harmonic analysis on real reductive groups. II. Wavepackets in the Schwartz space, *Invent. Math.* **36** (1976), 1–55.
55. Harish-Chandra, Harmonic analysis on real reductive groups. III. The Maass–Selberg relations and the Plancherel formula, *Ann. of Math. (2)* **104** (1976), 117–201.
56. P. de la Harpe and A. Valette, *La propriété (T) de Kazhdan pour les groupes localement compacts (avec un appendice de Marc Burger)*. *Astérisque* **175** 1989.
57. W. Hebisch, The subalgebra of $L^1(AN)$ generated by the Laplacian, *Proc. Amer. Math. Soc.* **117** (1993), 547–549.
58. S. Helgason, An analogue of the Paley-Wiener theorem for the Fourier transform on certain symmetric spaces, *Math. Ann.* **165** (1966), 297–308.
59. S. Helgason, *Differential Geometry, Lie Groups, and Symmetric Spaces*. Pure and Applied Math. Academic Press, New York, 1978.
60. S. Helgason, *Groups and Geometric Analysis: Integral Geometry, Invariant Differential Operators and Spherical Functions*. Pure and Applied Math. Academic Press, New York, 1984.
61. S. Helgason, Wave equations on homogeneous spaces, pages 254–287 in: *Lie Group Representations III (College Park, Md., 1982/1983)*. Lecture Notes in Math. **1077**. Springer, Berlin-New York, 1984.

62. S. Helgason, *Geometric Analysis on Symmetric Spaces*. Mathematical Surveys and Monographs **39**. American Mathematical Society, Providence, RI, 1994.

63. C.S. Herz, Sur le phénomène de Kunze–Stein, *C. R. Acad. Sci. Paris (Série A)* **271** (1970), 491–493.

64. B. Hoogenboom, Spherical functions and invariant differential operators on complex Grassmann manifolds, *Ark. Mat.* **20** (1982), 69–85.

65. R.E. Howe, On a notion of rank for unitary representations of the classical groups, pages 223–331 in: *Harmonic Analysis and Group Representations (C.I.M.E. II ciclo 1980)*. Liguori, Naples, 1982.

66. A. Ionescu, An endpoint estimate for the Kunze–Stein phenomenon and related maximal operators, *Ann. of Math.* **152** (2000), 259–275.

67. K.D. Johnson and N.R. Wallach, Composition series and intertwining operators for the spherical principal series. I, *Trans. Amer. Math. Soc.* **229** (1977), 137–173.

68. P. Julg, La conjecture de Baum-Connes à coefficients pour le groupe $\mathrm{Sp}(n,1)$, *C. R. Math. Acad. Sci. Paris* **334** (2002), 533–538.

69. M. Kanai, Rough isometries, and combinatorial approximations of geometries of noncompact Riemannian manifolds, *J. Math. Soc. Japan* **37** (1985), 391–413.

70. D.A. Kazhdan, Connection of the dual space of a group with the structure of its closed subgroups, *Funct. Anal. Appl.* **1** (1967), 63–65.

71. A.W. Knapp, *Representation Theory of Semisimple Groups. An Overview Based on Examples*. Princeton Mathematical Series **36**. Princeton University Press, Princeton, N.J., 1986.

72. T.H. Koornwinder, Jacobi functions and analysis on noncompact semisimple Lie groups, pages 1–85 in: *Special Functions: Group Theoretical Aspects and Applications*. R.A. Askey et al. (eds). Reidel, 1984.

73. B. Kostant, On the existence and irreducibility of certain series of representations, *Bull. Amer. Math. Soc.* **75** (1969), 627–642.

74. B. Kostant, On the existence and irreducibility of certain series of representations, pages 231–329 in: *Lie Groups and Their Representations (Proc. Summer School, Bolyai János Math. Soc., Budapest, 1971)*. Halsted, New York, 1975.

75. J.-S. Li, The minimal decay of matrix coefficients for classical groups, pages 146–169 in: *Harmonic Analysis in China*. M.D. Cheng et al. (eds), Math. Appl. **327**. Kluwer, 1995.

76. J.-S. Li, On the decay of matrix coefficients for exceptional groups, *Math. Ann.* **305** (1996), 249–270.

77. N. Lohoué and Th. Rychener, Die Resolvente von Δ auf symmetrischen Räumen vom nichtkompakten Typ, *Comment. Math. Helv.* **57** (1982), 445–468.

78. N. Lohoué and Th. Rychener, Some function spaces on symmetric spaces related to convolution operators, *J. Funct. Anal.* **55** (1984), 200–219.

79. A. Lubotzky, *Discrete Groups, Expanding Graphs and Invariant Measures*. Birkhäuser-Verlag, Basel, 1994.

80. I.G. Macdonald, *Spherical Functions on a Group of p-adic Type*. Publications of the Ramanujan Institute **2**. Ramanujan Institute for Advanced Study in Mathematics, University of Madras, Madras, 1971.

81. G.A. Margulis, Explicit constructions of expanders, *Problemy Peredachi Informatsii* **9** (1973), 71–80.

82. G.A. Margulis, Some remarks on invariant means, *Monatsh. Math.* **90** (1980), 233–235.

83. G.A. Margulis, Explicit group-theoretic constructions of combinatorial schemes and their applications in the construction of expanders and concentrators, *Problemy Peredachi Informatsii* **24** (1988), 51–60.

84. C.C. Moore, Exponential decay of correlation coefficients for geodesic flows, pages 163–181 in: *Group Representations, Ergodic Theory, Operator Algebras, and Mathematical Physics (Berkeley, Calif., 1984)*. Math. Sci. Res. Inst. Publ., 6. Springer, New York, 1987.

85. H. Oh, Tempered subgroups and representations with minimal decay of matrix coefficients, *Bull. Soc. Math. France* **126** (1998), 355–380.

86. H. Oh, Uniform pointwise bounds for matrix coefficients of unitary representations and applications to Kazhdan constants, *Duke Math. J.* **113** (2002), 133–192.

87. M. Ronan, *Lectures on Buildings*. Perspectives in Mathematics **7**. Academic Press Inc., Boston, MA, 1989.

88. J.M. Rosenblatt, Uniqueness of invariant means for measure-preserving transformations, *Trans. Amer. Math. Soc.* **265** (1981), 623–636.

89. P. Sarnak, *Some Applications of Modular Forms*. Cambridge Tracts in Mathematics **99**. Cambridge University Press, Cambridge, 1990.

90. I. Satake, Spherical functions and Ramanujan conjecture, pages 258–264 in: *Algebraic Groups and Discontinuous Subgroups*. Proc. Symp. Pure Math.,Boulder, Colo., 1965. Amer. Math. Soc., Providence, R.I., 1966.

91. P. Sawyer, The heat equation on the spaces of positive definite matrices, *Can. J. Math.* **44** (1992), 624–651.

92. P. Sawyer, On an upper bound for the heat kernel on $SU^*(2n)/Sp(n)$, *Can. Math. Bull.* **37** (1994), 408–418.

93. R. Scaramuzzi, A notion of rank for unitary representations of general linear groups, *Trans. Amer. Math. Soc.* **319** (1990), 349–379.

94. T.P. Schonbek, L^p-multipliers: a new proof of an old theorem, *Proc. Amer. Math. Soc.* **102** (1988), 361–364.

95. A.G. Setti, L^p and operator norm estimates for the complex time heat operator on homogeneous trees, *Trans. Amer. Math. Soc.* **350** (1998), 743–768.

96. R.J. Stanton and P. Tomas, Expansions for spherical functions on noncompact symmetric spaces, *Acta Math.* **140** (1978), 251–276.

97. J.-O. Strömberg, Weak type L^1 estimates for maximal functions on non-compact symmetric spaces, *Ann. of Math.* **114** (1981), 115–126.

98. D. Sullivan, For $n > 3$ there is only one finitely additive rotationally invariant measure on the n-sphere defined on all Lebesgue measurable subsets, *Bull. Amer. Math. Soc. (N.S.)* **4** (1981), 121–123.

99. M.E. Taylor, L^p-estimates on functions of the Laplace operator, *Duke Math. J.* **58** (1989), 773–793.

100. P.C. Trombi and V.S. Varadarajan, Spherical transforms of semisimple Lie groups, *Ann. of Math.* **94** (1971), 246–303.

101. L. Vretare, Elementary spherical functions on symmetric spaces, *Math. Scand.* **39** (1976), 343–358.

102. L. Vretare, On a recurrence formula for elementary spherical functions on symmetric spaces and its applications to multipliers for the spherical Fourier transform, *Math. Scand.* **41** (1977), 99–112.

103. S.P. Wang, The dual space of semi-simple Lie groups, *Amer. J. Math.* **91** (1969), 921–937.

Ramifications of the Geometric Langlands Program

Edward Frenkel *

Department of Mathematics, University of California, Berkeley, CA 94720, USA
frenkel@math.berkeley.edu

Introduction

The Langlands Program, conceived as a bridge between Number Theory and Automorphic Representations [L], has recently expanded into such areas as Geometry and Quantum Field Theory and exposed a myriad of unexpected connections and dualities between seemingly unrelated disciplines. There is something deeply mysterious in the ways the Langlands dualities manifest themselves and this is what makes their study so captivating.

In this review we will focus on the geometric Langlands correspondence for complex algebraic curves, which is a particular brand of the general theory.

* Supported by the DARPA grant HR0011-04-1-0031 and by the NSF grant DMS-0303529.
  November 2006. Based on the lectures given by the author at the CIME Summer School "Representation Theory and Complex Analysis", Venice, June 2004

Its origins and the connections with the classical Langlands correspondence are discussed in detail elsewhere (see, in particular, the reviews [F2, F6]), and we will not try to repeat this here. The general framework is the following: let X be a smooth projective curve over $\mathbb{C}$ and G be a simple Lie group over $\mathbb{C}$. Denote by $^L G$ the Langlands dual group of G (we recall this notion in Section 2.3). Suppose that we are given a principal $^L G$-bundle $\mathcal{F}$ on X equipped with a flat connection. This is equivalent to $\mathcal{F}$ being a holomorphic principal $^L G$-bundle equipped with a holomorphic connection ∇ (which is automatically flat as the complex dimension of X is equal to one). The pair $(\mathcal{F}, \nabla)$ may also be thought of as a $^L G$-local system on X, or as a homomorphism $\pi_1(X) \to {}^L G$ (corresponding to a base point in X and a trivialization of the fiber of $\mathcal{F}$ at this point).

The global Langlands correspondence is supposed to assign to $E = (\mathcal{F}, \nabla)$ an object Aut_E, called **Hecke eigensheaf with eigenvalue E**, on the moduli stack Bun_G of holomorphic G-bundles on X:

$$
\boxed{\begin{array}{c} \text{holomorphic } {}^L G\text{-bundles} \\ \text{with connection on } X \end{array}} \quad \longrightarrow \quad \boxed{\text{Hecke eigensheaves on } \mathrm{Bun}_G}
$$

$$E \mapsto \mathrm{Aut}_E$$

(see, e.g., [F6], Sect. 6.1, for the definition of Hecke eigensheaves). It is expected that there is a unique irreducible Hecke eigensheaf Aut_E (up to isomorphism) if E is sufficiently generic.

The Hecke eigensheaves Aut_E have been constructed, and the Langlands correspondence proved, in [FGV, Ga2] for $G = GL_n$ and an arbitrary irreducible GL_n-local system, and in [BD1] for an arbitrary simple Lie group G and those $^L G$-local systems which admit the structure of a $^L G$-oper (which is recalled below).

Recently, A. Kapustin and E. Witten [KW] have related the geometric Langlands correspondence to the S-duality of supersymmetric four-dimensional Yang-Mills theories, bringing into the realm of the Langlands correspondence new ideas and insights from quantum physics.

So far, we have considered the **unramified** $^L G$-local systems. In other words, the corresponding flat connection has no poles. But what should happen if we allow the connection to be singular at finitely many points of X?

This **ramified geometric Langlands correspondence** is the subject of this paper. Here are the most important adjustments that one needs to make in order to formulate this correspondence:

- The moduli stack Bun_G of G-bundles has to be replaced by the moduli stack of G-bundles together with the level structures at the ramification points. We call them the *enhanced* moduli stacks. Recall that a level structure of order N is a trivialization of the bundle on the Nth infinitesimal neighborhood of the point. The order of the level structure should be at least the order of the pole of the connection at this point.

- At the points at which the connection has regular singularity (pole of order 1) one can take instead of the level structure, a parabolic structure, i.e., a reduction of the fiber of the bundle to a Borel subgroup of G.
- The Langlands correspondence will assign to a flat $^L G$-bundle $E = (\mathcal{F}, \nabla)$ with ramification at the points $y_1, \ldots, y_n$ a **category** $\mathcal{A}ut_E$ of Hecke eigensheaves on the corresponding enhanced moduli stack with eigenvalue $E|_{X \setminus \{y_1, \ldots, y_n\}}$, which is a subcategory of the category of (twisted) $\mathcal{D}$-modules on this moduli stack.

If E is unramified, then we may consider the category $\mathcal{A}ut_E$ on the moduli stack Bun_G itself. We then expect that for generic E this category is equivalent to the category of vector spaces: its unique (up to isomorphism) irreducible object is Aut_E discussed above, and all other objects are direct sums of copies of Aut_E. Because this category is expected to have such a simple structure, it makes sense to say that the unramified geometric Langlands correspondence assigns to an unramified $^L G$-local system on X a single Hecke eigensheaf, rather than a category. This is not possible for general ramified local systems.

The questions that we are facing now are

(1) How to construct the categories of Hecke eigensheaves for ramified local systems?
(2) How to describe them in terms of the Langlands dual group $^L G$?

In this article I will review an approach to these questions which has been developed by D. Gaitsgory and myself in [FG2].

The idea goes back to the construction of A. Beilinson and V. Drinfeld [BD1] of the unramified geometric Langlands correspondence, which may be interpreted in terms of a localization functor. Functors of this type were introduced by A. Beilinson and J. Bernstein [BB] in representation theory of simple Lie algebras. In our situation this functor sends representations of the affine Kac-Moody algebra $\widehat{\mathfrak{g}}$ to twisted $\mathcal{D}$-modules on Bun_G, or its enhanced versions. As explained in [F6], these $\mathcal{D}$-modules may be viewed as sheaves of conformal blocks (or coinvariants) naturally arising in the framework of Conformal Field Theory.

The affine Kac-Moody algebra $\widehat{\mathfrak{g}}$ is the universal one-dimensional central extension of the loop algebra $\mathfrak{g}((t))$. The representation categories of $\widehat{\mathfrak{g}}$ have a parameter κ, called the level, which determines the scalar by which a generator of the one-dimensional center of $\widehat{\mathfrak{g}}$ acts on representations. We consider a particular value κ_c of this parameter, called the **critical level**. The completed enveloping algebra of an affine Kac-Moody algebra acquires an unusually large center at the critical level and this makes the structure of the corresponding category $\widehat{\mathfrak{g}}_{\kappa_c}$-mod very rich and interesting. B. Feigin and I have shown [FF3, F3] that this center is canonically isomorphic to the algebra of functions on the space of $^L G$-**opers** on $D^\times$. Opers are bundles on $D^\times$ with flat connection and an additional datum (as defined by Drinfeld-Sokolov [DS] and Beilinson-Drinfeld [BD1]; we recall the definition below). Remarkably, their structure group turns out to be not G, but the Langlands dual group ^LG, in agreement with the general Langlands philosophy.

This result means that the category $\widehat{\mathfrak{g}}_{\kappa_c}$-mod of (smooth) $\widehat{\mathfrak{g}}$-modules of critical level "lives" over the space $\mathrm{Op}_{^LG}(D^\times)$ of ^LG-opers on the punctured disc $D^\times$. For each $\chi \in \mathrm{Op}_{^LG}(D^\times)$ we have a "fiber" category $\widehat{\mathfrak{g}}_{\kappa_c}$-mod$_\chi$ whose objects are $\widehat{\mathfrak{g}}$-modules on which the center acts via the central character corresponding to χ. Applying the localization functors to these categories, and their K-equivariant subcategories $\widehat{\mathfrak{g}}_{\kappa_c}$-mod$_\chi^K$ for various subgroups $K \subset G[[t]]$, we obtain categories of Hecke eigensheaves on the moduli spaces of G-bundles on X with level (or parabolic) structures.

Thus, the localization functor gives us a powerful tool for converting **local** categories of representations of $\widehat{\mathfrak{g}}$ into **global** categories of Hecke eigensheaves. This is a new phenomenon which does not have any obvious analogues in the classical Langlands correspondence.

The simplest special case of this construction gives us the Beilinson-Drinfeld Hecke eigensheaves Aut_E on Bun_G corresponding to unramified ^LG-local systems admitting the oper structure. Motivated by this, we wish to apply the localization functors to more general categories $\widehat{\mathfrak{g}}_{\kappa_c}$-mod$_\chi^K$ of $\widehat{\mathfrak{g}}$-modules of critical level, corresponding to opers on X with singularities, or ramifications.

These categories $\widehat{\mathfrak{g}}_{\kappa_c}$-mod$_\chi$ are assigned to ^LG-opers χ on the punctured disc $D^\times$. It is important to realize that the formal loop group $G((t))$ naturally acts on each of these categories via its adjoint action on $\widehat{\mathfrak{g}}_{\kappa_c}$ (because the center is invariant under the adjoint action of $G((t))$). Thus, we assign to each oper χ a categorical representation of $G((t))$ on $\widehat{\mathfrak{g}}_{\kappa_c}$-mod$_\chi$.

This is analogous to the classical **local Langlands correspondence**. Let F be a local non-archimedian field, such as the field $\mathbb{F}_q((t))$ or the field of p-adic numbers. Let W_F' be the Weil-Deligne group of F, which is a version of the Galois group of F (we recall the definition in Section 2.1). The local Langlands correspondence relates the equivalence classes of irreducible (smooth) representations of the group $G(F)$ (or "L-packets" of such representations) and the equivalence classes of (admissible) homomorphisms $W_F' \to {}^LG$. In the geometric setting we replace these homomorphisms by flat ^LG-bundles on $D^\times$ (or by ^LG-opers), the group $G(F)$ by the loop group $G((t))$ and representations of $G(F)$ by categorical representations of $G((t))$.

This analogy is very suggestive, as it turns out that the structure of the categories $\widehat{\mathfrak{g}}_{\kappa_c}$-mod$_\chi$ (and their K-equivariant subcategories $\widehat{\mathfrak{g}}_{\kappa_c}$-mod$_\chi^K$) is similar to the structure of irreducible representations of $G(F)$ (and their subspaces of K-invariants). We will see examples of this parallelism in Sects. 7 and 8 below. This means that what we are really doing is developing a **local Langlands correspondence for loop groups**.

To summarize, our strategy [FG2] for constructing the global geometric Langlands correspondence has two parts:

(1) the local part: describing the structure of the categories of $\widehat{\mathfrak{g}}$-modules of critical level, and

(2) the global part: applying the localization functor to these categories to obtain the categories of Hecke eigensheaves on enhanced moduli spaces of G-bundles.

We expect that these localization functors are equivalences of categories (at least, in the generic situation), and therefore we can infer a lot of information about the global categories by studying the local categories $\widehat{\mathfrak{g}}_{\kappa_c}$-mod$_\chi$ of $\widehat{\mathfrak{g}}$-modules. Thus, the local categories $\widehat{\mathfrak{g}}_{\kappa_c}$-mod$_\chi$ take the center stage.

In this paper I review the results and conjectures of [FG1]–[FG6] with the emphasis on unramified and tamely ramified local systems. (I also discuss the case of irregular singularities at the end.) In particular, our study of the categories of $\widehat{\mathfrak{g}}_{\kappa_c}$-modules leads us to the following conjecture. (For related results, see [AB, ABG, Bez1, Bez2].)

Suppose that $E = (\mathcal{F}, \nabla)$, where $\mathcal{F}$ is a ^LG-bundle and ∇ is a connection on $\mathcal{F}$ with regular singularity at a single point $y \in X$ and unipotent monodromy (this is easy to generalize to multiple points). Let $M = \exp(2\pi i u)$, where $u \in {}^LG$ be a representative of the conjugacy class of the monodromy of ∇ around y. Denote by Sp_u the **Springer fiber** of u, the variety of Borel subalgebras of $^L\mathfrak{g}$ containing u. The category $\mathcal{A}ut_E$ of Hecke eigensheaves with eigenvalue E may then be realized as a subcategory of the category of $\mathcal{D}$-modules on the moduli stack of G-bundles on X with parabolic structure at the point y. We have the following conjectural description of the derived category of $\mathcal{A}ut_E$:

$$D^b(\mathcal{A}ut_E) \simeq D^b(\mathrm{QCoh}(\mathrm{Sp}_u^{\mathrm{DG}})),$$

where $\mathrm{QCoh}(\mathrm{Sp}_u^{\mathrm{DG}})$ is the category of quasicoherent sheaves on a suitable "DG enhancement" of $\mathrm{Sp}_u^{\mathrm{DG}}$. This is a category of differential graded (DG) modules over a sheaf of DG algebras whose zeroth cohomology is the structure sheaf of $\mathrm{Sp}_u^{\mathrm{DG}}$ (we discuss this in detail in Section 9).

Thus, we expect that the geometric Langlands correspondence attaches to a ^LG-local system on a Riemann surface with regular singularity at a puncture, a category which is closely related to the variety of Borel subgroups containing the monodromy around the puncture. We hope that further study of the categories of $\widehat{\mathfrak{g}}$-modules will help us to find a similar description of the Langlands correspondence for connections with irregular singularities.

The paper is organized as follows. In Sect. 1 we review the Beilinson-Drinfeld construction in the unramified case, in the framework of localization functors from representation categories of affine Kac-Moody algebras to $\mathcal{D}$-modules on on Bun$_G$. This will serve as a prototype for our construction of more general categories of Hecke eigensheaves, and it motivates us to study categories of $\widehat{\mathfrak{g}}$-modules of critical level. We wish to interpret these categories in the framework of the local geometric Langlands correspondence for loop groups. In order to do that, we first recall in Sect. 2 the setup of the classical Langlands correspondence. Then in Sect. 3 we explain the passage to the geometric context. In Sect. 4 we describe the structure of the center at the critical level and the isomorphism with functions on opers. In Sect. 5 we discuss the connection between the local Langlands parameters (^LG-local systems on the punctured disc) and opers. We introduce the categorical representations of loop groups corresponding to opers and the corresponding categories of Harish-Chandra modules in Sect. 6. We discuss these categories in detail in the unramified case in Sect. 7, paying particular attention to the analogies between the classical and the geometric settings. In Sect. 8 we do the same in the tamely ramified case. We then apply localization functor to these categories in Sect. 9 to obtain various results and conjectures on the global Langlands correspondence, both for regular and irregular singularities.

Much of the material of this paper is borrowed from my new book [F7], where I refer the reader for more details, in particular, for background on representation theory of affine Kac-Moody algebras of critical level.

Finally, I note that in a forthcoming paper [GW] the geometric Langlands correspondence with tame ramification is studied from the point of view of dimensional reduction of four-dimensional supersymmetric Yang-Mills theory.

Acknowledgments. I thank D. Gaitsgory for his collaboration on our joint works which are reviewed in this article. I am also grateful to R. Bezrukavnikov, V. Ginzburg, D. Kazhdan and E. Witten for useful discussions.

I thank the organizers of the CIME Summer School in Venice, especially, A. D'Agnolo, for the invitation to give lectures on this subject at this enjoyable conference.

1 The Unramified Global Langlands Correspondence

Our goal in this section is to construct Hecke eigensheaves Aut_E corresponding to unramified ^LG-local systems $E = (\mathcal{F}, \nabla)$ on X. By definition, Aut_E is a $\mathcal{D}$-module on Bun_G. We would like to construct Aut_E by applying a localization functor to representations of affine Kac-Moody algebra $\widehat{\mathfrak{g}}$.

Throughout this paper, unless specified otherwise, we let $\mathfrak{g}$ be a simple Lie algebra and G the corresponding connected and simply-connected algebraic group.

The key observation used in constructing the localization functor is that for a simple Lie group G the moduli stack Bun_G of G-bundles on X has a realization as a double quotient. Namely, let x be a point of X. Denote by $\mathcal{K}_x$ the completion of the field of rational functions on X at x, and by $\mathcal{O}_x$ its ring of integers. If we choose a coordinate t at x, then we may identify $\mathcal{K}_x \simeq \mathbb{C}((t)), \mathcal{O}_x \simeq \mathbb{C}[[t]]$. But in general there is no preferred coordinate, and so it is better not to use these identifications. Now let $G(\mathcal{K}_x) \simeq G((t))$ be the formal loop group corresponding to the punctured disc $D_x^\times$ around x.

It has two subgroups: one is $G(\mathcal{O}_x) \simeq G[[t]]$ and the other is G_{out}, the group of algebraic maps $X \backslash x \to G$. Then, according to [BeLa, DrSi], the algebraic stack Bun_G is isomorphic to the double quotient

$$(1.1) \qquad \text{Bun}_G \simeq G_{\text{out}} \backslash G(\mathcal{K}_x) / G(\mathcal{O}_x).$$

Intuitively, any G-bundle may be trivialized on the formal disc D_x and on $X \backslash x$. The transition function is then an element of $G(\mathcal{K}_x)$, which characterizes the bundle uniquely up to the right action of $G(\mathcal{O}_x)$ and the left action of G_{out} corresponding to changes of trivializations on D_x and $X \backslash x$, respectively.

The localization functor that we need is a special case of the following general construction. Let $\mathfrak{g}$ be a Lie algebra and K a Lie group $(\mathfrak{g}, K)$ whose Lie algebra is contained in $\mathfrak{g}$. The pair $(\mathfrak{g}, K)$ is called a Harish-Chandra pair. We will assume that K is connected. A $\mathfrak{g}$-module M is called K-equivariant if the action of the Lie subalgebra $\text{Lie}\, K \subset \mathfrak{g}$ on M may be exponentiated to an action of the Lie group K. Let $\mathfrak{g}\text{-mod}^K$ be the category of K-equivariant $\mathfrak{g}$-modules.

Now suppose that H is another subgroup of G. Let $\mathcal{D}_{H \backslash G / K}\text{-mod}$ be the category of $\mathcal{D}$-modules on $H \backslash G / K$. Then there is a localization functor [BB, BD1] (see also [F6, FB])

$$\Delta : \mathfrak{g}\text{-mod}^K \to \mathcal{D}_{H \backslash G / K}\text{-mod}.$$

Now let $\widehat{\mathfrak{g}}$ be a one-dimensional central extension of $\mathfrak{g}$ which becomes trivial when restricted to the Lie subalgebras $\text{Lie}\, K$ and $\text{Lie}\, H$. Suppose that this central extension can be exponentiated to a central extension $\widehat{G}$ of the corresponding Lie group G. Then we obtain a $\mathbb{C}^\times$-bundle $H \backslash \widehat{G} / K$ over $H \backslash G / K$. Let $\mathcal{L}$ be the corresponding line bundle and $\mathcal{D}_{\mathcal{L}}$ the sheaf of differential operators acting on $\mathcal{L}$. Then we have a functor

$$\Delta_{\mathcal{L}} : \widehat{\mathfrak{g}}\text{-mod}^K \to \mathcal{D}_{\mathcal{L}}\text{-mod}.$$

In our case we take the formal loop group $G(\mathcal{K}_x)$, and the subgroups $K = G(\mathcal{O}_x)$ and $H = G_{\text{out}}$ of $G(\mathcal{K}_x)$. We also consider the so-called critical central extension of $G(\mathcal{K}_x)$. Let us first discuss the corresponding central extension of the Lie algebra $\mathfrak{g} \otimes \mathcal{K}_x$. Choose a coordinate t at x and identify $\mathcal{K}_x \simeq \mathbb{C}((t))$. Then $\mathfrak{g} \otimes \mathcal{K}_x$ is identified with $\mathfrak{g}((t))$. Let κ be an invariant bilinear form on $\mathfrak{g}$. The **affine Kac-Moody algebra** $\widehat{\mathfrak{g}}_\kappa$ is defined as the central extension

$$0 \to \mathbb{C}\mathbf{1} \to \widehat{\mathfrak{g}}_\kappa \to \mathfrak{g}((t)) \to 0.$$

As a vector space, it is equal to the direct sum $\mathfrak{g}((t)) \oplus \mathbb{C}\mathbf{1}$, and the commutation relations read

$$(1.2) \qquad [A \otimes f(t), B \otimes g(t)] = [A, B] \otimes f(t)g(t) - (\kappa(A, B)\, \text{Res}\, f dg)\mathbf{1},$$

and $\mathbf{1}$ is a central element, which commutes with everything else. For a simple Lie algebra $\mathfrak{g}$ all invariant inner products are proportional to each other.

Therefore the Lie algebras $\widehat{\mathfrak{g}}_\kappa$ are isomorphic to each other for non-zero inner products κ.

Note that the restriction of the second term in (1.2) to the Lie subalgebra $\mathfrak{g} \otimes t^N \mathbb{C}[[t]]$, where $N \in \mathbb{Z}_+$, is equal to zero, and so it remains a Lie subalgebra of $\widehat{\mathfrak{g}}_\kappa$. A $\widehat{\mathfrak{g}}_\kappa$-module is called **smooth** if every vector in it is annihilated by this Lie subalgebra for sufficiently large N. We define the category $\widehat{\mathfrak{g}}_\kappa$-mod whose objects are smooth $\widehat{\mathfrak{g}}_\kappa$-modules on which the central element **1** acts as the identity. The morphisms are homomorphisms of representations of $\widehat{\mathfrak{g}}_\kappa$. Throughout this paper, unless specified otherwise, by a "$\widehat{\mathfrak{g}}_\kappa$-module" will always mean a module on which the central element **1** acts as the identity.[1] We will refer to κ as the **level**.

Now observe that formula (1.2) is independent of the choice of coordinate t at $x \in X$ and therefore defines a central extension of $\mathfrak{g} \otimes \mathcal{K}_x$, which we denote by $\widehat{\mathfrak{g}}_{\kappa,x}$. One can show that this central extension may be exponentiated to a central extension of the group $G(\mathcal{K}_x)$ if κ satisfies a certain integrality condition, namely, $\kappa = k\kappa_0$, where $k \in \mathbb{Z}$ and κ_0 is the inner product normalized by the condition that the square of the length of the maximal root is equal to 2. A particular example of the inner product which satisfies this condition is the **critical level** κ_c defined by the formula

$$(1.3) \qquad \kappa_c(A, B) = -\frac{1}{2} \operatorname{Tr}_\mathfrak{g} \operatorname{ad} A \operatorname{ad} B.$$

Thus, κ_c is equal to minus one half of the Killing form on $\mathfrak{g}$.[2] When $\kappa = \kappa_c$ representation theory of $\widehat{\mathfrak{g}}_\kappa$ changes dramatically because the completed enveloping algebra of $\widehat{\mathfrak{g}}_\kappa$ acquires a large center (see below).

Let $\widehat{G}_x$ be the corresponding critical central extension of $G(\mathcal{K}_x)$. It is known (see [BD1]) that in this case the corresponding line bundle $\mathcal{L}$ is the square root $K^{1/2}$ of the canonical line bundle on Bun_G.[3] Now we are ready to apply the localization functor in the situation where our group is $G(\mathcal{K}_x)$, with the two subgroups $K = G(\mathcal{O}_x)$ and $H = G_{\mathrm{out}}$, so that the double quotient $H \backslash G / K$ is Bun_G.[4] We choose $\mathcal{L} = K^{1/2}$. Then we have a localization functor

$$\Delta_{\kappa_c,x} : \widehat{\mathfrak{g}}_{\kappa_c,x}\text{-mod}^{G(\mathcal{O}_x)} \to \mathcal{D}_{\kappa_c}\text{-mod}.$$

We will apply this functor to a particular $\widehat{\mathfrak{g}}_{\kappa_c,x}$-module.

To construct this module, let us first define the vacuum module over $\widehat{\mathfrak{g}}_{\kappa_c,x}$ as the induced module

$$\mathbb{V}_{0,x} = \operatorname{Ind}_{\mathfrak{g} \otimes \mathcal{O}_x \oplus \mathbb{C}\mathbf{1}}^{\widehat{\mathfrak{g}}_{\kappa,x}} \mathbb{C},$$

---

[1] Note that we could have **1** act instead as λ times the identity for $\lambda \in \mathbb{C}^\times$; but the corresponding category would just be equivalent to the category $\widehat{\mathfrak{g}}_{\lambda\kappa}$-mod.

[2] It is also equal to $-h^\vee \kappa_0$, where $h^\vee$ is the dual Coxeter number of $\mathfrak{g}$.

[3] Recall that by our assumption G is simply-connected. In this case there is a unique square root.

[4] Since Bun_G is an algebraic stack, one needs to be careful in applying the localization functor. The appropriate formalism has been developed in [BD1].

where $\mathfrak{g} \otimes \mathcal{O}_x$ acts by 0 on $\mathbb{C}$ and $\mathbf{1}$ acts as the identity. According to the results of [FF3, F3], we have

$$\mathrm{End}_{\widehat{\mathfrak{g}}_{\kappa_c}} \mathbb{V}_{0,x} \simeq \mathrm{Fun} \, \mathrm{Op}_{{}^L G}(D_x),$$

where $\mathrm{Op}_{{}^L G}(D_x)$ is the space of ${}^L G$-opers on the formal disc $D_x = \mathrm{Spec} \, \mathcal{O}_x$ around x. We discuss this in detail in Section 4.

Now, given $\chi_x \in \mathrm{Op}_{{}^L G}(D_x)$, we obtain a maximal ideal $I(\chi_x)$ in the algebra $\mathrm{End}_{\widehat{\mathfrak{g}}_{\kappa_c}} \mathbb{V}_{0,x}$. Let $\mathbb{V}_0(\chi_x)$ be the $\widehat{\mathfrak{g}}_{\kappa_c,x}$-module which is the quotient of $\mathbb{V}_{0,x}$ by the image of $I(\chi_x)$ (it is non-zero, as explained in Section 7.3). The module $\mathbb{V}_{0,x}$ is clearly $G(\mathcal{O}_x)$-equivariant, and hence so is $\mathbb{V}_0(\chi_x)$. Therefore $\mathbb{V}_0(\chi_x)$ is an object of the category $\widehat{\mathfrak{g}}_{\kappa_c,x}$-$\mathrm{mod}^{G(\mathcal{O}_x)}$.

We now apply the localization functor $\Delta_{\kappa_c,x}$ to $\mathbb{V}_0(\chi_x)$. The following theorem is due to Beilinson and Drinfeld [BD1].

Theorem 1. (1) *The $\mathcal{D}_{\kappa_c}$-module $\Delta_{\kappa_c,x}(\mathbb{V}_0(\chi_x))$ is non-zero if and only if there exists a global ${}^L\mathfrak{g}$-oper on X, $\chi \in \mathrm{Op}_{{}^L G}(X)$ such that $\chi_x \in \mathrm{Op}_{{}^L G}(D_x)$ is the restriction of χ to D_x.*

(2) *If this holds, then $\Delta_{\kappa_c,x}(\mathbb{V}_0(\chi_x))$ depends only on χ and is independent of the choice of x in the sense that for any other point $y \in X$, if $\chi_y = \chi|_{D_y}$, then $\Delta_{\kappa_c,x}(\mathbb{V}_0(\chi_x)) \simeq \Delta_{\kappa_c,y}(\mathbb{V}_0(\chi_y))$.*

(3) *For any $\chi = (\mathcal{F}, \nabla, \mathcal{F}_{LB}) \in \mathrm{Op}_{{}^L G}(X)$ the $\mathcal{D}_{\kappa_c}$-module $\Delta_{\kappa_c,x}(\mathbb{V}_0(\chi_x))$ is a non-zero Hecke eigensheaf with the eigenvalue $E_\chi = (F, \nabla)$.*

Thus, for any $\chi \in \mathrm{Op}_{{}^L G}(X)$, the $\mathcal{D}_{\kappa_c}$-module $\Delta_{\kappa_c,x}(\mathbb{V}_0(\chi_x))$ is the sought-after Hecke eigensheaf Aut_{E_χ} corresponding to the ${}^L G$-local system E_χ under the global geometric Langlands correspondence.[5] For an outline of the proof of this theorem from [BD1], see [F6], Sect. 9.4.

A drawback of this construction is that not all ${}^L G$-local systems on X admit the structure of an oper. In fact, under our assumption that G is simply-connected (and so ${}^L G$ is of adjoint type), the local systems, or flat bundles $(\mathcal{F}, \nabla)$, on a smooth projective curve X that admit an oper structure correspond to a unique ${}^L G$-bundle on X described as follows (see [BD1]). Let $\Omega_X^{1/2}$ be a square root of the canonical line bundle Ω_X. There is a unique (up to an isomorphism) non-trivial extension

$$0 \to \Omega_X^{1/2} \to \mathcal{F}_0 \to \Omega_X^{-1/2} \to 0.$$

Let $\mathcal{F}_{PGL_2}$ be the PGL_2-bundle corresponding to the rank two vector bundle $\mathcal{F}_0$. Note that it does not depend on the choice of $\Omega_X^{1/2}$. This is the oper bundle for PGL_2. We define the oper bundle $\mathcal{F}_{LG}$ for a general simple Lie group ${}^L G$ of adjoint type as the push-forward of $\mathcal{F}_{PGL_2}$ with respect to a principal embedding $PGL_2 \hookrightarrow G$ (see Section 4.3).

---

[5] More precisely, Aut_{E_χ} is the $\mathcal{D}$-module $\Delta_{\kappa_c,x}(\mathbb{V}_0(\chi_x)) \otimes K^{-1/2}$, but here and below we will ignore the twist by $K^{1/2}$.

For each flat connection ∇ on the oper bundle $\mathcal{F}_{LG}$ there exists a unique ^LB-reduction $\mathcal{F}_{LB}$ satisfying the oper condition. Therefore $\operatorname{Op}_G(D)$ is a subset of $\operatorname{Loc}_{LG}(X)$, which is the fiber of the forgetful map $\operatorname{Loc}_{LG}(X) \to \operatorname{Bun}_{LG}$ over $\mathcal{F}_{LG}$.

Theorem 1 gives us a construction of Hecke eigensheaves for ^LG-local system that belong to the locus of opers. For a general ^LG-local system outside this locus, the above construction may be generalized as discussed at the end of Section 9.2 below.

Thus, Theorem 1, and its generalization to other unramified ^LG-local systems, give us an effective tool for constructing Hecke eigensheaves on Bun_G. It is natural to ask whether it can be generalized to the ramified case if we consider more general representations of $\widehat{\mathfrak{g}}_{\kappa_c,x}$. The goal of this paper is to explain how to do that.

We will see below that the completed universal enveloping algebra of $\widehat{\mathfrak{g}}_{\kappa_c,x}$ contains a large center. It is isomorphic to the algebra $\operatorname{Fun} \operatorname{Op}_{LG}(D_x^\times)$ of functions on the space $\operatorname{Op}_{LG}(D_x^\times)$ of ^LG-opers on the punctured disc $D_x^\times$. For $\chi_x \in \operatorname{Op}_{LG}(D_x^\times)$, let $\widehat{\mathfrak{g}}_{\kappa_c,x}$-$\operatorname{mod}_{\chi_x}$ be the full subcategory of $\widehat{\mathfrak{g}}_{\kappa_c,x}$-mod whose objects are $\widehat{\mathfrak{g}}_{\kappa_c,x}$-modules on which the center acts according to the character corresponding to χ_x.

The construction of Hecke eigensheaves now breaks into two steps:

(1) we study the Harish-Chandra categories $\widehat{\mathfrak{g}}_{\kappa_c,x}$-$\operatorname{mod}_{\chi_x}^K$ for various subgroups $K \subset G(\mathcal{O}_x)$;

(2) we apply the localization functors to these categories.

The simplest case of this construction is precisely the Beilinson-Drinfeld construction explained above. In this case we take χ_x to be a point in the subspace $\operatorname{Op}_{LG}(D_x) \subset \operatorname{Op}_{LG}(D_x^\times)$. Then the category $\widehat{\mathfrak{g}}_{\kappa_c,x}$-$\operatorname{mod}_{\chi_x}^{G(\mathcal{O}_x)}$ is equivalent to the category of vector spaces: its unique up to an isomorphism irreducible object is the above $\mathbb{V}_0(\chi_x)$, and all other objects are direct sums of copies of $\mathbb{V}_0(\chi_x)$ (see [FG1] and Theorem 3 below). Therefore the localization functor $\Delta_{\kappa_c,x}$ is determined by $\Delta_{\kappa_c,x}(\mathbb{V}_0(\chi_x))$, which is described in Theorem 1. It turns out to be the desired Hecke eigensheaf $\operatorname{Aut}_{E_\chi}$. Moreover, we expect that the functor $\Delta_{\kappa_c,x}$ sets up an equivalence between $\widehat{\mathfrak{g}}_{\kappa_c,x}$-$\operatorname{mod}_{\chi_x}^{G(\mathcal{O}_x)}$ and the category of Hecke eigensheaves on Bun_G with eigenvalue E_χ.

For general opers χ_x, with ramification, the (local) categories $\widehat{\mathfrak{g}}_{\kappa_c,x}$-$\operatorname{mod}_{\chi_x}^K$ are more complicated, as we will see below, and so are the corresponding (global) categories of Hecke eigensheaves. In order to understand the structure of the global categories, we need to study first of local categories of $\widehat{\mathfrak{g}}_{\kappa_c,x}$-modules. Using the localization functor, we can then understand the structure of the global categories. We will consider examples of the local categories in the following sections.

It is natural to view our study of the local categories $\widehat{\mathfrak{g}}_{\kappa_c,x}$-$\operatorname{mod}_{\chi_x}$ and $\widehat{\mathfrak{g}}_{\kappa_c,x}$-$\operatorname{mod}_{\chi_x}^K$ as a geometric analogue of the local Langlands correspondence. We will explain this point of view in the next section.

2 Classical Local Langlands Correspondence

The local Langlands correspondence relates smooth representations of reductive algebraic groups over local fields and representations of the Galois group of this field. In this section we define these objects and explain the main features of this correspondence. As the material of this section serves motivational purposes, we will only mention those aspects of this story that are most relevant for us. For a more detailed treatment, we refer the reader to the informative surveys [Vog, Ku] and references therein.

The local Langlands correspondence may be formulated for any local non-archimedian field. There are two possibilities: either F is the field $\mathbb{Q}_p$ of p-adic numbers or a finite extension of $\mathbb{Q}_p$, or F is the field $\mathbb{F}_q((t))$ of formal Laurent power series with coefficients in $\mathbb{F}_q$, the finite field with q elements (where q is a power of a prime number). For the sake of definiteness, in what follows we will restrict ourselves to the second case.

2.1 Langlands Parameters

Consider the group $GL_n(F)$, where $F = \mathbb{F}_q((t))$. A representation of $GL_n(F)$ on a complex vector space V is a homomorphism $\pi : GL_n(F) \to \operatorname{End} V$ such that $\pi(gh) = \pi(g)\pi(h)$ and $\pi(1) = \operatorname{Id}$. Define a topology on $GL_n(F)$ by stipulating that the base of open neighborhoods of $1 \in GL_n(F)$ is formed by the congruence subgroups

$$K_N = \{g \in GL_n(\mathbb{F}_q[[t]]) \mid g \equiv 1 \bmod t^N\}, \qquad N \in \mathbb{Z}_+.$$

For each $v \in V$ we obtain a map $\pi(\cdot)v : GL_n(F) \to V, g \mapsto \pi(g)v$. A representation (V, π) is called **smooth** if the map $\pi(\cdot)v$ is continuous for each v, where we give V the discrete topology. In other words, V is smooth if for any vector $v \in V$ there exists $N \in \mathbb{Z}_+$ such that

$$\pi(g)v = v, \qquad \forall g \in K_N.$$

We are interested in describing the equivalence classes of irreducible smooth representations of $GL_n(F)$. Surprisingly, those turn out to be related to objects of a different kind: n-dimensional representations of the Galois group of F.

Recall that the algebraic closure of F is a field obtained by adjoining to F the roots of all polynomials with coefficients in F. However, in the case when $F = \mathbb{F}_q((t))$ some of the extensions of F may be non-separable. We wish to avoid the non-separable extensions, because they do not contribute to the Galois group. Let $\overline{F}$ be the maximal separable extension inside a given algebraic closure of F. It is uniquely defined up to an isomorphism.

Let $\operatorname{Gal}(\overline{F}/F)$ be the **absolute Galois group** of F. Its elements are the automorphisms σ of the field $\overline{F}$ such that $\sigma(y) = y$ for all $y \in F$.

Now set $F = \mathbb{F}_q((t))$. Observe that we have a natural map $\mathrm{Gal}(\overline{F}/F) \to \mathrm{Gal}(\overline{\mathbb{F}}_q/\mathbb{F}_q)$ obtained by applying an automorphism of F to $\overline{\mathbb{F}}_q \subset F$. The group $\mathrm{Gal}(\overline{\mathbb{F}}_q/\mathbb{F}_q)$ is isomorphic to the profinite completion $\widehat{\mathbb{Z}}$ of $\mathbb{Z}$ (see, e.g., [F6], Sect. 1.3). Its subgroup $\mathbb{Z} \subset \widehat{\mathbb{Z}}$ is generated by the **geometric Frobenius element** which is inverse to the automorphism $x \mapsto x^q$ of $\overline{\mathbb{F}}_q$. Let W_F be the preimage of the subgroup $\mathbb{Z} \subset \mathrm{Gal}(\overline{\mathbb{F}}_q/\mathbb{F}_q)$. This is the **Weil group** of F. Denote by ν be the corresponding homomorphism $W_F \to \mathbb{Z}$.

Now let $W_F' = W_F \ltimes \mathbb{C}$ be the semi-direct product of W_F and the one-dimensional complex additive group $\mathbb{C}$, where W_F acts on $\mathbb{C}$ by the formula

$$(2.1) \qquad \sigma x \sigma^{-1} = q^{\nu(\sigma)} x, \qquad \sigma \in W_F, x \in \mathbb{C}.$$

This is the **Weil-Deligne group** of F.

An n-dimensional complex representation of W_F' is by definition a homomorphism $\rho' : W_F' \to GL_n(\mathbb{C})$ which may be described as a pair (ρ, N), where ρ is an n-dimensional representation of W_F, $N \in GL_n(\mathbb{C})$, and we have $\rho(\sigma) N \rho(\sigma)^{-1} = q^{\nu(\sigma)} \rho(N)$ for all $\sigma \in W_F$. The group W_F is topological, with respect to the Krull topology (in which the open neighborhoods of the identity are the normal subgroups of finite index). The representation (ρ, N) is called **admissible** if ρ is continuous (equivalently, factors through a finite quotient of W_F) and semisimple, and N is a unipotent element of $GL_n(\mathbb{C})$.

The group W_F' was introduced by P. Deligne [De2]. The idea is that by adjoining the unipotent element N to W_F one obtains a group whose complex admissible representations are the same as continuous ℓ-adic representations of W_F (where $\ell \neq p$ is a prime).

2.2 The Local Langlands Correspondence for GL_n

Now we are ready to state the local Langlands correspondence for the group GL_n over a local non-archimedean field F. It is a bijection between two different sorts of data. One is the set of the equivalence classes of irreducible smooth representations of $GL_n(F)$. The other is the set of equivalence classes of n-dimensional admissible representations of W_F'. We represent it schematically as follows:

$$\boxed{\begin{array}{c} n\text{-dimensional admissible} \\ \text{representations of } W_F' \end{array}} \quad \Longleftrightarrow \quad \boxed{\begin{array}{c} \text{irreducible smooth} \\ \text{representations of } GL_n(F) \end{array}}$$

This correspondence is supposed to satisfy an overdetermined system of constraints which we will not recall here (see, e.g., [Ku]).

The local Langlands correspondence for GL_n is a theorem. In the case when $F = \mathbb{F}_q((t))$ it has been proved in [LRS], and when $F = \mathbb{Q}_p$ or its finite extension in [HT] and also in [He].

2.3 Generalization to Other Reductive Groups

Let us replace the group GL_n by an arbitrary connected reductive group G over a local non-archimedian field F. The group $G(F)$ is also a topological group, and there is a notion of smooth representation of $G(F)$ on a complex vector space. It is natural to ask whether one can relate irreducible smooth representations of $G(F)$ to representations of the Weil-Deligne group W'_F. This question is addressed in the general local Langlands conjectures. It would take us too far afield to try to give here a precise formulation of these conjectures. So we will only indicate some of the objects involved referring the reader to the articles [Vog, Ku] where these conjectures are described in great detail.

Recall that in the case when $G = GL_n$ the irreducible smooth representations are parametrized by admissible homomorphisms $W'_F \to GL_n(\mathbb{C})$. In the case of a general reductive group G, the representations are conjecturally parametrized by admissible homomorphisms from W'_F to the so-called **Langlands dual group** ^LG, which is defined over $\mathbb{C}$.

In order to explain the notion of the Langlands dual group, consider first the group G over the closure $\overline{F}$ of the field F. All maximal tori T of this group are conjugate to each other and are necessarily split, i.e., we have an isomorphism $T(\overline{F}) \simeq (\overline{F}^\times)$. For example, in the case of GL_n, all maximal tori are conjugate to the subgroup of diagonal matrices. We associate to $T(\overline{F})$ two lattices: the weight lattice $X^*(T)$ of homomorphisms $T(\overline{F}) \to \overline{F}^\times$ and the coweight lattice $X_*(T)$ of homomorphisms $\overline{F}^\times \to T(\overline{F})$. They contain the sets of roots $\Delta \subset X^*(T)$ and coroots $\Delta^\vee \subset X_*(T)$, respectively. The quadruple $(X^*(T), X_*(T), \Delta, \Delta^\vee)$ is called the root datum for G over $\overline{F}$. The root datum determines G up to an isomorphism defined over $\overline{F}$. The choice of a Borel subgroup $B(\overline{F})$ containing $T(\overline{F})$ is equivalent to a choice of a basis in Δ, namely, the set of simple roots Δ_s, and the corresponding basis $\Delta_s^\vee$ in $\Delta^\vee$.

Now, given $\gamma \in \mathrm{Gal}(\overline{F}/F)$, there is $g \in G(\overline{F})$ such that $g(\gamma(T(\overline{F}))g^{-1} = T(\overline{F})$ and $g(\gamma(B(\overline{F}))g^{-1} = B(\overline{F})$. Then g gives rise to an automorphism of the based root data $(X^*(T), X_*(T), \Delta_s, \Delta_s^\vee)$. Thus, we obtain an action of $\mathrm{Gal}(\overline{F}/F)$ on the based root data.

Let us now exchange the lattices of weights and coweights and the sets of simple roots and coroots. Then we obtain the based root data

$$(X_*(T), X^*(T), \Delta_s^\vee, \Delta_s)$$

of a reductive algebraic group over $\mathbb{C}$ which is denoted by $^LG^\circ$. For instance, the group GL_n is self-dual, the dual of SO_{2n+1} is Sp_{2n}, the dual of Sp_{2n} is SO_{2n+1}, and SO_{2n} is self-dual.

The action of $\mathrm{Gal}(\overline{F}/F)$ on the based root data gives rise to its action on $^LG^\circ$. The semi-direct product $^LG = \mathrm{Gal}(\overline{F}/F) \ltimes {}^LG^\circ$ is called the **Langlands dual group** of G.

According to the local Langlands conjecture, the equivalence classes of irreducible smooth representations of $G(F)$ are, roughly speaking, parameterized by the equivalence classes of admissible homomorphisms $W'_F \to {}^L G$. In fact, the conjecture is more subtle: one needs to consider simultaneously representations of all inner forms of G, and a homomorphism $W'_F \to {}^L G$ corresponds in general not to a single irreducible representation of $G(F)$, but to a finite set of representations called an *L*-**packet**. To distinguish between them, one needs additional data (see [Vog] and Section 8.1 below for more details). But in the first approximation one can say that the essence of the local Langlands correspondence is that

Irreducible smooth representations of $G(F)$ are parameterized in terms of admissible homomorphisms $W'_F \to {}^L G$.

3 Geometric Local Langlands Correspondence over $\mathbb{C}$

We now wish to find a generalization of the local Langlands conjectures in which we replace the field $F = \mathbb{F}_q((t))$ by the field $\mathbb{C}((t))$. We would like to see how the ideas and patterns of the Langlands correspondence play out in this new context, with the hope of better understanding the deep underlying structures behind this correspondence.

So let G be a connected simply-connected algebraic group over $\mathbb{C}$, and $G(F)$ the loop group $G((t)) = G(\mathbb{C}((t)))$. Thus, we wish to study smooth representations of the loop group $G((t))$ and try to relate them to some "Langlands parameters", which we expect, by analogy with the case of local non-archimedian fields described above, to be related to the Galois group of $\mathbb{C}((t))$ and the Langlands dual group ${}^L G$.

3.1 Geometric Langlands Parameters

Unfortunately, the Galois group of $\mathbb{C}((t))$ is too small: it is isomorphic to the pro-finite completion $\widehat{\mathbb{Z}}$ of $\mathbb{Z}$. This is not surprising from the point of view of the analogy between the Galois groups and the fundamental groups (see, e.g., [F6], Sect. 3.1). The topological fundamental group of the punctured disc is $\mathbb{Z}$, and the algebraic fundamental group is its pro-finite completion.

However, we may introduce additional Langlands parameters by using a more geometric perspective on homomorphisms from the fundamental group to ${}^L G$. Those may be viewed as ${}^L G$-local systems. In general, ${}^L G$-local systems on a compact variety Z are the same as flat ${}^L G$-bundles $(\mathcal{F}, \nabla)$ on Z. If the variety is not compact (as in the case of $D^\times$), then we should impose the additional condition that the connection has **regular singularities** (pole of order at most 1) at infinity. In our case we obtain ${}^L G$-bundles on $D^\times$ with a connection that has regular singularity at the origin. Then the monodromy of the connection gives rise to a homomorphism from $\pi_1(D^\times)$ to ${}^L G$. Now we generalize this by allowing connections with **arbitrary**, that is regular and **irregular**, singularities at the origin. Thus, we want to use as the general Langlands parameters, the equivalence classes of pairs $(\mathcal{F}, \nabla)$, where $\mathcal{F}$ is a ^LG-bundle on $D^\times$ and ∇ is an arbitrary connection on $\mathcal{F}$.

Any bundle $\mathcal{F}$ on $D^\times$ may be trivialized. Then ∇ may be represented by the first-order differential operator

$$(3.1) \qquad \nabla = \partial_t + A(t), \qquad A(t) \in {}^L\mathfrak{g}((t)).$$

where $^L\mathfrak{g}$ is the Lie algebra of the Langlands dual group ^LG. Changing the trivialization of $\mathcal{F}$ amounts to a gauge transformation

$$\nabla \mapsto \nabla' = \partial_t + gAg^{-1} - (\partial_t g)g^{-1}$$

with $g \in {}^LG((t))$. Therefore the set of equivalence classes of ^LG-bundles with a connection on $D^\times$ is in bijection with the set of gauge equivalence classes of operators (3.1). We denote this set by $\mathrm{Loc}_{{}^LG}(D^\times)$. Thus, we have

$$(3.2) \qquad \mathrm{Loc}_{{}^LG}(D^\times) = \{\partial_t + A(t),\ A(t) \in {}^L\mathfrak{g}((t))\}/{}^LG((t)).$$

We declare that the local Langlands parameters in the complex setting should be the points of $\mathrm{Loc}_{{}^LG}(D^\times)$: the equivalence classes of flat ^LG-bundles on $D^\times$ or, more concretely, the gauge equivalence classes (3.2) of first-order differential operators.

Having settled the issue of the Langlands parameters, we have to decide what it is that we will be parameterizing. Recall that in the classical setting the homomorphism $W'_F \to {}^LG$ parameterized irreducible smooth representations of the group $G(F)$, $F = \mathbb{F}_q((t))$. We start by translating this notion to the representation theory of loop groups.

3.2 Representations of the Loop Group

The loop group $G((t))$ contains the congruence subgroups

$$(3.3) \qquad K_N = \{g \in G[[t]] \,|\, g \equiv 1 \bmod t^N\}, \qquad N \in \mathbb{Z}_+.$$

It is natural to call a representation of $G((t))$ on a complex vector space V **smooth** if for any vector $v \in V$ there exists $N \in \mathbb{Z}_+$ such that $K_N \cdot v = v$. This condition may be interpreted as the continuity condition, if we define a topology on $G((t))$ by taking as the base of open neighborhoods of the identity the subgroups $K_N, N \in \mathbb{Z}_+$, as before.

But our group G is now a complex Lie group (not a finite group), and so $G((t))$ is an infinite-dimensional Lie group. More precisely, we view $G((t))$ as an ind-group, i.e., as a group object in the category of ind-schemes. At first glance, it is natural to consider the algebraic representations of $G((t))$. We observe that $G((t))$ is generated by the "parahoric" algebraic groups P_i

corresponding to the affine simple roots. For these subgroups the notion of algebraic representation makes perfect sense. A representation of $G((t))$ is then said to be algebraic if its restriction to each of the P_i's is algebraic.

However, this naive approach leads us to the following discouraging fact: an irreducible smooth representation of $G((t))$, which is algebraic, is necessarily trivial (see [BD1], 3.7.11(ii)). Thus, we find that the class of algebraic representations of loop groups turns out to be too restrictive. We could relax this condition and consider differentiable representations, i.e., the representations of $G((t))$ considered as a Lie group. But it is easy to see that the result would be the same. Replacing $G((t))$ by its central extension $\widehat{G}$ would not help us much either: irreducible integrable representations of $\widehat{G}$ are parameterized by dominant integral weights, and there are no extensions between them [K2]. These representations are again too sparse to be parameterized by the geometric data considered above. Therefore we should look for other types of representations.

Going back to the original setup of the local Langlands correspondence, we recall that there we considered representations of $G(\mathbb{F}_q((t)))$ on $\mathbb{C}$-vector spaces, so we could not possibly use the algebraic structure of $G(\mathbb{F}_q((t)))$ as an ind-group over $\mathbb{F}_q$. Therefore we cannot expect the class of algebraic (or differentiable) representations of the complex loop group $G((t))$ to be meaningful from the point of view of the Langlands correspondence. We should view the loop group $G((t))$ as an abstract topological group, with the topology defined by means of the congruence subgroups, in other words, consider its smooth representations as an *abstract* group.

So we need to search for some geometric objects that encapsulate representations of our groups and make sense both over a finite field and over the complex field.

3.3 From Functions to Sheaves

We start by revisiting smooth representations of the group $G(F)$, where $F = \mathbb{F}_q((t))$. We realize such representations more concretely by considering their matrix coefficients. Let (V, π) be an irreducible smooth representation of $G(F)$. We define the **contragredient** representation $V^\vee$ as the linear span of all smooth vectors in the dual representation V^*. This span is stable under the action of $G(F)$ and so it admits a smooth representation $(V^\vee, \pi^\vee)$ of $G(F)$. Now let ϕ be a K_N-invariant vector in $V^\vee$. Then we define a linear map

$$V \to C(G(F)/K_N), \qquad v \mapsto f_v,$$

where

$$f_v(g) = \langle \pi^\vee(g)\phi, v \rangle.$$

Here $C(G(F)/K_N)$ denotes the vector space of $\mathbb{C}$-valued locally constant functions on $G(F)/K_N$. The group $G(F)$ naturally acts on this space by the formula $(g \cdot f)(h) = f(g^{-1}h)$, and the above map is a morphism of representations, which is non-zero, and hence injective, if (V, π) is irreducible.

Thus, we realize our representation in the space of functions on the quotient $G(F)/K_N$. More generally, we may realize representations in spaces of functions on the quotient $G((t))/K$ with values in a finite-dimensional vector space, by considering a finite-dimensional subrepresentation of K inside V rather than the trivial one.

An important observation here is that $G(F)/K$, where $F = \mathbb{F}_q((t))$ and K is a compact subgroup of $G(F)$, is not only a set, but it is a set of points of an algebraic variety (more precisely, an ind-scheme) defined over the field $\mathbb{F}_q$. For example, for $K_0 = G(\mathbb{F}_q[[t]])$, which is the maximal compact subgroup, the quotient $G(F)/K_0$ is the set of $\mathbb{F}_q$-points of the ind-scheme called the **affine Grassmannian**.

Next, we recall an important idea going back to Grothendieck that functions on the set of $\mathbb{F}_q$-points on an algebraic variety X defined over $\mathbb{F}_q$ can often be viewed as the "shadows" of the so-called ℓ-adic sheaves on X. We will not give the definition of these sheaves, referring the reader to [Mi, FK]. The Grothendieck **fonctions-faisceaux** dictionary (see, e.g., [La]) is formulated as follows. Let $\mathcal{F}$ be an ℓ-adic sheaf and x be an $\mathbb{F}_{q_1}$-point of X, where $q_1 = q^m$. Then one has the Frobenius conjugacy class Fr_x acting on the stalk $\mathcal{F}_x$ of $\mathcal{F}$ at x. Hence we can define a function $\mathbf{f}_{q_1}(\mathcal{F})$ on the set of $\mathbb{F}_{q_1}$-points of V, whose value at x is $\mathrm{Tr}(\mathrm{Fr}_x, \mathcal{F}_x)$. This function takes values in the algebraic closure $\overline{\mathbb{Q}}_\ell$ of $\mathbb{Q}_\ell$. But there is not much of a difference between $\overline{\mathbb{Q}}_\ell$-valued functions and $\mathbb{C}$-valued functions: since they have the same cardinality, $\overline{\mathbb{Q}}_\ell$ and $\mathbb{C}$ may be identified as abstract fields. Besides, in most interesting cases, the values actually belong to $\overline{\mathbb{Q}}$, which is inside both $\overline{\mathbb{Q}}_\ell$ and $\mathbb{C}$.

More generally, if $\mathcal{K}$ is a complex of ℓ-adic sheaves, one defines a function $\mathbf{f}_{q_1}(\mathcal{K})$ on $V(\mathbb{F}_{q_1})$ by taking the alternating sums of the traces of Fr_x on the stalk cohomologies of $\mathcal{K}$ at x. The map $\mathcal{K} \to \mathbf{f}_{q_1}(\mathcal{K})$ intertwines the natural operations on sheaves with natural operations on functions (see [La], Sect. 1.2).

Let $K_0(\mathcal{S}h_X)$ be the complexified Grothendieck group of the category of ℓ-adic sheaves on X. Then the above construction gives us a map

$$K_0(\mathcal{S}h_X) \to \prod_{m \geq 1} X(\mathbb{F}_{q^m}),$$

and it is known that this map is injective (see [La]).

Therefore we may hope that the functions on the quotients $G(F)/K_N$ which realize our representations come by this constructions from ℓ-adic sheaves, or more generally, from complexes of ℓ-adic sheaves, on X.

Now, the notion of constructible sheaf (unlike the notion of a function) has a transparent and meaningful analogue for a complex algebraic variety X, namely, those sheaves of $\mathbb{C}$-vector spaces whose restrictions to the strata of a stratification of the variety X are locally constant. The affine Grassmannian and more general ind-schemes underlying the quotients $G(F)/K_N$ may be defined both over $\mathbb{F}_q$ and $\mathbb{C}$. Thus, it is natural to consider the categories of such sheaves (or, more precisely, their derived categories) on these ind-schemes over $\mathbb{C}$ as the replacements for the vector spaces of functions on their points realizing smooth representations of the group $G(F)$.

We therefore naturally come to the idea, advanced in [FG2], that the representations of the loop group $G((t))$ that we need to consider are not realized on vector spaces, but on **categories**, such as the derived category of coherent sheaves on the affine Grassmannian. Of course, such a category has a Grothendieck group, and the group $G((t))$ will act on the Grothendieck group as well, giving us a representation of $G((t))$ on a vector space. But we obtain much more structure by looking at the categorical representation. The objects of the category, as well as the action, will have a geometric meaning, and thus we will be using the geometry as much as possible.

Let us summarize: to each local Langlands parameter $\chi \in \mathrm{Loc}_{{}^L G}(D^\times)$ we wish to attach a category $\mathcal{C}_\chi$ equipped with an action of the loop group $G((t))$. But what kind of categories should these $\mathcal{C}_\chi$ be and what properties do we expect them to satisfy?

To get closer to answering these questions, we wish to discuss two more steps that we can make in the above discussion to get to the types of categories with an action of the loop group that we will consider in this paper.

3.4 A Toy Model

At this point it is instructive to detour slightly and consider a toy model of our construction. Let G be a split reductive group over $\mathbb{Z}$, and B its Borel subgroup. A natural representation of $G(\mathbb{F}_q)$ is realized in the space of complex (or $\overline{\mathbb{Q}}_\ell$-) valued functions on the quotient $G(\mathbb{F}_q)/B(\mathbb{F}_q)$. It is natural to ask what is the "correct" analogue of this representation if we replace the field $\mathbb{F}_q$ by the complex field and the group $G(\mathbb{F}_q)$ by $G(\mathbb{C})$. This may be viewed as a simplified version of our quandary, since instead of considering $G(\mathbb{F}_q((t)))$ we now look at $G(\mathbb{F}_q)$.

The quotient $G(\mathbb{F}_q)/B(\mathbb{F}_q)$ is the set of $\mathbb{F}_q$-points of the algebraic variety defined over $\mathbb{Z}$ called the flag variety of G and defined by Fl. Our discussion in the previous section suggests that we first need to replace the notion of a function on $\mathrm{Fl}(\mathbb{F}_q)$ by the notion of an ℓ-adic sheaf on the variety $\mathrm{Fl}_{\mathbb{F}_q} = \mathrm{Fl} \underset{\mathbb{Z}}{\otimes} \mathbb{F}_q$.

Next, we replace the notion of an ℓ-adic sheaf on Fl considered as an algebraic variety over $\mathbb{F}_q$, by the notion of a constructible sheaf on $\mathrm{Fl}_\mathbb{C} = \mathrm{Fl} \underset{\mathbb{Z}}{\otimes} \mathbb{C}$ which is an algebraic variety over $\mathbb{C}$. The complex algebraic group $G_\mathbb{C}$ naturally acts on $\mathrm{Fl}_\mathbb{C}$ and hence on this category. Now we make two more reformulations of this category.

First of all, for a smooth complex algebraic variety X we have a **Riemann-Hilbert correspondence** which is an equivalence between the derived category of constructible sheaves on X and the derived category of $\mathcal{D}$-modules on X that are holonomic and have regular singularities.

Here we consider the sheaf of algebraic differential operators on X and sheaves of modules over it, which we simply refer to as $\mathcal{D}$-modules. The simplest example of a $\mathcal{D}$-module is the sheaf of sections of a vector bundle on V equipped with a flat connection. The flat connection enables us to multiply any section by a function and we can use the flat connection to act on sections by vector fields. The two actions generate an action of the sheaf of differential operators on the sections of our bundle. The sheaf of horizontal sections of this bundle is then a locally constant sheaf of X. We have seen above that there is a bijection between the set of isomorphism classes of rank n bundles on X with connection having regular singularities and the set of isomorphism classes of locally constant sheaves on X of rank n, or equivalently, n-dimensional representations of $\pi_1(X)$. This bijection may be elevated to an equivalence of the corresponding categories, and the general Riemann-Hilbert correspondence is a generalization of this equivalence of categories that encompasses more general $\mathcal{D}$-modules.

The Riemann-Hilbert correspondence allows us to associate to any holonomic $\mathcal{D}$-module on X a complex of constructible sheaves on X, and this gives us a functor between the corresponding derived categories which turns out to be an equivalence if we restrict ourselves to the holonomic $\mathcal{D}$-modules with regular singularities (see [B2, GM] for more details).

Thus, over $\mathbb{C}$ we may pass from constructible sheaves to $\mathcal{D}$-modules. In our case, we consider the category of (regular holonomic) $\mathcal{D}$-modules on the flag variety $\mathrm{Fl}_{\mathbb{C}}$. This category carries a natural action of $G_{\mathbb{C}}$.

Finally, let us observe that the Lie algebra $\mathfrak{g}$ of $G_{\mathbb{C}}$ acts on the flag variety infinitesimally by vector fields. Therefore, given a $\mathcal{D}$-module $\mathcal{F}$ on $\mathrm{Fl}_{\mathbb{C}}$, the space of its global sections $\Gamma(\mathrm{Fl}_{\mathbb{C}}, \mathcal{F})$ has the structure of $\mathfrak{g}$-module. We obtain a functor Γ from the category of $\mathcal{D}$-modules on $\mathrm{Fl}_{\mathbb{C}}$ to the category of $\mathfrak{g}$-modules. A. Beilinson and J. Bernstein have proved that this functor is an equivalence between the category of all $\mathcal{D}$-modules on $\mathrm{Fl}_{\mathbb{C}}$ (not necessarily regular holonomic) and the category $\mathcal{C}_0$ of $\mathfrak{g}$-modules on which the center of the universal enveloping algebra $U(\mathfrak{g})$ acts through the augmentation character.

Thus, we can now answer our question as to what is a meaningful geometric analogue of the representation of the finite group $G(\mathbb{F}_q)$ on the space of functions on the quotient $G(\mathbb{F}_q)/B(\mathbb{F}_q)$. The answer is the following: it is an **abelian category** equipped with an action of the algebraic group $G_{\mathbb{C}}$. This category has two incarnations: one is the category of $\mathcal{D}$-modules on the flag variety $\mathrm{Fl}_{\mathbb{C}}$, and the other is the category $\mathcal{C}_0$ of modules over the Lie algebra $\mathfrak{g}$ with the trivial central character. Both categories are equipped with natural actions of the group $G_{\mathbb{C}}$.

Let us pause for a moment and spell out what exactly we mean when we say that the group $G_{\mathbb{C}}$ acts on the category $\mathcal{C}_0$. For simplicity, we will describe the action of the corresponding group $G(\mathbb{C})$ of $\mathbb{C}$-points of $G_{\mathbb{C}}$.[6] This means the following: each element $g \in G$ gives rise to a functor F_g on $\mathcal{C}_0$ such that F_1 is the identity functor, and the functor $\mathcal{F}_{g^{-1}}$ is quasi-inverse to F_g. Moreover, for any pair $g, h \in G$ we have a fixed isomorphism of functors $i_{g,h} : F_{gh} \to F_g \circ F_h$ so that for any triple $g, h, k \in G$ we have the equality $i_{h,k} i_{g,hk} = i_{g,h} i_{gh,k}$ of isomorphisms $F_{ghk} \to F_g \circ F_h \circ F_k$.

The functors F_g are defined as follows. Given a representation (V, π) of $\mathfrak{g}$ and an element $g \in G(\mathbb{C})$, we define a new representation $F_g((V, \pi)) = (V, \pi_g)$, where by definition $\pi_g(x) = \pi(\mathrm{Ad}_g(x))$. Suppose that (V, π) is irreducible. Then it is easy to see that $(V, \pi_g) \simeq (V, \pi)$ if and only if (V, π) is integrable, i.e., is obtained from an algebraic representation of G.[7] This is equivalent to this representation being finite-dimensional. But a general representation (V, π) is infinite-dimensional, and so it will not be isomorphic to (V, π_g), at least for some $g \in G$.

Now we consider morphisms in $\mathcal{C}_0$, which are just $\mathfrak{g}$-homomorphisms. Given a $\mathfrak{g}$-homomorphism between representations (V, π) and (V', π'), i.e., a linear map $T : V \to V'$ such that $T\pi(x) = \pi'(x)T$ for all $x \in \mathfrak{g}$, we set $F_g(T) = T$. The isomorphisms $i_{g,h}$ are all identical in this case.

3.5 Back to Loop Groups

In our quest for a complex analogue of the local Langlands correspondence we need to decide what will replace the notion of a smooth representation of the group $G(F)$, where $F = \mathbb{F}_q((t))$. As the previous discussion demonstrates, we should consider representations of the complex loop group $G((t))$ on various categories of $\mathcal{D}$-modules on the ind-schemes $G((t))/K$, where K is a "compact" subgroup of $G((t))$, such as $G[[t]]$ or the Iwahori subgroup (the preimage of a Borel subgroup $B \subset G$ under the homomorphism $G[[t]] \to G$), or the categories of representations of the Lie algebra $\mathfrak{g}((t))$. Both scenarios are viable, and they lead to interesting results and conjectures which we will discuss in detail in Section 9, following [FG2]. In this paper we will concentrate on the second scenario and consider categories of modules over the loop algebra $\mathfrak{g}((t))$.

The group $G((t))$ acts on the category of representations of $\mathfrak{g}((t))$ in the way that we described in the previous section. An analogue of a smooth representation of $G(F)$ is a category of smooth representations of $\mathfrak{g}((t))$. Let us observe however that we could choose instead the category of smooth representations of the central extension of $\mathfrak{g}((t))$, namely, $\widehat{\mathfrak{g}}_\kappa$.

---

[6] More generally, for any $\mathbb{C}$-algebra R, we have an action of $G(R)$ on the corresponding base-changed category over R. Thus, we are naturally led to the notion of an algebraic group (or, more generally, a group scheme) acting on an abelian category, which is spelled out in [FG2], Sect. 20.

[7] In general, we could obtain a representation of a central extension of G, but if G is reductive, it does not have non-trivial central extensions.

The group $G((t))$ acts on the Lie algebra $\widehat{\mathfrak{g}}_\kappa$ for any κ, because the adjoint action of the central extension of $G((t))$ factors through the action of $G((t))$. We use the action of $G((t))$ on $\widehat{\mathfrak{g}}_\kappa$ to construct an action of $G((t))$ on the category $\widehat{\mathfrak{g}}_\kappa$-mod, in the same way as in Section 3.4.

Now recall the space $\mathrm{Loc}_{{}^L G}(D^\times)$ of the Langlands parameters that we defined in Section 3.1. Elements of $\mathrm{Loc}_{{}^L G}(D^\times)$ have a concrete description as gauge equivalence classes of first order operators $\partial_t + A(t), A(t) \in {}^L\mathfrak{g}((t))$, modulo the action of ${}^L G((t))$ (see formula (3.2)).

We can now formulate the local Langlands correspondence over $\mathbb{C}$ as the following problem:

To each local Langlands parameter $\chi \in \mathrm{Loc}_{{}^L G}(D^\times)$ associate a subcategory $\widehat{\mathfrak{g}}_\kappa$-mod$_\chi$ of $\widehat{\mathfrak{g}}_\kappa$-mod which is stable under the action of the loop group $G((t))$.

We wish to think of the category $\widehat{\mathfrak{g}}_\kappa$-mod as "fibering" over the space of local Langlands parameters $\mathrm{Loc}_{{}^L G}(D^\times)$, with the categories $\widehat{\mathfrak{g}}_\kappa$-mod$_\chi$ being the "fibers" and the group $G((t))$ acting along these fibers. From this point of view the categories $\widehat{\mathfrak{g}}_\kappa$-mod$_\chi$ should give us a "spectral decomposition" of the category $\widehat{\mathfrak{g}}_\kappa$-mod over $\mathrm{Loc}_{{}^L G}(D^\times)$.

In the next sections we will present a concrete proposal made in [FG2] describing these categories in the special case when $\kappa = \kappa_c$, the critical level.

4 Center and Opers

In Section 1 we have introduced the category $\widehat{\mathfrak{g}}_\kappa$-mod whose objects are smooth $\widehat{\mathfrak{g}}_\kappa$-modules on which the central element $\mathbf{1}$ acts as the identity. As explained at the end of the previous section, we wish to show that this category "fibers" over the space of the Langlands parameters, which are gauge equivalence classes of ${}^L G$-connections on the punctured disc $D^\times$ (or perhaps, something similar). Moreover, the loop group $G((t))$ should act on this category "along the fibers".

Any abelian category may be thought of as "fibering" over the spectrum of its center. Hence the first idea that comes to mind is to describe the center of the category $\widehat{\mathfrak{g}}_\kappa$-mod in the hope that its spectrum is related to the Langlands parameters. As we will see, this is indeed the case for a particular value of κ.

4.1 Center of an Abelian Category

Let us first recall what is the center of an abelian category. Let $\mathcal{C}$ be an abelian category over $\mathbb{C}$. The center $Z(\mathcal{C})$ is by definition the set of endomorphisms of the identity functor on $\mathcal{C}$. Let us recall such such an endomorphism is a system of endomorphisms $e_M \in \mathrm{Hom}_{\mathcal{C}}(M, M)$, for each object M of $\mathcal{C}$, which is compatible with the morphisms in $\mathcal{C}$: for any morphism $f : M \to N$ in $\mathcal{C}$ we have $f \circ e_M = e_N \circ f$. It is clear that $Z(\mathcal{C})$ has a natural structure of a commutative algebra over $\mathbb{C}$.

Let $S = \operatorname{Spec} Z(\mathcal{C})$. This is an affine algebraic variety such that $Z(\mathcal{C})$ is the algebra of functions on S. Each point $s \in S$ defines an algebra homomorphism (equivalently, a character) $\rho_s : Z(\mathcal{C}) \to \mathbb{C}$ (evaluation of a function at the point s). We define the full subcategory $\mathcal{C}_s$ of $\mathcal{C}$ whose objects are the objects of $\mathcal{C}$ on which $Z(\mathcal{C})$ acts according to the character ρ_s. It is instructive to think of the category $\mathcal{C}$ as "fibering" over S, with the fibers being the categories $\mathcal{C}_s$.

Now suppose that $\mathcal{C} = A$-mod is the category of left modules over an associative $\mathbb{C}$-algebra A. Then A itself, considered as a left A-module, is an object of $\mathcal{C}$, and so we obtain a homomorphism

$$Z(\mathcal{C}) \to Z(\operatorname{End}_A A) = Z(A^{\mathrm{opp}}) = Z(A),$$

where $Z(A)$ is the center of A. On the other hand, each element of $Z(A)$ defines an endomorphism of each object of A-mod, and so we obtain a homomorphism $Z(A) \to Z(\mathcal{C})$. It is easy to see that these maps set mutually inverse isomorphisms between $Z(\mathcal{C})$ and $Z(A)$.

If $\mathfrak{g}$ is a Lie algebra, then the category $\mathfrak{g}$-mod of $\mathfrak{g}$-modules coincides with the category $U(\mathfrak{g})$-mod of $U(\mathfrak{g})$-modules, where $U(\mathfrak{g})$ is the universal enveloping algebra of $\mathfrak{g}$. Therefore the center of the category $\mathfrak{g}$-mod is equal to the center of $U(\mathfrak{g})$, which by abuse of notation we denote by $Z(\mathfrak{g})$.

Now consider the category $\widehat{\mathfrak{g}}_\kappa$-mod. Let us recall from Section 1 that objects of $\widehat{\mathfrak{g}}_\kappa$-mod are $\widehat{\mathfrak{g}}_\kappa$-modules M on which the central element $\mathbf{1}$ acts as the identity and which are **smooth**, that is for any vector $v \in M$ we have

$$(4.1) \qquad (\mathfrak{g} \otimes t^N \mathbb{C}[[t]]) \cdot v = 0$$

for sufficiently large N.

Thus, we see that there are two properties that its objects satisfy. Therefore it does not coincide with the category of all modules over the universal enveloping algebra $U(\widehat{\mathfrak{g}}_\kappa)$ (which is the category of all $\widehat{\mathfrak{g}}_\kappa$-modules). We need to modify this algebra.

First of all, since $\mathbf{1}$ acts as the identity, the action of $U(\widehat{\mathfrak{g}}_\kappa)$ factors through the quotient

$$U_\kappa(\widehat{\mathfrak{g}}) \overset{\text{def}}{=} U_\kappa(\widehat{\mathfrak{g}})/(\mathbf{1} - 1).$$

Second, the smoothness condition (4.1) implies that the action of $U_\kappa(\widehat{\mathfrak{g}})$ extends to an action of its completion defined as follows.

Define a linear topology on $U_\kappa(\widehat{\mathfrak{g}})$ by using as the basis of neighborhoods for 0 the following left ideals:

$$I_N = U_\kappa(\widehat{\mathfrak{g}})(\mathfrak{g} \otimes t^N \mathbb{C}[[t]]), \qquad N \geqslant 0.$$

Let $\widetilde{U}_\kappa(\widehat{\mathfrak{g}})$ be the completion of $U_\kappa(\widehat{\mathfrak{g}})$ with respect to this topology. Note that, equivalently, we can write

$$\widetilde{U}_\kappa(\widehat{\mathfrak{g}}) = \varprojlim U_\kappa(\widehat{\mathfrak{g}})/I_N.$$

Even though the I_N's are only left ideals (and not two-sided ideals), one checks that the associative product structure on $U_\kappa(\widehat{\mathfrak{g}})$ extends by continuity to an associative product structure on $\widetilde{U}_\kappa(\widehat{\mathfrak{g}})$ (this follows from the fact that the Lie bracket on $U_\kappa(\widehat{\mathfrak{g}})$ is continuous in the above topology). Thus, $\widetilde{U}_\kappa(\widehat{\mathfrak{g}})$ is a complete topological algebra. It follows from the definition that the category $\widehat{\mathfrak{g}}_\kappa$-mod coincides with the category of discrete modules over $\widetilde{U}_\kappa(\widehat{\mathfrak{g}})$ on which the action of $\widetilde{U}_\kappa(\widehat{\mathfrak{g}})$ is pointwise continuous (this is precisely equivalent to the condition (4.1)).

It is now easy to see that the center of our category $\widehat{\mathfrak{g}}_\kappa$-mod is equal to the center of the algebra $\widetilde{U}_\kappa(\widehat{\mathfrak{g}})$, which we will denote by $Z_\kappa(\widehat{\mathfrak{g}})$. The argument is similar to the one we used above: though $\widetilde{U}_\kappa(\widehat{\mathfrak{g}})$ itself is not an object of $\widehat{\mathfrak{g}}_\kappa$-mod, we have a collection of objects $\widetilde{U}_\kappa(\widehat{\mathfrak{g}})/I_N$. Using this collection, we obtain an isomorphism between the center of category $\widehat{\mathfrak{g}}_\kappa$-mod and the inverse limit of the algebras $Z(\mathrm{End}_{\widehat{\mathfrak{g}}_\kappa} \widetilde{U}_\kappa(\widehat{\mathfrak{g}})/I_N)$, which, by definition, coincides with $Z_\kappa(\widehat{\mathfrak{g}})$.

Now we can formulate our first question:

$$\text{describe the center } Z_\kappa(\widehat{\mathfrak{g}}) \text{ for all levels } \kappa.$$

In order to answer this question we need to introduce the concept of G-opers.

4.2 Opers

Let G be a simple algebraic group of adjoint type, B its Borel subgroup and $N = [B, B]$ its unipotent radical, with the corresponding Lie algebras $\mathfrak{n} \subset \mathfrak{b} \subset \mathfrak{g}$.

Thus, $\mathfrak{g}$ is a simple Lie algebra, and as such it has the Cartan decomposition

$$\mathfrak{g} = \mathfrak{n}_- \oplus \mathfrak{h} \oplus \mathfrak{n}_+.$$

We will choose generators $e_1, \ldots, e_\ell$ (resp., $f_1, \ldots, f_\ell$) of $\mathfrak{n}_+$ (resp., $\mathfrak{n}_-$). We have $\mathfrak{n}_{\alpha_i} = \mathbb{C}e_i, \mathfrak{n}_{-\alpha_i} = \mathbb{C}f_i$. We take $\mathfrak{b} = \mathfrak{h} \oplus \mathfrak{n}_+$ as the Lie algebra of B. Then $\mathfrak{n}$ is the Lie algebra of N. In what follows we will use the notation $\mathfrak{n}$ for $\mathfrak{n}_+$.

Let $[\mathfrak{n}, \mathfrak{n}]^\perp \subset \mathfrak{g}$ be the orthogonal complement of $[\mathfrak{n}, \mathfrak{n}]$ with respect to a non-degenerate invariant bilinear form κ_0. We have

$$[\mathfrak{n}, \mathfrak{n}]^\perp / \mathfrak{b} \simeq \bigoplus_{i=1}^{\ell} \mathfrak{n}_{-\alpha_i}.$$

Clearly, the group B acts on $\mathfrak{n}^\perp / \mathfrak{b}$. Our first observation is that there is an open B-orbit $\mathbf{O} \subset \mathfrak{n}^\perp / \mathfrak{b} \subset \mathfrak{g}/\mathfrak{b}$, consisting of vectors whose projection on each subspace $\mathfrak{n}_{-\alpha_i}$ is non-zero. This orbit may also be described as the B-orbit of the sum of the projections of the generators $f_i, i = 1, \ldots, \ell$, of any possible subalgebra $\mathfrak{n}_-$, onto $\mathfrak{g}/\mathfrak{b}$. The action of B on $\mathbf{O}$ factors through an action of $H = B/N$. The latter is simply transitive and makes $\mathbf{O}$ into an H-torsor.

Let X be a smooth curve and x a point of X. As before, we denote by $\mathcal{O}_x$ the completed local ring and by $\mathcal{K}_x$ its field of fractions. The ring $\mathcal{O}_x$ is isomorphic, but not canonically, to $\mathbb{C}[[t]]$. Then $D_x = \operatorname{Spec} \mathcal{O}_x$ is the disc without a coordinate and $D_x^\times = \operatorname{Spec} \mathcal{K}_x$ is the corresponding punctured disc.

Suppose now that we are given a principal G-bundle $\mathcal{F}$ on a smooth curve X, or D_x, or $D_x^\times$, together with a connection ∇ (automatically flat) and a reduction $\mathcal{F}_B$ to the Borel subgroup B of G. Then we define the relative position of ∇ and $\mathcal{F}_B$ (i.e., the failure of ∇ to preserve $\mathcal{F}_B$) as follows. Locally, choose any flat connection ∇' on $\mathcal{F}$ preserving $\mathcal{F}_B$, and take the difference $\nabla - \nabla'$, which is a section of $\mathfrak{g}_{\mathcal{F}_B} \otimes \Omega_X$. We project it onto $(\mathfrak{g}/\mathfrak{b})_{\mathcal{F}_B} \otimes \Omega_X$. It is clear that the resulting local section of $(\mathfrak{g}/\mathfrak{b})_{\mathcal{F}_B} \otimes \Omega_X$ are independent of the choice ∇'. These sections patch together to define a global $(\mathfrak{g}/\mathfrak{b})_{\mathcal{F}_B}$-valued one-form on X, denoted by $\nabla/\mathcal{F}_B$.

Let X be a smooth curve, or D_x, or $D_x^\times$. Suppose we are given a principal G-bundle $\mathcal{F}$ on X, a connection ∇ on $\mathcal{F}$ and a B-reduction $\mathcal{F}_B$. We will say that $\mathcal{F}_B$ is **transversal** to ∇ if the one-form $\nabla/\mathcal{F}_B$ takes values in $\mathbf{O}_{\mathcal{F}_B} \subset (\mathfrak{g}/\mathfrak{b})_{\mathcal{F}_B}$. Note that $\mathbf{O}$ is $\mathbb{C}^\times$-invariant, so that $\mathbf{O} \otimes \Omega_X$ is a well-defined subset of $(\mathfrak{g}/\mathfrak{b})_{\mathcal{F}_B} \otimes \Omega_X$.

Now, a G-**oper** on X is by definition a triple $(\mathcal{F}, \nabla, \mathcal{F}_B)$, where $\mathcal{F}$ is a principal G-bundle $\mathcal{F}$ on X, ∇ is a connection on $\mathcal{F}$ and $\mathcal{F}_B$ is a B-reduction of $\mathcal{F}$, such that $\mathcal{F}_B$ is transversal to ∇.

This definition is due to A. Beilinson and V. Drinfeld [BD1] (in the case when X is the punctured disc opers were introduced earlier by V. Drinfeld and V. Sokolov in [DS]).

Equivalently, the transversality condition may be reformulated as saying that if we choose a local trivialization of $\mathcal{F}_B$ and a local coordinate t then the connection will be of the form

$$(4.2) \qquad \nabla = \partial_t + \sum_{i=1}^{\ell} \psi_i(t) f_i + \mathbf{v}(t),$$

where each $\psi_i(t)$ is a nowhere vanishing function, and $\mathbf{v}(t)$ is a $\mathfrak{b}$-valued function.

If we change the trivialization of $\mathcal{F}_B$, then this operator will get transformed by the corresponding B-valued gauge transformation. This observation allows us to describe opers on the disc $D_x = \operatorname{Spec} \mathcal{O}_x$ and the punctured disc $D_x^\times = \operatorname{Spec} \mathcal{K}_x$ in a more explicit way. The same reasoning will work on any sufficiently small analytic subset U of any curve, equipped with a local coordinate t, or on a Zariski open subset equipped with an étale coordinate. For the sake of definiteness, we will consider now the case of the base $D_x^\times$.

Let us choose a coordinate t on D_x, i.e., an isomorphism $\mathcal{O}_x \simeq \mathbb{C}[[t]]$. Then we identify D_x with $D = \operatorname{Spec} \mathbb{C}[[t]]$ and $D_x^\times$ with $D^\times = \operatorname{Spec} \mathbb{C}((t))$.

The space $\mathrm{Op}_G(D^\times)$ of G-opers on $D^\times$ is then the quotient of the space of all operators of the form (4.2), where $\psi_i(t) \in \mathbb{C}((t)), \psi_i(0) \neq 0, i = 1, \ldots, \ell$, and $\mathbf{v}(t) \in \mathfrak{b}((t))$, by the action of the group $B((t))$ of gauge transformations:

$$g \cdot (\partial_t + A(t)) = \partial_t + gA(t)g^{-1} - g^{-1}\partial_t g.$$

Let us choose a splitting $\imath : H \to B$ of the homomorphism $B \to H$. Then B becomes the semi-direct product $B = H \ltimes N$. The B-orbit $\mathbf{O}$ is an H-torsor, and so we can use H-valued gauge transformations to make all functions $\psi_i(t)$ equal to 1. In other words, there is a unique element of $H((t))$, namely, the element $\prod_{i=1}^\ell \check{\omega}_i(\psi_i(t))$, where $\check{\omega}_i : \mathbb{C}^\times \to H$ is the ith fundamental coweight of G, such that the corresponding gauge transformation brings our connection operator to the form

$$(4.3) \qquad \nabla = \partial_t + \sum_{i=1}^\ell f_i + \mathbf{v}(t), \qquad \mathbf{v}(t) \in \mathfrak{b}((t)).$$

What remains is the group of N-valued gauge transformations. Thus, we obtain that $\mathrm{Op}_G(D^\times)$ is equal to the quotient of the space $\widetilde{\mathrm{Op}}_G(D^\times)$ of operators of the form (4.3) by the action of the group $N((t))$ by gauge transformations:

$$\mathrm{Op}_G(D^\times) = \widetilde{\mathrm{Op}}_G(D^\times)/N((t)).$$

Lemma 1 ([DS]). *The action of $N((t))$ on $\widetilde{\mathrm{Op}}_G(D^\times)$ is free.*

4.3 Canonical Representatives

Now we construct canonical representatives in the $N((t))$-gauge classes of connections of the form (4.3), following [BD1]. Observe that the operator $\mathrm{ad}\,\check{\rho}$ defines a gradation on $\mathfrak{g}$, called the **principal gradation**, with respect to which we have a direct sum decomposition $\mathfrak{g} = \bigoplus_i \mathfrak{g}_i$. In particular, we have $\mathfrak{b} = \bigoplus_{i \geq 0} \mathfrak{b}_i$, where $\mathfrak{b}_0 = \mathfrak{h}$.

Let now

$$p_{-1} = \sum_{i=1}^\ell f_i.$$

The operator $\mathrm{ad}\,p_{-1}$ acts from $\mathfrak{b}_{i+1}$ to $\mathfrak{b}_i$ injectively for all $i \geq 0$. Hence we can find for each $i \geq 0$ a subspace $V_i \subset \mathfrak{b}_i$, such that $\mathfrak{b}_i = [p_{-1}, \mathfrak{b}_{i+1}] \oplus V_i$. It is well-known that $V_i \neq 0$ if and only if i is an **exponent** of $\mathfrak{g}$, and in that case $\dim V_i$ is equal to the multiplicity of the exponent i. In particular, $V_0 = 0$.

Let $V = \bigoplus_{i \in E} V_i \subset \mathfrak{n}$, where $E = \{d_1, \ldots, d_\ell\}$ is the set of exponents of $\mathfrak{g}$ counted with multiplicity. They are equal to the orders of the generators of the center of $U(\mathfrak{g})$ minus one. We note that the multiplicity of each exponent is equal to one in all cases except the case $\mathfrak{g} = D_{2n}, d_n = 2n$, when it is equal to two.

There is a special choice of the transversal subspace $V = \bigoplus_{i \in E} V_i$. Namely, there exists a unique element p_1 in $\mathfrak{n}$, such that $\{p_{-1}, 2\check{\rho}, p_1\}$ is an $\mathfrak{sl}_2$-triple. This means that they have the same relations as the generators $\{e, h, f\}$ of $\mathfrak{sl}_2$. We have $p_1 = \sum_{i=1}^{\ell} m_i e_i$, where e_i's are generators of $\mathfrak{n}_+$ and m_i are certain coefficients uniquely determined by the condition that $\{p_{-1}, 2\check{\rho}, p_1\}$ is an $\mathfrak{sl}_2$-triple.

Let $V^{\mathrm{can}} = \bigoplus_{i \in E} V_i^{\mathrm{can}}$ be the space of $\operatorname{ad} p_1$-invariants in $\mathfrak{n}$. Then p_1 spans V_1^{can}. Let p_j be a linear generator of $V_{d_j}^{\mathrm{can}}$. If the multiplicity of d_j is greater than one, then we choose linearly independent vectors in $V_{d_j}^{\mathrm{can}}$.

Each $N((t))$-equivalence class contains a unique operator of the form $\nabla = \partial_t + p_{-1} + \mathbf{v}(t)$, where $\mathbf{v}(t) \in V^{\mathrm{can}}[[t]]$, so that we can write

$$\mathbf{v}(t) = \sum_{j=1}^{\ell} v_j(t) \cdot p_j, \qquad v_j(t) \in \mathbb{C}[[t]].$$

It is easy to find (see, e.g., [F6], Sect. 8.3) that under changes of coordinate t, v_1 transforms as a projective connection, and $v_j, j > 1$, transforms as a $(d_j + 1)$-differential on D_x. Thus, we obtain an isomorphism

$$(4.4) \qquad \operatorname{Op}_G(D^\times) \simeq \mathcal{P}roj(D^\times) \times \bigoplus_{j=2}^{\ell} \Omega_{\mathcal{K}}^{\otimes(d_j+1)},$$

where $\Omega_{\mathcal{K}}^{\otimes n}$ is the space of n-differentials on $D^\times$ and $\mathcal{P}roj(D^\times)$ is the $\Omega_{\mathcal{K}}^{\otimes 2}$-torsor of projective connections on $D^\times$.

We have an analogous isomorphism with $D^\times$ replaced by formal disc D or any smooth algebraic curve X.

4.4 Description of the Center

Now we are ready to describe the center of the completed universal enveloping algebra $\widetilde{U}_{\kappa_c}(\widehat{\mathfrak{g}})$. The following assertion is proved in [F7], using results of [K1]:

Proposition 1. *The center of $\widetilde{U}_\kappa(\widehat{\mathfrak{g}})$ consists of the scalars for $\kappa \neq \kappa_c$.*

Let us denote the center of $\widetilde{U}_{\kappa_c}(\widehat{\mathfrak{g}})$ by $Z(\widehat{\mathfrak{g}})$. The following theorem was proved in [FF3, F3] (it was conjectured by V. Drinfeld).

Theorem 2. *The center $Z(\widehat{\mathfrak{g}})$ is isomorphic to the algebra $\operatorname{Fun} \operatorname{Op}_{{}^L G}(D^\times)$ in a way compatible with the action of the group of coordinate changes.*

This implies the following result. Let x be a point of a smooth curve X. Then we have the affine algebra $\widehat{\mathfrak{g}}_{\kappa_c,x}$ as defined in Section 1 and the corresponding completed universal enveloping algebra of critical level. We denote its center by $Z(\widehat{\mathfrak{g}}_x)$.

Corollary 1. *The center $Z(\widehat{\mathfrak{g}}_x)$ is isomorphic to the algebra $\operatorname{Fun} \operatorname{Op}_{{}^L G}(D_x^\times)$ of functions on the space of ${}^L G$-opers on $D_x^\times$.*

5 Opers vs. Local Systems

We now go back to the question posed at the end of Section 3: let

$$(5.1) \qquad \mathrm{Loc}_{L_G}(D^\times) = \{\partial_t + A(t), \ A(t) \in {}^L\mathfrak{g}((t))\} \, / \, {}^L G((t))$$

be the set of gauge equivalence classes of ${}^L G$-connections on the punctured disc $D^\times = \mathrm{Spec}\,\mathbb{C}((t))$. We had argued in Section 3 that $\mathrm{Loc}_{L_G}(D^\times)$ should be taken as the space of Langlands parameters for the loop group $G((t))$. Recall that the loop group $G((t))$ acts on the category $\widehat{\mathfrak{g}}_\kappa$-mod of (smooth) $\widehat{\mathfrak{g}}$-modules of level κ (see Section 1 for the definition of this category). We asked the following question:

Associate to each local Langlands parameter $\sigma \in \mathrm{Loc}_{L_G}(D^\times)$ a subcategory $\widehat{\mathfrak{g}}_\kappa$-mod$_\sigma$ of $\widehat{\mathfrak{g}}_\kappa$-mod which is stable under the action of the loop group $G((t))$.

Even more ambitiously, we wish to represent the category $\widehat{\mathfrak{g}}_\kappa$-mod as "fibering" over the space of local Langlands parameters $\mathrm{Loc}_{L_G}(D^\times)$, with the categories $\widehat{\mathfrak{g}}_\kappa$-mod$_\sigma$ being the "fibers" and the group $G((t))$ acting along these fibers. If we could do that, then we would think of this fibration as a "spectral decomposition" of the category $\widehat{\mathfrak{g}}_\kappa$-mod over $\mathrm{Loc}_{L_G}(D^\times)$.

At the beginning of Section 4 we proposed a possible scenario for solving this problem. Namely, we observed that any abelian category may be thought of as "fibering" over the spectrum of its center. Hence our idea was to describe the center of the category $\widehat{\mathfrak{g}}_\kappa$-mod (for each value of κ) and see if its spectrum is related to the space $\mathrm{Loc}_{L_G}(D^\times)$ of Langlands parameters.

We have identified the center of the category $\widehat{\mathfrak{g}}_\kappa$-mod with the center $Z_\kappa(\widehat{\mathfrak{g}})$ of the associative algebra $\widetilde{U}_\kappa(\widehat{\mathfrak{g}})$, the completed enveloping algebra of $\widehat{\mathfrak{g}}$ of level κ, defined in Section 4. Next, we described the algebra $Z_\kappa(\widehat{\mathfrak{g}})$. According to Proposition 1, if $\kappa \neq \kappa_c$, the critical level, then $Z_\kappa(\widehat{\mathfrak{g}}) = \mathbb{C}$. Therefore our approach cannot work for $\kappa \neq \kappa_c$. However, we found that the center $Z_{\kappa_c}(\widehat{\mathfrak{g}})$ at the critical level is highly non-trivial and indeed related to ${}^L G$-connections on the punctured disc.

Now, following the works [FG1]–[FG6] of D. Gaitsgory and myself, I will use these results to formulate more precise conjectures on the local Langlands correspondence for loop groups and to provide some evidence for these conjectures. I will then discuss the implications of these conjectures for the global geometric Langlands correspondence.[8]

According to Theorem 2, $Z_{\kappa_c}(\widehat{\mathfrak{g}})$ is isomorphic to $\mathrm{Fun}\,\mathrm{Op}_{L_G}(D^\times)$, the algebra of functions on the space of ${}^L G$-opers on the punctured disc $D^\times$. This isomorphism is compatible with various symmetries and structures on both

---

[8] Note that A. Beilinson has another proposal [Bei] for local geometric Langlands correspondence, using representations of affine Kac-Moody algebras of levels *less than critical*. It would be interesting to understand the connection between his proposal and ours.

algebras, such as the action of the group of coordinate changes. There is a one-to-one correspondence between points $\chi \in \mathrm{Op}_{L_G}(D^\times)$ and homomorphisms (equivalently, characters)

$$\mathrm{Fun}\,\mathrm{Op}_{L_G}(D^\times) \to \mathbb{C},$$

corresponding to evaluating a function at χ. Hence points of $\mathrm{Op}_{L_G}(D^\times)$ parametrize **central characters** $Z_{\kappa_c}(\widehat{\mathfrak{g}}) \to \mathbb{C}$.

Given a $^L G$-oper $\chi \in \mathrm{Op}_{L_G}(D^\times)$, define the category

$$\widehat{\mathfrak{g}}_{\kappa_c}\text{-mod}_\chi$$

as a full subcategory of $\widehat{\mathfrak{g}}_{\kappa_c}$-mod whose objects are $\widehat{\mathfrak{g}}$-modules of critical level (hence $\widetilde{U}_{\kappa_c}(\widehat{\mathfrak{g}})$-modules) on which the center $Z_{\kappa_c}(\widehat{\mathfrak{g}}) \subset \widetilde{U}_{\kappa_c}(\widehat{\mathfrak{g}})$ acts according to the central character corresponding to χ. More generally, for any closed algebraic subvariety $Y \subset \mathrm{Op}_{L_G}(D^\times)$ (not necessarily a point), we have an ideal

$$I_Y \subset \mathrm{Fun}\,\mathrm{Op}_{L_G}(D^\times) \simeq Z_{\kappa_c}(\widehat{\mathfrak{g}})$$

of those functions that vanish on Y. We then have a full subcategory $\widehat{\mathfrak{g}}_{\kappa_c}$-mod$_Y$ of $\widehat{\mathfrak{g}}_{\kappa_c}$-mod whose objects are $\widehat{\mathfrak{g}}$-modules of critical level on which I_Y acts by 0. This category is an example of a "base change" of the category $\widehat{\mathfrak{g}}_{\kappa_c}$-mod with respect to the morphism $Y \to \mathrm{Op}_{L_G}(D^\times)$. It is easy to generalize this definition to an arbitrary affine scheme Y equipped with a morphism $Y \to \mathrm{Op}_{L_G}(D^\times)$.[9]

Since the algebra $\mathrm{Op}_{L_G}(D^\times)$ acts on the category $\widehat{\mathfrak{g}}_{\kappa_c}$-mod, one can say that the category $\widehat{\mathfrak{g}}_\kappa$-mod "fibers" over the space $\mathrm{Op}_{L_G}(D^\times)$, in such a way that the fiber-category corresponding to $\chi \in \mathrm{Op}_{L_G}(D^\times)$ is the category $\widehat{\mathfrak{g}}_{\kappa_c}$-mod$_\chi$.[10]

Recall that the group $G(\!(t)\!)$ acts on $\widetilde{U}_{\kappa_c}(\widehat{\mathfrak{g}})$ and on the category $\widehat{\mathfrak{g}}_{\kappa_c}$-mod. One can show (see [BD1], Remark 3.7.11(iii)) that the action of $G(\!(t)\!)$ on $Z_{\kappa_c}(\widehat{\mathfrak{g}}) \subset \widetilde{U}_{\kappa_c}(\widehat{\mathfrak{g}})$ is trivial. Therefore the subcategories $\widehat{\mathfrak{g}}_{\kappa_c}$-mod$_\chi$ (and, more generally, $\widehat{\mathfrak{g}}_{\kappa_c}$-mod$_Y$) are stable under the action of $G(\!(t)\!)$. Thus, the group $G(\!(t)\!)$ acts "along the fibers" of the "fibration" $\widehat{\mathfrak{g}}_{\kappa_c}$-mod $\to \mathrm{Op}_{L_G}(D^\times)$ (see [FG2], Sect. 20, for more details).

The fibration $\widehat{\mathfrak{g}}_{\kappa_c}$-mod $\to \mathrm{Op}_{L_G}(D^\times)$ almost gives us the desired local Langlands correspondence for loop groups. But there is one important difference: we asked that the category $\widehat{\mathfrak{g}}_{\kappa_c}$-mod fiber over the space $\mathrm{Loc}_{L_G}(D^\times)$ of local systems on $D^\times$. We have shown, however, that $\widehat{\mathfrak{g}}_{\kappa_c}$-mod fibers over the space $\mathrm{Op}_{L_G}(D^\times)$ of $^L G$-opers.

---

[9] The corresponding base changed categories $\widehat{\mathfrak{g}}_{\kappa_c}$-mod$_Y$ may then be "glued" together, which allows us to define the base changed category $\widehat{\mathfrak{g}}_{\kappa_c}$-mod$_Y$ for any scheme Y mapping to $\mathrm{Op}_{L_G}(D^\times)$.

[10] The precise notion of an abelian category fibering over a scheme is spelled out in [Ga3].

What is the difference between the two spaces? While a ^LG-local system is a pair $(\mathcal{F}, \nabla)$, where $\mathcal{F}$ is an ^LG-bundle and ∇ is a connection on $\mathcal{F}$, an ^LG-oper is a triple $(\mathcal{F}, \nabla, \mathcal{F}_{^LB})$, where $\mathcal{F}$ and ∇ are as before, and $\mathcal{F}_{^LB}$ is an additional piece of structure, namely, a reduction of $\mathcal{F}$ to a (fixed) Borel subgroup $^LB \subset {}^LG$ satisfying the transversality condition explained in Section 4.2. Thus, for any curve X we clearly have a forgetful map

$$\mathrm{Op}_{^LG}(X) \to \mathrm{Loc}_{^LG}(X).$$

The fiber of this map over $(\mathcal{F}, \nabla) \in \mathrm{Loc}_{^LG}(X)$ consists of all ^LB-reductions of $\mathcal{F}$ satisfying the transversality condition with respect to ∇.

For a general X it may well be that this map is not surjective, i.e., that the fiber of this map over a particular local system $(\mathcal{F}, \nabla)$ is empty. For example, if X is a projective curve and ^LG is a group of adjoint type, then there is a unique ^LG-bundle $\mathcal{F}_{^LG}$ such that the fiber over $(\mathcal{F}_{^LG}, \nabla)$ is non-empty, as we saw in Section 1.

The situation is quite different when $X = D^\times$. In this case any ^LG-bundle $\mathcal{F}$ may be trivialized. A connection ∇ therefore may be represented as a first order operator $\partial_t + A(t)$, $A(t) \in {}^L\mathfrak{g}((t))$. However, the trivialization of $\mathcal{F}$ is not unique; two trivializations differ by an element of $^LG((t))$. Therefore the set of equivalence classes of pairs $(\mathcal{F}, \nabla)$ is identified with the quotient (5.1).

Suppose now that $(\mathcal{F}, \nabla)$ carries an oper reduction $\mathcal{F}_{^LB}$. Then we consider only those trivializations of $\mathcal{F}$ which come from trivializations of $\mathcal{F}_{^LB}$. There are fewer of those, since two trivializations now differ by an element of $^LB((t))$ rather than $^LG((t))$. Due to the oper transversality condition, the connection ∇ must have a special form with respect to any of those trivializations, namely,

$$\nabla = \partial_t + \sum_{i=1}^{\ell} \psi_i(t) f_i + \mathbf{v}(t),$$

where each $\psi_i(t) \neq 0$ and $\mathbf{v}(t) \in {}^L\mathfrak{b}((t))$ (see Section 4.2). Thus, we obtain a concrete realization of the space of opers as a space of gauge equivalence classes

(5.2)

$$\mathrm{Op}_{^LG}(D^\times) = \left\{ \partial_t + \sum_{i=1}^{\ell} \psi_i(t) f_i + \mathbf{v}(t),\ \psi_i \neq 0, \mathbf{v}(t) \in {}^L\mathfrak{b}((t)) \right\} \Big/ {}^LB((t)).$$

Now the map

$$\alpha : \mathrm{Op}_{^LG}(D^\times) \to \mathrm{Loc}_{^LG}(D^\times)$$

simply takes a $^LB((t))$-equivalence class of operators of the form (5.2) to its $^LG((t))$-equivalence class.

Unlike the case of projective curves X discussed above, we expect that the map α is **surjective** for any simple Lie group ^LG. In the case of $G = SL_n$ this follows from the results of P. Deligne [Del], and we conjecture it to be true in general.

Conjecture 1. *The map α is surjective for any simple Lie group $^L G$.*

Now we find ourselves in the following situation: we *expect* that there exists a category $\mathcal{C}$ fibering over the space $\mathrm{Loc}_{^L G}(D^\times)$ of "true" local Langlands parameters, equipped with a fiberwise action of the loop group $G((t))$. The fiber categories $\mathcal{C}_\sigma$ corresponding to various $\sigma \in \mathrm{Loc}_{^L G}(D^\times)$ should satisfy various, not yet specified, properties. This should be the ultimate form of the local Langlands correspondence. On the other hand, we have *constructed* a category $\widehat{\mathfrak{g}}_{\kappa_c}$-mod which fibers over a close cousin of the space $\mathrm{Loc}_{^L G}(D^\times)$, namely, the space $\mathrm{Op}_{^L G}(D^\times)$ of $^L G$-opers, and is equipped with a fiberwise action of the loop group $G((t))$.

What should be the relationship between the two?

The idea of [FG2] is that the second fibration is a "base change" of the first one, that is we have a Cartesian diagram

(5.3)

$$
\begin{array}{ccc}
\widehat{\mathfrak{g}}_{\kappa_c}\text{-mod} & \longrightarrow & \mathcal{C} \\
\downarrow & & \downarrow \\
\mathrm{Op}_{^L G}(D^\times) & \xrightarrow{\ \alpha\ } & \mathrm{Loc}_{^L G}(D^\times)
\end{array}
$$

that commutes with the action of $G((t))$ along the fibers of the two vertical maps. In other words,

$$
\widehat{\mathfrak{g}}_{\kappa_c}\text{-mod} \simeq \mathcal{C} \underset{\mathrm{Loc}_{^L G}(D^\times)}{\times} \mathrm{Op}_{^L G}(D^\times).
$$

At present, we do not have a definition of $\mathcal{C}$, and therefore we cannot make this isomorphism precise. But we will use it as our guiding principle. We will now discuss various corollaries of this conjecture and various pieces of evidence that make us believe that it is true.

In particular, let us fix a Langlands parameter $\sigma \in \mathrm{Loc}_{^L G}(D^\times)$ that is in the image of the map α (according to Conjecture 1, all Langlands parameters are). Let χ be a $^L G$-oper in the preimage of σ, $\alpha^{-1}(\sigma)$. Then, according to the above conjecture, the category $\widehat{\mathfrak{g}}_{\kappa_c}$-mod$_\chi$ is equivalent to the "would be" Langlands category $\mathcal{C}_\sigma$ attached to σ. Hence we may take $\widehat{\mathfrak{g}}_{\kappa_c}$-mod$_\chi$ as the **definition** of $\mathcal{C}_\sigma$.

The caveat is, of course, that we need to ensure that this definition is independent of the choice of χ in $\alpha^{-1}(\sigma)$. This means that for any two $^L G$-opers, χ and χ', in the preimage of σ, the corresponding categories, $\widehat{\mathfrak{g}}_{\kappa_c}$-mod$_\chi$ and $\widehat{\mathfrak{g}}_{\kappa_c}$-mod$_{\chi'}$, should be equivalent to each other, and this equivalence should commute with the action of the loop group $G((t))$. Moreover, we should expect that these equivalences are compatible with each other as we move along the fiber $\alpha^{-1}(\sigma)$. We will not try to make this condition more precise here (however, we will explain below in Conjecture 4 what this means for regular opers).

Even putting the questions of compatibility aside, we arrive at the following rather non-trivial conjecture (see [FG2]).

Conjecture 2. *Suppose that* $\chi, \chi' \in \text{Op}_{L_G}(D^\times)$ *are such that* $\alpha(\chi) = \alpha(\chi')$, *i.e., that the flat* $^L G$*-bundles on* $D^\times$ *underlying the* $^L G$*-opers* χ *and* χ' *are isomorphic to each other. Then there is an equivalence between the categories* $\widehat{\mathfrak{g}}_{\kappa_c}$*-mod*$_\chi$ *and* $\widehat{\mathfrak{g}}_{\kappa_c}$*-mod*$_{\chi'}$ *which commutes with the actions of the group* $G(\!(t)\!)$ *on the two categories.*

Thus, motivated by our quest for the local Langlands correspondence, we have found an unexpected symmetry in the structure of the category $\widehat{\mathfrak{g}}_{\kappa_c}$-mod of $\widehat{\mathfrak{g}}$-modules of critical level.

6 Harish–Chandra Categories

As explained in Section 3, the local Langlands correspondence for the loop group $G(\!(t)\!)$ should be viewed as a categorification of the local Langlands correspondence for the group $G(F)$, where F is a local non-archimedean field. This means that the categories $\mathcal{C}_\sigma$, equipped with an action of $G(\!(t)\!)$, that we are trying to attach to the Langlands parameters $\sigma \in \text{Loc}_{L_G}(D^\times)$ should be viewed as categorifications of the smooth representations of $G(F)$ on complex vector spaces attached to the corresponding local Langlands parameters discussed in Section 2.3. Here we use the term "categorification" to indicate that we expect the Grothendieck groups of the categories $\mathcal{C}_\sigma$ to "look like" irreducible smooth representations of $G(F)$. We begin by taking a closer look at the structure of these representations.

6.1 Spaces of K-Invariant Vectors

It is known that an irreducible smooth representation (R, π) of $G(F)$ is automatically **admissible**, in the sense that for any open compact subgroup K, such as the Nth congruence subgroup K_N defined in Section 2.1, the space $R^{\pi(K)}$ of K-invariant vectors in R is finite-dimensional. Thus, while most of the irreducible smooth representations (R, π) of $G(F)$ are infinite-dimensional, they are filtered by the finite-dimensional subspaces $R^{\pi(K)}$ of K-invariant vectors, where K are smaller and smaller open compact subgroups. The space $R^{\pi(K)}$ does not carry an action of $G(F)$, but it carries an action of the **Hecke algebra** $H(G(F), K)$.

By definition, $H(G(F), K)$ is the space of compactly supported K bi-invariant functions on $G(F)$. It is given an algebra structure with respect to the **convolution product**

$$(6.1) \qquad (f_1 \star f_2)(g) = \int_{G(F)} f_1(gh^{-1}) f_2(h)\, dh,$$

where dh is the Haar measure on $G(F)$ normalized in such a way that the volume of the subgroup $K_0 = G(\mathcal{O})$ is equal to 1 (here $\mathcal{O}$ is the ring of integers of F; e.g., for $F = \mathbb{F}_q((t))$ we have $\mathcal{O} = \mathbb{F}_q[[t]]$). The algebra $H(G(F), K)$ acts on the space $R^{\pi(K)}$ by the formula

$$(6.2) \qquad f \star v = \int_{G(F)} f_1(gh^{-1})(\pi(h) \cdot v) \, dh, \qquad v \in R^{\pi(K)}.$$

Studying the spaces of K-invariant vectors and their $\mathcal{H}(G(F), K)$-module structure gives us an effective tool for analyzing representations of the group $G(F)$, where $F = \mathbb{F}_q((t))$.

Can we find a similar structure in the categorical local Langlands correspondence for loop groups?

6.2 Equivariant Modules

In the categorical setting a representation (R, π) of the group $G(F)$ is replaced by a category equipped with an action of $G((t))$, such as $\widehat{\mathfrak{g}}_{\kappa_c}$-mod$_\chi$. The open compact subgroups of $G(F)$ have obvious analogues for the loop group $G((t))$ (although they are, of course, not compact with respect to the usual topology on $G((t))$). For instance, we have the "maximal compact subgroup" $K_0 = G[[t]]$, or, more generally, the Nth congruence subgroup K_N, whose elements are congruent to 1 modulo $t^N \mathbb{C}[[t]]$. Another important example is the analogue of the **Iwahori subgroup**. This is the subgroup of $G[[t]]$, which we denote by I, whose elements $g(t)$ have the property that their value at 0, that is $g(0)$, belong to a fixed Borel subgroup $B \subset G$.

Now, for a subgroup $K \subset G((t))$ of this type, an analogue of a K-invariant vector in the categorical setting is an object of our category, i.e., a smooth $\widehat{\mathfrak{g}}_{\kappa_c}$-module (M, ρ), where $\rho : \widehat{\mathfrak{g}}_{\kappa_c} \to \operatorname{End} M$, which is stable under the action of K. Recall from Section 3.5 that for any $g \in G((t))$ we have a new $\widehat{\mathfrak{g}}_{\kappa_c}$-module (M, ρ_g), where $\rho_g(x) = \rho(\operatorname{Ad}_g(x))$. We say that (M, ρ) is stable under K, or that (M, ρ) is **weakly K-equivariant**, if there is a compatible system of isomorphisms between (M, ρ) and (M, ρ_k) for all $k \in K$. More precisely, this means that for each $k \in K$ there exists a linear map $T_k^M : M \to M$ such that

$$T_k^M \rho(x)(T_k^M)^{-1} = \rho(\operatorname{Ad}_k(x))$$

for all $x \in \widehat{\mathfrak{g}}_{\kappa_c}$, and we have

$$T_1^M = \operatorname{Id}_M, \qquad T_{k_1}^M T_{k_2}^M = T_{k_1 k_2}^M.$$

Thus, M becomes a representation of the group K.[11] Consider the corresponding representation of the Lie algebra $\mathfrak{k} = \operatorname{Lie} K$ on M. Let us assume that the embedding $\mathfrak{k} \hookrightarrow \mathfrak{g}((t))$ lifts to $\mathfrak{k} \hookrightarrow \widehat{\mathfrak{g}}_{\kappa_c}$ (i.e., that the central extension cocycle is trivial on $\mathfrak{k}$). This is true, for instance, for any subgroup contained

---

[11] In general, it is reasonable to modify the last condition to allow for a non-trivial two-cocycle and hence a non-trivial central extension of K; however, in the case of interest K does not have any non-trivial central extensions.

in $K_0 = G[[t]]$, or its conjugate. Then we also have a representation of $\mathfrak{k}$ on M obtained by restriction of ρ. In general, the two representations do not have to coincide. If they do coincide, then the module M is called **strongly K-equivariant**, or simply **K-equivariant**.

The pair $(\widehat{\mathfrak{g}}_{\kappa_c}, K)$ is an example of **Harish-Chandra pair**, that is a pair $(\mathfrak{g}, H)$ consisting of a Lie algebra $\mathfrak{g}$ and a Lie group H whose Lie algebra is contained in $\mathfrak{g}$. The K-equivariant $\widehat{\mathfrak{g}}_{\kappa_c}$-modules are therefore called $(\widehat{\mathfrak{g}}_{\kappa_c}, K)$ **Harish-Chandra modules**. These are (smooth) $\widehat{\mathfrak{g}}_{\kappa_c}$-modules on which the action of the Lie algebra $\mathrm{Lie}\, K \subset \widehat{\mathfrak{g}}_{\kappa_c}$ may be exponentiated to an action of K (we will assume that K is connected). We denote by $\widehat{\mathfrak{g}}_{\kappa_c}\text{-mod}^K$ and $\widehat{\mathfrak{g}}_{\kappa_c}\text{-mod}^K_\chi$ the full subcategories of $\widehat{\mathfrak{g}}_{\kappa_c}$-mod and $\widehat{\mathfrak{g}}_{\kappa_c}$-mod$_\chi$, respectively, whose objects are $(\widehat{\mathfrak{g}}_{\kappa_c}, K)$ Harish-Chandra modules.

We will stipulate that the analogues of K-invariant vectors in the category $\widehat{\mathfrak{g}}_{\kappa_c}$-mod$_\chi$ are $(\widehat{\mathfrak{g}}_{\kappa_c}, K)$ Harish-Chandra modules. Thus, while the categories $\widehat{\mathfrak{g}}_{\kappa_c}$-mod$_\chi$ should be viewed as analogues of smooth irreducible representations (R, π) of the group $G(F)$, the categories $\widehat{\mathfrak{g}}_{\kappa_c}$-mod$^K_\chi$ are analogues of the spaces of K-invariant vectors $R^{\pi(K)}$.

Next, we discuss the categorical analogue of the Hecke algebra $H(G(F), K)$.

6.3 Categorical Hecke Algebras

We recall that $H(G(F), K)$ is the algebra of compactly supported K bi-invariant functions on $G(F)$. We realize it as the algebra of left K-invariant compactly supported functions on $G(F)/K$. In Section 3.4 we have already discussed the question of categorification of the algebra of functions on a homogeneous space like $G(F)/K$. Our conclusion was that the categorical analogue of this algebra, when $G(F)$ is replaced by the complex loop group $G((t))$, is the category of $\mathcal{D}$-modules on $G((t))/K$. More precisely, this quotient has the structure of an ind-scheme which is a direct limit of finite-dimensional algebraic varieties with respect to closed embeddings. The appropriate notion of (right) $\mathcal{D}$-modules on such ind-schemes is formulated in [BD1] (see also [FG1, FG2]). As the categorical analogue of the algebra of left K-invariant functions on $G(F)/K$, we take the category $\mathcal{H}(G((t)), K)$ of K-equivariant $\mathcal{D}$-modules on the ind-scheme $G((t))/K$ (with respect to the left action of K on $G((t))/K$). We call it the **categorical Hecke algebra** associated to K.

It is easy to define the convolution of two objects of $\mathcal{H}(G((t)), K)$ by imitating formula (6.1). Namely, we interpret this formula as a composition of the operations of pulling back and integrating functions. Then we apply the same operations to $\mathcal{D}$-modules, thinking of the integral as push-forward. However, here one encounters two problems. The first problem is that for a general group K the morphisms involved will not be proper, and so we have to choose between the $*$- and !-push-forward. This problem does not arise, however, if K is such that $I \subset K \subset G[[t]]$, which will be our main case of interest. The second, and more serious, issue is that in general the push-forward is not an exact functor, and so the convolution of two $\mathcal{D}$-modules will not be a $\mathcal{D}$-module, but a complex, more precisely, an object of the corresponding K-equivariant (bounded) derived category $D^b(G((t))/K)^K$ of $\mathcal{D}$-modules on $G((t))/K$. We will not spell out the exact definition of this category here, referring the interested reader to [BD1] and [FG2]. The exception is the case of the subgroup $K_0 = G[[t]]$, when the convolution functor is exact and so we may restrict ourselves to the abelian category of K_0-equivariant $\mathcal{D}$-modules on $G((t))/K_0$.

Now the category $D^b(G((t))/K)^K$ has a monoidal structure, and as such it acts on the derived category of $(\widehat{\mathfrak{g}}_{\kappa_c}, K)$ Harish-Chandra modules (again, we refer the reader to [BD1, FG2] for the precise definition). In the special case when $K = K_0$, we may restrict ourselves to the corresponding abelian categories. This action should be viewed as the categorical analogue of the action of $H(G(F), K)$ on the space $R^{\pi(K)}$ of K-invariant vectors discussed above.

Our ultimate goal is understanding the "local Langlands categories" $\mathcal{C}_\sigma$ associated to the "local Langlands parameters $\sigma \in \mathrm{Loc}_{{}^L G}(D^\times)$. We now have a candidate for the category $\mathcal{C}_\sigma$, namely, the category $\widehat{\mathfrak{g}}_{\kappa_c}$-$\mathrm{mod}_\chi$, where $\sigma = \alpha(\chi)$. Therefore $\widehat{\mathfrak{g}}_{\kappa_c}$-$\mathrm{mod}_\chi$ should be viewed as a categorification of a smooth representation (R, π) of $G(F)$. The corresponding category $\widehat{\mathfrak{g}}_{\kappa_c}$-$\mathrm{mod}_\chi^K$ of $(\widehat{\mathfrak{g}}_{\kappa_c}, K)$ Harish-Chandra modules should therefore be viewed as a categorification of $R^{\pi(K)}$. This category (or, more precisely, its derived category) is acted upon by the categorical Hecke algebra $\mathcal{H}(G((t)), K)$. We summarize this analogy in the following table.

Classical Theory	Geometric Theory
Representation of $G(F)$ on a vector space R	Representation of $G((t))$ on a category $\widehat{\mathfrak{g}}_{\kappa_c}$-$\mathrm{mod}_\chi$
A vector in R	An object of $\widehat{\mathfrak{g}}_{\kappa_c}$-$\mathrm{mod}_\chi$
The subspace $R^{\pi(K)}$ of K-invariant vectors of R	The subcategory $\widehat{\mathfrak{g}}_{\kappa_c}$-$\mathrm{mod}_\chi^K$ of $(\widehat{\mathfrak{g}}_{\kappa_c}, K)$ Harish-Chandra modules
Hecke algebra $H(G(F), K)$ acts on $R^{\pi(K)}$	Categorical Hecke algebra $\mathcal{H}(G((t)), K)$ acts on $\widehat{\mathfrak{g}}_{\kappa_c}$-$\mathrm{mod}_\chi^K$

Now we may test our proposal for the local Langlands correspondence by studying the categories $\widehat{\mathfrak{g}}_{\kappa_c}$-$\mathrm{mod}_\chi^K$ of Harish-Chandra modules and comparing their structure to the structure of the spaces $R^{\pi(K)}$ of K-invariant vectors of smooth representations of $G(F)$ in the known cases. Another possibility is to test Conjecture 2 when applied to the categories of Harish-Chandra modules.

In the next section we consider the case of the "maximal compact subgroup" $K_0 = G[[t]]$ and find perfect agreement with the classical results about unramified representations of $G(F)$. We then take up the more complicated case of the Iwahori subgroup I. There we also find the conjectures and results of [FG2] to be consistent with the known results about representations of $G(F)$ with Iwahori fixed vectors.

7 Local Langlands Correspondence: Unramified Case

We first take up the case of the "maximal compact subgroup" $K_0 = G[[t]]$ of $G((t))$ and consider the categories $\widehat{\mathfrak{g}}_{\kappa_c}$-mod$_\chi$ which contain non-trivial K_0-equivariant objects.

7.1 Unramified Representations of $G(F)$

These categories are analogues of smooth representations of the group $G(F)$, where F is a local non-archimedean field (such as $\mathbb{F}_q((t))$) that contain non-zero K_0-invariant vectors. Such representations are called **unramified**. The classification of the irreducible unramified representations of $G(F)$ is the simplest case of the local Langlands correspondence discussed in Sections 2.2 and 2.3. Namely, we have a bijection between the sets of equivalence classes of the following objects:

$$(7.1) \quad \boxed{\begin{array}{c} \text{unramified admissible} \\ \text{homomorphisms } W'_F \to {}^L G \end{array}} \Longleftrightarrow \boxed{\begin{array}{c} \text{irreducible unramified} \\ \text{representations of } G(F) \end{array}}$$

where W'_F is the Weil-Deligne group introduced in Section 2.1.

By definition, unramified homomorphisms $W'_F \longrightarrow {}^L G$ are those which factor through the quotient

$$W'_F \to W_F \to \mathbb{Z}$$

(see Section 2.1 for the definitions of these groups and homomorphisms). It is admissible if its image in ${}^L G$ consists of semi-simple elements. Therefore the set on the left hand side of (7.1) is just the set of conjugacy classes of semi-simple elements of ${}^L G$. Thus, the above bijection may be reinterpreted as follows:

$$(7.2) \quad \boxed{\begin{array}{c} \text{semi-simple conjugacy} \\ \text{classes in } {}^L G \end{array}} \Longleftrightarrow \boxed{\begin{array}{c} \text{irreducible unramified} \\ \text{representations of } G(F) \end{array}}$$

To construct this bijection, we look at the Hecke algebra $H(G(F), K_0)$. According to the Satake isomorphism [Sat], in the interpretation of Langlands [L], this algebra is commutative and isomorphic to the representation ring of the Langlands dual group $^L G$:

$$(7.3) \qquad\qquad H(G(F), K_0) \simeq \mathrm{Rep}\, {}^L G.$$

We recall that $\mathrm{Rep}\, {}^L G$ consists of finite linear combinations $\sum_i a_i[V_i]$, where the V_i are finite-dimensional representations of $^L G$ (without loss of generality we may assume that they are irreducible) and $a_i \in \mathbb{C}$, with respect to the multiplication

$$[V] \cdot [W] = [V \otimes W].$$

Since $\mathrm{Rep}\, {}^L G$ is commutative, its irreducible modules are all one-dimensional. They correspond to characters $\mathrm{Rep}\, {}^L G \to \mathbb{C}$. We have a bijection

$$(7.4) \qquad \boxed{\begin{array}{c} \text{semi-simple conjugacy} \\ \text{classes in } {}^L G \end{array}} \quad \Longleftrightarrow \quad \boxed{\begin{array}{c} \text{characters} \\ \text{of } \mathrm{Rep}\, {}^L G \end{array}}$$

where the character ϕ_γ corresponding to the conjugacy class γ is given by the formula[12]

$$\phi_\gamma : [V] \mapsto \mathrm{Tr}(\gamma, V).$$

Now, if (R, π) is a representation of $G(F)$, then the space $R^{\pi(K_0)}$ of K_0-invariant vectors in V is a module over $H(G(F), K_0)$. It is easy to show that this sets up a one-to-one correspondence between equivalence classes of irreducible unramified representations of $G(F)$ and irreducible $H(G(F), K_0)$-modules. Combining this with the bijection (7.4) and the isomorphism (7.3), we obtain the sought-after bijections (7.1) and (7.2).

In particular, we find that, because the Hecke algebra $H(G(F), K_0)$ is commutative, the space $R^{\pi(K_0)}$ of K_0-invariants of an irreducible representation, which is an irreducible $H(G(F), K_0)$-module, is either zero or one-dimensional. If it is one-dimensional, then $H(G(F), K_0)$ acts on it by the character ϕ_γ for some γ:

$$(7.5) \qquad H_V \star v = \mathrm{Tr}(\gamma, V) v, \qquad v \in R^{\pi(K_0)}, [V] \in \mathrm{Rep}\, {}^L G,$$

where H_V is the element of $H(G(F), K_0)$ corresponding to $[V]$ under the isomorphism (7.3) (see formula (6.2) for the definition of the convolution action). We now discuss the categorical analogues of these statements.

---

[12] It is customary to multiply the right hand side of this formula, for irreducible representation V, by a scalar depending on q and the highest weight of V, but this is not essential for our discussion.

7.2 Unramified Categories $\widehat{\mathfrak{g}}_{\kappa_c}$-Modules

In the categorical setting, the role of an irreducible representation (R, π) of $G(F)$ is played by the category $\widehat{\mathfrak{g}}_{\kappa_c}$-mod$_\chi$ for some $\chi \in \operatorname{Op}_{LG}(D^\times)$. The analogue of an unramified representation is a category $\widehat{\mathfrak{g}}_{\kappa_c}$-mod$_\chi$ which contains non-zero $(\widehat{\mathfrak{g}}_{\kappa_c}, G[[t]])$ Harish-Chandra modules. This leads us to the following question: for what $\chi \in \operatorname{Op}_{LG}(D^\times)$ does the category $\widehat{\mathfrak{g}}_{\kappa_c}$-mod$_\chi$ contain non-zero $(\widehat{\mathfrak{g}}_{\kappa_c}, G[[t]])$ Harish-Chandra modules?

We saw in the previous section that (R, π) is unramified if and only if it corresponds to an unramified Langlands parameter, which is a homomorphism $W'_F \to {}^LG$ that factors through $W'_F \to \mathbb{Z}$. Recall that in the geometric setting the Langlands parameters are ${}^LG$-local systems on $D^\times$. The analogues of unramified homomorphisms $W'_F \to {}^LG$ are those local systems on $D^\times$ which extend to the disc D, in other words, have no singularity at the origin $0 \in D$. Note that there is a unique, up to isomorphism local system on D. Indeed, suppose that we are given a regular connection on a ${}^LG$-bundle $\mathcal{F}$ on D. Let us trivialize the fiber $\mathcal{F}_0$ of $\mathcal{F}$ at $0 \in D$. Then, because D is contractible, the connection identifies $\mathcal{F}$ with the trivial bundle on D. Under this identification the connection itself becomes trivial, i.e., represented by the operator $\nabla = \partial_t$.

Therefore all regular ${}^LG$-local systems (i.e., those which extend to D) correspond to a single point of the set $\operatorname{Loc}_{LG}(D^\times)$, namely, the equivalence class of the trivial local system σ_0.[13] From the point of view of the realization of $\operatorname{Loc}_{LG}(D^\times)$ as the quotient (3.2) this simply means that there is a unique ${}^LG((t))$ gauge equivalence class containing all regular connections of the form $\partial_t + A(t)$, where $A(t) \in {}^L\mathfrak{g}[[t]]$.

The gauge equivalence class of regular connections is the unique local Langlands parameter that we may view as unramified in the geometric setting. Therefore, by analogy with the unramified Langlands correspondence for $G(F)$, we expect that the category $\widehat{\mathfrak{g}}_{\kappa_c}$-mod$_\chi$ contains non-zero $(\widehat{\mathfrak{g}}_{\kappa_c}, G[[t]])$ Harish-Chandra modules if and only if the ${}^LG$-oper $\chi \in \operatorname{Op}_{LG}(D^\times)$ is ${}^LG((t))$ gauge equivalent to the trivial connection, or, in other words, χ belongs to the fiber $\alpha^{-1}(\sigma_0)$ over σ_0.

What does this fiber look like? Let P^+ be the set of dominant integral weights of G (equivalently, dominant integral coweights of ${}^LG$). In [FG2] we defined, for each $\lambda \in P^+$, the space $\operatorname{Op}^\lambda_{LG}$ of ${}^LB[[t]]$-equivalence classes of operators of the form

$$(7.6) \qquad \nabla = \partial_t + \sum_{i=1}^{\ell} t^{\langle \check{\alpha}_i, \lambda \rangle} \psi_i(t) f_i + \mathbf{v}(t),$$

---

[13] Note however that the trivial ${}^LG$-local system on D has a non-trivial group of automorphisms, namely, the group ${}^LG$ itself (it may be realized as the group of automorphisms of the fiber at $0 \in D$). Therefore if we think of $\operatorname{Loc}_{LG}(D^\times)$ as a stack rather than as a set, then the trivial local system corresponds to a substack $\operatorname{pt}/{}^LG$.

where $\psi_i(t) \in \mathbb{C}[[t]], \psi_i(0) \neq 0, \mathbf{v}(t) \in {}^L\mathfrak{b}[[t]]$.

Lemma 2. *Suppose that the local system underlying an oper* $\chi \in \mathrm{Op}_{{}^L G}(D^\times)$ *is trivial. Then* χ *belongs to the disjoint union of the subsets* $\mathrm{Op}_{{}^L G}^\lambda \subset \mathrm{Op}_{{}^L G}(D^\times), \lambda \in P^+$.

Proof. It is clear from the definition that any oper in $\mathrm{Op}_{{}^L G}^\lambda$ is regular on the disc D and is therefore ${}^L G((t))$ gauge equivalent to the trivial connection.

Now suppose that we have an oper $\chi = (\mathcal{F}, \nabla, \mathcal{F}_{{}^L B})$ such that the underlying ${}^L G$-local system is trivial. Then ∇ is ${}^L G((t))$ gauge equivalent to a regular connection, that is one of the form $\partial_t + A(t)$, where $A(t) \in {}^L\mathfrak{g}[[t]]$. We have the decomposition ${}^L G((t)) = {}^L G[[t]]{}^L B((t))$. The gauge action of ${}^L G[[t]]$ clearly preserves the space of regular connections. Therefore if an oper connection ∇ is ${}^L G((t))$ gauge equivalent to a regular connection, then its ${}^L B((t))$ gauge class already must contain a regular connection. The oper condition then implies that this gauge class contains a connection operator of the form (7.6) for some dominant integral weight λ of ${}^L G$. Therefore $\chi \in \mathrm{Op}_{{}^L G}^\lambda$. $\square$

Thus, we see that the set of opers corresponding to the (unique) unramified Langlands parameter is the disjoint union $\bigsqcup_{\lambda \in P^+} \mathrm{Op}_{{}^L G}^\lambda$. We call such opers "unramified". The following result then confirms our expectation that the category $\widehat{\mathfrak{g}}_{\kappa_c}\text{-mod}_\chi$ is "unramified", that is contains non-zero $G[[t]]$-equivariant objects, if and only if χ is unramified (see [FG3] for a proof).

Lemma 3. *The category* $\widehat{\mathfrak{g}}_{\kappa_c}\text{-mod}_\chi$ *contains a non-zero* $(\widehat{\mathfrak{g}}_{\kappa_c}, G[[t]])$ *Harish-Chandra module if and only if*

$$(7.7) \qquad \chi \in \bigsqcup_{\lambda \in P^+} \mathrm{Op}_{{}^L G}^\lambda.$$

The next question is to describe the category $\widehat{\mathfrak{g}}_{\kappa_c}\text{-mod}_\chi^{G[[t]]}$ of $(\widehat{\mathfrak{g}}_{\kappa_c}, G[[t]])$ modules for $\chi \in \mathrm{Op}_{{}^L G}^\lambda$.

7.3 Categories of $G[[t]]$-Equivariant Modules

Let us recall from Section 7.1 that the space of K_0-invariant vectors in an unramified irreducible representation of $G(F)$ is always one-dimensional. We have proposed that the category $\widehat{\mathfrak{g}}_{\kappa_c}\text{-mod}_\chi^{G[[t]]}$ should be viewed as a categorical analogue of this space. Therefore we expect it to be the simplest possible abelian category: the category of $\mathbb{C}$-vector spaces. Here we assume that χ belongs to the union of the spaces $\mathrm{Op}_{{}^L G}^\lambda$, where $\lambda \in P^+$, for otherwise the category $\widehat{\mathfrak{g}}_{\kappa_c}\text{-mod}_\chi^{G[[t]]}$ would be trivial (zero object is the only object).

In this subsection we will prove, following [FG1] (see also [BD1]), that our expectation is in fact correct provided that $\lambda = 0$, in which case $\mathrm{Op}_{{}^L G}^0 = \mathrm{Op}_{{}^L G}(D)$, and so

$$\chi \in \mathrm{Op}_{{}^L G}(D) \subset \mathrm{Op}_{{}^L G}(D^\times).$$

We will also conjecture that this is true for $\chi \in \mathrm{Op}_{^LG}^\lambda$ for all $\lambda \in P^+$.

Recall the vacuum module $\mathbb{V}_0 = V_{\kappa_c}(\mathfrak{g})$. According to [FF3, F3], we have

$$(7.8) \qquad \mathrm{End}_{\widehat{\mathfrak{g}}_{\kappa_c}} \mathbb{V}_0 \simeq \mathrm{Fun}\,\mathrm{Op}_{^LG}(D).$$

Let $\chi \in \mathrm{Op}_{^LG}(D) \subset \mathrm{Op}_{^LG}(D^\times)$. Then χ defines a character of the algebra $\mathrm{End}_{\widehat{\mathfrak{g}}_{\kappa_c}} \mathbb{V}_0$. Let $\mathbb{V}_0(\chi)$ be the quotient of $\mathbb{V}_0$ by the kernel of this character. Then we have the following result.

Theorem 3. *Let* $\chi \in \mathrm{Op}_{^LG}(D) \subset \mathrm{Op}_{^LG}(D^\times)$. *The category* $\widehat{\mathfrak{g}}_{\kappa_c}$-$\mathrm{mod}_\chi^{G[[t]]}$ *is equivalent to the category of vector spaces: its unique, up to isomorphism, irreducible object is* $\mathbb{V}_0(\chi)$ *and any other object is isomorphic to the direct sum of copies of* $\mathbb{V}_0(\chi)$.

This theorem provides the first piece of evidence for Conjecture 2: we see that the categories $\widehat{\mathfrak{g}}_{\kappa_c}$-$\mathrm{mod}_\chi^{G[[t]]}$ are equivalent to each other for all $\chi \in \mathrm{Op}_{^LG}(D)$.

It is more convenient to consider, instead of an individual regular ^LG-oper χ, the entire family $\mathrm{Op}_{^LG}^0 = \mathrm{Op}_{^LG}(D)$ of regular opers on the disc D. Let $\widehat{\mathfrak{g}}_{\kappa_c}$-$\mathrm{mod}_{\mathrm{reg}}$ be the full subcategory of the category $\widehat{\mathfrak{g}}_{\kappa_c}$-mod whose objects are $\widehat{\mathfrak{g}}_{\kappa_c}$-modules on which the action of the center $Z(\widehat{\mathfrak{g}})$ factors through the homomorphism

$$Z(\widehat{\mathfrak{g}}) \simeq \mathrm{Fun}\,\mathrm{Op}_{^LG}(D^\times) \to \mathrm{Fun}\,\mathrm{Op}_{^LG}(D).$$

Note that the category $\widehat{\mathfrak{g}}_{\kappa_c}$-$\mathrm{mod}_{\mathrm{reg}}$ is an example of a category $\widehat{\mathfrak{g}}_{\kappa_c}$-$\mathrm{mod}_V$ introduced in Section 5, in the case when $V = \mathrm{Op}_{^LG}(D)$.

Let $\widehat{\mathfrak{g}}_{\kappa_c}$-$\mathrm{mod}_{\mathrm{reg}}^{G[[t]]}$ be the corresponding $G[[t]]$-equivariant category. It is instructive to think of $\widehat{\mathfrak{g}}_{\kappa_c}$-$\mathrm{mod}_{\mathrm{reg}}$ and $\widehat{\mathfrak{g}}_{\kappa_c}$-$\mathrm{mod}_{\mathrm{reg}}^{G[[t]]}$ as categories fibered over $\mathrm{Op}_{^LG}(D)$, with the fibers over $\chi \in \mathrm{Op}_{^LG}(D)$ being $\widehat{\mathfrak{g}}_{\kappa_c}$-$\mathrm{mod}_\chi$ and $\widehat{\mathfrak{g}}_{\kappa_c}$-$\mathrm{mod}_\chi^{G[[t]]}$, respectively.

We will now describe the category $\widehat{\mathfrak{g}}_{\kappa_c}$-$\mathrm{mod}_{\mathrm{reg}}^{G[[t]]}$. This description will in particular imply Theorem 3.

In order to simplify our formulas, in what follows we will use the following notation for $\mathrm{Fun}\,\mathrm{Op}_{^LG}(D)$:

$$\mathfrak{z} = \mathfrak{z}(\widehat{\mathfrak{g}}) = \mathrm{Fun}\,\mathrm{Op}_{^LG}(D).$$

Let $\mathfrak{z}$-mod be the category of modules over the commutative algebra $\mathfrak{z}$. Equivalently, this is the category of quasicoherent sheaves on the space $\mathrm{Op}_{^LG}(D)$.

By definition, any object of $\widehat{\mathfrak{g}}_{\kappa_c}$-$\mathrm{mod}_{\mathrm{reg}}^{G[[t]]}$ is a $\mathfrak{z}$-module. Introduce the functors

$$\mathsf{F} : \widehat{\mathfrak{g}}_{\kappa_c}\text{-}\mathrm{mod}_{\mathrm{reg}}^{G[[t]]} \to \mathfrak{z}\text{-mod}, \qquad M \mapsto \mathrm{Hom}_{\widehat{\mathfrak{g}}_{\kappa_c}}(\mathbb{V}_0, M),$$

$$\mathsf{G} : \mathfrak{z}\text{-mod} \to \widehat{\mathfrak{g}}_{\kappa_c}\text{-}\mathrm{mod}_{\mathrm{reg}}^{G[[t]]}, \qquad \mathcal{F} \mapsto \mathbb{V}_0 \underset{\mathfrak{z}}{\otimes} \mathcal{F}.$$

The following theorem has been proved in [FG1], Theorem 6.3 (important results in this direction were obtained earlier in [BD1]).

Theorem 4. *The functors* F *and* G *are mutually inverse equivalences of categories*

$$(7.9) \qquad \widehat{\mathfrak{g}}_{\kappa_c}\text{-mod}_{\text{reg}}^{G[[t]]} \simeq \mathfrak{z}\text{-mod}.$$

This immediately implies Theorem 3. Indeed, for each $\chi \in \text{Op}_{^LG}(D)$ the category $\widehat{\mathfrak{g}}_{\kappa_c}\text{-mod}_\chi^{G[[t]]}$ is the full subcategory of $\widehat{\mathfrak{g}}_{\kappa_c}\text{-mod}_{\text{reg}}^{G[[t]]}$ which are annihilated, as $\mathfrak{z}$-modules, by the maximal ideal I_χ of χ. By Theorem 4, this category is equivalent to the category of $\mathfrak{z}$-modules annihilated by I_χ. But this is the category of $\mathfrak{z}$-modules supported (scheme-theoretically) at the point χ, which is equivalent to the category of vector spaces.

7.4 The Action of the Spherical Hecke Algebra

In Section 7.1 we discussed irreducible unramified representations of the group $G(F)$, where F is a local non-archimedian field. We have seen that such representations are parameterized by conjugacy classes of the Langlands dual group ^LG. Given such a conjugacy class γ, we have an irreducible unramified representation (R_γ, π_γ), which contains a one-dimensional subspace $(R_\gamma)^{\pi_\gamma(K_0)}$ of K_0-invariant vectors. The spherical Hecke algebra $H(G(F), K_0)$, which is isomorphic to $\text{Rep}\,^LG$ via the Satake isomorphism, acts on this space by a character ϕ_γ, see formula (7.5).

In the geometric setting, we have argued that for any $\chi \in \text{Op}_{^LG}(D)$ the category $\widehat{\mathfrak{g}}_{\kappa_c}\text{-mod}_\chi$, equipped with an action of the loop group $G((t))$, should be viewed as a categorification of (R_γ, π_γ). Furthermore, its subcategory $\widehat{\mathfrak{g}}_{\kappa_c}\text{-mod}_\chi^{G[[t]]}$ of $(\widehat{\mathfrak{g}}_{\kappa_c}, G[[t]])$ Harish-Chandra modules should be viewed as a categorification of the one-dimensional space $(R_\gamma)^{\pi_\gamma(K_0)}$. According to Theorem 3, the latter category is equivalent to the category of vector spaces, which is consistent with our expectations.

We now discuss the categorical analogue of the action of the spherical Hecke algebra.

As explained in Section 6.3, the categorical analogue of the spherical Hecke algebra is the category of $G[[t]]$-equivariant $\mathcal{D}$-modules on the **affine Grassmannian** $\text{Gr} = G((t))/G[[t]]$. We refer the reader to [BD1, FG2] for the precise definition of Gr and this category. There is an important property that is satisfied in the unramified case: the convolution functors with these $\mathcal{D}$-modules are exact, which means that we do not need to consider the derived category; the abelian category of such $\mathcal{D}$-modules will do. Let us denote this abelian category by $\mathcal{H}(G((t)), G[[t]])$.

According to the results of [MV], this category carries a natural structure of tensor category, which is equivalent to the tensor category $\mathcal{R}ep\,^LG$ of representations of ^LG. This should be viewed as a categorical analogue of the

Satake isomorphism. Thus, for each object V of $Rep\,{}^L G$ we have an object of $\mathcal{H}(G((t)), G[[t]])$ which we denote by $\mathcal{H}_V$. What should be the analogue of the Hecke eigenvector property (7.5)?

As we explained in Section 6.3, the category $\mathcal{H}(G((t)), G[[t]])$ naturally acts on the category $\widehat{\mathfrak{g}}_{\kappa_c}\text{-mod}_\chi^{G[[t]]}$, and this action should be viewed as a categorical analogue of the action of $H(G(F), K_0)$ on $(R_\gamma)^{\pi_\gamma(K_0)}$.

Now, by Theorem 3, any object of $\widehat{\mathfrak{g}}_{\kappa_c}\text{-mod}_\chi^{G[[t]]}$ is a direct sum of copies of $\mathbb{V}_0(\chi)$. Therefore it is sufficient to describe the action of $\mathcal{H}(G((t)), G[[t]])$ on $\mathbb{V}_0(\chi)$. This action is described by the following statement, which follows from [BD1]: there exists a family of isomorphisms

$$(7.10) \qquad \alpha_V : \mathcal{H}_V \star \mathbb{V}_0(\chi) \xrightarrow{\sim} \underline{V} \otimes \mathbb{V}_0(\chi), \qquad V \in Rep\,{}^L G,$$

where $\underline{V}$ is the vector space underlying the representation V. Moreover, these isomorphisms are compatible with the tensor product structure on $\mathcal{H}_V$ (given by the convolution) and on $\underline{V}$ (given by tensor product of vector spaces).

In view of Theorem 3, this is not surprising. Indeed, it follows from the definition that $\mathcal{H}_V \star \mathbb{V}_0(\chi)$ is again an object of the category $\widehat{\mathfrak{g}}_{\kappa_c}\text{-mod}_\chi^{G[[t]]}$. Therefore it must be isomorphic to $U_V \otimes_{\mathbb{C}} \mathbb{V}_0(\chi)$, where U_V is a vector space. But then we obtain a functor $\mathcal{H}(G((t)), G[[t]]) \to Vect, \mathcal{H}_V \mapsto U_V$. It follows from the construction that this is a tensor functor. Therefore the standard Tannakian formalism implies that U_V is isomorphic to $\underline{V}$.

The isomorphisms (7.10) should be viewed as the categorical analogues of the Hecke eigenvector conditions (7.5). The difference is that while in (7.5) the action of elements of the Hecke algebra on a K_0-invariant vector in R_γ amounts to multiplication by a scalar, the action of an object of the Hecke category $\mathcal{H}(G((t)), G[[t]])$ on the $G[[t]]$-equivariant object $\mathbb{V}_0(\chi)$ of $\widehat{\mathfrak{g}}_{\kappa_c}\text{-mod}_\chi$ amounts to multiplication by a *vector space*, namely, the vector space underlying the corresponding representation of ${}^L G$. It is natural to call a module satisfying this property a **Hecke eigenmodule**. Thus, we obtain that $\mathbb{V}_0(\chi)$ is a Hecke eigenmodule. This is in agreement with our expectation that the category $\widehat{\mathfrak{g}}_{\kappa_c}\text{-mod}_\chi^{G[[t]]}$ is a categorical version of the space of K_0-invariant vectors in R_γ.

One ingredient that is missing in the geometric case is the conjugacy class γ of ${}^L G$. We recall that in the classical Langlands correspondence this was the image of the Frobenius element of the Galois group $\mathrm{Gal}(\overline{\mathbb{F}}_q/\mathbb{F}_q)$, which does not have an analogue in the geometric setting where our ground field is $\mathbb{C}$, which is algebraically closed. So while unramified local systems in the classical case are parameterized by the conjugacy classes γ, there is only one, up to an isomorphism, unramified local system in the geometric case. However, this local system has a large group of automorphisms, namely, ${}^L G$ itself. One can argue that what replaces γ in the geometric setting is the action of this group ${}^L G$ by automorphisms of the category $\widehat{\mathfrak{g}}_{\kappa_c}\text{-mod}_\chi$, which we will discuss in the next two sections.

7.5 Categories of Representations and $\mathcal{D}$-Modules

When we discussed the procedure of categorification of representations in Section 3.5, we saw that there are two possible scenarios for constructing categories equipped with an action of the loop group $G((t))$. In the first one we consider categories of $\mathcal{D}$-modules on the ind-schemes $G((t))/K$, where K is a "compact" subgroup of $G((t))$, such as $G[[t]]$ or the Iwahori subgroup. In the second one we consider categories of representations $\widehat{\mathfrak{g}}_{\kappa_c}$-mod$_\chi$. So far we have focused exclusively on the second scenario, but it is instructive to also discuss categories of the first type.

In the toy model considered in Section 3.4 we discussed the category of $\mathfrak{g}$-modules with fixed central character and the category of $\mathcal{D}$-modules on the flag variety G/B. We have argued that both could be viewed as categorifications of the representation of the group $G(\mathbb{F}_q)$ on the space of functions on $(G/B)(\mathbb{F}_q)$. These categories are equivalent, according to the Beilinson-Bernstein theory, with the functor of global sections connecting the two. Could something like this be true in the case of affine Kac-Moody algebras as well?

The affine Grassmannian $\mathrm{Gr} = G((t))/G[[t]]$ may be viewed as the simplest possible analogue of the flag variety G/B for the loop group $G((t))$. Consider the category of $\mathcal{D}$-modules on $G((t))/G[[t]]$ (see [BD1, FG2] for the precise definition). We have a functor of global sections from this category to the category of $\mathfrak{g}((t))$-modules. In order to obtain $\widehat{\mathfrak{g}}_{\kappa_c}$-modules, we need to take instead the category $\mathcal{D}_{\kappa_c}$-mod of $\mathcal{D}$-modules twisted by a line bundle $\mathcal{L}_{\kappa_c}$. This is the unique line bundle $\mathcal{L}_{\kappa_c}$ on Gr which carries an action of $\widehat{\mathfrak{g}}_{\kappa_c}$ (such that the central element $\mathbf{1}$ is mapped to the identity) lifting the natural action of $\mathfrak{g}((t))$ on Gr. Then for any object $\mathcal{M}$ of $\mathcal{D}_{\kappa_c}$-mod, the space of global sections $\Gamma(\mathrm{Gr}, \mathcal{M})$ is a $\widehat{\mathfrak{g}}_{\kappa_c}$-module. Moreover, it is known (see [BD1, FG1]) that $\Gamma(\mathrm{Gr}, \mathcal{M})$ is in fact an object of $\widehat{\mathfrak{g}}_{\kappa_c}$-mod$_{\mathrm{reg}}$. Therefore we have a functor of global sections

$$\Gamma : \mathcal{D}_{\kappa_c}\text{-mod} \to \widehat{\mathfrak{g}}_{\kappa_c}\text{-mod}_{\mathrm{reg}} .$$

We note that the categories $\mathcal{D}$-mod and $\mathcal{D}_{\kappa_c}$-mod are equivalent under the functor $\mathcal{M} \mapsto \mathcal{M} \otimes \mathcal{L}_{\kappa_c}$. But the corresponding global sections functors are very different.

However, unlike in the Beilinson-Bernstein scenario, the functor Γ cannot possibly be an equivalence of categories. There are two reasons for this. First of all, the category $\widehat{\mathfrak{g}}_{\kappa_c}$-mod$_{\mathrm{reg}}$ has a large center, namely, the algebra $\mathfrak{z} = \mathrm{Fun}\,\mathrm{Op}_{{}^L G}(D)$, while the center of the category $\mathcal{D}_{\kappa_c}$-mod is trivial.[14] The second, and more serious, reason is that the category $\mathcal{D}_{\kappa_c}$-mod carries an

---

[14] Recall that we are under the assumption that G is a connected simply-connected algebraic group, and in this case Gr has one connected component. In general, the center of the category $\mathcal{D}_{\kappa_c}$-mod has a basis enumerated by the connected components of Gr and is isomorphic to the group algebra of the finite group $\pi_1(G)$.

additional symmetry, namely, an action of the tensor category $Rep^L G$ of representations of the Langlands dual group $^L G$, and this action trivializes under the functor Γ as we explain presently.

Over $\mathrm{Op}_{L_G}(D)$ there exists a canonical principal $^L G$-bundle, which we will denote by $\mathcal{P}$. By definition, the fiber of $\mathcal{P}$ at $\chi = (\mathcal{F}, \nabla, \mathcal{F}_{L_B}) \in \mathrm{Op}_{L_G}(D)$ is $\mathcal{F}_0$, the fiber at $0 \in D$ of the $^L G$-bundle $\mathcal{F}$ underlying χ. For an object $V \in Rep^L G$ let us denote by $\mathcal{V}$ the associated vector bundle over $\mathrm{Op}_{L_G}(D)$, i.e.,

$$\mathcal{V} = \mathcal{P} \underset{L_G}{\times} V.$$

Next, consider the category $\mathcal{D}_{\kappa_c}\text{-mod}^{G[[t]]}$ of $G[[t]]$-equivariant $\mathcal{D}_{\kappa_c}$-modules on Gr. It is equivalent to the category

$$\mathcal{D}\text{-mod}^{G[[t]]} = \mathcal{H}(G((t)), G[[t]])$$

considered above. This is a tensor category, with respect to the convolution functor, which is equivalent to the category $Rep^L G$. We will use the same notation $\mathcal{H}_V$ for the object of $\mathcal{D}_{\kappa_c}\text{-mod}^{G[[t]]}$ corresponding to $V \in Rep^L G$. The category $\mathcal{D}_{\kappa_c}\text{-mod}^{G[[t]]}$ acts on $\mathcal{D}_{\kappa_c}$-mod by convolution functors

$$\mathcal{M} \mapsto \mathcal{H}_V \star \mathcal{M}$$

which are exact. This amounts to a tensor action of the category $Rep^L G$ on $\mathcal{D}_{\kappa_c}$-mod.

Now, A. Beilinson and V. Drinfeld have proved in [BD1] that there are functorial isomorphisms

$$\Gamma(\mathrm{Gr}, \mathcal{H}_V \star \mathcal{M}) \simeq \Gamma(\mathrm{Gr}, \mathcal{M}) \underset{3}{\otimes} \mathcal{V}, \qquad V \in Rep^L G,$$

compatible with the tensor structure. Thus, we see that there are non-isomorphic objects of $\mathcal{D}_{\kappa_c}$-mod, which the functor Γ sends to isomorphic objects of $\widehat{\mathfrak{g}}_{\kappa_c}\text{-mod}_{\mathrm{reg}}$. Therefore the category $\mathcal{D}_{\kappa_c}$-mod and the functor Γ need to be modified in order to have a chance to obtain a category equivalent to $\widehat{\mathfrak{g}}_{\kappa_c}\text{-mod}_{\mathrm{reg}}$.

In [FG2] it was shown how to modify the category $\mathcal{D}_{\kappa_c}$-mod, by simultaneously "adding" to it $\mathfrak{z}$ as a center, and "dividing" it by the above $Rep^L G$-action. As the result, we obtain a candidate for a category that can be equivalent to $\widehat{\mathfrak{g}}_{\kappa_c}\text{-mod}_{\mathrm{reg}}$. This is the category of **Hecke eigenmodules** on Gr, denoted by $\mathcal{D}_{\kappa_c}^{\mathrm{Hecke}}\text{-mod}_{\mathrm{reg}}$.

By definition, an object of $\mathcal{D}_{\kappa_c}^{\mathrm{Hecke}}\text{-mod}_{\mathrm{reg}}$ is an object of $\mathcal{D}_{\kappa_c}$-mod, equipped with an action of the algebra $\mathfrak{z}$ by endomorphisms and a system of isomorphisms

$$\alpha_V : \mathcal{H}_V \star \mathcal{M} \xrightarrow{\sim} \mathcal{V} \underset{3}{\otimes} \mathcal{M}, \qquad V \in Rep^L G,$$

compatible with the tensor structure.

The above functor Γ naturally gives rise to a functor

$$(7.11) \qquad \Gamma^{\text{Hecke}} : \mathcal{D}^{\text{Hecke}}_{\kappa_c}\text{-mod}_{\text{reg}} \to \widehat{\mathfrak{g}}_{\kappa_c}\text{-mod}_{\text{reg}} .$$

This is in fact a general property. Suppose for simplicity that we have an abelian category $\mathcal{C}$ which is acted upon by the tensor category $\mathcal{R}ep\, H$, where H is an algebraic group; we denote this functor by $\mathcal{M} \mapsto \mathcal{M} \star V, V \in \mathcal{R}ep\, H$. Let $\mathcal{C}^{\text{Hecke}}$ be the category whose objects are collections $(\mathcal{M}, \{\alpha_V\}_{V \in \mathcal{R}ep\, H})$, where $\mathcal{M} \in \mathcal{C}$ and $\{\alpha_V\}$ is a compatible system of isomorphisms

$$\alpha_V : \mathcal{M} \star V \xrightarrow{\sim} \underline{V} \underset{\mathbb{C}}{\otimes} \mathcal{M}, \qquad V \in \mathcal{R}ep\, H,$$

where $\underline{V}$ is the vector space underlying V. One may think of $\mathcal{C}^{\text{Hecke}}$ as the "de-equivariantized" category $\mathcal{C}$ with respect to the action of H. It carries a natural action of the group H: for $h \in H$, we have

$$h \cdot (\mathcal{M}, \{\alpha_V\}_{V \in \mathcal{R}ep\, H}) = (\mathcal{M}, \{(h \otimes \text{id}_{\mathcal{M}}) \circ \alpha_V\}_{V \in \mathcal{R}ep\, H}).$$

In other words, $\mathcal{M}$ remains unchanged, but the isomorphisms α_V get composed with h.

The category $\mathcal{C}$ may in turn be reconstructed as the category of H-equivariant objects of $\mathcal{C}^{\text{Hecke}}$ with respect to this action, see [Ga3].

Suppose that we have a functor $\mathsf{G} : \mathcal{C} \to \mathcal{C}'$, such that we have functorial isomorphisms

$$(7.12) \qquad \mathsf{G}(\mathcal{M} \star V) \simeq \mathsf{G}(\mathcal{M}) \underset{\mathbb{C}}{\otimes} \underline{V}, \qquad V \in \mathcal{R}ep\, H,$$

compatible with the tensor structure. Then, according to [AG], there exists a functor $\mathsf{G}^{\text{Hecke}} : \mathcal{C}^{\text{Hecke}} \to \mathcal{C}'$ such that $\mathsf{G} \simeq \mathsf{G}^{\text{Hecke}} \circ \text{Ind}$, where the functor $\text{Ind} : \mathcal{C} \to \mathcal{C}^{\text{Hecke}}$ sends $\mathcal{M}$ to $\mathcal{M} \star \mathcal{O}_H$, where $\mathcal{O}_H$ is the regular representation of H. The functor $\mathsf{G}^{\text{Hecke}}$ may be explicitly described as follows: the isomorphisms α_V and (7.12) give rise to an action of the algebra $\mathcal{O}_H$ on $\mathsf{G}(\mathcal{M})$, and $\mathsf{G}^{\text{Hecke}}(\mathcal{M})$ is obtained by taking the fiber of $\mathsf{G}(\mathcal{M})$ at $1 \in H$.

We take $\mathcal{C} = \mathcal{D}_{\kappa_c}\text{-mod}$, $\mathcal{C}' = \widehat{\mathfrak{g}}_{\kappa_c}\text{-mod}_{\text{reg}}$, and $\mathsf{G} = \Gamma$. The only difference is that now we are working over the base $\text{Op}_{L_G}(D)$, which we have to take into account. Thus, we obtain a functor (7.11) (see [FG2, FG4] for more details). Moreover, the left action of the group $G(\!(t)\!)$ on Gr gives rise to its action on the category $\mathcal{D}^{\text{Hecke}}_{\kappa_c}\text{-mod}_{\text{reg}}$, and the functor Γ^{Hecke} intertwines this action with the action of $G(\!(t)\!)$ on $\widehat{\mathfrak{g}}_{\kappa_c}\text{-mod}_{\text{reg}}$.

The following was conjectured in [FG2]:

Conjecture 3. *The functor Γ^{Hecke} in formula (7.11) defines an equivalence of the categories $\mathcal{D}^{\text{Hecke}}_{\kappa_c}\text{-mod}_{\text{reg}}$ and $\widehat{\mathfrak{g}}_{\kappa_c}\text{-mod}_{\text{reg}}$.*

It was proved in [FG2] that the functor Γ^{Hecke}, when extended to the derived categories, is fully faithful. Furthermore, it was proved in [FG4] that it sets up an equivalence of the corresponding I^0-equivariant categories, where $I^0 = [I, I]$ is the radical of the Iwahori subgroup.

Let us specialize Conjecture 3 to a point $\chi = (\mathcal{F}, \nabla, \mathcal{F}_{LB}) \in \mathrm{Op}_{LG}(D)$. Then on the right hand side we consider the category $\widehat{\mathfrak{g}}_{\kappa_c}\text{-mod}_\chi$, and on the left hand side we consider the category $\mathcal{D}^{\mathrm{Hecke}}_{\kappa_c}\text{-mod}_\chi$. Its objects consist of a $\mathcal{D}_{\kappa_c}$-module $\mathcal{M}$ and a collection of isomorphisms

$$(7.13) \qquad \alpha_V : \mathcal{H}_V \star \mathcal{M} \overset{\sim}{\longrightarrow} V_{\mathcal{F}_0} \otimes \mathcal{M}, \qquad V \in \mathcal{R}ep\,^L G.$$

Here $V_{\mathcal{F}_0}$ is the twist of the representation V by the $^L G$-torsor $\mathcal{F}_0$. These isomorphisms have to be compatible with the tensor structure on the category $\mathcal{H}(G((t)), G[[t]])$.

Conjecture 3 implies that there is a canonical equivalence of categories

$$(7.14) \qquad \mathcal{D}^{\mathrm{Hecke}}_{\kappa_c}\text{-mod}_\chi \simeq \widehat{\mathfrak{g}}_{\kappa_c}\text{-mod}_\chi.$$

It is this conjectural equivalence that should be viewed as an analogue of the Beilinson-Bernstein equivalence.

From this point of view, one can think of each of the categories $\mathcal{D}^{\mathrm{Hecke}}_{\kappa_c}\text{-mod}_\chi$ as the second incarnation of the sought-after Langlands category $\mathcal{C}_{\sigma_0}$ corresponding to the trivial $^L G$-local system.

Now we give another explanation why it is natural to view the category $\mathcal{D}^{\mathrm{Hecke}}_{\kappa_c}\text{-mod}_\chi$ as a categorification of an unramified representation of the group $G(F)$. First of all, observe that these categories are all equivalent to each other and to the category $\mathcal{D}^{\mathrm{Hecke}}_{\kappa_c}\text{-mod}$, whose objects are $\mathcal{D}_{\kappa_c}$-modules $\mathcal{M}$ together with a collection of isomorphisms

$$(7.15) \qquad \alpha_V : \mathcal{H}_V \star \mathcal{M} \overset{\sim}{\longrightarrow} \underline{V} \otimes \mathcal{M}, \qquad V \in \mathcal{R}ep\,^L G.$$

Comparing formulas (7.13) and (7.15), we see that there is an equivalence

$$\mathcal{D}^{\mathrm{Hecke}}_{\kappa_c}\text{-mod}_\chi \simeq \mathcal{D}^{\mathrm{Hecke}}_{\kappa_c}\text{-mod},$$

for each choice of trivialization of the $^L G$-torsor $\mathcal{F}_0$ (the fiber at $0 \in D$ of the principal $^L G$-bundle $\mathcal{F}$ on D underlying the oper χ).

Now recall from Section 7.1 that to each semi-simple conjugacy class γ in $^L G$ corresponds an irreducible unramified representation (R_γ, π_γ) of $G(F)$ via the Satake correspondence (7.2). It is known that there is a non-degenerate pairing

$$\langle,\rangle : R_\gamma \times R_{\gamma^{-1}} \to \mathbb{C},$$

in other words, $R_{\gamma^{-1}}$ is the representation of $G(F)$ which is contragredient to R_γ (it may be realized in the space of smooth vectors in the dual space to R_γ).

Let $v \in R_{\gamma^{-1}}$ be a non-zero vector such that $K_0 v = v$ (this vector is unique up to a scalar). It then satisfies the Hecke eigenvector property (7.5) (in which we need to replace γ by γ^{-1}). This allows us to embed R_γ into the space of locally constant right K_0-invariant functions on $G(F)$ (equivalently, functions on $G(F)/K_0$), by using matrix coefficients, as follows:

$$u \in R_\gamma \mapsto f_u, \qquad f_u(g) = \langle u, gv \rangle.$$

The Hecke eigenvector property (7.5) implies that the functions f_u are right K_0-invariant and satisfy the condition

$$(7.16) \qquad\qquad f \star H_V = \mathrm{Tr}(\gamma^{-1}, V)f,$$

where $\star$ denotes the convolution product (6.1). Let $C(G(F)/K_0)_\gamma$ be the space of locally constant functions on $G(F)/K_0$ satisfying (7.16). It carries a representation of $G(F)$ induced by its left action on $G(F)/K_0$. We have constructed an injective map $R_\gamma \to C(G(R)/G(R))_\gamma$, and one can show that for generic γ it is an isomorphism.

Thus, we obtain a realization of an irreducible unramified representation of $G(F)$ in the space of functions on the quotient $G(F)/K_0$ satisfying the Hecke eigenfunction condition (7.16). The Hecke eigenmodule condition (7.15) may be viewed as a categorical analogue of (7.16). Therefore the category $\mathcal{D}^{\mathrm{Hecke}}_{\kappa_c}$-mod of twisted $\mathcal{D}$-modules on $\mathrm{Gr} = G((t))/K_0$ satisfying the Hecke eigenmodule condition (7.15), equipped with a $G((t))$-action appears to be a natural categorification of the irreducible unramified representations of $G(F)$.

7.6 Equivalences Between Categories of Modules

All opers in $\mathrm{Op}_{LG}(D)$ correspond to one and the same ^LG-local system, namely, the trivial local system. Therefore, according to Conjecture 2, we expect that the categories $\widehat{\mathfrak{g}}_{\kappa_c}$-mod$_\chi$ are equivalent to each other. More precisely, for each isomorphism between the underlying local systems of any two opers in $\mathrm{Op}_{LG}(D)$ we wish to have an equivalence of the corresponding categories, and these equivalences should be compatible with respect to the operation of composition of these isomorphisms.

Let us spell this out in detail. Let $\chi = (\mathcal{F}, \nabla, \mathcal{F}_{LB})$ and $\chi' = (\mathcal{F}', \nabla', \mathcal{F}'_{LB})$ be two opers in $\mathrm{Op}_{LG}(D)$. Then an isomorphism between the underlying local systems $(\mathcal{F}, \nabla) \xrightarrow{\sim} (\mathcal{F}', \nabla')$ is the same as an isomorphism $\mathcal{F}_0 \xrightarrow{\sim} \mathcal{F}'_0$ between the ^LG-torsors $\mathcal{F}_0$ and $\mathcal{F}'_0$, which are the fibers of the ^LG-bundles $\mathcal{F}$ and $\mathcal{F}'$, respectively, at $0 \in D$. Let us denote this set of isomorphisms by $\mathrm{Isom}_{\chi, \chi'}$. Then we have

$$\mathrm{Isom}_{\chi, \chi'} = \mathcal{F}_0 \underset{LG}{\times} {}^LG \underset{LG}{\times} \mathcal{F}'_0,$$

where we twist ^LG by $\mathcal{F}_0$ with respect to the left action and by $\mathcal{F}'_0$ with respect to the right action. In particular,

$$\mathrm{Isom}_{\chi, \chi} = {}^LG_{\mathcal{F}_0} = \mathcal{F}_0 \underset{LG}{\times} \mathrm{Ad}\, {}^LG$$

is just the group of automorphisms of $\mathcal{F}_0$.

It is instructive to combine the sets $\mathrm{Isom}_{\chi,\chi'}$ into a groupoid Isom over $\mathrm{Op}_{L_G}(D)$. Thus, by definition Isom consists of triples (χ, χ', ϕ), where $\chi, \chi' \in \mathrm{Op}_{L_G}(D)$ and $\phi \in \mathrm{Isom}_{\chi,\chi}$ is an isomorphism of the underlying local systems. The two morphisms Isom $\to \mathrm{Op}_{L_G}(D)$ correspond to sending such a triple to χ and χ'. The identity morphism $\mathrm{Op}_{L_G}(D) \to$ Isom sends χ to $(\chi, \chi, \mathrm{Id})$, and the composition morphism

$$\mathrm{Isom} \underset{\mathrm{Op}_{L_G}(D)}{\times} \mathrm{Isom} \to \mathrm{Isom}$$

corresponds to composing two isomorphisms.

Conjecture 2 has the following more precise formulation for regular opers:

Conjecture 4. *For each $\phi \in \mathrm{Isom}_{\chi,\chi'}$ there exists an equivalence*

$$E_\phi : \widehat{\mathfrak{g}}_{\kappa_c}\text{-mod}_\chi \to \widehat{\mathfrak{g}}_{\kappa_c}\text{-mod}_{\chi'},$$

which intertwines the actions of $G(\!(t)\!)$ on the two categories, such that $E_{\mathrm{Id}} = \mathrm{Id}$ and there exist isomorphisms $\beta_{\phi,\phi'} : E_{\phi \circ \phi'} \simeq E_\phi \circ E_{\phi'}$ satisfying

$$\beta_{\phi \circ \phi', \phi''} \beta_{\phi,\phi'} = \beta_{\phi,\phi' \circ \phi''} \beta_{\phi',\phi''}$$

for all isomorphisms ϕ, ϕ', ϕ'', whenever they may be composed in the appropriate order.

In other words, the groupoid Isom over $\mathrm{Op}_{L_G}(D)$ acts on the category $\widehat{\mathfrak{g}}_{\kappa_c}\text{-mod}_{\mathrm{reg}}$ fibered over $\mathrm{Op}_{L_G}(D)$, preserving the action of $G(\!(t)\!)$ along the fibers.

In particular, this conjecture implies that the group $^L G_{\mathcal{F}_0}$ acts on the category $\widehat{\mathfrak{g}}_{\kappa_c}\text{-mod}_\chi$ for any $\chi \in \mathrm{Op}_{L_G}(D)$.

Now we observe that Conjecture 3 implies Conjecture 4. Indeed, by Conjecture 3, there is a canonical equivalence of categories (7.14),

$$\mathcal{D}_{\kappa_c}^{\mathrm{Hecke}}\text{-mod}_\chi \simeq \widehat{\mathfrak{g}}_{\kappa_c}\text{-mod}_\chi.$$

It follows immediately from the definition of the category $\mathcal{D}_{\kappa_c}^{\mathrm{Hecke}}\text{-mod}_\chi$ (namely, formula (7.13)) that for each isomorphism $\phi \in \mathrm{Isom}_{\chi,\chi'}$, i.e., an isomorphism of the $^L G$-torsors $\mathcal{F}_0$ and $\mathcal{F}_0'$ underlying the opers χ and χ', there is a canonical equivalence

$$\mathcal{D}_{\kappa_c}^{\mathrm{Hecke}}\text{-mod}_\chi \simeq \mathcal{D}_{\kappa_c}^{\mathrm{Hecke}}\text{-mod}_{\chi'}.$$

Therefore we obtain the sought-after equivalence $E_\phi : \widehat{\mathfrak{g}}_{\kappa_c}\text{-mod}_\chi \to \widehat{\mathfrak{g}}_{\kappa_c}\text{-mod}_{\chi'}$. Furthermore, it is clear that these equivalences satisfy the conditions of Conjecture 4. In particular, they intertwine the actions of $G(\!(t)\!)$, which affects the $\mathcal{D}$-module $\mathcal{M}$ underlying an object of $\mathcal{D}_{\kappa_c}^{\mathrm{Hecke}}\text{-mod}_\chi$, but does not affect the isomorphisms α_V.

Equivalently, we can express this by saying that the groupoid Isom naturally acts on the category $\mathcal{D}^{\mathrm{Hecke}}_{\kappa_c}$-mod$_{\mathrm{reg}}$. By Conjecture 3, this gives rise to an action of Isom on $\widehat{\mathfrak{g}}_{\kappa_c}$-mod$_{\mathrm{reg}}$.

In particular, we construct an action of the group $(^L G)_{\mathcal{F}_0}$, the twist of $^L G$ by the $^L G$-torsor $\mathcal{F}_0$ underlying a particular oper χ, on the category $\mathcal{D}^{\mathrm{Hecke}}_{\kappa_c}$-mod$_\chi$. Indeed, each element $g \in (^L G)_{\mathcal{F}_0}$ acts on the $\mathcal{F}_0$-twist $V_{\mathcal{F}_0}$ of any finite-dimensional representation V of $^L G$. Given an object $(\mathcal{M}, (\alpha_V))$ of $\mathcal{D}^{\mathrm{Hecke}}_{\kappa_c}$-mod$_{\chi'}$, we construct a new object, namely, $(\mathcal{M}, ((g \otimes \mathrm{Id}_{\mathcal{M}}) \circ \alpha_V))$. Thus, we do not change the $\mathcal{D}$-module $\mathcal{M}$, but we change the isomorphisms α_V appearing in the Hecke eigenmodule condition (7.13) by composing them with the action of g on $V_{\mathcal{F}_0}$. According to Conjecture 3, the category $\mathcal{D}^{\mathrm{Hecke}}_{\kappa_c}$-mod$_\chi$ is equivalent to $\widehat{\mathfrak{g}}_{\kappa_c}$-mod$_\chi$. Therefore this gives rise to an action of the group $(^L G)_{\mathcal{F}_0}$ on $\widehat{\mathfrak{g}}_{\kappa_c}$-mod$_\chi$. But this action is much more difficult to describe in terms of $\widehat{\mathfrak{g}}_{\kappa_c}$-modules.

7.7 Generalization to other Dominant Integral Weights

We have extensively studied above the categories $\widehat{\mathfrak{g}}_{\kappa_c}$-mod$_\chi$ and $\widehat{\mathfrak{g}}_{\kappa_c}$-mod$_\chi^{G[[t]]}$ associated to regular opers $\chi \in \mathrm{Op}_{^L G}(D)$. However, according to Lemma 2, the (set-theoretic) fiber of the map $\alpha : \mathrm{Op}_{^L G}(D^\times) \to \mathrm{Loc}_{^L G}(D^\times)$ over the trivial local system σ_0 is the disjoint union of the subsets $\mathrm{Op}^\lambda_{^L G}, \lambda \in P^+$. Here we discuss briefly the categories $\widehat{\mathfrak{g}}_{\kappa_c}$-mod$_\chi$ and $\widehat{\mathfrak{g}}_{\kappa_c}$-mod$_\chi^{G[[t]]}$ for $\chi \in \mathrm{Op}^\lambda_{^L G}$, where $\lambda \neq 0$.

Consider the Weyl module $\mathbb{V}_\lambda$ with highest weight λ,

$$\mathbb{V}_\lambda = U(\widehat{\mathfrak{g}}_{\kappa_c}) \underset{U(\mathfrak{g}[[t]] \oplus \mathbb{C}\mathbf{1})}{\otimes} V_\lambda.$$

According to [FG6], we have

$$(7.17) \qquad \mathrm{End}_{\widehat{\mathfrak{g}}_{\kappa_c}} \mathbb{V}_\lambda \simeq \mathrm{Fun}\,\mathrm{Op}^\lambda_{^L G}.$$

Let $\chi \in \mathrm{Op}^\lambda_{^L G} \subset \mathrm{Op}_{^L G}(D^\times)$. Then χ defines a character of the algebra $\mathrm{End}_{\widehat{\mathfrak{g}}_{\kappa_c}} \mathbb{V}_\lambda$. Let $\mathbb{V}_\lambda(\chi)$ be the quotient of $\mathbb{V}_\lambda$ by the kernel of this character. The following conjecture of [FG6] is an analogue of Theorem 3:

Conjecture 5. *Let* $\chi \in \mathrm{Op}^\lambda_{^L G} \subset \mathrm{Op}_{^L G}(D^\times)$. *Then the category* $\widehat{\mathfrak{g}}_{\kappa_c}$-mod$_\chi^{G[[t]]}$ *is equivalent to the category of vector spaces: its unique, up to isomorphism, irreducible object is* $\mathbb{V}_\lambda(\chi)$ *and any other object is isomorphic to the direct sum of copies of* $\mathbb{V}_\lambda(\chi)$.

Note that this is consistent with Conjecture 2, which tells us that the categories $\widehat{\mathfrak{g}}_{\kappa_c}$-mod$_\chi^{G[[t]]}$ should be equivalent to each other for all opers which are gauge equivalent to the trivial local system on D.

8 Local Langlands Correspondence: Tamely Ramified Case

In the previous section we have considered categorical analogues of the irreducible unramified representations of a reductive group $G(F)$ over a local non-archimedian field F. We recall that these are the representations containing non-zero vectors fixed by the maximal compact subgroup $K_0 \subset G(F)$. The corresponding Langlands parameters are unramified admissible homomorphisms from the Weil-Deligne group W_F' to $^L G$, i.e., those which factor through the quotient

$$W_F' \to W_F \to \mathbb{Z},$$

and whose image in $^L G$ is semi-simple. Such homomorphisms are parameterized by semi-simple conjugacy classes in $^L G$.

We have seen that the categorical analogues of unramified representations of $G(F)$ are the categories $\widehat{\mathfrak{g}}_{\kappa_c}$-mod$_\chi$ (equipped with an action of the loop group $G((t))$), where χ is a $^L G$-oper on $D^\times$ whose underlying $^L G$-local system is trivial. These categories can be called unramified in the sense that they contain non-zero $G[[t]]$-equivariant objects. The corresponding Langlands parameter is the trivial $^L G$-local system σ_0 on $D^\times$, which should be viewed as an analogue of an unramified homomorphism $W_F' \to {}^L G$. However, the local system σ_0 is realized by many different opers, and this introduces an additional complication into our picture: at the end of the day we need to show that the categories $\widehat{\mathfrak{g}}_{\kappa_c}$-mod$_\chi$, where χ is of the above type, are equivalent to each other. In particular, Conjecture 4, which describes what we expect to happen when $\chi \in \mathrm{Op}_{{}^L G}(D)$.

The next natural step is to consider categorical analogues of representations of $G(F)$ that contain vectors invariant under the Iwahori subgroup $I \subset G[[t]]$, the preimage of a fixed Borel subgroup $B \subset G$ under the evaluation homomorphism $G[[t]] \to G$. We begin this section by recalling a classification of these representations, due to D. Kazhdan and G. Lusztig [KL] and V. Ginzburg [CG]. We then discuss the categorical analogues of these representations following [FG2]–[FG5] and the intricate interplay between the classical and the geometric pictures.

8.1 Tamely Ramified Representations

The Langlands parameters corresponding to irreducible representations of $G(F)$ with I-invariant vectors are **tamely ramified** homomorphisms $W_F' \to {}^L G$. Recall from Section 2.1 that $W_F' = W_F \ltimes \mathbb{C}$. A homomorphism $W_F' \to {}^L G$ is called tamely ramified if it factors through the quotient

$$W_F' \to \mathbb{Z} \ltimes \mathbb{C}.$$

According to the relation (2.1), the group $\mathbb{Z} \ltimes \mathbb{C}$ is generated by two elements $F = 1 \in \mathbb{Z}$ (Frobenius) and $M = 1 \in \mathbb{C}$ (monodromy) satisfying the relation

(8.1) $$FMF^{-1} = qM.$$

Under an admissible tamely ramified homomorphism the generator F goes to a semi-simple element $\gamma \in {}^L G$ and the generator M goes to a unipotent element $N \in {}^L G$. According to formula (8.1), they have to satisfy the relation

(8.2) $$\gamma N \gamma^{-1} = N^q.$$

Alternatively, we may write $N = \exp(u)$, where u is a nilpotent element of ${}^L \mathfrak{g}$. Then this relation becomes

$$\gamma u \gamma^{-1} = qu.$$

Thus, we have the following bijection between the sets of equivalence classes
(8.3)

tamely ramified admissible homomorphisms $W_F' \to {}^L G$	$\Longleftrightarrow$	pairs $\gamma \in {}^L G$, semi-simple, $u \in {}^L \mathfrak{g}$, nilpotent, $\gamma u \gamma^{-1} = qu$

In both cases equivalence relation amounts to conjugation by an element of ${}^L G$.

Now to each Langlands parameter of this type we wish to attach an irreducible representation of $G(F)$ which contains non-zero I-invariant vectors. It turns out that if $G = GL_n$ there is indeed a bijection, proved in [BZ], between the sets of equivalence classes of the following objects:
(8.4)

tamely ramified admissible homomorphisms $W_F' \to GL_n$	$\Longleftrightarrow$	irreducible representations (R, π) of $GL_n(F)$, $R^{\pi(I)} \neq 0$

However, such a bijection is no longer true for other reductive groups: two new phenomena appear, which we discuss presently.

The first one is the appearance of L-**packets**. One no longer expects to be able to assign to a particular admissible homomorphism $W_F' \to {}^L G$ a single irreducible smooth representations of $G(F)$. Instead, a finite collection of such representations (more precisely, a collection of equivalence classes of representations) is assigned, called an L-packet. In order to distinguish representations in a given L-packet, one needs to introduce an additional parameter. We will see how this is done in the case at hand shortly. However, and this is the second subtlety alluded to above, it turns out that not all irreducible representations of $G(F)$ within the L-packet associated to a given tamely ramified homomorphism $W_F' \to {}^L G$ contain non-zero I-invariant vectors. Fortunately, there is a certain property of the extra parameter used to distinguish representations inside the L-packet that tells us whether the corresponding representation of $G(F)$ has I-invariant vectors.

In the case of tamely ramified homomorphisms $W_F' \to {}^L G$ this extra parameter is an irreducible representation ρ of the finite group $C(\gamma, u)$ of components of the simultaneous centralizer of γ and u in ${}^L G$, on which the center of $^L G$ acts trivially (see [Lu1]). In the case of $G = GL_n$ these centralizers are always connected, and so this parameter never appears. But for other reductive groups G this group of components is often non-trivial. The simplest example is when $^L G = G_2$ and u is a subprincipal nilpotent element of the Lie algebra $^L\mathfrak{g}$.[15] In this case for some γ satisfying $\gamma u \gamma^{-1} = qu$ the group of components $C(\gamma, u)$ is the symmetric group S_3, which has three irreducible representations (up to equivalence). Each of them corresponds to a particular member of the L-packet associated with the tamely ramified homomorphism $W'_F \to {}^L G$ defined by (γ, u). Thus, the L-packet consists of three (equivalence classes of) irreducible smooth representations of $G(F)$. However, not all of them contain non-zero I-invariant vectors.

The representations ρ of the finite group $C(\gamma, u)$ which correspond to representations of $G(F)$ with I-invariant vectors are distinguished by the following property. Consider the **Springer fiber** Sp_u. We recall that

$$(8.5) \qquad \mathrm{Sp}_u = \{\mathfrak{b}' \in {}^L G/{}^L B \mid u \in \mathfrak{b}'\}.$$

The group $C(\gamma, u)$ acts on the homology of the variety Sp_u^γ of γ-fixed points of Sp_u. A representation ρ of $C(\gamma, u)$ corresponds to a representation of $G(F)$ with non-zero I-invariant vectors if and only if ρ occurs in the homology of Sp_u^γ, $H_\bullet(\mathrm{Sp}_u^\gamma)$.

In the case of G_2 the Springer fiber Sp_u of the subprincipal element u is a union of four projective lines connected with each other as in the Dynkin diagram of D_4. For some γ the set Sp_u^γ is the union of a projective line (corresponding to the central vertex in the Dynkin diagram of D_4) and three points (each in one of the remaining three projective lines). The corresponding group $C(\gamma, u) = S_3$ on Sp_u^γ acts trivially on the projective line and by permutation of the three points. Therefore the trivial and the two-dimensional representations of S_3 occur in $H_\bullet(\mathrm{Sp}_u^\gamma)$, but the sign representation does not. The irreducible representations of $G(F)$ corresponding to the first two contain non-zero I-invariant vectors, whereas the one corresponding to the sign representation of S_3 does not.

The ultimate form of the local Langlands correspondence for representations of $G(F)$ with I-invariant vectors is then as follows (here we assume, as in [KL, CG]), that the group G is split and has connected center):
(8.6)

triples (γ, u, ρ), $\gamma u \gamma^{-1} = qu$, $\rho \in \mathcal{R}ep\, C(\gamma, u)$ occurs in $H_\bullet(\mathrm{Sp}_u^\gamma, \mathbb{C})$	$\Longleftrightarrow$	irreducible representations (R, π) of $G(F)$, $R^{\pi(I)} \neq 0$

Again, this should be understood as a bijection between two sets of equivalence classes of the objects listed. This bijection is due to [KL] (see also [CG]). It was conjectured by Deligne and Langlands, with a subsequent modification (addition of ρ) made by Lusztig.

---

[15] The term "subprincipal" means that the adjoint orbit of this element has codimension 2 in the nilpotent cone.

How to set up this bijection? The idea is to replace irreducible representations of $G(F)$ appearing on the right hand side of (8.6) with irreducible modules over the corresponding Hecke algebra $H(G(F), I)$. Recall from Section 6.1 that this is the algebra of compactly supported I bi-invariant functions on $G(F)$, with respect to convolution. It naturally acts on the space of I-invariant vectors of any smooth representation of $G(F)$ (see formula (6.2)). Thus, we obtain a functor from the category of smooth representations of $G(F)$ to the category of $H(G(F), I)$. According to a theorem of A. Borel [B1], it induces a bijection between the set of equivalence classes of irreducible representations of $G(F)$ with non-zero I-invariant vectors and the set of equivalence classes of irreducible $H(G(F), I)$-modules.

The algebra $H(G(F), I)$ is known as the **affine Hecke algebra** and has the standard description in terms of generators and relations. However, for our purposes we need another description, due to [KL, CG], which identifies it with the equivariant K-theory of the **Steinberg variety**

$$\mathrm{St} = \widetilde{\mathcal{N}} \times_{\mathcal{N}} \widetilde{\mathcal{N}},$$

where $\mathcal{N} \subset {}^L\mathfrak{g}$ is the nilpotent cone and $\widetilde{\mathcal{N}}$ is the **Springer resolution**

$$\widetilde{\mathcal{N}} = \{x \in \mathcal{N}, \mathfrak{b}' \in {}^LG/{}^LB \mid x \in \mathfrak{b}'\}.$$

Thus, a point of St is a triple consisting of a nilpotent element of ${}^L\mathfrak{g}$ and two Borel subalgebras containing it. The group ${}^LG \times \mathbb{C}^\times$ naturally acts on St, with ${}^LG$ conjugating members of the triple and $\mathbb{C}^\times$ acting by multiplication on the nilpotent elements,

$$(8.7) \qquad a \cdot (x, \mathfrak{b}', \mathfrak{b}'') = (a^{-1}x, \mathfrak{b}', \mathfrak{b}'').$$

According to a theorem of [KL, CG], there is an isomorphism

$$(8.8) \qquad H(G(F), I) \simeq K^{{}^LG \times \mathbb{C}^\times}(\mathrm{St}).$$

The right hand side is the ${}^LG \times \mathbb{C}^\times$-equivariant K-theory of St. It is an algebra with respect to a natural operation of convolution (see [CG] for details). It is also a free module over its center, isomorphic to

$$K^{{}^LG \times \mathbb{C}^\times}(\mathrm{pt}) = \mathrm{Rep}\,{}^LG \otimes \mathbb{C}[\mathbf{q}, \mathbf{q}^{-1}].$$

Under the isomorphism (8.8) the element $\mathbf{q}$ goes to the standard parameter $\mathbf{q}$ of the affine Hecke algebra $H(G(F), I)$ (here we consider $H(G(F), I)$ as a $\mathbb{C}[\mathbf{q}, \mathbf{q}^{-1}]$-module).

Now, the algebra $K^{{}^LG \times \mathbb{C}^\times}(\mathrm{St})$, and hence the algebra $H(G(F), I)$, has a natural family of modules which are parameterized precisely by the conjugacy classes of pairs (γ, u) as above. On these modules $H(G(F), I)$ acts via a central character corresponding to a point in $\mathrm{Spec}\,\mathrm{Rep}\,{}^LG \underset{\mathbb{C}}{\otimes} \mathbb{C}[\mathbf{q}, \mathbf{q}^{-1}]$, which is just a pair (γ, q), where γ is a semi-simple conjugacy class in $^L G$ and $q \in \mathbb{C}^\times$. In our situation q is the cardinality of the residue field of F (hence a power of a prime), but in what follows we will allow a larger range of possible values of q: all non-zero complex numbers except for the roots of unity. Consider the quotient of $H(G(F), I)$ by the central character defined by (γ, u). This is just the algebra $K^{^L G \times \mathbb{C}^\times}(\mathrm{St})$, specialized at (γ, q). We denote it by $K^{^L G \times \mathbb{C}^\times}(\mathrm{St})_{(\gamma, q)}$.

Now for a nilpotent element $u \in \mathcal{N}$ consider the Springer fiber Sp_u. The condition that $\gamma u \gamma^{-1} = qu$ means that u, and hence Sp_u, is stabilized by the action of $(\gamma, q) \in {^L G} \times \mathbb{C}^\times$ (see formula (8.7)). Let A be the smallest algebraic subgroup of $^L G \times \mathbb{C}^\times$ containing (γ, q). The algebra $K^{^L G \times \mathbb{C}^\times}(\mathrm{St})_{(\gamma, q)}$ naturally acts on the equivariant K-theory $K^A(\mathrm{Sp}_u)$ specialized at (γ, q),

$$ K^A(\mathrm{Sp}_u)_{(\gamma, q)} = K^A(\mathrm{Sp}_u) \underset{\mathrm{Rep}\, A}{\otimes} \mathbb{C}_{(\gamma, q)}. $$

It is known that $K^A(\mathrm{Sp}_u)_{(\gamma, q)}$ is isomorphic to the homology $H_\bullet(\mathrm{Sp}_u^\gamma)$ of the γ-fixed subset of Sp_u (see [KL, CG]). Thus, we obtain that $K^A(\mathrm{Sp}_u)_{(\gamma, q)}$ is a module over $H(G(F), I)$.

Unfortunately, these $H(G(F), I)$-modules are not irreducible in general, and one needs to work harder to describe the irreducible modules over $H(G(F), I)$. For $G = GL_n$ one can show that each of these modules has a unique irreducible quotient, and this way one recovers the bijection (8.4). But for a general groups G the finite groups $C(\gamma, u)$ come into play. Namely, the group $C(\gamma, u)$ acts on $K^A(\mathrm{Sp}_u)_{(\gamma, q)}$, and this action commutes with the action of $K^{^L G \times \mathbb{C}^\times}(\mathrm{St})_{(\gamma, q)}$. Therefore we have a decomposition

$$ K^A(\mathrm{Sp}_u)_{(\gamma, q)} = \bigoplus_{\rho \in \mathrm{Irrep}\, C(\gamma, u)} \rho \otimes K^A(\mathrm{Sp}_u)_{(\gamma, q, \rho)}, $$

of $K^A(\mathrm{Sp}_u)_{(\gamma, q)}$ as a representation of $C(\gamma, u) \times H(G(F), I)$. One shows (see [KL, CG] for details) that each $H(G(F), I)$-module $K^A(\mathrm{Sp}_u)_{(\gamma, q, \rho)}$ has a unique irreducible quotient, and this way one obtains a parameterization of irreducible modules by the triples appearing in the left hand side of (8.6). Therefore we obtain that the same set is in bijection with the right hand side of (8.6). This is how the tame local Langlands correspondence (8.6), also known as the Deligne–Langlands conjecture, is proved.

8.2 Categories Admitting $(\widehat{\mathfrak{g}}_{\kappa_c}, I)$ Harish-Chandra Modules

We now wish to find categorical analogues of the above results in the framework of the categorical Langlands correspondence for loop groups.

As we explained in Section 6.2, in the categorical setting a representation of $G(F)$ is replaced by a category $\widehat{\mathfrak{g}}_{\kappa_c}\text{-mod}_\chi$ equipped with an action of $G((t))$, and the space of I-invariant vectors is replaced by the subcategory of $(\widehat{\mathfrak{g}}_{\kappa_c}, I)$ Harish-Chandra modules in $\widehat{\mathfrak{g}}_{\kappa_c}\text{-mod}_\chi$. Hence the analogue of the question which representations of $G(F)$ admit non-zero I-invariant vectors becomes the following question: for what χ does the category $\widehat{\mathfrak{g}}_{\kappa_c}$ -mod$_\chi$ contain non-zero $(\widehat{\mathfrak{g}}_{\kappa_c}, I)$ Harish-Chandra modules?

To answer this question, we introduce the space $\text{Op}_{{}^L G}^{\text{RS}}(D)$ of **opers with regular singularity**. By definition (see [BD1], Sect. 3.8.8), an element of this space is an ${}^L N[[t]]$-conjugacy class of operators of the form

$$(8.9) \qquad \nabla = \partial_t + t^{-1}(p_{-1} + \mathbf{v}(t)),$$

where $\mathbf{v}(t) \in {}^L\mathfrak{b}[[t]]$. One can show that a natural map $\text{Op}_{{}^L G}^{\text{RS}}(D) \to \text{Op}_{{}^L G}(D^\times)$ is an embedding.

Following [BD1], we associate to an oper with regular singularity its **residue**. For an operator (8.9) the residue is by definition equal to $p_{-1} + \mathbf{v}(0)$. Clearly, under gauge transformations by an element $x(t)$ of ${}^L N[[t]]$ the residue gets conjugated by $x(0) \in N$. Therefore its projection onto

$$ {}^L\mathfrak{g}/{}^L G = \text{Spec}(\text{Fun}\, {}^L\mathfrak{g})^{{}^L G} = \text{Spec}(\text{Fun}\, {}^L\mathfrak{h})^W = \mathfrak{h}^*/W $$

is well-defined.

Given $\mu \in \mathfrak{h}^*$, we write $\varpi(\mu)$ for the projection of μ onto $\mathfrak{h}^*/W$. Finally, let P be the set of integral (not necessarily dominant) weights of $\mathfrak{g}$, viewed as a subset of $\mathfrak{h}^*$. The next result follows from [F3, FG2].

Lemma 4. *The category* $\widehat{\mathfrak{g}}_{\kappa_c}$ -mod$_\chi$ *contains a non-zero* $(\widehat{\mathfrak{g}}_{\kappa_c}, I)$ *Harish-Chandra module if and only if*

$$(8.10) \qquad \chi \in \bigsqcup_{\nu \in P/W} \text{Op}_{{}^L G}^{\text{RS}}(D)_{\varpi(\nu)}.$$

Thus, the opers χ for which the corresponding category $\widehat{\mathfrak{g}}_{\kappa_c}$ -mod$_\chi$ contain non-trivial I-equivariant objects are precisely the points of the subscheme (8.10) of $\text{Op}_{{}^L G}(D^\times)$. The next question is what are the corresponding ${}^L G$-local systems.

Let $\text{Loc}_{{}^L G}^{\text{RS,uni}} \subset \text{Loc}_{{}^L G}(D^\times)$ be the locus of ${}^L G$-local systems on $D^\times$ with regular singularity and unipotent monodromy. Such a local system is determined, up to an isomorphism, by the conjugacy class of its monodromy (see, e.g., [BV], Sect. 8). Therefore $\text{Loc}_{{}^L G}^{\text{RS,uni}}$ is an algebraic stack isomorphic to $\mathcal{N}/{}^L G$. The following result is proved in a way similar to the proof of Lemma 2.

Lemma 5. *If the local system underlying an oper* $\chi \in \text{Op}_{{}^L G}(D^\times)$ *belongs to* $\text{Loc}_{{}^L G}^{\text{RS,uni}}$, *then* χ *belongs to the subset* (8.10) *of* $\text{Op}_{{}^L G}(D^\times)$.

Indeed, the subscheme (8.10) is precisely the (set-theoretic) preimage of $\text{Loc}_{{}^L G}^{\text{RS,uni}} \subset \text{Loc}_{{}^L G}(D^\times)$ under the map $\alpha : \text{Op}_{{}^L G}(D^\times) \to \text{Loc}_{{}^L G}(D^\times)$.

This hardly comes as a surprise. Indeed, by analogy with the classical Langlands correspondence we expect that the categories $\widehat{\mathfrak{g}}_{\kappa_c}$-$\mathrm{mod}_\chi$ containing non-trivial I-equivariant objects correspond to the Langlands parameters which are the geometric counterparts of tamely ramified homomorphisms $W_F' \to {}^L G$. The most obvious candidates for those are precisely the ${}^L G$-local systems on $D^\times$ with regular singularity and unipotent monodromy. For this reason we will call such local systems **tamely ramified**.

Let us summarize: suppose that σ is a tamely ramified ${}^L G$-local system on $D^\times$, and let χ be a ${}^L G$-oper that is in the gauge equivalence class of σ. Then χ belongs to the subscheme (8.10), and the corresponding category $\widehat{\mathfrak{g}}_{\kappa_c}$-$\mathrm{mod}_\chi$ contains non-zero I-equivariant objects, by Lemma 4. Let $\widehat{\mathfrak{g}}_{\kappa_c}$-$\mathrm{mod}_\chi^I$ be the corresponding category of I-equivariant (or, equivalently, $(\widehat{\mathfrak{g}}_{\kappa_c}, I)$ Harish-Chandra) modules. Note that according to Conjecture 2, the categories $\widehat{\mathfrak{g}}_{\kappa_c}$-$\mathrm{mod}_\chi$ (resp., $\widehat{\mathfrak{g}}_{\kappa_c}$-$\mathrm{mod}_\chi^I$) should be equivalent to each other for all χ which are gauge equivalent to each other as ${}^L G$-local systems.

In the next section, following [FG2], we will give a conjectural description of the categories $\widehat{\mathfrak{g}}_{\kappa_c}$-$\mathrm{mod}_\chi^I$ for $\chi \in \mathrm{Op}_{LG}^{RS}(D)_{\varpi(-\rho)}$ in terms of the category of coherent sheaves on the Springer fiber corresponding to the residue of χ. This description in particular implies that at least the derived categories of these categories are equivalent to each other for the opers corresponding to the same local system. We have a similar conjecture for $\chi \in \mathrm{Op}_{LG}^{RS}(D)_{\varpi(\nu)}$ for other $\nu \in P$, which the reader may easily reconstruct from our discussion of the case $\nu = -\rho$.

8.3 Conjectural Description of the Categories of $(\widehat{\mathfrak{g}}_{\kappa_c}, I)$ Harish-Chandra Modules

Let us consider one of the connected components of the subscheme (8.10), namely, $\mathrm{Op}_{LG}^{RS}(D)_{\varpi(-\rho)}$. Here it will be convenient to use a different realization of this space, as the space $\mathrm{Op}_{LG}^{\mathrm{nilp}}$ of **nilpotent opers** introduced in [FG2]. By definition, an element of this space is an ${}^L N[[t]]$-gauge equivalence class of operators of the form

$$(8.11) \qquad \nabla = \partial_t + p_{-1} + \mathbf{v}(t) + \frac{v}{t},$$

where $\mathbf{v}(t) \in {}^L\mathfrak{b}[[t]]$ and $v \in {}^L\mathfrak{n}$. It is shown in [FG2] that $\mathrm{Op}_{LG}^{\mathrm{nilp}} \simeq \mathrm{Op}_{LG}^{RS}(D)_{\varpi(-\rho)}$. In particular, $\mathrm{Op}_{LG}^{\mathrm{nilp}}$ is a subspace of $\mathrm{Op}_{LG}(D^\times)$.

We have the (secondary) residue map

$$\mathrm{Res} : \mathrm{Op}_{LG}^{\mathrm{nilp}} \to {}^L\mathfrak{n}_{\mathcal{F}_{LB,0}} = \mathcal{F}_{LB,0} \underset{LB}{\times} {}^L\mathfrak{n},$$

sending a gauge equivalence class of operators (8.11) to v. By abuse of notation, we will denote the corresponding map

$$\mathrm{Op}_{LG}^{\mathrm{nilp}} \to {}^L\mathfrak{n}/{}^L B = \widetilde{\mathcal{N}}/{}^L G$$

also by Res.

For any $\chi \in \mathrm{Op}^{\mathrm{nilp}}_{^L G}$ the $^L G$-gauge equivalence class of the corresponding connection is a tamely ramified $^L G$-local system on $D^\times$. Moreover, its monodromy conjugacy class is equal to $\exp(2\pi i \, \mathrm{Res}(\chi))$.

We wish to describe the category $\widehat{\mathfrak{g}}_{\kappa_c}\text{-mod}^I_\chi$ of $(\widehat{\mathfrak{g}}_{\kappa_c}, I)$ Harish-Chandra modules with the central character $\chi \in \mathrm{Op}^{\mathrm{nilp}}_{^L G}$. However, here we face the first major complication as compared to the unramified case. While in the ramified case we worked with the abelian category $\widehat{\mathfrak{g}}_{\kappa_c}\text{-mod}^{G[[t]]}_\chi$, this does not seem to be possible in the tamely ramified case. So from now on we will work with the appropriate derived category $D^b(\widehat{\mathfrak{g}}_{\kappa_c}\text{-mod}_\chi)^I$. By definition, this is the full subcategory of the bounded derived category $D^b(\widehat{\mathfrak{g}}_{\kappa_c}\text{-mod}_\chi)$ whose objects are complexes with cohomologies in $\widehat{\mathfrak{g}}_{\kappa_c}\text{-mod}^I_\chi$.

Roughly speaking, the conjecture of [FG2] is that $D^b(\widehat{\mathfrak{g}}_{\kappa_c}\text{-mod}_\chi)^I$ is equivalent to $D^b(\mathrm{QCoh}(\mathrm{Sp}_{\mathrm{Res}(\chi)}))$, where $\mathrm{QCoh}(\mathrm{Sp}_{\mathrm{Res}(\chi)})$ is the category of quasi-coherent sheaves on the Springer fiber of $\mathrm{Res}(\chi)$. However, we need to make some adjustments to this statement. These adjustments are needed to arrive at a "nice" statement, Conjecture 6 below. We now explain what these adjustments are the reasons behind them.

The first adjustment is that we need to consider a slightly larger category of representations than $D^b(\widehat{\mathfrak{g}}_{\kappa_c}\text{-mod}_\chi)^I$. Namely, we wish to include extensions of I-equivariant $\widehat{\mathfrak{g}}_{\kappa_c}$-modules which are not necessarily I-equivariant, but only I^0-equivariant, where $I^0 = [I, I]$. To explain this more precisely, let us choose a Cartan subgroup $H \subset B \subset I$ and the corresponding Lie subalgebra $\mathfrak{h} \subset \mathfrak{b} \subset \mathrm{Lie}\, I$. We then have an isomorphism $I = H \ltimes I^0$. An I-equivariant $\widehat{\mathfrak{g}}_{\kappa_c}$-module is the same as a module on which $\mathfrak{h}$ acts diagonally with eigenvalues given by integral weights and the Lie algebra $\mathrm{Lie}\, I^0$ acts locally nilpotently. However, there may exist extensions between such modules on which the action of $\mathfrak{h}$ is no longer semi-simple. Such modules are called I-**monodromic**. More precisely, an I-monodromic $\widehat{\mathfrak{g}}_{\kappa_c}$-module is a module that admits an increasing filtration whose consecutive quotients are I-equivariant. It is natural to include such modules in our category. However, it is easy to show that an I-monodromic object of $\widehat{\mathfrak{g}}_{\kappa_c}\text{-mod}_\chi$ is the same as an I^0-equivariant object of $\widehat{\mathfrak{g}}_{\kappa_c}\text{-mod}_\chi$ for any $\chi \in \mathrm{Op}^{\mathrm{nilp}}_{^L G}$ (see [FG2]). Therefore instead of I-monodromic modules we will use I^0-equivariant modules. Denote by $D^b(\widehat{\mathfrak{g}}_{\kappa_c}\text{-mod}_\chi)^{I^0}$ the the full subcategory of $D^b(\widehat{\mathfrak{g}}_{\kappa_c}\text{-mod}_\chi)$ whose objects are complexes with the cohomologies in $\widehat{\mathfrak{g}}_{\kappa_c}\text{-mod}^{I^0}_\chi$.

The second adjustment has to do with the non-flatness of the Springer resolution $\widetilde{\mathcal{N}} \to \mathcal{N}$. By definition, the Springer fiber Sp_u is the fiber product $\widetilde{\mathcal{N}} \underset{\mathcal{N}}{\times} \mathrm{pt}$, where pt is the point $u \in \mathcal{N}$. This means that the structure sheaf of Sp_u is given by

$$(8.12) \qquad \mathcal{O}_{\mathrm{Sp}_u} = \mathcal{O}_{\widetilde{\mathcal{N}}} \underset{\mathcal{O}_{\mathcal{N}}}{\otimes} \mathbb{C}.$$

However, because the morphism $\widetilde{\mathcal{N}} \to \mathcal{N}$ is not flat, this tensor product functor is not left exact, and there are non-trivial higher derived tensor products (the Tor's). Our (conjectural) equivalence is not going to be an exact functor: it sends a general object of the category $\widehat{\mathfrak{g}}_{\kappa_c}$-mod$_{\chi}^{I^0}$ not to an object of the category of quasicoherent sheaves, but to a complex of sheaves, or, more precisely, an object of the corresponding derived category. Hence we are forced to work with derived categories, and so the higher derived tensor products need to be taken into account.

To understand better the consequences of this non-exactness, let us consider the following model example. Suppose that we have established an equivalence between the derived category $D^b(\mathrm{QCoh}(\widetilde{\mathcal{N}}))$ and another derived category $D^b(\mathcal{C})$. In particular, this means that both categories carry an action of the algebra $\mathrm{Fun}\,\mathcal{N}$ (recall that $\mathcal{N}$ is an affine algebraic variety). Let us suppose that the action of $\mathrm{Fun}\,\mathcal{N}$ on $D^b(\mathcal{C})$ comes from its action on the abelian category $\mathcal{C}$. Thus, $\mathcal{C}$ fibers over $\mathcal{N}$, and let $\mathcal{C}_u$ the fiber category corresponding to $u \in \mathcal{N}$. This is the full subcategory of $\mathcal{C}$ whose objects are objects of $\mathcal{C}$ on which the ideal of u in $\mathrm{Fun}\,\mathcal{N}$ acts by 0.[16] What is the category $D^b(\mathcal{C}_u)$ equivalent to?

It is tempting to say that it is equivalent to $D^b(\mathrm{QCoh}(\mathrm{Sp}_u))$. However, this does not follow from the equivalence of $D^b(\mathrm{QCoh}(\mathcal{N}))$ and $D^b(\mathcal{C})$ because of the tensor product (8.12) having non-trivial higher derived functors. The correct answer is that $D^b(\mathcal{C}_u)$ is equivalent to the category $D^b(\mathrm{QCoh}(\mathrm{Sp}_u^{\mathrm{DG}}))$, where $\mathrm{Sp}_u^{\mathrm{DG}}$ is the "DG fiber" of $\widetilde{\mathcal{N}} \to \mathcal{N}$ at u. By definition, a quasicoherent sheaf on $\mathrm{Sp}_u^{\mathrm{DG}}$ is a DG module over the DG algebra

$$(8.13) \qquad \mathcal{O}_{\mathrm{Sp}_u^{\mathrm{DG}}} = \mathcal{O}_{\widetilde{\mathcal{N}}} \overset{L}{\underset{\mathcal{O}_{\mathcal{N}}}{\otimes}} \mathcal{C}_u,$$

where we now take the full derived functor of tensor product. Thus, the category $D^b(\mathrm{QCoh}(\mathrm{Sp}_u^{\mathrm{DG}}))$ may be thought of as the derived category of quasicoherent sheaves on the "DG scheme" $\mathrm{Sp}_u^{\mathrm{DG}}$ (see [CK] for a precise definition of DG scheme).

Finally, the last adjustment is that we should consider the non-reduced Springer fibers. This means that instead of the Springer resolution $\widetilde{\mathcal{N}}$ we should consider the "thickened" Springer resolution

$$\widetilde{\widetilde{\mathcal{N}}} = {}^L\widetilde{\mathfrak{g}} \underset{{}^L\mathfrak{g}}{\times} \mathcal{N},$$

where ${}^L\widetilde{\mathfrak{g}}$ is the so-called **Grothendieck alteration**,

$${}^L\widetilde{\mathfrak{g}} = \{x \in {}^L\mathfrak{g}, \mathfrak{b}' \in {}^LG/{}^LB \mid x \in \mathfrak{b}'\}.$$

The variety $\widetilde{\widetilde{\mathcal{N}}}$ is non-reduced, and the underlying reduced variety is the Springer resolution $\widetilde{\mathcal{N}}$. For instance, the fiber of $\widetilde{\widetilde{\mathcal{N}}}$ over a regular element

---

[16] The relationship between $\mathcal{C}$ and $\mathcal{C}_u$ is similar to the relationship between $\widehat{\mathfrak{g}}_{\kappa_c}$-mod and and $\widehat{\mathfrak{g}}_{\kappa_c}$-mod$_{\chi}$, where $\chi \in \mathrm{Op}_{{}^LG}(D^{\times})$.

in $\mathcal{N}$ consists of a single point, but the corresponding fiber of $\widetilde{\mathcal{N}}$ is the spectrum of the Artinian ring $h_0 = \mathrm{Fun}\,{}^L\mathfrak{h}/(\mathrm{Fun}\,{}^L\mathfrak{h})^W_+$. Here $(\mathrm{Fun}\,{}^L\mathfrak{h})^W_+$ is the ideal in $\mathrm{Fun}\,{}^L\mathfrak{h}$ generated by the augmentation ideal of the subalgebra of W-invariants. Thus, $\mathrm{Spec}\,h_0$ is the scheme-theoretic fiber of $\varpi : {}^L\mathfrak{h} \to {}^L\mathfrak{h}/W$ at 0. It turns out that in order to describe the category $D^b(\widehat{\mathfrak{g}}_{\kappa_c}\text{-mod}_\chi)^{I^0}$ we need to use the "thickened" Springer resolution.

Let us summarize: in order to construct the sought-after equivalence of categories we take, instead of individual Springer fibers, the whole Springer resolution, and we further replace it by the "thickened" Springer resolution $\widetilde{\mathcal{N}}$ defined above. In this version we will be able to formulate our equivalence in such a way that we avoid DG schemes.

This means that instead of considering the categories $\widehat{\mathfrak{g}}_{\kappa_c}\text{-mod}_\chi$ for individual nilpotent opers χ, we should consider the "universal" category $\widehat{\mathfrak{g}}_{\kappa_c}\text{-mod}_{\mathrm{nilp}}$ which is the "family version" of all of these categories. By definition, the category $\widehat{\mathfrak{g}}_{\kappa_c}\text{-mod}_{\mathrm{nilp}}$ is the full subcategory of $\widehat{\mathfrak{g}}_{\kappa_c}\text{-mod}$ whose objects have the property that the action of $Z(\widehat{\mathfrak{g}}) = \mathrm{Fun}\,\mathrm{Op}_{{}^LG}(D)$ on them factors through the quotient $\mathrm{Fun}\,\mathrm{Op}_{{}^LG}(D) \to \mathrm{Fun}\,\mathrm{Op}_{{}^LG}^{\mathrm{nilp}}$. Thus, the category $\widehat{\mathfrak{g}}_{\kappa_c}\text{-mod}_{\mathrm{nilp}}$ is similar to the category $\widehat{\mathfrak{g}}_{\kappa_c}\text{-mod}_{\mathrm{reg}}$ that we have considered above. While the former fibers over $\mathrm{Op}_{{}^LG}^{\mathrm{nilp}}$, the latter fibers over $\mathrm{Op}_{{}^LG}(D)$. The individual categories $\widehat{\mathfrak{g}}_{\kappa_c}\text{-mod}_\chi$ are now realized as fibers of these categories over particular opers χ.

Our naive idea was that for each $\chi \in \mathrm{Op}_{{}^LG}^{\mathrm{nilp}}$ the category $D^b(\widehat{\mathfrak{g}}_{\kappa_c}\text{-mod}_\chi)^{I^0}$ is equivalent to $\mathrm{QCoh}(\mathrm{Sp}_{\mathrm{Res}(\chi)})$. We would like to formulate now a "family version" of such an equivalence. To this end we form the fiber product

$$^L\widetilde{\mathfrak{n}} = {}^L\widetilde{\mathfrak{g}} \underset{{}^L\mathfrak{g}}{\times} {}^L\mathfrak{n}.$$

It turns out that this fiber product does not suffer from the problem of the individual Springer fibers, as the following lemma shows:

Lemma 6 ([FG2],Lemma 6.4). *The derived tensor product*

$$\mathrm{Fun}\,{}^L\widetilde{\mathfrak{g}} \overset{L}{\underset{\mathrm{Fun}\,{}^L\mathfrak{g}}{\otimes}} \mathrm{Fun}\,{}^L\mathfrak{n}$$

is concentrated in cohomological dimension 0.

The variety $^L\widetilde{\mathfrak{n}}$ may be thought of as the family of (non-reduced) Springer fibers parameterized by $^L\mathfrak{n} \subset {}^L\mathfrak{g}$. It is important to note that it is singular, reducible and non-reduced. For example, if $\mathfrak{g} = \mathfrak{sl}_2$, it has two components, one of which is $\mathbb{P}^1$ (the Springer fiber at 0) and the other is the doubled affine line (i.e., $\mathrm{Spec}\,\mathbb{C}[x,y]/(y^2)$).

We note that the corresponding reduced scheme is

$$(8.14) \qquad\qquad {}^L\widetilde{\mathfrak{n}} = \widetilde{\mathcal{N}} \underset{\mathcal{N}}{\times} {}^L\mathfrak{n}.$$

However, the derived tensor product corresponding to (8.14) is not concentrated in cohomological dimension 0, and this is the reason why we prefer to use $^L\widetilde{\mathfrak{n}}$ rather than $^L\widetilde{\mathfrak{n}}$.

Now we set

$$\mathrm{MOp}_{^LG}^{\mathrm{nilp}} = \mathrm{Op}_{^LG}^{\mathrm{nilp}} \underset{^L\mathfrak{n}/^LB}{\times} {}^L\widetilde{\mathfrak{n}}/^LB,$$

where we use the residue morphism $\mathrm{Res} : \mathrm{Op}_{^LG}^{\mathrm{nilp}} \to {}^L\mathfrak{n}/^LB$. Thus, informally $\mathrm{MOp}_{^LG}^{\mathrm{nilp}}$ may be thought as the family over $\mathrm{Op}_{^LG}^{\mathrm{nilp}}$ whose fiber over $\chi \in \mathrm{Op}_{^LG}^{\mathrm{nilp}}$ is the (non-reduced) Springer fiber of $\mathrm{Res}(\chi)$.

The space $\mathrm{MOp}_{^LG}^{\mathrm{nilp}}$ is the space of **Miura opers** whose underlying opers are nilpotent, introduced in [FG2].

We also introduce the category $\widehat{\mathfrak{g}}_{\kappa_c}\text{-mod}_{\mathrm{nilp}}^{I^0}$ which is a full subcategory of $\widehat{\mathfrak{g}}_{\kappa_c}\text{-mod}_{\mathrm{nilp}}$ whose objects are I^0-equivariant. Let $D^b(\widehat{\mathfrak{g}}_{\kappa_c}\text{-mod}_{\mathrm{nilp}})^{I^0}$ be the corresponding derived category.

Now we can formulate the Main Conjecture of [FG2]:

Conjecture 6. *There is an equivalence of categories*

$$(8.15) \qquad D^b(\widehat{\mathfrak{g}}_{\kappa_c}\text{-mod}_{\mathrm{nilp}})^{I^0} \simeq D^b(\mathrm{QCoh}(\mathrm{MOp}_{^LG}^{\mathrm{nilp}}))$$

which is compatible with the action of the algebra $\mathrm{Fun}\,\mathrm{Op}_{^LG}^{\mathrm{nilp}}$ *on both categories.*

Note that the action of $\mathrm{Fun}\,\mathrm{Op}_{^LG}^{\mathrm{nilp}}$ on the first category comes from the action of the center $Z(\widehat{\mathfrak{g}})$, and on the second category it comes from the fact that $\mathrm{MOp}_{^LG}^{\mathrm{nilp}}$ is a scheme over $\mathrm{Op}_{^LG}^{\mathrm{nilp}}$.

Another important remark is that the equivalence (8.15) does not preserve the t-structures on the two categories. In other words, (8.15) is expected in general to map objects of the abelian category $\widehat{\mathfrak{g}}_{\kappa_c}\text{-mod}_{\mathrm{nilp}}^{I^0}$ to complexes in $D^b(\mathrm{QCoh}(\mathrm{MOp}_{^LG}^{\mathrm{nilp}}))$, and vice versa.

There are similar conjectures for the categories corresponding to the spaces $\mathrm{Op}_{^LG}^{\mathrm{nilp},\lambda}$ of nilpotent opers with dominant integral weights $\lambda \in P^+$.

In the next section we will discuss the connection between Conjecture 6 and the classical tamely ramified Langlands correspondence. We then present some evidence for this conjecture.

8.4 Connection between the Classical and the Geometric Settings

Let us discuss the connection between the equivalence (8.15) and the realization of representations of affine Hecke algebras in terms of K-theory of the Springer fibers. As we have explained, we would like to view the category $D^b(\widehat{\mathfrak{g}}_{\kappa_c}\text{-mod}_\chi)^{I^0}$ for $\chi \in \mathrm{Op}_{^LG}^{\mathrm{nilp}}$ as, roughly, a categorification of the space $R^{\pi(I)}$ of I-invariant vectors in an irreducible representation (R, π) of

$G(F)$. Therefore, we expect that the Grothendieck group of the category $D^b(\widehat{\mathfrak{g}}_{\kappa_c}\text{-mod}_\chi)^{I^0}$ is somehow related to the space $R^{\pi(I)}$.

Let us try to specialize the statement of Conjecture 6 to a particular oper

$$\chi = (\mathcal{F}, \nabla, \mathcal{F}_{L_B}) \in \mathrm{Op}^{\mathrm{nilp}}_{L_G}.$$

Let $\widetilde{\mathrm{Sp}}^{\mathrm{DG}}_{\mathrm{Res}(\chi)}$ be the DG fiber of $\mathrm{MOp}^{\mathrm{nilp}}_{L_G}$ over χ. By definition (see Section 8.3), the residue $\mathrm{Res}(\chi)$ of χ is a vector in the twist of $^L\mathfrak{n}$ by the $^L B$-torsor $\mathcal{F}_{L_{B,0}}$. It follows that $\widetilde{\mathrm{Sp}}^{\mathrm{DG}}_{\mathrm{Res}(\chi)}$ is the DG fiber over $\mathrm{Res}(\chi)$ of the $\mathcal{F}_{L_{B,0}}$-twist of the Grothendieck alteration.

If we trivialize $\mathcal{F}_{L_{B,0}}$, then $u = \mathrm{Res}(\chi)$ becomes an element of $^L\mathfrak{n}$. By definition, the (non-reduced) DG Springer fiber $\widetilde{\mathrm{Sp}}^{\mathrm{DG}}_u$ is the DG fiber of the Grothendieck alteration $^L\widetilde{\mathfrak{g}} \to {}^L\mathfrak{g}$ at u. In other words, the corresponding structure sheaf is the DG algebra

$$\mathcal{O}_{\widetilde{\mathrm{Sp}}^{\mathrm{DG}}_u} = \mathcal{O}_{L\widetilde{\mathfrak{g}}} \overset{L}{\underset{\mathcal{O}_{L_\mathfrak{g}}}{\otimes}} \mathbb{C}_u$$

(compare with formula (8.13)).

To see what these DG fibers look like, let $u = 0$. Then the naive Springer fiber is just the flag variety $^L G/^L B$ (it is reduced in this case), and $\mathcal{O}_{\widetilde{\mathrm{Sp}}_0}$ is the structure sheaf of $^L G/^L B$. But the sheaf $\mathcal{O}_{\widetilde{\mathrm{Sp}}^{\mathrm{DG}}_0}$ is a sheaf of DG algebras, which is quasi-isomorphic to the complex of differential forms on $^L G/^L B$, with the zero differential. In other words, $\widetilde{\mathrm{Sp}}^{\mathrm{DG}}_0$ may be viewed as a "$\mathbb{Z}$-graded manifold" such that the corresponding supermanifold, obtained by replacing the $\mathbb{Z}$-grading by the corresponding $\mathbb{Z}/2\mathbb{Z}$-grading, is $\Pi T(^L G/^L B)$, the tangent bundle to $^L G/^L B$ with the parity of the fibers changed from even to odd.

We expect that the category $\widehat{\mathfrak{g}}_{\kappa_c}\text{-mod}^{I^0}_{\mathrm{nilp}}$ is flat over $\mathrm{Op}^{\mathrm{nilp}}_{L_G}$. Therefore, specializing Conjecture 6 to a particular oper $\chi \in \mathrm{Op}^{\mathrm{nilp}}_{L_G}$, we obtain as a corollary an equivalence of categories

$$(8.16) \qquad D^b(\widehat{\mathfrak{g}}_{\kappa_c}\text{-mod}_\chi)^{I^0} \simeq D^b(\mathrm{QCoh}(\widetilde{\mathrm{Sp}}^{\mathrm{DG}}_{\mathrm{Res}(\chi)})).$$

This bodes well with Conjecture 2 saying that the categories $\widehat{\mathfrak{g}}_{\kappa_c}\text{-mod}_{\chi_1}$ and $\widehat{\mathfrak{g}}_{\kappa_c}\text{-mod}_{\chi_2}$ (and hence $D^b(\widehat{\mathfrak{g}}_{\kappa_c}\text{-mod}_{\chi_1})^{I^0}$ and $D^b(\widehat{\mathfrak{g}}_{\kappa_c}\text{-mod}_{\chi_2})^{I^0}$) should be equivalent if the underlying local systems of the opers χ_1 and χ_2 are isomorphic. For nilpotent opers χ_1 and χ_2 this is so if and only if their monodromies are conjugate to each other. Since their monodromies are obtained by exponentiating their residues, this is equivalent to saying that the residues, $\mathrm{Res}(\chi_1)$ and $\mathrm{Res}(\chi_2)$, are conjugate with respect to the $\mathcal{F}_{L_{B,0}}$-twist of $^L G$. But in this case the DG Springer fibers corresponding to χ_1 and χ_2 are also isomorphic, and so $D^b(\widehat{\mathfrak{g}}_{\kappa_c}\text{-mod}_{\chi_1})^{I^0}$ and $D^b(\widehat{\mathfrak{g}}_{\kappa_c}\text{-mod}_{\chi_2})^{I^0}$ are equivalent to each other, by (8.16).

The Grothendieck group of the category $D^b(\mathrm{QCoh}(\widetilde{\mathrm{Sp}}_u^{\mathrm{DG}}))$, where u is any nilpotent element, is the same as the Grothendieck group of $\mathrm{QCoh}(\mathrm{Sp}_u)$. In other words, the Grothendieck group does not "know" about the DG or the non-reduced structure of $\widetilde{\mathrm{Sp}}_u^{\mathrm{DG}}$. Hence it is nothing but the algebraic K-theory $K(\mathrm{Sp}_u)$. As we explained at the end of Section 8.1, equivariant variants of this algebraic K-theory realize the "standard modules" over the affine Hecke algebra $H(G(F), I)$. Moreover, the spaces of I-invariant vectors $R^{\pi(I)}$ as above, which are naturally modules over the affine Hecke algebra, may be realized as subquotients of $K(\mathrm{Sp}_u)$. This indicates that the equivalences (8.16) and (8.15) are compatible with the classical results.

However, at first glance there are some important differences between the classical and the categorical pictures, which we now discuss in more detail.

In the construction of $H(G(F), I)$-modules outlined in Section 8.1 we had to pick a semi-simple element γ of $^L G$ such that $\gamma u \gamma^{-1} = qu$, where q is the number of elements in the residue field of F. Then we consider the specialized A-equivariant K-theory $K^A(\mathrm{Sp}_u)_{(\gamma, q)}$ where A is the the smallest algebraic subgroup of $^L G \times \mathbb{C}^\times$ containing (γ, q). This gives $K(\mathrm{Sp}_u)$ the structure of an $H(G(F), I)$-module. But this module carries a residual symmetry with respect to the group $C(\gamma, u)$ of components of the centralizer of γ and u in $^L G$, which commutes with the action of $H(G(F), I)$. Hence we consider the $H(G(F), I)$-module

$$K^A(\mathrm{Sp}_u)_{(\gamma, q, \rho)} = \mathrm{Hom}_{C(\gamma, u)}(\rho, K(\mathrm{Sp}_u)),$$

corresponding to an irreducible representation ρ of $C(\gamma, u)$. Finally, each of these components has a unique irreducible quotient, and this is an irreducible representation of $H(G(F), I)$ which is realized on the space $R^{\pi(I)}$, where (R, π) is an irreducible representation of $G(F)$ corresponding to (γ, u, ρ) under the bijection (8.6). How is this intricate structure reflected in the categorical setting?

Our category $D^b(\mathrm{QCoh}(\widetilde{\mathrm{Sp}}_u^{\mathrm{DG}}))$, where $u = \mathrm{Res}(\chi)$, is a particular categorification of the (non-equivariant) K-theory $K(\mathrm{Sp}_u)$. Note that in the classical local Langlands correspondence (8.6) the element u of the triple (γ, u, ρ) is interpreted as the logarithm of the monodromy of the corresponding representation of the Weil-Deligne group W_F'. This is in agreement with the interpretation of $\mathrm{Res}(\chi)$ as the logarithm of the monodromy of the $^L G$-local system on $D^\times$ corresponding to χ, which plays the role of the local Langlands parameter for the category $\widehat{\mathfrak{g}}_{\kappa_c}$-$\mathrm{mod}_\chi$ (up to the inessential factor $2\pi i$).

But what about the other parameters, γ and ρ? And why does our category correspond to the non-equivariant K-theory of the Springer fiber, and not the equivariant K-theory, as in the classical setting?

The element γ corresponding to the Frobenius in W_F' does not seem to have an analogue in the geometric setting. We have already seen this above in the unramified case: while in the classical setting unramified local Langlands parameters are the semi-simple conjugacy classes γ in $^L G$, in the geometric setting we have only one unramified local Langlands parameter, namely, the trivial local system.

To understand better what's going on here, we revisit the unramified case. Recall that the spherical Hecke algebra $H(G(F), K_0)$ is isomorphic to the representation ring $\text{Rep}\,^L G$. The one-dimensional space of K_0-invariants in an irreducible unramified representation (R, π) of $G(F)$ realizes a one-dimensional representation of $H(G(F), K_0)$, i.e., a homomorphism $\text{Rep}\,^L G \to \mathbb{C}$. The unramified Langlands parameter γ of (R, π), which is a semi-simple conjugacy class in $^L G$, is the point in $\text{Spec}(\text{Rep}\,^L G)$ corresponding to this homomorphism. What is a categorical analogue of this homomorphism? The categorification of $\text{Rep}\,^L G$ is the category $\mathcal{R}ep\,^L G$. The product structure on $\text{Rep}\,^L G$ is reflected in the structure of tensor category on $\mathcal{R}ep\,^L G$. On the other hand, the categorification of the algebra $\mathbb{C}$ is the category $\mathcal{V}ect$ of vector spaces. Therefore a categorical analogue of a homomorphism $\text{Rep}\,^L G \to \mathbb{C}$ is a functor $\mathcal{R}ep\,^L G \to \mathcal{V}ect$ respecting the tensor structures on both categories. Such functors are called the fiber functors. The fiber functors form a category of their own, which is equivalent to the category of $^L G$-torsors. Thus, any two fiber functors are isomorphic, but not canonically. In particular, the group of automorphisms of each fiber functor is isomorphic to $^L G$. (Incidentally, this is how $^L G$ is reconstructed from a fiber functor in the Tannakian formalism.) Thus, we see that while in the categorical world we do not have analogues of semi-simple conjugacy classes γ (the points of $\text{Spec}(\text{Rep}\,^L G)$), their role is in some sense played by the group of automorphisms of a fiber functor.

This is reflected in the fact that while in the categorical setting we have a unique unramified Langlands parameter, namely, the trivial $^L G$-local system σ_0 on $D^\times$, this local system has a non-trivial group of automorphisms, namely, $^L G$. We therefore expect that the group $^L G$ should act by automorphisms of the Langlands category $\mathcal{C}_{\sigma_0}$ corresponding to σ_0, and this action should commute with the action of the loop group $G((t))$ on $\mathcal{C}_{\sigma_0}$. It is this action of $^L G$ that is meant to compensate for the lack of unramified Langlands parameters, as compared to the classical setting.

We have argued in Section 7 that the category $\widehat{\mathfrak{g}}_{\kappa_c}\text{-mod}_\chi$, where $\chi = (\mathcal{F}, \nabla, \mathcal{F}_{L_B}) \in \text{Op}_{L_G}(D)$, is a candidate for the Langlands category $\mathcal{C}_{\sigma_0}$. Therefore we expect that the group $^L G$ (more precisely, its twist $^L G_{\mathcal{F}}$) acts on the category $\widehat{\mathfrak{g}}_{\kappa_c}\text{-mod}_\chi$. In Section 7.6 we showed how to obtain this action using the conjectural equivalence between $\widehat{\mathfrak{g}}_{\kappa_c}\text{-mod}_\chi$ and the category $\mathcal{D}_{\kappa_c}^{\text{Hecke}}\text{-mod}_\chi$ of Hecke eigenmodules on the affine Grassmannian Gr (see Conjecture 3). The category $\mathcal{D}_{\kappa_c}^{\text{Hecke}}\text{-mod}_\chi$ was defined in Section 7.5 as a "de-equivariantization" of the category $\mathcal{D}_{\kappa_c}\text{-mod}$ of twisted $\mathcal{D}$-modules on Gr with respect to the monoidal action of the category $\mathcal{R}ep\,^L G$.

Now comes a crucial observation which will be useful for understanding the way things work in the tamely ramified case: the category $\mathcal{R}ep\,^L G$ may be interpreted as the category of $^L G$-equivariant quasicoherent sheaves on the variety $\text{pt} = \text{Spec}\,\mathbb{C}$. In other words, $\mathcal{R}ep\,^L G$ may be interpreted as the category of quasicoherent sheaves on the stack $\text{pt}\,/^L G$. The existence of monoidal action of the category $\mathcal{R}ep\,{}^L G$ on $\mathcal{D}_{\kappa_c}$-mod should be viewed as the statement that the category $\mathcal{D}_{\kappa_c}$-mod "lives" over the stack pt $/{}^L G$. The statement of Conjecture 3 may then be interpreted as saying that

$$\widehat{\mathfrak{g}}_{\kappa_c}\text{-mod}_\chi \simeq \mathcal{D}_{\kappa_c}\text{-mod} \underset{\mathrm{pt}\,/{}^L G}{\times} \mathrm{pt}.$$

In other words, if $\mathcal{C}$ is the conjectural Langlands category fibering over the stack $\mathrm{Loc}_{LG}(D^\times)$ of all ${}^L G$-local systems on $D^\times$, then

$$\mathcal{D}_{\kappa_c}\text{-mod} \simeq \mathcal{C} \underset{\mathrm{Loc}_{LG}(D^\times)}{\times} \mathrm{pt}\,/{}^L G,$$

whereas

$$\widehat{\mathfrak{g}}_{\kappa_c}\text{-mod}_\chi \simeq \mathcal{C} \underset{\mathrm{Loc}_{LG}(D^\times)}{\times} \mathrm{pt},$$

where the morphism pt $\to \mathrm{Loc}_{LG}(D^\times)$ corresponds to the oper χ.

Thus, in the categorical setting there are two different ways to think about the trivial local system σ_0: as a point (defined by a particular ${}^L G$-bundle on D with connection, such as a regular oper χ), or as a stack pt $/{}^L G$. The base change of the Langlands category in the first case gives us a category with an action of ${}^L G$, such as the categories $\widehat{\mathfrak{g}}_{\kappa_c}$-mod$_\chi$ or $\mathcal{D}_{\kappa_c}^{\mathrm{Hecke}}$-mod. The base change in the second case gives us a category with a monoidal action of $\mathcal{R}ep\,{}^L G$, such as the category $\mathcal{D}_{\kappa_c}$-mod. We can go back and forth between the two by applying the procedures of equivariantization and de-equivariantization with respect to ${}^L G$ and $\mathcal{R}ep\,{}^L G$, respectively.

Now we return to the tamely ramified case. The semi-simple element γ appearing in the triple (γ, u, ρ) plays the same role as the unramified Langlands parameter γ. However, now it must satisfy the identity $\gamma u \gamma^{-1} = qu$. Recall that the center Z of $H(G(F), I)$ is isomorphic to Rep ${}^L G$, and so Spec Z is the set of all semi-simple elements in ${}^L G$. For a fixed nilpotent element u the equation $\gamma u \gamma^{-1} = qu$ cuts out a locus C_u in Spec Z corresponding to those central characters which may occur on irreducible $H(G(F), I)$-modules corresponding to u. In the categorical setting (where we set $q = 1$) the analogue of C_u is the centralizer $Z(u)$ of u in ${}^L G$, which is precisely the group $\mathrm{Aut}(\sigma)$ of automorphisms of a tame local system σ on $D^\times$ with monodromy $\exp(2\pi i u)$. On general grounds we expect that the group $\mathrm{Aut}(\sigma)$ acts on the Langlands category $\mathcal{C}_\sigma$, just as we expect the group ${}^L G$ of automorphisms of the trivial local system σ_0 to act on the category $\mathcal{C}_{\sigma_0}$. It is this action that replaces the parameter γ in the geometric setting.

In the classical setting we also have one more parameter, ρ. Let us recall that ρ is a representation of the group $C(\gamma, u)$ of connected components of the centralizer $Z(\gamma, u)$ of γ and u. But the group $Z(\gamma, u)$ is a subgroup of $Z(u)$, which becomes the group $\mathrm{Aut}(\sigma)$ in the geometric setting. Therefore one can argue that the parameter ρ is also absorbed into the action of $\mathrm{Aut}(\sigma)$ on the category $\mathcal{C}_\sigma$.

If we have an action of $\text{Aut}(\sigma)$ on the category $\mathcal{C}_\sigma$, or on one of its many incarnations $\widehat{\mathfrak{g}}_{\kappa_c}\text{-mod}_\chi, \chi \in \text{Op}_{L_G}^{\text{nilp}}$, it means that these categories must be "de-equivariantized", just like the categories $\widehat{\mathfrak{g}}_{\kappa_c}\text{-mod}_\chi, \chi \in \text{Op}_{L_G}(D)$, in the unramified case. This is the reason why in the equivalence (8.16) (and in Conjecture 6) we have the non-equivariant categories of quasicoherent sheaves (whose Grothendieck groups correspond to the non-equivariant K-theory of the Springer fibers).

However, there is also an equivariant version of these categories. Consider the substack of tamely ramified local systems in $\text{Loc}_{L_G}(D^\times)$ introduced in Section 8.2. Since a tamely ramified local system is completely determined by the logarithm of its (unipotent) monodromy, this substack is isomorphic to $\mathcal{N}/{}^L G$. This substack plays the role of the substack $\text{pt}/{}^L G$ corresponding to the trivial local system. Let us set

$$\mathcal{C}_{\text{tame}} = \mathcal{C} \underset{\text{Loc}_{L_G}(D^\times)}{\times} \mathcal{N}/{}^L G.$$

Then, according to our general conjecture expressed by the Cartesian diagram (5.3), we expect to have

$$(8.17) \qquad \widehat{\mathfrak{g}}_{\kappa_c}\text{-mod}_{\text{nilp}} \simeq \mathcal{C}_{\text{tame}} \underset{\mathcal{N}/{}^L G}{\times} \text{Op}_{L_G}^{\text{nilp}}.$$

Let $D^b(\mathcal{C}_{\text{tame}})^{I^0}$ be the I^0-equivariant derived category corresponding to $\mathcal{C}_{\text{tame}}$. Combining (8.17) with Conjecture 6, and noting that

$$\text{MOp}_{L_G}^{\text{nilp}} \simeq \text{Op}_{L_G}^{\text{nilp}} \underset{\mathcal{N}/{}^L G}{\times} \widetilde{\widetilde{\mathcal{N}}}/{}^L G,$$

we obtain the following conjecture (see [FG2]):

$$(8.18) \qquad D^b(\mathcal{C}_{\text{tame}})^{I^0} \simeq D^b(\text{QCoh}(\widetilde{\widetilde{\mathcal{N}}}/{}^L G)).$$

The category on the right hand side may be interpreted as the derived category of ${}^L G$-equivariant quasicoherent sheaves on the "thickened" Springer resolution $\widetilde{\widetilde{\mathcal{N}}}$.

Together, the conjectural equivalences (8.16) and (8.18) should be thought of as the categorical versions of the realizations of modules over the affine Hecke algebra in the K-theory of the Springer fibers.

One corollary of the equivalence (8.16) is the following: the classes of irreducible objects of the category $\widehat{\mathfrak{g}}_{\kappa_c}\text{-mod}_\chi^{I^0}$ in the Grothendieck group of $\widehat{\mathfrak{g}}_{\kappa_c}\text{-mod}_\chi^{I^0}$ give rise to a basis in the algebraic K-theory $K(\text{Sp}_u)$, where $u = \text{Res}(\chi)$. Presumably, this basis is closely related to the bases in (equivariant version of) this K-theory constructed by G. Lusztig in [Lu2] (from the perspective of unrestricted $\mathfrak{g}$-modules in positive characteristic).

8.5 Evidence for the Conjecture

We now describe some evidence for Conjecture 6. It consists of the following four groups of results:

- Interpretation of the Wakimoto modules as $\widehat{\mathfrak{g}}_{\kappa_c}$-modules corresponding to the skyscraper sheaves on $\mathrm{MOp}_{L_G}^{\mathrm{nilp}}$;
- Connection to R. Bezrukavnikov's theory;
- Proof of the equivalence of certain quotient categories of $D^b(\widehat{\mathfrak{g}}_{\kappa_c}\text{-mod}_{\mathrm{nilp}})^{I^0}$ and $D^b(\mathrm{QCoh}(\mathrm{MOp}_{L_G}^{\mathrm{nilp}}))$, [FG2].
- Proof of the restriction of the equivalence (8.15) to regular opers, [FG4].

We start with the discussion of Wakimoto modules.

Suppose that we have proved the equivalence of categories (8.15). Then each quasicoherent sheaf on $\mathrm{MOp}_{L_G}^{\mathrm{nilp}}$ should correspond to an object of the derived category $D^b(\widehat{\mathfrak{g}}_{\kappa_c}\text{-mod}_{\mathrm{nilp}})^{I^0}$. The simplest quasicoherent sheaves on $\mathrm{MOp}_{L_G}^{\mathrm{nilp}}$ are the **skyscraper sheaves** supported at the $\mathbb{C}$-points of $\mathrm{MOp}_{L_G}^{\mathrm{nilp}}$. It follows from the definition that a $\mathbb{C}$-point of $\mathrm{MOp}_{L_G}^{\mathrm{nilp}}$, which is the same as a $\mathbb{C}$-point of the reduced scheme MOp_G^0, is a pair $(\chi, \mathfrak{b}')$, where $\chi = (\mathcal{F}, \nabla, \mathcal{F}_{L_B})$ is a nilpotent $^L G$-oper in $\mathrm{Op}_{L_G}^{\mathrm{nilp}}$ and $\mathfrak{b}'$ is a point of the Springer fiber corresponding to $\mathrm{Res}(\chi)$, which is the variety of Borel subalgebras in $^L \mathfrak{g}_{\mathcal{F}_0}$ that contain $\mathrm{Res}(\chi)$. Thus, if Conjecture 6 is true, we should have a family of objects of the category $D^b(\widehat{\mathfrak{g}}_{\kappa_c}\text{-mod}_{\mathrm{nilp}})^{I^0}$ parameterized by these data. What are these objects?

The answer is that these are the **Wakimoto modules**. These modules were originally introduced by M. Wakimoto [Wak] for $\mathfrak{g} = \mathfrak{sl}_2$ and by B. Feigin and myself in general in [FF1, FF2] (see also [F3]). We recall from [F3] that Wakimoto modules of critical level are parameterized by the space $\mathrm{Conn}(\Omega^{-\rho})_{D^\times}$ of connections on the $^L H$-bundle $\Omega^{-\rho}$ over $D^\times$. This is the push-forward of the $\mathbb{C}^\times$-bundle corresponding to the canonical line bundle Ω with respect to the homomorphism $\rho : \mathbb{C}^\times \to {}^L H$. Let us denote the Wakimoto module corresponding to $\overline{\nabla} \in \mathrm{Conn}(\Omega^{-\rho})_{D^\times}$ by $W_{\overline{\nabla}}$. According to [F3], Theorem 12.6, the center $Z(\widehat{\mathfrak{g}})$ acts on $W_{\overline{\nabla}}$ via the central character $\mu(\overline{\nabla})$, where

$$\mu : \mathrm{Conn}(\Omega^{-\rho})_{D^\times} \to \mathrm{Op}_{L_G}(D^\times)$$

is the **Miura transformation**.

It is not difficult to show that if $\chi \in \mathrm{Op}_{L_G}^{\mathrm{nilp}}$, then $W_{\overline{\nabla}}$ is an object of the category $\widehat{\mathfrak{g}}_{\kappa_c}\text{-mod}_\chi^I$ for any $\overline{\nabla} \in \mu^{-1}(\chi)$. Now, according to the results presented in [FG2], the points of the fiber $\mu^{-1}(\chi)$ of the Miura transformation over χ are in bijection with the points of the Springer fiber $\mathrm{Sp}_{\mathrm{Res}(\chi)}$ corresponding to the nilpotent element $\mathrm{Res}(\chi)$. Therefore to each point of $\mathrm{Sp}_{\mathrm{Res}(\chi)}$ we may assign a Wakimoto module, which is an object of the category $\widehat{\mathfrak{g}}_{\kappa_c}\text{-mod}_\chi^{I^0}$ (and hence of the corresponding derived category). In other words, Wakimoto modules are objects of the category $\widehat{\mathfrak{g}}_{\kappa_c}\text{-mod}_{\mathrm{nilp}}^I$ parameterized by the $\mathbb{C}$-points of $\mathrm{MOp}_{LG}^{\mathrm{nilp}}$. It is natural to assume that they correspond to the skyscraper sheaves on $\mathrm{MOp}_{LG}^{\mathrm{nilp}}$ under the equivalence (8.15). This was in fact one of our motivations for this conjecture.

Incidentally, this gives us a glimpse into how the group of automorphisms of the ^LG-local system underlying the oper χ acts on the category $\widehat{\mathfrak{g}}_{\kappa_c}\text{-mod}_\chi$. This group is $Z(\mathrm{Res}(\chi))$, the centralizer of the residue $\mathrm{Res}(\chi)$, and it acts on the Springer fiber $\mathrm{Sp}_{\mathrm{Res}(\chi)}$. Therefore $g \in Z(\mathrm{Res}(\chi))$ sends the skyscraper sheaf supported at a point $p \in \mathrm{Sp}_{\mathrm{Res}(\chi)}$ to the skyscraper sheaf supported at $g \cdot p$. Thus, we expect that g sends the Wakimoto module corresponding to p to the Wakimoto module corresponding to $g \cdot p$.

If the Wakimoto modules indeed correspond to the skyscraper sheaves, then the equivalence (8.15) may be thought of as a kind of "spectral decomposition" of the category $D^b(\widehat{\mathfrak{g}}_{\kappa_c}\text{-mod}_{\mathrm{nilp}})^{I^0}$, with the basic objects being the Wakimoto modules $W_{\overline{\nabla}}$, where $\overline{\nabla}$ runs over the locus in $\mathrm{Conn}(\Omega^{-\rho})_{D^\times}$ which is isomorphic, pointwise, to $\mathrm{MOp}_{LG}^{\mathrm{nilp}}$ (see [FG5] for more details).

Now we discuss the second piece of evidence, connection with Bezrukanikov's theory.

To motivate it, let us recall that in Section 7.4 we discussed the action of the categorical spherical algebra $\mathcal{H}(G((t)), G[[t]])$ on the category $\widehat{\mathfrak{g}}_{\kappa_c}\text{-mod}_\chi$, where χ is a regular oper. The affine Hecke algebra $H(G(F), I)$ also has a categorical analogue. Consider the **affine flag variety** $\mathrm{Fl} = G((t))/I$. The categorical Hecke algebra is the category $\mathcal{H}(G((t)), I)$ which is the full subcategory of the derived category of $\mathcal{D}$-modules on $\mathrm{Fl} = G((t))/I$ whose objects are complexes with I-equivariant cohomologies. This category naturally acts on the derived category $D^b(\widehat{\mathfrak{g}}_{\kappa_c}\text{-mod}_\chi)^I$. What does this action correspond to on the other side of the equivalence (8.15)?

The answer is given by a theorem of R. Bezrukavnikov [Bez2], which may be viewed as a categorification of the isomorphism (8.8):

$$(8.19) \qquad D^b(\mathcal{D}_{\kappa_c}^{\mathrm{Fl}}\text{-mod})^{I^0} \simeq D^b(\mathrm{QCoh}(\widetilde{\mathrm{St}})),$$

where $\mathcal{D}_{\kappa_c}^{\mathrm{Fl}}$-mod is the category of twisted $\mathcal{D}$-modules on Fl and $\widetilde{\mathrm{St}}$ is the "thickened" Steinberg variety

$$\widetilde{\mathrm{St}} = \widetilde{\mathcal{N}} \underset{\mathcal{N}}{\times} \widetilde{\widetilde{\mathcal{N}}} = \widetilde{\mathcal{N}} \underset{L\mathfrak{g}}{\times} {}^L\widetilde{\mathfrak{g}}.$$

Morally, we expect that the two categories in (8.19) act on the two categories in (8.15) in a compatible way. However, strictly speaking, the left hand side of (8.19) acts like this:

$$D^b(\widehat{\mathfrak{g}}_{\kappa_c}\text{-mod}_{\mathrm{nilp}})^I \to D^b(\widehat{\mathfrak{g}}_{\kappa_c}\text{-mod}_{\mathrm{nilp}})^{I^0},$$

and the right hand side of (8.19) acts like this:

$$D^b(\mathrm{QCoh}(\mathrm{MOp}_{LG}^0)) \to D^b(\mathrm{QCoh}(\mathrm{MOp}_{LG}^{\mathrm{nilp}})).$$

So one needs a more precise statement, which may be found in [Bez2], Sect. 4.2. Alternatively, one can consider the corresponding actions of the affine braid group of ^LG, as in [Bez2].

A special case of this compatibility concerns some special objects of the category $D^b(\mathcal{D}^{\mathrm{Fl}}_{\kappa_c}\text{-mod})^I$, the central sheaves introduced in [Ga1]. They correspond to the central elements of the affine Hecke algebra $H(G(F), I)$. These central elements act as scalars on irreducible $H(G(F), I)$-modules, as well as on the standard modules $K^A(\mathrm{Sp}_u)_{(\gamma,q,\rho)}$ discussed above. We have argued that the categories $\widehat{\mathfrak{g}}_{\kappa_c}\text{-mod}^{I^0}_\chi, \chi \in \mathrm{Op}^{\mathrm{nilp}}_{LG}$, are categorical versions of these representations. Therefore it is natural to expect that its objects are "eigenmodules" with respect to the action of the central sheaves from $D^b(\mathcal{D}^{\mathrm{Fl}}_{\kappa_c}\text{-mod})^I$ (in the sense of Section 7.4). This has indeed been proved in [FG3].

This discussion indicates an intimate connection between the category $D^b(\widehat{\mathfrak{g}}_{\kappa_c}\text{-mod}_{\mathrm{nilp}})$ and the category of twisted $\mathcal{D}$-modules on the affine flag variety, which is similar to the connection between $\widehat{\mathfrak{g}}_{\kappa_c}\text{-mod}_{\mathrm{reg}}$ and the category of twisted $\mathcal{D}$-modules on the affine Grassmannian which we discussed in Section 7.5. A more precise conjecture relating $D^b(\widehat{\mathfrak{g}}_{\kappa_c}\text{-mod}_{\mathrm{nilp}})$ and $D^b(\mathcal{D}^{\mathrm{Fl}}_{\kappa_c}\text{-mod})$ was formulated in [FG2] (see the Introduction and Sect. 6), where we refer the reader for more details. This conjecture may be viewed as an analogue of Conjecture 3 for nilpotent opers. As explained in [FG2], this conjecture is supported by the results of [AB, ABG] (see also [Bez1, Bez2]). Together, these results and conjectures provide additional evidence for the equivalence (8.15).

9 Ramified Global Langlands Correspondence

We now discuss the implications of the local Langlands correspondence for the global geometric Langlands correspondence.

We begin by briefly discussing the setting of the classical global Langlands correspondence.

9.1 The Classical Setting

Let X be a smooth projective curve over $\mathbb{F}_q$. Denote by F the field $\mathbb{F}_q(X)$ of rational functions on X. For any closed point x of X we denote by F_x the completion of F at x and by $\mathcal{O}_x$ its ring of integers. If we choose a local coordinate t_x at x (i.e., a rational function on X which vanishes at x to order one), then we obtain isomorphisms $F_x \simeq \mathbb{F}_{q_x}((t_x))$ and $\mathcal{O}_x \simeq \mathbb{F}_{q_x}[[t_x]]$, where $\mathbb{F}_{q_x}$ is the residue field of x; in general, it is a finite extension of $\mathbb{F}_q$ containing $q_x = q^{\mathrm{ord}_x}$ elements.

Thus, we now have a local field attached to each point of X. The ring $\mathbb{A} = \mathbb{A}_F$ of **adèles** of F is by definition the **restricted** product of the fields F_x, where x runs over the set $|X|$ of all closed points of X. The word "restricted" means that we consider only the collections $(f_x)_{x \in |X|}$ of elements of F_x in which $f_x \in \mathcal{O}_x$ for all but finitely many x. The ring $\mathbb{A}$ contains the field F, which is embedded into $\mathbb{A}$ diagonally, by taking the expansions of rational functions on X at all points.

While in the local Langlands correspondence we considered irreducible smooth representations of the group GL_n over a local field, in the global Langlands correspondence we consider irreducible **automorphic representations** of the group $GL_n(\mathbb{A})$. The word "automorphic" means, roughly, that the representation may be realized in a reasonable space of functions on the quotient $GL_n(F)\backslash GL_n(\mathbb{A})$ (on which the group $GL_n(\mathbb{A})$ acts from the right).

On the other side of the correspondence we consider n-dimensional representations of the Galois group $\mathrm{Gal}(\overline{F}/F)$, or, more precisely, the Weil group W_F, which is a subgroup of $\mathrm{Gal}(\overline{F}/F)$ defined in the same way as in the local case.

Roughly speaking, the global Langlands correspondence is a bijection between the set of equivalence classes of n-dimensional representations of W_F and the set of equivalence classes of irreducible automorphic representations of $GL_n(\mathbb{A})$:

The precise statement is more subtle. For example, we should consider the so-called ℓ-adic representations of the Weil group (while in the local case we considered the admissible complex representations of the Weil-Deligne group; the reason is that in the local case those are equivalent to the ℓ-adic representations). Moreover, under this correspondence important invariants attached to the objects appearing on both sides (Frobenius eigenvalues on the Galois side and the Hecke eigenvalues on the other side) are supposed to match. We refer the reader to Part I of the review [F6] for more details.

The global Langlands correspondence has been proved for GL_2 in the 80's by V. Drinfeld [Dr1]–[Dr4] and more recently by L. Lafforgue [Laf] for GL_n with an arbitrary n.

Like in the local story, we may also wish to replace the group GL_n by an arbitrary reductive algebraic group defined over F. Then on one side of the global Langlands correspondence we have homomorphisms $\sigma : W_F \to {}^L G$ satisfying some properties (or perhaps, some more refined data, as in [A]). We expect to be able to attach to each σ an **automorphic representation** π of $GL_n(\mathbb{A}_F)$.[17] The word "automorphic" again means, roughly, that the representation may be realized in a reasonable space of functions on the

---

[17] In this section, by abuse of notation, we will use the same symbol to denote a representation of the group and the vector space underlying this representation.

quotient $GL_n(F)\backslash GL_n(\mathbb{A})$ (on which the group $GL_n(\mathbb{A})$ acts from the right). We will not try to make this precise. In general, we expect not one but several automorphic representations assigned to σ which are the global analogues of the L-packets discussed above (see [A]). Another complication is that the multiplicity of a given irreducible automorphic representation in the space of functions on $GL_n(F)\backslash GL_n(\mathbb{A})$ may be greater than one. We will mostly ignore all of these issues here, as our main interest is in the geometric theory (note also that these issues do not arise if $G = GL_n$).

An irreducible automorphic representation may always be decomposed as the restricted tensor product $\bigotimes'_{x\in X} \pi_x$, where each π_x is an irreducible representation of $G(F_x)$. Moreover, for all by finitely many $x \in X$ the factor π_x is an **unramified** representation of $G(F_x)$: it contains a non-zero vector invariant under the maximal compact subgroup $K_{0,x} = G(\mathcal{O}_x)$ (see Section 7.1). Let us choose such a vector $v_x \in \pi_x$ (it is unique up to a scalar). The word "restricted" means that we consider the span of vectors of the form $\otimes_{x\in X} u_x$, where $u_x \in \pi_x$ and $u_x = v_x$ for all but finitely many $x \in X$.

An important property of the global Langlands correspondence is its compatibility with the local one. We can embed the Weil group W_{F_x} of each of the local fields F_x into the global Weil group W_F. Such an embedding is not unique, but it is well-defined up to conjugation in W_F. Therefore an equivalence class of $\sigma : W_F \to {}^L G$ gives rise to a well-defined equivalence class of $\sigma_x : W_{F_x} \to {}^L G$. We will impose the condition on σ that for all but finitely many $x \in X$ the homomorphism σ_x is unramified (see Section 7.1).

By the local Langlands correspondence, to σ_x one can attach an equivalence class of irreducible smooth representations π_x of $G(F_x)$.[18] Moreover, an unramified σ_x will correspond to an unramified irreducible representation π_x. The compatibility between local and global correspondences is the statement that the automorphic representation of $G(\mathbb{A})$ corresponding to σ should be isomorphic to the restricted tensor product $\bigotimes'_{x\in X} \pi_x$. Schematically, this is represented as follows:

$$\sigma \overset{\text{global}}{\longleftrightarrow} \pi = \bigotimes_{x\in X}{}' \pi_x$$

$$\sigma_x \overset{\text{local}}{\longleftrightarrow} \pi_x.$$

In this section we discuss an analogue of this local-to-global principle in the geometric setting and the implications of our local results and conjectures for the global geometric Langlands correspondence. We focus in particular on the unramified and tamely ramified Langlands parameters. At the end of the section we also discuss connections with irregular singularities.

---

[18] Here we are considering ℓ-adic homomorphisms from the Weil group W_{F_x} to ${}^L G$, and therefore we do not need to pass from the Weil group to the Weil-Deligne group.

9.2 The Unramified Case, Revisited

An important special case is when $\sigma : W_F \to {}^L G$ is everywhere unramified. Then for each $x \in X$ the corresponding homomorphism $\sigma_x : W_{F_x} \to {}^L G$ is unramified, and hence corresponds, as explained in Section 7.1, to a semi-simple conjugacy class γ_x in ${}^L G$, which is the image of the Frobenius element under σ_x. This conjugacy class in turn gives rise to an unramified irreducible representation π_x of $G(F_x)$ with a unique, up to a scalar, vector v_x such that $G(\mathcal{O}_x)v_x = v_x$. The spherical Hecke algebra $H(G(F_x), G(\mathcal{O}_x)) \simeq \mathrm{Rep}\,{}^L G$ acts on this vector according to formula (7.5):

$$(9.1) \qquad H_{V,x} \star v_x = \mathrm{Tr}(\gamma_x, V)v_x, \qquad [V] \in \mathrm{Rep}\,{}^L G.$$

The tensor product $v = \otimes_{x \in X} v_x$ of this vectors is a $G(\mathcal{O})$-invariant vector in $\pi = \bigotimes'_{x \in X} \pi_x$, which, according to the global Langlands conjecture is automorphic. This means that π is realized in the space of functions on $G(F) \backslash G(\mathbb{A}_F)$. In this realization vector v corresponds to a right $G(\mathcal{O})$-invariant function on $G(F) \backslash G(\mathbb{A}_F)$, or equivalently, a function on the double quotient

$$(9.2) \qquad G(F) \backslash G(\mathbb{A}_F)/G(\mathcal{O}).$$

Thus, an unramified global Langlands parameter σ gives rise to a function on (9.2). This function is the **automorphic function** corresponding to σ. We denote it by f_π. Since it corresponds to a vector in an irreducible representation π of $G(\mathbb{A}_F)$, the entire representation π may be reconstructed from this function. Thus, we do not lose any information by passing from π to f_π.

Since $v \in \pi$ is an eigenvector of the Hecke operators, according to formula (9.1), we obtain that the function f_π is a **Hecke eigenfunction** on the double quotient (9.2). In fact, the local Hecke algebras $H(G(F_x), G(\mathcal{O}_x))$ act naturally (from the right) on the space of functions on (9.2), and f_π is an eigenfunction of this action. It satisfies the same property (9.1).

To summarize, the unramified global Langlands correspondence in the classical setting may be viewed as a correspondence between unramified homomorphisms $\sigma : W_F \to {}^L G$ and Hecke eigenfunctions on (9.2) (some irreducibility condition on σ needs to be added to make this more precise, but we will ignore this).

What should be the geometric analogue of this correspondence, when X is a complex algebraic curve?

As explained in Section 3.1, the geometric analogue of an unramified homomorphism $W_F \to {}^L G$ is a homomorphism $\pi_1(X) \to {}^L G$, or equivalently, since X is assumed to be compact, a holomorphic ${}^L G$-bundle on X with a holomorphic connection (it automatically gives rise to a flat connection). The global geometric Langlands correspondence should therefore associate to a flat holomorphic ${}^L G$-bundle on X a geometric object on a geometric version of the double quotient (9.2). As we argued in Section 3.3, this should be a $\mathcal{D}$-module on an algebraic variety whose set of points is (9.2).

Now, it is known that (9.2) is in bijection with the set of isomorphism classes of G-bundles on X. This key result is due to A. Weil (see, e.g., [F6], Sect. 3.2). This suggests that (9.2) is the set of points of the moduli space of G-bundles on X. Unfortunately, in general this is not an algebraic variety, but an algebraic stack, which locally looks like the quotient of an algebraic variety by an action of an algebraic group. We denote it by Bun_G. The theory of $\mathcal{D}$-modules has been developed in the setting of algebraic stacks like Bun_G in [BD1], and so we can use it for our purposes. Thus, we would like to attach to a flat holomorphic ^LG-bundle E on X a $\mathcal{D}$-module Aut_E on Bun_G. This $\mathcal{D}$-module should satisfy an analogue of the Hecke eigenfunction condition, which makes it into a **Hecke eigensheaf** with eigenvalue E. This notion is spelled out in [F6], Sect. 6.1 (following [BD1]), where we refer the reader for details.

This brings us to the following question:

How to relate this global correspondence to the local geometric Langlands correspondence discussed above?

As we have already seen in Section 1, the key element in answering this question is a **localization functor** $\Delta_{\kappa_c,x}$ from $(\widehat{\mathfrak{g}}_{\kappa_c,x}, G(\mathcal{O}_x))$-modules to (twisted) $\mathcal{D}$-modules on Bun_G. In Section 1 we have applied this functor to the object $\mathbb{V}_0(\chi_x)$ of the Harish-Chandra category $\widehat{\mathfrak{g}}_{\kappa_c,x}\text{-mod}^{G(\mathcal{O}_x)}$, where $\chi_x \in \mathrm{Op}_{^LG}(D_x)$. For an oper χ_x which extends from D_x to the entire curve X we have obtained this way the Hecke eigensheaf associated to the underlying ^LG-local system (see Theorem 1).

For a ^LG-local system $E = (\mathcal{F}, \nabla)$ on X which does not admit the structure of a regular oper on X, the above construction may be modified as follows (see the discussion in [F6], Sect. 9.6, based on an unpublished work of Beilinson and Drinfeld). In this case one can choose an ^LB-reduction $\mathcal{F}_{^LB}$ satisfying the oper condition away from a finite set of points $y_1, \ldots, y_n$ and such that the restriction χ_{y_i} of the corresponding oper χ on $X \setminus \{y_1, \ldots, y_n\}$ to $D_{y_i}^\times$ belongs to $\mathrm{Op}_{^LG}^{\lambda_i}(D_{y_i}) \subset \mathrm{Op}_{^LG}(D_{y_i}^\times)$ for some $\lambda_i \in P^+$. Then one can construct a Hecke eigensheaf corresponding to E by applying a multi-point version of the localization functor to the tensor product of the quotients $\mathbb{V}_{\lambda_i}(\chi_{y_i})$ of the Weyl modules $\mathbb{V}_{\lambda_i,y_i}$ (see [F6], Sect. 9.6).

The main lesson of this construction is that in the geometric setting the localization functor gives us a powerful tool for converting local Langlands categories, such as $\widehat{\mathfrak{g}}_{\kappa_c,x}\text{-mod}^{G(\mathcal{O}_x)}_{\chi_x}$, into global categories of Hecke eigensheaves. The category $\widehat{\mathfrak{g}}_{\kappa_c,x}\text{-mod}^{G(\mathcal{O}_x)}_{\chi_x}$ turns out to be very simple: it has a unique irreducible object, $\mathbb{V}_0(\chi_x)$. That is why it is sufficient to consider its image under the localization functor, which turns out to be the desired Hecke eigensheaf Aut_{E_χ}. For general opers, with ramification, the corresponding local categories are more complicated, as we have seen above, and so are the corresponding categories of Hecke eigensheaves. We will consider examples of these categories in the next section.

9.3 Classical Langlands Correspondence with Ramification

Let us first consider ramified global Langlands correspondence in the classical setting. Suppose that we are given a homomorphism $\sigma : W_F \to {}^L G$ that is ramified at finitely many points $y_1, \ldots, y_n$ of X. Then we expect that to such σ corresponds an automorphic representation $\bigotimes'_{x \in X} \pi_x$ (more precisely, an L-packet of representations). Here π_x is still unramified for all $x \in X \backslash \{y_1, \ldots, y_n\}$, but is *ramified* at $y_1, \ldots, y_n$, i.e., the space of $G(\mathcal{O}_{y_i})$-invariant vectors in π_{y_i} is zero. In particular, consider the special case when each $\sigma_{y_i} : W_{F_{y_i}} \to {}^L G$ is tamely ramified (see Section 8.1 for the definition). Then, according to the results presented in Section 8.1, the corresponding L-packet of representations of $G(F_{y_i})$ contains an irreducible representation π_{y_i} with non-zero invariant vectors with respect to the Iwahori subgroup I_{y_i}. Let us choose such a representation for each point y_i.

Consider the subspace

$$(9.3) \qquad \bigotimes_{i=1}^{n} \pi_{y_i}^{I_{y_i}} \otimes \bigotimes_{x \neq y_i} v_x \subset \bigotimes'_{x \in X} \pi_x,$$

where v_x is a $G(\mathcal{O}_x)$-vector in $\pi_x, x \neq y_i, i = 1, \ldots, n$. Then, because $\bigotimes'_{x \in X} \pi_x$ is realized in the space of functions on $G(F) \backslash G(\mathbb{A}_F)$, we obtain that the subspace (9.3) is realized in the space of functions on the double quotient

$$(9.4) \qquad G(F) \backslash G(\mathbb{A}_F) / \prod_{i=1}^{n} I_{y_i} \times \prod_{x \neq y_i} G(\mathcal{O}_x).$$

The spherical Hecke algebras $H(G(F_x), G(\mathcal{O}_x)), x \neq y_i$, act on the subspace (9.3), and all elements of (9.3) are eigenfunctions of these algebras (they satisfy formula (9.1)). At the points y_i we have, instead of the action of the commutative spherical Hecke algebra $H(G(F_{y_i}), G(\mathcal{O}_{y_i})$, the action of the non-commutative affine Hecke algebra $H(G(F_{y_i}), I_{y_i})$. Thus, we obtain a subspace of the space of functions on (9.4), which consists of Hecke eigenfunctions with respect to the spherical Hecke algebras $H(G(F_x), G(\mathcal{O}_x)), x \neq y_i$, and which realize a module over $\bigotimes_{i=1}^{n} H(G(F_{y_i}), I_{y_i})$ (which is irreducible, since each π_{y_i} is irreducible).

This subspace encapsulates the automorphic representation $\bigotimes'_{x \in X} \pi_x$ the way the automorphic function f_π encapsulates an unramified automorphic representation. The difference is that in the unramified case the function f_π spans the one-dimensional space of invariants of the maximal compact subgroup $G(\mathcal{O})$ in $\bigotimes'_{x \in X} \pi_x$, whereas in the tamely ramified case the subspace (9.3) is in general a multi-dimensional vector space.

9.4 Geometric Langlands Correspondence in the Tamely Ramified Case

Now let us see how this plays out in the geometric setting. As we discussed before, the analogue of a homomorphism $\sigma : W_F \to {}^L G$ tamely ramified at the points $y_1, \ldots, y_n \in X$ is now a local system $E = (\mathcal{F}, \nabla)$, where $\mathcal{F}$ a $^L G$-bundle $\mathcal{F}$ on X with a connection ∇ that has regular singularities at $y_1, \ldots, y_n$ and unipotent monodromies around these points. We will call such a local system **tamely ramified** at $y_1, \ldots, y_n$. What should the global geometric Langlands correspondence attach to such a local system? It is clear that we need to find a geometric object replacing the finite-dimensional vector space (9.3) realized in the space of functions on (9.4).

Just as (9.2) is the set of points of the moduli stack Bun_G of G-bundles, the double quotient (9.4) is the set of points of the moduli stack $\mathrm{Bun}_{G,(y_i)}$ of G-bundles on X with the **parabolic structures** at $y_i, i = 1, \ldots, n$. By definition, a parabolic structure of a G-bundle $\mathcal{P}$ at $y \in X$ is a reduction of the fiber $\mathcal{P}_y$ of $\mathcal{P}$ at y to a Borel subgroup $B \subset G$. Therefore, as before, we obtain that a proper replacement for (9.3) is a category of $\mathcal{D}$-modules on $\mathrm{Bun}_{G,(y_i)}$. As in the unramified case, we have the notion of a Hecke eigensheaf on $\mathrm{Bun}_{G,(y_i)}$. But because the Hecke functors are now defined using the Hecke correspondences over $X \backslash \{y_1, \ldots, y_n\}$ (and not over X as before), an "eigenvalue" of the Hecke operators is now an $^L G$-local system on $X \backslash \{y_1, \ldots, y_n\}$ (rather than on X). Thus, we obtain that the global geometric Langlands correspondence now should assign to a $^L G$-local system E on X, tamely ramified at the points $y_1, \ldots, y_n$, a **category** $\mathcal{A}ut_E$ of $\mathcal{D}$-modules on $\mathrm{Bun}_{G,(y_i)}$ with the eigenvalue $E|_{X \backslash \{y_1, \ldots, y_n\}}$,

$$E \mapsto \mathcal{A}ut_E.$$

We now construct these categories using a generalization of the localization functor we used in the unramified case (see [FG2]). For the sake of notational simplicity, let us assume that our $^L G$-local system $E = (\mathcal{F}, \nabla)$ is tamely ramified at a single point $y \in X$. Suppose that this local system on $X \backslash y$ admits the structure of a $^L G$-oper $\chi = (\mathcal{F}, \nabla, \mathcal{F}_{L B})$ whose restriction χ_y to the punctured disc $D_y^\times$ belongs to the subspace $\mathrm{Op}_{L G}^{\mathrm{nilp}}(D_y)$ of nilpotent $^L G$-opers.

For a simple Lie group G, the moduli stack $\mathrm{Bun}_{G,y}$ has a realization analogous to (1.1):

$$\mathrm{Bun}_{G,y} \simeq G_{\mathrm{out}} \backslash G(\mathcal{K}_y)/I_y.$$

Let $\mathcal{D}_{\kappa_c, I_y}$ be the sheaf of twisted differential operators on $\mathrm{Bun}_{G,y}$ acting on the line bundle corresponding to the critical level (it is the pull-back of the square root of the canonical line bundle $K^{1/2}$ on Bun_G under the natural projection $\mathrm{Bun}_{G,y} \to \mathrm{Bun}_G$). Applying the formalism of the previous section, we obtain a localization functor

$$\Delta_{\kappa_c, I_y} : \widehat{\mathfrak{g}}_{\kappa_c, y}\text{-mod}^{I_y} \to \mathcal{D}_{\kappa_c, I_y}\text{-mod}.$$

However, in order to make contact with the results obtained above we also consider the larger category $\widehat{\mathfrak{g}}_{\kappa_c, y}\text{-mod}^{I_y^0}$ of I_y^0-equivariant modules, where $I_y^0 = [I_y, I_y]$.

Set

$$\mathrm{Bun}'_{G,y} = G_{\mathrm{out}} \backslash G(\mathcal{K}_y)/I_y^0,$$

and let $\mathcal{D}_{\kappa_c, I_y^0}$ be the sheaf of twisted differential operators on $\mathrm{Bun}'_{G,y}$ acting on the pull-back of the line bundle $K^{1/2}$ on Bun_G. Let $\mathcal{D}_{\kappa_c, I_y^0}$-mod be the category of $\mathcal{D}_{\kappa_c, I_y^0}$-modules. Applying the general formalism, we obtain a localization functor

$$(9.5) \qquad \Delta_{\kappa_c, I_y^0} : \widehat{\mathfrak{g}}_{\kappa_c, y}\text{-mod}^{I_y^0} \to \mathcal{D}_{\kappa_c, I_y^0}\text{-mod}.$$

We note that a version of the categorical affine Hecke algebra $\mathcal{H}(G(\mathcal{K}_y), I_y)$ discussed in Section 8.5 naturally acts on the derived categories of the above categories, and the functors Δ_{κ_c, I_y} and Δ_{κ_c, I_y^0} intertwine these actions. Equivalently, one can say that this functor intertwines the corresponding actions of the affine braid group associated to $^L G$ on the two categories (as in [Bez2]).

We now restrict the functors Δ_{κ_c, I_y} and Δ_{κ_c, I_y^0} to the subcategories $\widehat{\mathfrak{g}}_{\kappa_c, y}\text{-mod}_{\chi_y}^{I_y}$ and $\widehat{\mathfrak{g}}_{\kappa_c, y}\text{-mod}_{\chi_y}^{I_y^0}$, respectively. By using the same argument as in [BD1], we obtain the following analogue of Theorem 1.

Theorem 5. *Fix $\chi_y \in \mathrm{Op}_{^L G}^{\mathrm{nilp}}(D_y)$ and let M be an object of the category $\widehat{\mathfrak{g}}_{\kappa_c, y}\text{-mod}_{\chi_y}^{I_y}$ (resp. $\widehat{\mathfrak{g}}_{\kappa_c, y}\text{-mod}_{\chi_y}^{I_y^0}$). Then*
(1) $\Delta_{\kappa_c, I_y}(M) = 0$ (resp., $\Delta_{\kappa_c, I_y^0}(M) = 0$) unless χ_y is the restriction of a regular oper $\chi = (\mathcal{F}, \nabla, \mathcal{F}_{L_B})$ on $X \backslash y$ to $D_y^\times$.
(2) In that case $\Delta_{\kappa_c, y}(M)$ (resp., $\Delta_{\kappa_c, I_y^0}(M)$) is a Hecke eigensheaf with the eigenvalue $E_\chi = (\mathcal{F}, \nabla)$.

Thus, we obtain that if $\chi_y = \chi|_{D_y^\times}$, then the image of any object of $\widehat{\mathfrak{g}}_{\kappa_c, y}\text{-mod}_{\chi_y}^{I_y}$ under the functor Δ_{κ_c, I_y} belongs to the category $\mathcal{A}ut_{E_\chi}^{I_y}$ of Hecke eigensheaves on $\mathrm{Bun}_{G,y}$. Now consider the restriction of the functor Δ_{κ_c, I_y^0} to $\widehat{\mathfrak{g}}_{\kappa_c, y}\text{-mod}_{\chi_y}^{I_y^0}$. As discussed in Section 8.3, the category $\widehat{\mathfrak{g}}_{\kappa_c, y}\text{-mod}_{\chi_y}^{I_y^0}$ coincides with the corresponding category $\widehat{\mathfrak{g}}_{\kappa_c, y}\text{-mod}_{\chi_y}^{I_y, m}$ of I_y-monodromic modules. Therefore the image of any object of $\widehat{\mathfrak{g}}_{\kappa_c, y}\text{-mod}_{\chi_y}^{I_y^0}$ under the functor Δ_{κ_c, I_y^0} belongs to the subcategory $\mathcal{D}_{\kappa_c, I_y}^m\text{-mod}$ of $\mathcal{D}_{\kappa_c, I_y^0}\text{-mod}$ whose objects admit an increasing filtration such that the consecutive quotients are pull-backs of $\mathcal{D}_{\kappa_c, I_y}$-modules from $\mathrm{Bun}_{G,y}$. Such $\mathcal{D}_{\kappa_c, I_y^0}$-modules are called **monodromic**.

Let $\mathcal{A}ut_{E_\chi}^{I_y, m}$ be the subcategory of $\mathcal{D}_{\kappa_c, I_y^0}^m\text{-mod}$ whose objects are Hecke eigensheaves with eigenvalue E_χ.

Thus, we obtain the functors

$$(9.6) \quad \Delta_{\kappa_c, I_y} : \widehat{\mathfrak{g}}_{\kappa_c, y}\text{-mod}_{\chi_y}^{I_y} \to \mathcal{A}ut_{E_\chi}^{I_y}, \qquad \Delta_{\kappa_c, I_y^0} : \widehat{\mathfrak{g}}_{\kappa_c, y}\text{-mod}_{\chi_y}^{I_y^0} \to \mathcal{A}ut_{E_\chi}^{I_y, m}$$

It is tempting to conjecture (see [FG2]) that these functors are equivalences of categories, at least for generic χ. Suppose that this is true. Then we may identify the *global* categories $\mathcal{A}ut_{E_\chi}^{I_y}$ and $\mathcal{A}ut_{E_\chi}^{I_y, m}$ of Hecke eigensheaves on

Bun_{G,I_y} and Bun'_{G,I_y^0} with the *local* categories $\widehat{\mathfrak{g}}_{\kappa_c,y}$-$\mathrm{mod}_{\chi_y}^{I_y}$ and $\widehat{\mathfrak{g}}_{\kappa_c,y}$-$\mathrm{mod}_{\chi_y}^{I_y^0}$, respectively. Therefore we can use our results and conjectures on the local Langlands categories, such as $\widehat{\mathfrak{g}}_{\kappa_c,y}$-$\mathrm{mod}_{\chi_y}^{I_y^0}$, to describe the global categories of Hecke eigensheaves on the moduli stacks of G-bundles on X with parabolic structures.

We have the following conjectural description of the derived category of I_y^0-equivariant modules, $D^b(\widehat{\mathfrak{g}}_{\kappa_c,y}$-$\mathrm{mod}_{\chi_y})^{I_y^0}$ (see formula (8.16)):

$$(9.7) \qquad D^b(\widehat{\mathfrak{g}}_{\kappa_c,y}\text{-}\mathrm{mod}_{\chi_y})^{I_y^0} \simeq D^b(\mathrm{QCoh}(\widetilde{\mathrm{Sp}}_{\mathrm{Res}(\chi_y)}^{\mathrm{DG}})).$$

The corresponding I_y-equivariant version is

$$(9.8) \qquad D^b(\widehat{\mathfrak{g}}_{\kappa_c,y}\text{-}\mathrm{mod}_{\chi_y})^{I_y} \simeq D^b(\mathrm{QCoh}(\mathrm{Sp}_{\mathrm{Res}(\chi_y)}^{\mathrm{DG}})),$$

where we replace the non-reduced DG Springer fiber by the reduced one: it is defined as the DG fiber of the morphism $\widetilde{\mathcal{N}} \to \mathfrak{g}$ over u.

If the functors (9.6) are equivalences, then by combining them with (9.7) and (9.8), we obtain the following conjectural equivalences of categories: (9.9)

$$D^b(\mathcal{A}ut_{E_\chi}^{I_y}) \simeq D^b(\mathrm{QCoh}(\mathrm{Sp}_{\mathrm{Res}(\chi_y)}^{\mathrm{DG}})), \quad D^b(\mathcal{A}ut_{E_\chi}^{I_y,m}) \simeq D^b(\mathrm{QCoh}(\widetilde{\mathrm{Sp}}_{\mathrm{Res}(\chi_y)}^{\mathrm{DG}})).$$

In other words, the derived category of a global Langlands category (monodromic or not) corresponding to a local system tamely ramified at $y \in X$ is equivalent to the derived category of quasicoherent sheaves on the DG Springer fiber of its residue at y (non-reduced or reduced).

Again, these equivalences are supposed to intertwine the natural actions on the above categories of the categorical affine Hecke algebra $\mathcal{H}(G(\mathcal{K}_y), I_y)$ (or, equivalently, the affine braid group associated to $^L G$).

The categories appearing in (9.9) actually make sense for an arbitrary $^L G$-local system E on X tamely ramified at y. It is therefore tempting to conjecture that these equivalences still hold in general: (9.10)

$$D^b(\mathcal{A}ut_E^{I_y}) \simeq D^b(\mathrm{QCoh}(\mathrm{Sp}_{\mathrm{Res}(E)}^{\mathrm{DG}})), \quad D^b(\mathcal{A}ut_E^{I_y,m}) \simeq D^b(\mathrm{QCoh}(\widetilde{\mathrm{Sp}}_{\mathrm{Res}(E)}^{\mathrm{DG}})).$$

The corresponding localization functors are constructed as follows: we represent a general local system E on X with tame ramification at y by an oper χ on the complement of finitely many points $y_1, \ldots, y_n$, whose restriction to $D_{y_i}^\times$ belongs to $\mathrm{Op}_{^L G}^{\lambda_i}(D_{y_i}) \subset \mathrm{Op}_{^L G}(D_{y_i}^\times)$ for some $\lambda_i \in P^+$. Then, in the same way as in the unramified case, we construct localization functors from $\widehat{\mathfrak{g}}_{\kappa_c,y}$-$\mathrm{mod}_{\chi_y}^{I_y}$ to $\mathcal{A}ut_E^{I_y}$ and from $\widehat{\mathfrak{g}}_{\kappa_c,y}$-$\mathrm{mod}_{\chi_y}^{I_y^0}$ to $\mathcal{A}ut_E^{I_y,m}$ (here, as before, $\chi_y = \chi|_{D_y^\times}$), and this leads us to the conjectural equivalences (9.10).

The equivalences (9.10) also have family versions in which we allow E to vary. It is analogous to the family version (8.15) of the local equivalences. As in the local case, in a family version we can avoid using DG schemes.

The above construction may be generalized to allow local systems tamely ramified at finitely many points $y_1, \ldots, y_n$. The corresponding Hecke eigensheaves are then $\mathcal{D}$-modules on the moduli stack of G-bundles on X with parabolic structures at $y_1, \ldots, y_n$. Non-trivial examples of these Hecke eigensheaves arise already in genus zero. These sheaves were constructed explicitly in [F1] (see also [F4, F5]), and they are closely related to the Gaudin integrable system.

9.5 Connections with Regular Singularities

So far we have only considered the categories of $\widehat{\mathfrak{g}}_{\kappa_c}$-modules corresponding to ^LG-opers on X which are regular everywhere except at a point $y \in X$ (or perhaps, at several points) and whose restriction to $D_y^\times$ is a nilpotent oper χ_y in $\mathrm{Op}_{LG}^{\mathrm{nilp}}(D_y)$. In other words, χ_y is an oper with regular singularity at y with residue $\varpi(-\rho)$ (where $\varpi : \mathfrak{h}^* \to \mathfrak{h}^*/W$). However, we can easily generalize the localization functor to the categories of $\widehat{\mathfrak{g}}_{\kappa_c}$-modules corresponding to ^LG-opers which have regular singularity at y with *arbitrary* residue.

So suppose we are given an oper $\chi \in \mathrm{Op}_{LG}^{\mathrm{RS}}(D)_{\varpi(-\lambda-\rho)}$ with regular singularity and residue $\varpi(-\lambda - \rho)$, where $\lambda \in \mathfrak{h}^*$. In this case the monodromy of this oper around y is conjugate to

$$M = \exp(2\pi i(\lambda + \rho)) = \exp(2\pi i\lambda).$$

We then have the category $\widehat{\mathfrak{g}}_{\kappa_c}\text{-mod}_\chi^{I^0}$ of I^0-equivariant $\widehat{\mathfrak{g}}_{\kappa_c}$-modules with central character χ. The case of $\lambda = 0$ is an "extremal" case when the category $\widehat{\mathfrak{g}}_{\kappa_c}\text{-mod}_\chi^{I^0}$ is most complicated. On the other "extreme" is the case of generic opers χ, corresponding to a generic λ. In this case one can show that the category $\widehat{\mathfrak{g}}_{\kappa_c}\text{-mod}_\chi^{I^0}$ is quite simple: it contains irreducible objects $\mathbb{M}_{w(\lambda+\rho)-\rho}(\chi)$ labeled by the Weyl group of $\mathfrak{g}$, and each object of $\widehat{\mathfrak{g}}_{\kappa_c}\text{-mod}_\chi^{I^0}$ is a direct sum of these irreducible modules. Here $\mathbb{M}_{w(\lambda+\rho)-\rho}(\chi)$ is the quotient of the Verma module

$$\mathbb{M}_{w(\lambda+\rho)-\rho} = \mathrm{Ind}_{\mathfrak{b}_+ \oplus \mathbb{C}\mathbf{1}}^{\widehat{\mathfrak{g}}_{\kappa_c}} \mathbb{C}_{w(\lambda+\rho)-\rho}, \qquad w \in W,$$

by the central character corresponding to χ.

For other values of λ the structure of $\widehat{\mathfrak{g}}_{\kappa_c}\text{-mod}_\chi^{I^0}$ is somewhere in-between these two extreme cases.

Recall that we have a localization functor (9.5)

$$\Delta_{\kappa_c, I_y^0}^\lambda : \widehat{\mathfrak{g}}_{\kappa_c, y}\text{-mod}^{I_y^0} \to \mathcal{D}_{\kappa_c, I_y^0}\text{-mod}.$$

from $\widehat{\mathfrak{g}}_{\kappa_c, y}\text{-mod}_{\chi_y}^{I_y^0}$ to a category of $\mathcal{D}$-modules on Bun_{G, I_y}' twisted by the pull-back of the line bundle $K^{1/2}$ on Bun_G. We now restrict this functor to the subcategory $\widehat{\mathfrak{g}}_{\kappa_c, y}\text{-mod}_{\chi_y}^{I_y^0}$ where χ_y is a ^LG-oper on D_y with regular singularity at y and residue $\varpi(-\lambda - \rho)$.

Consider first the case when $\lambda \in \mathfrak{h}^*$ is generic. Suppose that χ_y extends to a regular oper χ on $X \backslash y$. One then shows in the same way as in Theorem 5 that for any object M of $\widehat{\mathfrak{g}}_{\kappa_c,y}$-$\mathrm{mod}_{\chi_y}^{I_y^0}$ the corresponding $\mathcal{D}_{\kappa_c,I_y^0}$-module $\Delta_{\kappa_c,I_y^0}(M)$ is a Hecke eigensheaf with eigenvalue E_χ, which is the ^LG-local system on X with regular singularity at y underlying χ (if χ_y cannot be extended to $X \backslash y$, then $\Delta_{\kappa_c,I_y}^\lambda(M) = 0$, as before). Therefore we obtain a functor

$$\Delta_{\kappa_c,I_y^0} : \widehat{\mathfrak{g}}_{\kappa_c,y}\text{-}\mathrm{mod}_{\chi_y}^{I_y^0} \to \mathcal{A}ut_{E_\chi}^{I_y^0},$$

where $\mathcal{A}ut_{E_\chi}^{I_y^0}$ is the category of Hecke eigensheaves on Bun_{G,I_y}' with eigenvalue E_χ.

Since we have assumed that the residue of the oper χ_y is generic, the monodromy of E_χ around y belongs to a regular semi-simple conjugacy class of ^LG containing $\exp(2\pi i\lambda)$. In this case the category $\widehat{\mathfrak{g}}_{\kappa_c,y}$-$\mathrm{mod}_{\chi_y}^{I_y^0}$ is particularly simple, as we have discussed above. We expect that the functor Δ_{κ_c,I_y^0} sets up an equivalence between $\widehat{\mathfrak{g}}_{\kappa_c,y}$-$\mathrm{mod}_{\chi_y}^{I_y^0}$ and $\mathcal{A}ut_{E_\chi}^{I_y^0}$.

We can formulate this more neatly as follows. For $M \in {}^LG$ let $\mathcal{B}_M$ be the variety of Borel subgroups containing M. Observe that if M is regular semi-simple, then $\mathcal{B}_M$ is a set of points which is in bijection with W. Therefore our conjecture is that $\mathcal{A}ut_{E_\chi}^{I_y^0}$ is equivalent to the category $\mathrm{QCoh}(\mathcal{B}_M)$ of quasicoherent sheaves on $\mathcal{B}_M$, where M is a representative of the conjugacy class of the monodromy of E_χ.

Consider now an arbitrary ^LG-local system E on X with regular singularity at $y \in X$ whose monodromy around y is regular semi-simple. It is then tempting to conjecture that, at least if E is generic, this category has the same structure as in the case when E has the structure of an oper, i.e., it is equivalent to the category $\mathrm{QCoh}(\mathcal{B}_M)$, where M is a representative of the conjugacy class of the monodromy of E around y.

On the other hand, if the monodromy around y is unipotent, then $\mathcal{B}_M$ is nothing but the Springer fiber Sp_u, where $M = \exp(2\pi i u)$. The corresponding category $\mathcal{A}ut_E^{I_y^0}$ was discussed in Section 9.4 (we expect that it coincides with $\mathcal{A}ut_E^{I_y,m}$). Thus, we see that in both "extreme" cases: unipotent monodromy and regular semi-simple monodromy, our conjectures identify the derived category of $\mathcal{A}ut_E^{I_y^0}$ with the derived category of the category $\mathrm{QCoh}(\mathcal{B}_M)$ (where $\mathcal{B}_M$ should be viewed as a DG scheme $\widetilde{\mathrm{Sp}}_u^{DG}$ in the unipotent case). One is then led to conjecture, most ambitiously, that for any ^LG-local system E on X with regular singularity at $y \in X$ the derived category of $\mathcal{A}ut_E^{I_y^0}$ is equivalent to the derived category of quasicoherent sheaves on a suitable DG version of the scheme $\mathcal{B}_M$, where M is a representative of the conjugacy class of the monodromy of E around y:

$$D^b(\mathcal{A}ut_E^{I_y^0}) \simeq D^b(\mathrm{QCoh}(\mathcal{B}_M^{DG})).$$

This has an obvious generalization to the case of multiple ramification points, where on the right hand side we take the Cartesian product of the varieties $\mathcal{B}_{M_i}^{DG}$ corresponding to the monodromies. Thus, we obtain a conjectural realization of the categories of Hecke eigensheaves, whose eigenvalues are local systems with regular singularities, in terms of categories of quasicoherent sheaves.

It is useful to note that the Hecke eigensheaves on Bun'_{G,I_y} obtained above via the localization functors may be viewed as pull-backs of twisted $\mathcal{D}$-modules on Bun_{G,I_y} (or, more generally, extensions of such pull-backs).

More precisely, for each $\lambda \in \mathfrak{h}^*$ we have the sheaf of twisted differential operators on $\text{Bun}_{G,y}$ acting on a "line bundle" $\widetilde{\mathcal{L}}_\lambda$. If λ were an integral weight, this would be an actual line bundle, which is constructed as follows: note that the map $p : \text{Bun}_{G,I_y} \to \text{Bun}_G$, corresponding to forgetting the parabolic structure, is a fibration with the fibers isomorphic to the flag manifold G/B. For each integral weight λ we have the G-equivariant line bundle $\ell_\lambda = G \underset{B}{\times} \mathbb{C}_\lambda$ on G/B. The line bundle $\mathcal{L}_\lambda$ on Bun_{G,I_y} is defined in such a way that its restriction to each fiber of the projection p is isomorphic to ℓ_λ. We then set $\widetilde{\mathcal{L}}_\lambda = \mathcal{L}_\lambda \otimes p^*(K^{1/2})$, where $K^{1/2}$ is the square root of the canonical line bundle on Bun_G corresponding to the critical level. Now, it is well-known (see, e.g., [BB]) that even though the line bundle $\widetilde{\mathcal{L}}_\lambda$ does not exist if λ is not an integral weight, the corresponding sheaf $\mathcal{D}_{\kappa_c,I_y}^\lambda$ of $\widetilde{\mathcal{L}}_\lambda$-twisted differential operators on Bun_{G,I_y} is still well-defined.

Observe that we have an equivalence between the category $\mathcal{D}_{\kappa_c,I_y}^\lambda$-mod and the category of weakly H-equivariant $\mathcal{D}_{\kappa_c,I_y^0}$-module on $\text{Bun}'_{G,y}$ on which $\mathfrak{h}$ acts via the character $\lambda : \mathfrak{h} \to \mathbb{C}$. If $\mathcal{F}$ is an object of $\mathcal{D}_{\kappa_c,I_y}^\lambda$-mod, then the corresponding weakly H-equivariant $\mathcal{D}_{\kappa_c,I_y^0}$-module on $\text{Bun}'_{G,y}$ is $\pi^*(\mathcal{F})$, where $\pi : \text{Bun}'_{G,y} \to \text{Bun}_{G,I_y}$.

Now, it is easy to see that the $\mathcal{D}_{\kappa_c,I_y^0}$-modules on $\text{Bun}'_{G,y}$ obtained by applying the localization functor Δ_{κ_c,I_y^0} to objects of $\widehat{\mathfrak{g}}_{\kappa_c,y}\text{-mod}_{\chi_y}^{I_y^0}$ are always weakly H-equivariant. Consider, for example, the case when χ_y is a generic oper with regular singularity at y. Then its residue is equal to $\varpi(-\lambda - \rho)$, where λ is a regular element of $\mathfrak{h}^*$, and so its monodromy is $M = \exp(2\pi i \lambda)$. The corresponding category $\widehat{\mathfrak{g}}_{\kappa_c,y}\text{-mod}_{\chi_y}^{I_y^0}$ has objects $\mathbb{M}_{w(\lambda+\rho)-\rho}(\chi_y)$ that we introduced above. The Cartan subalgebra $\mathfrak{h}$ of $\widehat{\mathfrak{g}}_{\kappa_c,y}$ acts on $\mathbb{M}_{w(\lambda+\rho)-\rho}(\chi_y)$ semi-simply with the eigenvalues given by the weights of the form $w(\lambda + \rho) - \rho + \mu$, where μ is an integral weight. In other words,

$$\mathbb{M}_{w(\lambda+\rho)-\rho}(\chi_y) \otimes \mathbb{C}_{-w(\lambda+\rho)+\rho}$$

is I_y-equivariant. Therefore we find that $\Delta_{\kappa_c,I_y^0}(\mathbb{M}_{w(\lambda+\rho)-\rho}(\chi_y))$ is weakly H-equivariant, and the corresponding action of $\mathfrak{h}$ is given by $w(\lambda + \rho) - \rho :$ $\mathfrak{h} \to \mathbb{C}$. Thus, $\Delta_{\kappa_c,I_y^0}(\mathbb{M}_{w(\lambda+\rho)-\rho}(\chi_y))$ is the pull-back of a $\mathcal{D}_{\kappa_c,I_y}^{w(\lambda+\rho)-\rho}$-module on $\text{Bun}_{G,y}$. This $\mathcal{D}^{w(\lambda+\rho)-\rho}_{\kappa_c,I_y}$-module is a Hecke eigensheaf with eigenvalue E_χ provided that $\chi_y = \chi|_{D_y^\times}$, where χ is a regular oper on $X\backslash y$.

Thus, for a given generic oper χ_y we have $|W|$ different Hecke eigensheaves

$$\Delta_{\kappa_c,I_y^0}(\mathbb{M}_{w(\lambda+\rho)-\rho}(\chi_y)), \qquad w \in W,$$

on $\text{Bun}'_{G,y}$. However, each of them is the pull-back of a twisted $\mathcal{D}$-module on $\text{Bun}_{G,y}$ corresponding to a particular twist: namely, by a "line bundle" $\widetilde{\mathcal{L}}_{w(\lambda+\rho)-\rho}$. (Since we have assumed that λ is generic, all of these twists are different; note also that if $\mu = w(\lambda+\rho) - \rho$, then $\exp(2\pi i\mu)$ is in the conjugacy class of the monodromy $\exp(2\pi i\lambda)$.) It is therefore natural to conjecture that there is a unique Hecke eigensheaf on $\text{Bun}_{G,y}$ with eigenvalue E_χ, which is a twisted $\mathcal{D}$-module with the twisting given by $\widetilde{\mathcal{L}}_{w(\lambda+\rho)-\rho}$.

More generally, suppose that E is a local system on X with regular singularity at y and generic regular semi-simple monodromy. Let us choose a representative M of the monodromy which belongs to the Cartan subgroup $^LH \subset {}^LG$. Choose $\mu \in \mathfrak{h}^* \simeq {}^L\mathfrak{h}$ to be such that $M = \exp(2\pi i\mu)$. Note that there are exactly $|W|$ such choices up to a shift by an integral weight ν. Let $\text{Aut}_E^{I_y,\mu}$ be the category of Hecke eigensheaves with eigenvalue E in the category of twisted $\mathcal{D}$-modules on Bun_{G,I_y} with the twisting given by $\widetilde{\mathcal{L}}_\mu$. Then we expect that for generic E the category $\text{Aut}_E^{I_y,\mu}$ has a unique irreducible object. Its pull-back to $\text{Bun}'_{G,y}$ is one of the $|W|$ irreducible objects of $\text{Aut}_E^{I_y^0}$. (Note that tensoring with the line bundle $\mathcal{L}_\nu$, where ν is an integral weight, we identify the categories $\text{Aut}_E^{I_y,\mu}$ and $\text{Aut}_E^{I_y,\mu'}$ if $\mu' = \mu + \nu$.)

Similarly, one can describe the Hecke eigensheaves on $\text{Bun}'_{G,y}$ obtained by applying Δ_{κ_c,I_y^0} to the categories $\widehat{\mathfrak{g}}_{\kappa_c,y}$-$\text{mod}_{\chi_y}^{I_y^0}$ for other opers χ_y in terms of twisted $\mathcal{D}$-modules on $\text{Bun}_{G,y}$. In the opposite extreme case, when the residue of χ_y is 0 (and so χ_y is a nilpotent oper), this is explained in Section 9.4. (In this case one may choose to consider monodromic $\mathcal{D}$-modules; this is not necessary if λ is generic, because in this case there are no non-trivial extensions.)

Finally, it is natural to ask whether these equivalences for individual local systems may be combined into a family version encompassing all of them. The global geometric Langlands correspondence in the unramified case may be viewed as a kind of non-abelian Fourier-Mukai transform relating the (derived) category of $\mathcal{D}$-modules on Bun_G and the (derived) category of quasicoherent sheaves on $\text{Loc}_{{}^LG}(X)$, the stack of ^LG-local systems on the curve X. Under this correspondence, the skyscraper sheaf supported at a ^LG-local system E is supposed to go to the Hecke eigensheaf Aut_E on Bun_G. Thus, one may think of $\text{Loc}_{{}^LG}(X)$ as a parameter space of a "spectral decomposition" of the derived category of $\mathcal{D}$-modules on Bun_G (see, e.g., [F6], Sect. 6.2, for more details).

The above results and conjectures suggest that one may also view the geometric Langlands correspondence in the tamely ramified case in a similar way. Now the role of $\mathrm{Loc}_{{}^L G}(X)$ should be played by the stack $\mathrm{Loc}_{{}^L G, y}(X)$ of parabolic $^L G$-local systems with regular singularity at $y \in X$ (or, more generally, multiple points) and unipotent monodromy. This stack classifies triples $(\mathcal{F}, \nabla, \mathcal{F}_{{}^L B, y})$, where $\mathcal{F}$ is a $^L G$-bundle on X, ∇ is a connection on $\mathcal{F}$ with regular singularity at y and unipotent monodromy, and $\mathcal{F}_{{}^L B, y}$ is a $^L B$-reduction of the fiber $\mathcal{F}_y$ of $\mathcal{F}$ at y, which is preserved by ∇. This stack is now a candidate for a parameter space of a "spectral decomposition" of the derived category of $\mathcal{D}$-modules on the moduli stack $\mathrm{Bun}_{G, y}$ of G-bundles with parabolic structure at y.[19]

9.6 Irregular Connections

We now generalize the above results to the case of connections with irregular singularities. Let $\mathcal{F}$ be a $^L G$-bundle on X with connection ∇ that is regular everywhere except for a point $y \in X$, where it has a pole of order greater than 1. As before, we assume first that $(\mathcal{F}, \nabla)$ admits the structure of a $^L G$-oper on $X \backslash y$, which we denote by χ. Let χ_y be the the restriction of χ to $D_y^\times$. A typical example of such an oper is a $^L G$-oper with pole of order $\leq n$ on the disc D_y, which is, by definition (see [BD1], Sect. 3.8.8), an $^L N[[t]]$-conjugacy class of operators of the form

$$(9.11) \qquad \nabla = \partial_t + \frac{1}{t^n}\left(p_{-1} + \mathbf{v}(t)\right), \qquad \mathbf{v}(t) \in {}^L \mathfrak{b}[[t]].$$

We denote the space of such opers by $\mathrm{Op}_{{}^L G}^{\leq n}(D_y)$.

One can show that for $\chi_y \in \mathrm{Op}_{{}^L G}^{\leq n}(D_y)$ the category $\widehat{\mathfrak{g}}_{\kappa_c, y}\text{-mod}_{\chi_y}^K$ is non-trivial if K is the congruence subgroup $K_{m,y} \subset G(\mathcal{O}_y)$ with $m \geq n$. (We recall that for $m > 0$ we have $K_{m,\bar{y}} = \exp(\mathfrak{g} \otimes (\mathfrak{m}_y)^m)$, where $\mathfrak{m}_y$ is the maximal ideal of $\mathcal{O}_y$.) Let us take the category $\widehat{\mathfrak{g}}_{\kappa_c, y}\text{-mod}_{\chi_y}^{K_{n,y}}$. Then our general formalism gives us a localization functor

$$\Delta_{\kappa_c, K_{n,y}} : \widehat{\mathfrak{g}}_{\kappa_c, y}\text{-mod}_{\chi_y}^{K_{n,y}} \to \mathcal{D}_{\kappa_c, K_{n,y}}\text{-mod},$$

where $\mathcal{D}_{\kappa_c, K_{n,y}}\text{-mod}$ is the category of critically twisted[20] $\mathcal{D}$-modules on

$$\mathrm{Bun}_{G, y, n} \simeq G_{\mathrm{out}} \backslash G(\mathcal{K}_y) / K_{n,y}.$$

This is the moduli stack of G-bundles on X with a level n structure at $y \in X$ (which is a trivialization of the restriction of the G-bundle to the nth infinitesimal neighborhood of y).

---

[19] One may also try to extend this "spectral decomposition" to the case of all connections with regular singularities, but here the situation is more subtle, as can already be seen in the abelian case.

[20] this refers to the twisting by the line bundle on $\mathrm{Bun}_{G, y, n}$ obtained by pull-back of the line bundle $K^{1/2}$ on Bun_G, as before

In the same way as above, one shows that the $\mathcal{D}$-modules obtained by applying $\Delta_{\kappa_c, K_{n,y}}$ to objects of $\widehat{\mathfrak{g}}_{\kappa_c,y}$-mod$^{K_{n,y}}_{\chi_y}$ are Hecke eigensheaves with the eigenvalue $E_\chi|_{X\backslash y}$, where E_χ is the $^L G$-local system underlying the oper χ. Let $\mathcal{A}ut^{K_{n,y}}_{E_\chi}$ be the category of these eigensheaves. Thus, we really obtain a functor

$$\widehat{\mathfrak{g}}_{\kappa_c,y}\text{-mod}^{K_{n,y}}_{\chi_y} \to \mathcal{A}ut^{K_{n,y}}_{E_\chi}.$$

By analogy with the case of regular connections, we expect that this functor is an equivalence of categories.

As before, this functor may be generalized to an arbitrary flat bundle $E = (\mathcal{F}, \nabla)$, where ∇ has singularity at y, by representing it as an oper with mild ramification at additional points $y_1, \ldots, y_m$ on X. Let χ_y be the restriction of this oper to $D_y^\times$. Then it belongs to $\operatorname{Op}^{\leq n}_{L_G}(D_y)$ for some n, and we obtain a functor

$$\widehat{\mathfrak{g}}_{\kappa_c,y}\text{-mod}^{K_{n,y}}_{\chi_y} \to \mathcal{A}ut^{K_{n,y}}_{E},$$

which we expect to be an equivalence of categories for generic E. This also has an obvious multi-point generalization.

This way we obtain a conjectural description of the categories of Hecke eigensheaves corresponding to (generic) connections on X with arbitrary singularities at finitely many points in terms of categories of Harish-Chandra modules of critical level over $\widehat{\mathfrak{g}}$. However, in the case of regular singularities, we also have an alternative description of these categories: in terms of (derived) categories of quasicoherent sheaves on the varieties $\mathcal{B}^{DG}_M$. It would be desirable to obtain such a description for irregular connections as well.

Finally, we remark that the above construction has a kind of limiting version where we take the infinite level structure at y, i.e., a trivialization of the restriction of a G-bundle to the disc D_y. Let $\operatorname{Bun}_{G,y,\infty}$ be the moduli stack of G-bundles on X with an infinite level structure at y. Then

$$\operatorname{Bun}_{G,y,\infty} \simeq G_{\text{out}}\backslash G(\mathcal{K}_y).$$

We now have a localization functor

$$\widehat{\mathfrak{g}}_{\kappa_c,y}\text{-mod}_{\chi_y} \to \mathcal{A}ut^\infty_E,$$

where E and χ_y are as above, and $\mathcal{A}ut^\infty_E$ is the category of Hecke eigensheaves on $\operatorname{Bun}_{G,y,\infty}$ with eigenvalue $E|_{X\backslash y}$. Thus, instead of the category $\widehat{\mathfrak{g}}_{\kappa_c,y}$-mod$^{K_{n,y}}_{\chi_y}$ of Harish-Chandra modules we now have the category $\widehat{\mathfrak{g}}_{\kappa_c,y}$-mod$_{\chi_y}$ of all (smooth) $\widehat{\mathfrak{g}}_{\kappa_c,y}$-modules with fixed central character (corresponding to χ).

According to our general local conjecture, this is precisely the local Langlands category associated to the restriction of the local system E to $D_y^\times$ (equipped with an action of the loop group $G(\mathcal{K}_y)$). It is natural to assume that for generic E this functor establishes an equivalence between this category and the category $\mathcal{A}ut^\infty_E$ of Hecke eigensheaves on $\operatorname{Bun}_{G,y,\infty}$ (also equipped with an action of the loop group $G(\mathcal{K}_y)$. This may be thought of as the ultimate form of the local–to–global compatibility in the geometric Langlands Program:

$$
\begin{array}{ccc}
E & \longrightarrow & \mathcal{A}ut_E^\infty \\
\downarrow & & \uparrow \\
E|_{D_y^\times} & \longrightarrow & \widehat{\mathfrak{g}}_{\kappa_c}\text{-mod}_{\chi_y}\,.
\end{array}
$$

Let us summarize: by using representation theory of affine Kac-Moody algebras at the critical level we have constructed the local Langlands categories corresponding to the local Langlands parameters: $^L G$-local systems on the punctured disc. We then applied the technique of localization functors to produce from these local categories, the global categories of Hecke eigensheaves on the moduli stacks of G-bundles on a curve X with parabolic (or level) structures. These global categories correspond to the global Langlands parameters: $^L G$-local systems on X with ramification. We have used our results and conjectures on the structure of the local categories to investigate these global categories. We hope that in this way representation theory of affine Kac-Moody algebras may one day fulfill the dream of uncovering the mysteries of the geometric Langlands correspondence.

References

[AB]   A. Arkhipov and R. Bezrukavnikov, *Perverse sheaves on affine flags and Langlands dual group*, Preprint math.RT/0201073.

[ABG]   S. Arkhipov, R. Bezrukavnikov and V. Ginzburg, *Quantum groups, the loop Grassmannian, and the Springer resolution*, Journal of AMS **17** (2004) 595–678.

[AG]   S. Arkhipov and D. Gaitsgory, *Another realization of the category of modules over the small quantum group*, Adv. Math. **173** (2003) 114–143.

[A]   J. Arthur, *Unipotent automorphic representations: conjectures*, Asterisque **171-172** (1989) 13–71.

[BV]   D.G. Babbitt and V.S. Varadarajan, *Formal reduction theory of meromorphic differential equations: a group theoretic view*, Pacific J. Math. **109** (1983) 1–80.

[BeLa] A. Beauville and Y. Laszlo, *Un lemme de descente*, C.R. Acad. Sci. Paris, Sér. I Math. **320** (1995) 335–340.

[Bei]   A. Beilinson, *Langlands parameters for Heisenberg modules*, Preprint math.QA/0204020.

[BB]   A. Beilinson and J. Bernstein, *A proof of Jantzen conjectures*, Advances in Soviet Mathematics **16**, Part 1, pp. 1–50, AMS, 1993.

[BD1]   A. Beilinson and V. Drinfeld, *Quantization of Hitchin's integrable system and Hecke eigensheaves*, Preprint, available at www.math.uchicago.edu/~arinkin

[BD2]   A. Beilinson and V. Drinfeld, *Chiral algebras*, Colloq. Publ. **51**, AMS, 2004.

[BD3]   A. Beilinson and V. Drinfeld, *Opers*, Preprint math.AG/0501398.

[BeLu]   J. Bernstein and V. Lunts, *Localization for derived categories of* $(\mathfrak{g}, K)$-*modules*, Journal of AMS **8** (1995) 819–856.

[BZ]   J. Bernstein and A. Zelevinsky, *Induced representations of reductive p-adic groups*, I, Ann. Sci. ENS **10** (1977) 441–472.

[Bez1]   R. Bezrukavnikov, *Perverse sheaves on affine flags and nilpotent cone of the Langlands dual group*, Preprint math.RT/0201256.

[Bez2]   R. Bezrukavnikov, *Noncommutative counterparts of the Springer resolution*, Preprint math.RT/0604445.

[B1]   A. Borel, *Admissible representations of a semi-simple group over a local field with vectors fixed under an Iwahori subgroup*, Inv. Math. **35** (1976) 233–259.

[B2]   A. Borel, e.a., *Algebraic D-modules*, Academic Press, 1987.

[CG]   N. Chriss and V. Ginzburg, *Representation theory and complex geometry*, Birkhäuser 1997.

[CK]   I. Ciocan-Fountanine and M. Kapranov, *Derived Quot schemes*, Ann. Sci. ENS **34** (2001) 403–440.

[De1]   P. Deligne, *Equations différentielles á points singuliers réguliers*, Lect. Notes in Math. **163**, Springer, 1970.

[De2]   P. Deligne, *Les constantes des équations fonctionnelles des fonctions L*, in *Modular Functions one Variable II*, Proc. Internat. Summer School, Univ. Antwerp 1972, Lect. Notes Math. **349**, pp. 501–597, Springer 1973.

[Dr1]   V.G. Drinfeld, *Langlands conjecture for* $GL(2)$ *over function field*, Proc. of Int. Congress of Math. (Helsinki, 1978), pp. 565–574.

[Dr2]   V.G. Drinfeld, *Two-dimensional ℓ-adic representations of the fundamental group of a curve over a finite field and automorphic forms on* $GL(2)$, Amer. J. Math. **105** (1983) 85–114.

[Dr3]   V.G. Drinfeld, *Moduli varieties of F-sheaves*, Funct. Anal. Appl. **21** (1987) 107–122.

[Dr4]   V.G. Drinfeld, *The proof of Petersson's conjecture for* $GL(2)$ *over a global field of characteristic p*, Funct. Anal. Appl. **22** (1988) 28–43.

[DS]   V. Drinfeld and V. Sokolov, *Lie algebras and KdV type equations*, J. Sov. Math. **30** (1985) 1975–2036.

[DrSi]   V. Drinfeld and C. Simpson, *B-structures on G-bundles and local triviality*, Math. Res. Lett. **2** (1995) 823–829.

[EF]   D. Eisenbud and E. Frenkel, Appendix to M. Mustata, *Jet schemes of locally complete intersection canonical singularities*, Invent. Math. **145** (2001) 397–424.

[FF1]   B. Feigin and E. Frenkel, *A family of representations of affine Lie algebras*, Russ. Math. Surv. **43**, N 5 (1988) 221–222.

[FF2]   B. Feigin and E. Frenkel, *Affine Kac-Moody Algebras and semi-infinite flag manifolds*, Comm. Math. Phys. **128**, 161–189 (1990).

[FF3]   B. Feigin and E. Frenkel, *Affine Kac-Moody algebras at the critical level and Gelfand-Dikii algebras*, Int. Jour. Mod. Phys. **A7**, Supplement 1A (1992) 197–215.

[FK]   E. Freitag and R. Kiehl, *Etale Cohomology and the Weil conjecture*, Springer, 1988.

[F1]   E. Frenkel, *Affine algebras, Langlands duality and Bethe Ansatz*, in Proceedings of the International Congress of Mathematical Physics, Paris, 1994, ed. D. Iagolnitzer, pp. 606–642, International Press, 1995; arXiv: q-alg/9506003.

[F2]   E. Frenkel, *Recent advances in the Langlands Program*, Bull. Amer. Math. Soc. **41** (2004) 151–184 (math.AG/0303074).

[F3]   E. Frenkel, *Wakimoto modules, opers and the center at the critical level*, Advances in Math. **195** (2005) 297–404.

[F4]   E. Frenkel, *Gaudin model and opers*, in Infinite Dimensional Algebras and Quantum Integrable Systems, eds. P. Kulish, e.a., Progress in Math. **237**, pp. 1–60, Birkhäuser, 2005.

[F5]   E. Frenkel, *Opers on the projective line, flag manifolds and Bethe Ansatz*, Mosc. Math. J. **4** (2004) 655–705.

[F6]   E. Frenkel, *Lectures on the Langlands Program and conformal field theory*, Preprint hep-th/0512172.

[F7]   E. Frenkel, *Langlands Correspondence for Loop Groups. An Introduction*, to be published by Cambridge University Press; draft available at http://math.berkeley.edu/∼frenkel/

[FB]   E. Frenkel and D. Ben-Zvi, *Vertex algebras and algebraic curves*, Second Edition, Mathematical Surveys and Monographs, vol. 88. AMS 2004.

[FG1]   E. Frenkel and D. Gaitsgory, *D-modules on the affine Grassmannian and representations of affine Kac-Moody algebras*, Duke Math. J. **125** (2004) 279–327.

[FG2]   E. Frenkel and D. Gaitsgory, *Local geometric Langlands correspondence and affine Kac-Moody algebras*, Preprint math.RT/0508382.

[FG3]   E. Frenkel and D. Gaitsgory, *Fusion and convolution: applications to affine Kac-Moody algebras at the critical level*, Preprint math.RT/0511284.

[FG4]   E. Frenkel and D. Gaitsgory, *Localization of $\widehat{\mathfrak{g}}$-modules on the affine Grassmannian*, Preprint math.RT/0512562.

[FG5]   E. Frenkel and D. Gaitsgory, *Geometric realizations of Wakimoto modules at the critical level*, Preprint math.RT/0603524.

[FG6]   E. Frenkel and D. Gaitsgory, *Weyl modules and opers without monodromy*, to appear.

[FGV]   E. Frenkel, D. Gaitsgory and K. Vilonen, *On the geometric Langlands conjecture*, Journal of AMS **15** (2001) 367–417.

[FT]   E. Frenkel and C. Teleman, *Self-extensions of Verma modules and differential forms on opers*, Compositio Math. **142** (2006) 477–500.

[Ga1]   D. Gaitsgory, *Construction of central elements in the Iwahori Hecke algebra via nearby cycles*, Inv. Math. **144** (2001) 253–280.

[Ga2]   D. Gaitsgory, *On a vanishing conjecture appearing in the geometric Langlands correspondence*, Ann. Math. **160** (2004) 617–682.

[Ga3]   D. Gaitsgory, *The notion of category over an algebraic stack*, Preprint math.AG/0507192.

[GM]   S.I. Gelfand, Yu.I. Manin, *Homological algebra*, Encyclopedia of Mathematical Sciences **38**, Springer, 1994.

[GW]   S. Gukov and E. Witten, *Gauge theory, ramification, and the geometric Langlands Program*, to appear.

[HT]   M. Harris and R. Taylor, *The geometry and cohomology of some simple Shimura varieties*, Annals of Mathematics Studies **151**, Princeton University Press, 2001.

[He]   G. Henniart, *Une preuve simple des conjectures de Langlands pour* GL(n) *sur un corps p-adique*, Invent. Math. **139** (2000) 439–455.

[K1]   V. Kac, *Laplace operators of infinite-dimensional Lie algebras and theta functions*, Proc. Nat. Acad. Sci. U.S.A. **81** (1984) no. 2, Phys. Sci., 645–647.

[K2]   V.G. Kac, *Infinite-dimensional Lie Algebras*, 3rd Edition, Cambridge University Press, 1990.

[KW]   A. Kapustin and E. Witten, *Electric-magnetic duality and the geometric Langlands Program*, Preprint hep-th/0604151.

[KL]   D. Kazhdan and G. Lusztig, *Proof of the Deligne–Langlands conjecture for Hecke algebras*, Inv. Math. **87** (1987) 153–215.

[Ku]   S. Kudla, *The local Langlands correspondence: the non-Archimedean case*, in *Motives* (Seattle, 1991), pp. 365–391, Proc. Sympos. Pure Math. **55**, Part 2, AMS, 1994.

[Laf]   L. Lafforgue, *Chtoucas de Drinfeld et correspondance de Langlands*, Invent. Math. **147** (2002) 1–241.

[L]   R.P. Langlands, *Problems in the theory of automorphic forms*, in Lect. Notes in Math. **170**, pp. 18–61, Springer Verlag, 1970.

[La]   G. Laumon, *Transformation de Fourier, constantes d'équations fonctionelles et conjecture de Weil*, Publ. IHES **65** (1987) 131–210.

[LRS]   G. Laumon, M. Rapoport and U. Stuhler, *D-elliptic sheaves and the Langlands correspondence*, Invent. Math. **113** (1993) 217–338.

[Lu1]   G. Lusztig, *Classification of unipotent representations of simple p-adic groups*, Int. Math. Res. Notices (1995) no. 11, 517–589.

[Lu2]   G. Lusztig, *Bases in K-theory*, Represent. Theory **2** (1998) 298–369; **3** (1999) 281–353.

[Mi]   J.S. Milne, *Étale cohomology*, Princeton University Press, 1980.

[MV]   I. Mirković and K. Vilonen, *Geometric Langlands duality and representations of algebraic groups over commutative rings*, Preprint math.RT/0401222.

[Sat]   I. Satake, *Theory of spherical functions on reductive algebraic groups over p-adic fields*, IHES Publ. Math. **18** (1963) 5–69.

[Vog]   D.A. Vogan, *The local Langlands conjecture*, Contemporary Math. **145**, pp. 305–379, AMS, 1993.

[Wak]   M. Wakimoto, *Fock representations of affine Lie algebra $A_1^{(1)}$*, Comm. Math. Phys. **104** (1986) 605–609.

Equivariant Derived Category and Representation of Real Semisimple Lie Groups

Masaki Kashiwara

Research Institute for Mathematical Sciences, Kyoto University, Kyoto 606–8502, Japan
masaki@kurims.kyoto-u.ac.jp

1 Introduction

This note is based on five lectures on the geometry of flag manifolds and the representation theory of real semisimple Lie groups, delivered at the CIME summer school "Representation theory and Complex Analysis", June 10–17, 2004, Venezia.

The study of the relation between the geometry of flag manifolds and the representation theory of complex algebraic groups has a long history. However, it was rather recently that we realized the close relation between the representation theory of real semisimple Lie groups and the geometry of the flag manifold and its cotangent bundle. In these relations, there are two facets, complex geometry and real geometry. The Matsuki correspondence is an example: it is a correspondence between the orbits of the real semisimple group on the flag manifold and the orbits of the complexification of its maximal compact subgroup.

Among these relations, we focus on the diagram below.

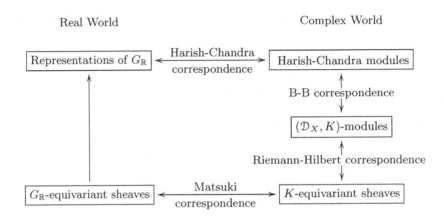

Fig. 1. Correspondences

The purpose of this note is to explain this diagram.

In the Introduction, we give an overview of this diagram, and we will explain more details in the subsequent sections. In order to simplify the arguments, in the Introduction we restrict ourselves to the case of the trivial infinitesimal character. In order to treat the general case, we need the "twisting" of sheaves and of the ring of differential operators. For them, see the subsequent sections.

Considerable parts of this note are joint work with W. Schmid, and were announced in [21].

Acknowledgement. The author would like to thank Andrea D'Agnolo for the organization of the Summer School and his help during the preparation of this note. He also thanks Kyo Nishiyama, Toshiyuki Kobayashi and Akira Kono for valuable advices.

1.1 Harish-Chandra Correspondence

Let $G_{\mathbb{R}}$ be a connected real semisimple Lie group with a finite center, and $K_{\mathbb{R}}$ a maximal compact subgroup of $G_{\mathbb{R}}$. Let $\mathfrak{g}_{\mathbb{R}}$ and $\mathfrak{k}_{\mathbb{R}}$ be the Lie algebras of $G_{\mathbb{R}}$ and $K_{\mathbb{R}}$, respectively. Let $\mathfrak{g}$ and $\mathfrak{k}$ be their complexifications. Let K be the complexification of $K_{\mathbb{R}}$.

We consider a representation of $G_{\mathbb{R}}$. Here, it means a complete locally convex topological space E with a continuous action of $G_{\mathbb{R}}$. A vector v in E

is called $K_{\mathbb{R}}$-finite if v is contained in a finite-dimensional $K_{\mathbb{R}}$-submodule of E. Harish-Chandra considered

$$\mathrm{HC}(E) := \{v \in E \,; \, v \text{ is } K_{\mathbb{R}}\text{-finite}\} .$$

If E has finite $K_{\mathbb{R}}$-multiplicities, i.e., $\dim \mathrm{Hom}_{K_{\mathbb{R}}}(V, E) < \infty$ for any finite-dimensional irreducible representation V of $K_{\mathbb{R}}$, he called E an *admissible* representation. The action of $G_{\mathbb{R}}$ on an admissible representation E can be differentiated on $\mathrm{HC}(E)$, and $\mathfrak{g}$ acts on $\mathrm{HC}(E)$. Since any continuous $K_{\mathbb{R}}$-action on a finite-dimensional vector space extends to a K-action, $\mathrm{HC}(E)$ has a $(\mathfrak{g}, K)$-module structure (see Definition 3.1.1).

Definition 1.1.1. A $(\mathfrak{g}, K)$-module M is called a *Harish-Chandra module* if it satisfies the conditions:

(a) M is $\mathfrak{z}(\mathfrak{g})$-finite,
(b) M has finite K-multiplicities,
(c) M is finitely generated over $U(\mathfrak{g})$.

Here, $U(\mathfrak{g})$ is the universal enveloping algebra of $\mathfrak{g}$ and $\mathfrak{z}(\mathfrak{g})$ is the center of $U(\mathfrak{g})$. The condition (a) means that the image of $\mathfrak{z}(\mathfrak{g}) \to \mathrm{End}(M)$ is finite-dimensional over $\mathbb{C}$.

In fact, if two of the three conditions (a)–(c) are satisfied, then all of the three are satisfied.

An admissible representation E is of finite length if and only if $\mathrm{HC}(E)$ is a Harish-Chandra module.

The $(\mathfrak{g}, K)$-module $\mathrm{HC}(E)$ is a dense subspace of E, and hence E is the completion of $\mathrm{HC}(E)$ with the induced topology on $\mathrm{HC}(E)$. However, for a Harish-Chandra module M, there exist many representations E such that $\mathrm{HC}(E) \simeq M$. Among them, there exist the smallest one $\mathrm{mg}(M)$ and the largest one $\mathrm{MG}(M)$.

More precisely, we have the following results ([24, 25]). Let $\mathcal{T}_{G_{\mathbb{R}}}^{\mathrm{adm}}$ be the category of admissible representations of $G_{\mathbb{R}}$ of finite length. Let $\mathrm{HC}(\mathfrak{g}, K)$ be the category of Harish-Chandra modules. Then, for any $M \in \mathrm{HC}(\mathfrak{g}, K)$, there exist $\mathrm{mg}(M)$ and $\mathrm{MG}(M)$ in $\mathcal{T}_{G_{\mathbb{R}}}^{\mathrm{adm}}$ satisfying:

(1.1.1)
$$\mathrm{Hom}_{\mathrm{HC}(\mathfrak{g}, K)}(M, \mathrm{HC}(E)) \simeq \mathrm{Hom}_{\mathcal{T}_{G_{\mathbb{R}}}^{\mathrm{adm}}}(\mathrm{mg}(M), E),$$
$$\mathrm{Hom}_{\mathrm{HC}(\mathfrak{g}, K)}(\mathrm{HC}(E), M) \simeq \mathrm{Hom}_{\mathcal{T}_{G_{\mathbb{R}}}^{\mathrm{adm}}}(E, \mathrm{MG}(M))$$

for any $E \in \mathcal{T}_{G_{\mathbb{R}}}^{\mathrm{adm}}$. In other words, $M \mapsto \mathrm{mg}(M)$ (resp. $M \mapsto \mathrm{MG}(M)$) is a left adjoint functor (resp. right adjoint functor) of the functor $\mathrm{HC} \colon \mathcal{T}_{G_R}^{\mathrm{adm}} \to \mathrm{HC}(\mathfrak{g}, K)$. Moreover we have

$$M \xrightarrow{\sim} \mathrm{HC}(\mathrm{mg}(M)) \xrightarrow{\sim} \mathrm{HC}(\mathrm{MG}(M)) \quad \text{for any } M \in \mathrm{HC}(\mathfrak{g}, K).$$

For a Harish-Chandra module M and a representation E such that $\mathrm{HC}(E) \simeq M$, we have

$$M \subset \mathrm{mg}(M) \subset E \subset \mathrm{MG}(M).$$

We call $\mathrm{mg}(M)$ the *minimal globalization* of M and $\mathrm{MG}(M)$ the *maximal globalization* of M. The space $\mathrm{mg}(M)$ is a dual Fréchet nuclear space and $\mathrm{MG}(M)$ is a Fréchet nuclear space (see Example 2.1.2 (ii)).

Example 1.1.2. Let $P_{\mathbb{R}}$ be a parabolic subgroup of $G_{\mathbb{R}}$ and $Y = G_{\mathbb{R}}/P_{\mathbb{R}}$. Then Y is compact. The space $\mathscr{A}(Y)$ of real analytic functions, the space $\mathrm{C}^{\infty}(Y)$ of C^{∞}-functions, the space $\mathrm{L}^2(Y)$ of L^2-functions, the space $\mathscr{D}ist(Y)$ of distributions, and the space $\mathscr{B}(Y)$ of hyperfunctions are admissible representations of $G_{\mathbb{R}}$, and they have the same Harish-Chandra module M. We have

$$\mathrm{mg}(M) = \mathscr{A}(Y) \subset \mathrm{C}^{\infty}(Y) \subset \mathrm{L}^2(Y) \subset \mathscr{D}ist(Y) \subset \mathscr{B}(Y) = \mathrm{MG}(M).$$

The representation $\mathrm{MG}(M)$ can be explicitly constructed as follows. Let us set

$$M^* = \mathrm{Hom}_{\mathbb{C}}(M, \mathbb{C})^{K\text{-fini}}.$$

Here, the superscript "K-fini" means the set of K-finite vectors. Then M^* is again a Harish-Chandra module, and we have

$$\mathrm{MG}(M) \simeq \mathrm{Hom}_{U(\mathfrak{g})}(M^*, \mathrm{C}^{\infty}(G_{\mathbb{R}})).$$

Here, $\mathrm{C}^{\infty}(G_{\mathbb{R}})$ is a $U(\mathfrak{g})$-module with respect to the right action of $G_{\mathbb{R}}$ on $G_{\mathbb{R}}$. The module $\mathrm{Hom}_{U(\mathfrak{g})}(M^*, \mathrm{C}^{\infty}(G_{\mathbb{R}}))$ is calculated with respect to this structure. Since the left $G_{\mathbb{R}}$-action on $G_{\mathbb{R}}$ commutes with the right action, $\mathrm{Hom}_{U(\mathfrak{g})}(M^*, \mathrm{C}^{\infty}(G_{\mathbb{R}}))$ is a representation of $G_{\mathbb{R}}$ by the left action of $G_{\mathbb{R}}$ on $G_{\mathbb{R}}$. We endow $\mathrm{Hom}_{U(\mathfrak{g})}(M^*, \mathrm{C}^{\infty}(G_{\mathbb{R}}))$ with the topology induced from the Fréchet nuclear topology of $\mathrm{C}^{\infty}(G_{\mathbb{R}})$. The minimal globalization $\mathrm{mg}(M)$ is the dual representation of $\mathrm{MG}(M^*)$.

In §10, we shall give a proof of the fact that $M \mapsto \mathrm{mg}(M)$ and $M \mapsto \mathrm{MG}(M)$ are exact functors, and $\mathrm{mg}(M) \simeq \Gamma_c(G_{\mathbb{R}}; \mathscr{D}ist_{G_{\mathbb{R}}}) \otimes_{U(\mathfrak{g})} M$. Here, $\Gamma_c(G_{\mathbb{R}}; \mathscr{D}ist_{G_{\mathbb{R}}})$ is the space of distributions on $G_{\mathbb{R}}$ with compact support.

1.2 Beilinson-Bernstein Correspondence

Beilinson and Bernstein established the correspondence between $U(\mathfrak{g})$-modules and D-modules on the flag manifold.

Let G be a semisimple algebraic group with $\mathfrak{g}$ as its Lie algebra. Let X be the flag manifold of G, i.e., the space of all Borel subgroups of G.

For a $\mathbb{C}$-algebra homomorphism $\chi \colon \mathfrak{z}(\mathfrak{g}) \to \mathbb{C}$ and a $\mathfrak{g}$-module M, we say that M has an *infinitesimal character* χ if $a \cdot u = \chi(a)u$ for any $a \in \mathfrak{z}(\mathfrak{g})$ and $u \in M$. In Introduction, we restrict ourselves to the case of the trivial infinitesimal character, although we treat the general case in the body of this note. Let $\chi_{\mathrm{triv}} \colon \mathfrak{z}(\mathfrak{g}) \to \mathbb{C}$ be the trivial infinitesimal character (the infinitesimal character of the trivial representation). We set $U_{\chi_{\mathrm{triv}}}(\mathfrak{g}) = U(\mathfrak{g})/U(\mathfrak{g}) \mathrm{Ker}(\chi_{\mathrm{triv}})$.

Then $U_{\chi_{\mathrm{triv}}}(\mathfrak{g})$-modules are nothing but $\mathfrak{g}$-modules with the trivial infinitesimal character.

Let $\mathscr{D}_X$ be the sheaf of differential operators on X. Then we have the following theorem due to Beilinson-Bernstein [1].

Theorem 1.2.1. (i) *The Lie algebra homomorphism* $\mathfrak{g} \to \Gamma(X; \mathscr{D}_X)$ *induces an isomorphism*

$$U_{\chi_{\mathrm{triv}}}(\mathfrak{g}) \xrightarrow{\sim} \Gamma(X; \mathscr{D}_X).$$

(ii) $H^n(X; \mathscr{M}) = 0$ *for any quasi-coherent* $\mathscr{D}_X$*-module* $\mathscr{M}$ *and* $n \neq 0$.
(iii) *The category* $\mathrm{Mod}(\mathscr{D}_X)$ *of quasi-coherent* $\mathscr{D}_X$*-modules and the category* $\mathrm{Mod}(U_{\chi_{\mathrm{triv}}}(\mathfrak{g}))$ *of* $U_{\chi_{\mathrm{triv}}}(\mathfrak{g})$*-modules are equivalent by*

$$\mathrm{Mod}(\mathscr{D}_X) \ni \mathscr{M} \longmapsto \Gamma(X; \mathscr{M}) \in \mathrm{Mod}(U_{\chi_{\mathrm{triv}}}(\mathfrak{g})),$$

$$\mathrm{Mod}(\mathscr{D}_X) \ni \mathscr{D}_X \otimes_{U(\mathfrak{g})} M \longleftarrow M \in \mathrm{Mod}(U_{\chi_{\mathrm{triv}}}(\mathfrak{g})).$$

In particular, we have the following corollary.

Corollary 1.2.2. *The category* $\mathrm{HC}_{\chi_{\mathrm{triv}}}(\mathfrak{g}, K)$ *of Harish-Chandra modules with the trivial infinitesimal character and the category* $\mathrm{Mod}_{K, \mathrm{coh}}(\mathscr{D}_X)$ *of coherent* K*-equivariant* $\mathscr{D}_X$*-modules are equivalent.*

The K-equivariant $\mathscr{D}_X$-modules are, roughly speaking, $\mathscr{D}_X$-modules with an action of K. (For the precise definition, see § 3.) We call this equivalence the B-B correspondence.

The set of isomorphism classes of irreducible K-equivariant $\mathscr{D}_X$-modules is isomorphic to the set of pairs (O, L) of a K-orbit O in X and an isomorphism class L of an irreducible representation of the finite group $K_x/(K_x)^\circ$. Here K_x is the isotropy subgroup of K at a point x of O, and $(K_x)^\circ$ is its connected component containing the identity. Hence the set of isomorphism classes of irreducible Harish-Chandra modules with the trivial infinitesimal character corresponds to the set of such pairs (O, L).

1.3 Riemann-Hilbert Correspondence

The flag manifold X has finitely many K-orbits. Therefore any coherent K-equivariant $\mathscr{D}_X$-module is a regular holonomic $\mathscr{D}_X$-module (see [15]). Let $\mathrm{D}^{\mathrm{b}}(\mathscr{D}_X)$ be the bounded derived category of $\mathscr{D}_X$-modules, and let $\mathrm{D}^{\mathrm{b}}_{\mathrm{rh}}(\mathscr{D}_X)$ be the full subcategory of $\mathrm{D}^{\mathrm{b}}(\mathscr{D}_X)$ consisting of bounded complexes of $\mathscr{D}_X$-modules with regular holonomic cohomology groups.

Let $Z \longmapsto Z^{\mathrm{an}}$ be the canonical functor from the category of complex algebraic varieties to the one of complex analytic spaces. Then there exists a morphism of ringed space $\pi \colon Z^{\mathrm{an}} \to Z$. For an $\mathscr{O}_Z$-module $\mathscr{F}$, let $\mathscr{F}^{\mathrm{an}} := \mathscr{O}_{Z^{\mathrm{an}}} \otimes_{\pi^{-1}\mathscr{O}_Z} \pi^{-1}\mathscr{F}$ be the corresponding $\mathscr{O}_{Z^{\mathrm{an}}}$-module. Similarly, for a $\mathscr{D}_Z$-module $\mathscr{M}$, let $\mathscr{M}^{\mathrm{an}} := \mathscr{D}_{Z^{\mathrm{an}}} \otimes_{\pi^{-1}\mathscr{D}_Z} \pi^{-1}\mathscr{M} \simeq \mathscr{O}_{Z^{\mathrm{an}}} \otimes_{\pi^{-1}\mathscr{O}_Z} \pi^{-1}\mathscr{M}$ be the corresponding $\mathscr{D}_{Z^{\mathrm{an}}}$-module. For a $\mathscr{D}_Z$-module $\mathscr{M}$ and a $\mathscr{D}_{Z^{\mathrm{an}}}$-module $\mathscr{N}$, we write

$\mathscr{H}om_{\mathscr{D}_Z}(\mathscr{M},\mathscr{N})$ instead of $\mathscr{H}om_{\pi^{-1}\mathscr{D}_Z}(\pi^{-1}\mathscr{M},\mathscr{N}) \simeq \mathscr{H}om_{\mathscr{D}_{Z^{an}}}(\mathscr{M}^{an},\mathscr{N})$ for short.

Let us denote by $D^b(\mathbb{C}_{X^{an}})$ the bounded derived category of sheaves of $\mathbb{C}$-vector spaces on X^{an}. Then the de Rham functor $DR_X: D^b(\mathscr{D}_X) \rightarrow D^b(\mathbb{C}_{X^{an}})$, given by $DR_X(\mathscr{M}) = \mathbf{R}\mathscr{H}om_{\mathscr{D}_X}(\mathscr{O}_X,\mathscr{M}^{an})$, induces an equivalence of triangulated categories, called the *Riemann-Hilbert correspondence* ([12])

$$DR_X: D^b_{rh}(\mathscr{D}_X) \xrightarrow{\sim} D^b_{\mathbb{C}\text{-}c}(\mathbb{C}_{X^{an}}).$$

Here $D^b_{\mathbb{C}\text{-}c}(\mathbb{C}_{X^{an}})$ is the full subcategory of $D^b(\mathbb{C}_{X^{an}})$ consisting of bounded complexes of sheaves of $\mathbb{C}$-vector spaces on X^{an} with constructible cohomologies (see [18] and also §4.4).

Let $RH(\mathscr{D}_X)$ be the category of regular holonomic $\mathscr{D}_X$-modules. Then it may be regarded as a full subcategory of $D^b_{rh}(\mathscr{D}_X)$. Its image by DR_X is a full subcategory of $D^b_{\mathbb{C}\text{-}c}(\mathbb{C}_{X^{an}})$ and denoted by $\text{Perv}(\mathbb{C}_{X^{an}})$. Since $RH(\mathscr{D}_X)$ is an abelian category, $\text{Perv}(\mathbb{C}_{X^{an}})$ is also an abelian category. An object of $\text{Perv}(\mathbb{C}_{X^{an}})$ is called a *perverse sheaf* on X^{an}.

Then the functor DR_X induces an equivalence between $\text{Mod}_{K,\text{coh}}(\mathscr{D}_X)$ and the category $\text{Perv}_{K^{an}}(\mathbb{C}_{X^{an}})$ of K^{an}-equivariant perverse sheaves on X^{an}:

$$DR_X: \text{Mod}_{K,\text{coh}}(\mathscr{D}_X) \xrightarrow{\sim} \text{Perv}_{K^{an}}(\mathbb{C}_{X^{an}}).$$

1.4 Matsuki Correspondence

The following theorem is due to Matsuki ([22]).

Proposition 1.4.1.   (i) *There are only finitely many K-orbits in X and also finitely many $G_{\mathbb{R}}$-orbits in X^{an}.*

(ii) *There is a one-to-one correspondence between the set of K-orbits and the set of $G_{\mathbb{R}}$-orbits.*

(iii) *A K-orbit U and a $G_{\mathbb{R}}$-orbit V correspond by the correspondence in (ii) if and only if $U^{an} \cap V$ is a $K_{\mathbb{R}}$-orbit.*

Its sheaf-theoretical version is conjectured by Kashiwara [14] and proved by Mirković-Uzawa-Vilonen [23].

In order to state the results, we have to use the equivariant derived category (see [4], and also §4). Let H be a real Lie group, and let Z be a topological space with an action of H. We assume that Z is locally compact with a finite cohomological dimension. Then we can define the equivariant derived category $D^b_H(\mathbb{C}_Z)$, which has the following properties:

(a) there exists a forgetful functor $D^b_H(\mathbb{C}_Z) \rightarrow D^b(\mathbb{C}_Z)$,
(b) for any $F \in D^b_H(\mathbb{C}_Z)$, its cohomology group $H^n(F)$ is an H-equivariant sheaf on Z for any n,

(c) for any H-equivariant morphism $f\colon Z \to Z'$, there exist canonical functors $f^{-1}, f^!\colon \mathrm{D}_H^{\mathrm{b}}(\mathbb{C}_{Z'}) \to \mathrm{D}_H^{\mathrm{b}}(\mathbb{C}_Z)$ and $f_*, f_!\colon \mathrm{D}_H^{\mathrm{b}}(\mathbb{C}_Z) \to \mathrm{D}_H^{\mathrm{b}}(\mathbb{C}_{Z'})$ which commute with the forgetful functors in (a), and satisfy the usual properties (see § 4),

(d) if H acts freely on Z, then $\mathrm{D}_H^{\mathrm{b}}(\mathbb{C}_Z) \simeq \mathrm{D}^{\mathrm{b}}(\mathbb{C}_{Z/H})$.

(e) if H is a closed subgroup of H', then we have an equivalence

$$\mathrm{Ind}_H^{H'}\colon \mathrm{D}_H^{\mathrm{b}}(\mathbb{C}_Z) \xrightarrow{\sim} \mathrm{D}_{H'}^{\mathrm{b}}(\mathbb{C}_{(Z\times H')/H}).$$

Now let us come back to the case of real semisimple groups. We have an equivalence of categories:

$$(1.4.1) \qquad \mathrm{Ind}_{K^{\mathrm{an}}}^{G^{\mathrm{an}}}\colon \mathrm{D}_{K^{\mathrm{an}}}^{\mathrm{b}}(\mathbb{C}_{X^{\mathrm{an}}}) \xrightarrow{\sim} \mathrm{D}_{G^{\mathrm{an}}}^{\mathrm{b}}(\mathbb{C}_{(X^{\mathrm{an}}\times G^{\mathrm{an}})/K^{\mathrm{an}}}).$$

Let us set $S = G/K$ and $S_{\mathbb{R}} = G_{\mathbb{R}}/K_{\mathbb{R}}$. Then $S_{\mathbb{R}}$ is a Riemannian symmetric space and $S_{\mathbb{R}} \subset S$. Let $i\colon S_{\mathbb{R}} \hookrightarrow S^{\mathrm{an}}$ be the closed embedding. Since $(X \times G)/K \simeq X \times S$, we obtain an equivalence of categories

$$\mathrm{Ind}_{K^{\mathrm{an}}}^{G^{\mathrm{an}}}\colon \mathrm{D}_{K^{\mathrm{an}}}^{\mathrm{b}}(\mathbb{C}_{X^{\mathrm{an}}}) \xrightarrow{\sim} \mathrm{D}_{G^{\mathrm{an}}}^{\mathrm{b}}(\mathbb{C}_{X^{\mathrm{an}}\times S^{\mathrm{an}}}).$$

Let $p_1\colon X^{\mathrm{an}} \times S^{\mathrm{an}} \to X^{\mathrm{an}}$ be the first projection and $p_2\colon X^{\mathrm{an}} \times S^{\mathrm{an}} \to S^{\mathrm{an}}$ the second projection. We define the functor

$$\Phi\colon \mathrm{D}_{K^{\mathrm{an}}}^{\mathrm{b}}(\mathbb{C}_{X^{\mathrm{an}}}) \to \mathrm{D}_{G_{\mathbb{R}}}^{\mathrm{b}}(\mathbb{C}_{X^{\mathrm{an}}})$$

by

$$\Phi(F) = \mathbf{R}p_{1!}(\mathrm{Ind}_{K^{\mathrm{an}}}^{G^{\mathrm{an}}}(F) \otimes p_2^{-1}i_*\mathbb{C}_{S_{\mathbb{R}}})[\mathrm{d}_S].$$

Here, we use the notation

$$(1.4.2) \qquad\qquad\qquad \mathrm{d}_S = \dim S.$$

Theorem 1.4.2 ([23]). $\Phi\colon \mathrm{D}_{K^{\mathrm{an}}}^{\mathrm{b}}(\mathbb{C}_{X^{\mathrm{an}}}) \to \mathrm{D}_{G_{\mathbb{R}}}^{\mathrm{b}}(\mathbb{C}_{X^{\mathrm{an}}})$ *is an equivalence of triangulated categories.*

Roughly speaking, there is a correspondence between K^{an}-equivariant sheaves on X^{an} and $G_{\mathbb{R}}$-equivariant sheaves on X^{an}. We call it the (sheaf-theoretical) *Matsuki correspondence*.

1.5 Construction of Representations of $G_{\mathbb{R}}$

Let H be an affine algebraic group, and let Z be an algebraic manifold with an action of H. We can in fact define two kinds of H-equivariance on $\mathscr{D}_Z$-modules: a *quasi-equivariance* and an *equivariance*. (For their definitions, see Definition 3.1.3.) Note that $\mathscr{D}_Z \otimes_{\mathscr{O}_Z} \mathscr{F}$ is quasi-H-equivariant for any H-equivariant $\mathscr{O}_Z$-module $\mathscr{F}$, but it is not H-equivariant in general. The $\mathscr{D}_Z$-module $\mathscr{O}_Z$ is

H-equivariant. Let us denote by $\mathrm{Mod}(\mathscr{D}_Z, H)$ (resp. $\mathrm{Mod}_H(\mathscr{D}_Z)$) the category of quasi-$H$-equivariant (resp. H-equivariant) $\mathscr{D}_Z$-modules. Then $\mathrm{Mod}_H(\mathscr{D}_Z)$ is a full abelian subcategory of $\mathrm{Mod}(\mathscr{D}_Z, H)$.

Let $G_{\mathbb{R}}$ be a real semisimple Lie group contained in a semisimple algebraic group G as a real form. Let $\mathbf{FN}$ be the category of Fréchet nuclear spaces (see Example 2.1.2 (ii)), and let $\mathbf{FN}_{G_{\mathbb{R}}}$ be the category of Fréchet nuclear spaces with a continuous $G_{\mathbb{R}}$-action. It is an additive category but not an abelian category. However it is a quasi-abelian category and we can define its bounded derived category $\mathrm{D}^{\mathrm{b}}(\mathbf{FN}_{G_{\mathbb{R}}})$ (see §2).

Let Z be an algebraic manifold with a G-action. Let $\mathrm{D}^{\mathrm{b}}_{\mathrm{coh}}(\mathrm{Mod}(\mathscr{D}_Z, G))$ be the full subcategory of $\mathrm{D}^{\mathrm{b}}(\mathrm{Mod}(\mathscr{D}_Z, G))$ consisting of objects with coherent cohomologies. Let $\mathrm{D}^{\mathrm{b}}_{G_{\mathbb{R}}, \mathbb{R}\text{-c}}(\mathbb{C}_{Z^{\mathrm{an}}})$ be the full subcategory of the $G_{\mathbb{R}}$-equivariant derived category $\mathrm{D}^{\mathrm{b}}_{G_{\mathbb{R}}}(\mathbb{C}_{Z^{\mathrm{an}}})$ consisting of objects with $\mathbb{R}$-constructible cohomologies (see §4.4). Then for $\mathscr{M} \in \mathrm{D}^{\mathrm{b}}_{\mathrm{coh}}(\mathrm{Mod}(\mathscr{D}_Z, G))$ and $F \in \mathrm{D}^{\mathrm{b}}_{G_{\mathbb{R}}, \mathbb{R}\text{-c}}(\mathbb{C}_{Z^{\mathrm{an}}})$, we can define

$$\mathbf{R}\mathrm{Hom}^{\mathrm{top}}_{\mathscr{D}_Z}(\mathscr{M} \otimes F, \mathscr{O}_{Z^{\mathrm{an}}})$$

as an object of $\mathrm{D}^{\mathrm{b}}(\mathbf{FN}_{G_{\mathbb{R}}})$.

Roughly speaking, it is constructed as follows. (For a precise construction, see §5.) We can take a bounded complex $\mathscr{D}_Z \otimes \mathcal{V}^{\bullet}$ of quasi-G-equivariant $\mathscr{D}_Z$-modules which is isomorphic to $\mathscr{M}$ in the derived category, where each $\mathcal{V}^n$ is a G-equivariant vector bundle on Z. On the other hand, we can represent F by a complex $K^{\bullet}$ of $G_{\mathbb{R}}$-equivariant sheaves such that each K^n has a form $\oplus_{a \in I_n} L_a$ for an index set I_n, where L_a is a $G_{\mathbb{R}}$-equivariant locally constant sheaf of finite rank on a $G_{\mathbb{R}}$-invariant open subset U_a of Z^{an}.[1] Let $\mathscr{E}^{(0, \bullet)}_{Z^{\mathrm{an}}}$ be the Dolbeault resolution of $\mathscr{O}_{Z^{\mathrm{an}}}$ by differential forms with C^{∞} coefficients. Then, $\mathrm{Hom}_{\mathscr{D}_Z}((\mathscr{D}_Z \otimes \mathcal{V}^{\bullet}) \otimes K^{\bullet}, \mathscr{E}^{(0, \bullet)}_{Z^{\mathrm{an}}})$ represents $\mathbf{R}\mathrm{Hom}_{\mathscr{D}_Z}(\mathscr{M} \otimes F, \mathscr{O}_{Z^{\mathrm{an}}}) \in \mathrm{D}^{\mathrm{b}}(\mathrm{Mod}(\mathbb{C}))$. On the other hand, $\mathrm{Hom}_{\mathscr{D}_Z}((\mathscr{D}_Z \otimes \mathcal{V}^n) \otimes L_a, \mathscr{E}^{(0, q)}_{Z^{\mathrm{an}}}) = \mathrm{Hom}_{\mathscr{O}_Z}(\mathcal{V}^n \otimes L_a, \mathscr{E}^{(0, q)}_{Z^{\mathrm{an}}})$ carries a natural topology of Fréchet nuclear spaces and is endowed with a continuous $G_{\mathbb{R}}$-action. Hence $\mathrm{Hom}_{\mathscr{D}_Z}((\mathscr{D}_Z \otimes \mathcal{V}^{\bullet}) \otimes K^{\bullet}, \mathscr{E}^{(0, \bullet)}_{Z^{\mathrm{an}}})$ is a complex of objects in $\mathbf{FN}_{G_{\mathbb{R}}}$. It is $\mathbf{R}\mathrm{Hom}^{\mathrm{top}}_{\mathscr{D}_Z}(\mathscr{M} \otimes F, \mathscr{O}_{Z^{\mathrm{an}}}) \in \mathrm{D}^{\mathrm{b}}(\mathbf{FN}_{G_{\mathbb{R}}})$.

Dually, we can consider the category $\mathbf{DFN}_{G_{\mathbb{R}}}$ of dual Fréchet nuclear spaces with a continuous $G_{\mathbb{R}}$-action and its bounded derived category $\mathrm{D}^{\mathrm{b}}(\mathbf{DFN}_{G_{\mathbb{R}}})$. Then, we can construct $\mathbf{R}\Gamma^{\mathrm{top}}_c(Z^{\mathrm{an}}; F \otimes \Omega_{Z^{\mathrm{an}}} \overset{\mathrm{L}}{\otimes}_{\mathscr{D}_Z} \mathscr{M})$, which is an object of $\mathrm{D}^{\mathrm{b}}(\mathbf{DFN}_{G_{\mathbb{R}}})$. Here, $\Omega_{Z^{\mathrm{an}}}$ is the sheaf of holomorphic differential forms with the maximal degree. Let $\mathscr{D}\!\mathit{ist}^{(\mathrm{d}z, \bullet)}$ be the Dolbeault resolution of $\Omega_{Z^{\mathrm{an}}}$ by differential forms with distribution coefficients. Then, the complex $\Gamma_c(Z^{\mathrm{an}}; K^{\bullet} \otimes \mathscr{D}\!\mathit{ist}^{(\mathrm{d}z, \bullet)} \otimes_{\mathscr{D}_Z} (\mathscr{D}_Z \otimes \mathcal{V}^{\bullet}))$ represents $\mathbf{R}\Gamma_c(Z^{\mathrm{an}}; F \otimes \Omega_{Z^{\mathrm{an}}} \otimes_{\mathscr{D}_Z} \mathscr{M}) \in \mathrm{D}^{\mathrm{b}}(\mathrm{Mod}(\mathbb{C}))$. On the other hand, since $\Gamma_c(Z^{\mathrm{an}}; K^{\bullet} \otimes \mathscr{D}\!\mathit{ist}^{(\mathrm{d}z, \bullet)} \otimes_{\mathscr{D}_Z}$

---

[1] In fact, it is not possible to represent F by such a $K^{\bullet}$ in general. We overcome this difficulty by a resolution of the base space Z (see §5).

$(\mathscr{D}_Z \otimes \mathcal{V}^\bullet))$ is a complex in $\mathbf{DFN}_{G_\mathbb{R}}$, it may be regarded as an object of $\mathrm{D}^b(\mathbf{DFN}_{G_\mathbb{R}})$. It is $\mathbf{R}\,\Gamma_c^{\mathrm{top}}(Z^{\mathrm{an}}; F \otimes \Omega_{Z^{\mathrm{an}}} \overset{\mathbf{L}}{\otimes}_{\mathscr{D}_Z} \mathcal{M})$. We have

$$\mathbf{R}\,\Gamma_c^{\mathrm{top}}(Z^{\mathrm{an}}; F \otimes \Omega_{Z^{\mathrm{an}}} \overset{\mathbf{L}}{\otimes}_{\mathscr{D}_Z} \mathcal{M}) \simeq \big(\mathbf{R}\mathrm{Hom}_{\mathscr{D}_Z}^{\mathrm{top}}(\mathcal{M} \otimes F, \mathscr{O}_{Z^{\mathrm{an}}})\big)^*.$$

Let us apply it to the flag manifold X with the action of G. Let F be an object of $\mathrm{D}^b_{G_\mathbb{R}, \mathbb{R}\text{-}c}(\mathbb{C}_{X^{\mathrm{an}}})$. Then $\mathbf{R}\mathrm{Hom}_{\mathbb{C}}^{\mathrm{top}}(F, \mathscr{O}_{X^{\mathrm{an}}}) := \mathbf{R}\mathrm{Hom}_{\mathscr{D}_X}^{\mathrm{top}}(\mathscr{D}_X \otimes F, \mathscr{O}_{X^{\mathrm{an}}})$ is an object of $\mathrm{D}^b(\mathbf{FN}_{G_\mathbb{R}})$. This is strict, i.e., if we represent $\mathbf{R}\mathrm{Hom}_{\mathbb{C}}^{\mathrm{top}}(F, \mathscr{O}_{X^{\mathrm{an}}})$ as a complex in $\mathbf{FN}_{G_\mathbb{R}}$, the differentials of such a complex have closed ranges. Moreover, its cohomology group $H^n(\mathbf{R}\mathrm{Hom}_{\mathbb{C}}^{\mathrm{top}}(F, \mathscr{O}_{X^{\mathrm{an}}}))$ is the maximal globalization of some Harish-Chandra module (see § 10). Similarly, $\mathbf{R}\mathrm{Hom}_{\mathbb{C}}^{\mathrm{top}}(F, \Omega_{X^{\mathrm{an}}}) := \mathbf{R}\mathrm{Hom}_{\mathscr{D}_X}^{\mathrm{top}}\big((\mathscr{D}_X \otimes \Omega_X^{\otimes -1}) \otimes F, \mathscr{O}_{X^{\mathrm{an}}}\big)$ is a strict object of $\mathrm{D}^b(\mathbf{FN}_{G_\mathbb{R}})$ and its cohomology groups are the maximal globalization of a Harish-Chandra module. Here Ω_X is the sheaf of differential forms with degree d_X on X.

Dually, we can consider $\mathbf{R}\,\Gamma_c^{\mathrm{top}}(X^{\mathrm{an}}; F \otimes \mathscr{O}_{X^{\mathrm{an}}})$ as an object of $\mathrm{D}^b(\mathbf{DFN}_{G_\mathbb{R}})$, whose cohomology groups are the minimal globalization of a Harish-Chandra module.

This is the left vertical arrow in Fig. 1.

Remark 1.5.1. Note the works by Hecht-Taylor [11] and Smithies-Taylor [27] which are relevant to this note. They considered the $\mathscr{D}_{X^{\mathrm{an}}}$-module $\mathscr{O}_{X^{\mathrm{an}}} \otimes F$ instead of F, and construct the left vertical arrow in Fig. 1 in a similar way to the Beilinson-Bernstein correspondence.

Let us denote by $\mathrm{Mod}_f(\mathfrak{g}, K)$ the category of $(\mathfrak{g}, K)$-modules finitely generated over $U(\mathfrak{g})$. Then, $\mathrm{Mod}_f(\mathfrak{g}, K)$ has enough projectives. Indeed, $U(\mathfrak{g}) \otimes_{U(\mathfrak{k})} N$ is a projective object of $\mathrm{Mod}_f(\mathfrak{g}, K)$ for any finite-dimensional K-module N. Hence there exists a right derived functor

$$\mathbf{R}\mathrm{Hom}_{U(\mathfrak{g})}^{\mathrm{top}}(\bullet, \mathrm{C}^\infty(G_\mathbb{R})) : \mathrm{D}^b(\mathrm{Mod}_f(\mathfrak{g}, K))^{\mathrm{op}} \to \mathrm{D}^b(\mathbf{FN}_{G_\mathbb{R}})$$

of the functor $\mathrm{Hom}_{U(\mathfrak{g})}(\bullet, \mathrm{C}^\infty(G_\mathbb{R})) : \mathrm{Mod}_f(\mathfrak{g}, K)^{\mathrm{op}} \to \mathbf{FN}_{G_\mathbb{R}}$. Similarly, there exists a left derived functor

$$\Gamma_c(G_\mathbb{R}; \mathscr{D}ist_{G_\mathbb{R}}) \overset{\mathbf{L}}{\otimes}_{U(\mathfrak{g})} \bullet : \mathrm{D}^b(\mathrm{Mod}_f(\mathfrak{g}, K)) \to \mathrm{D}^b(\mathbf{DFN}_{G_\mathbb{R}})$$

of the functor $\Gamma_c(G_\mathbb{R}; \mathscr{D}ist_{G_\mathbb{R}}) \otimes_{U(\mathfrak{g})} \bullet : \mathrm{Mod}_f(\mathfrak{g}, K) \to \mathbf{DFN}_{G_\mathbb{R}}$.[2] In § 10, we prove $H^n(\mathbf{R}\mathrm{Hom}_{U(\mathfrak{g})}^{\mathrm{top}}(M, \mathrm{C}^\infty(G_\mathbb{R}))) = 0$, $H^n(\Gamma_c(G_\mathbb{R}; \mathscr{D}ist_{G_\mathbb{R}}) \overset{\mathbf{L}}{\otimes}_{U(\mathfrak{g})} M) = 0$ for $n \neq 0$, and

---

[2] They are denoted by $\mathbf{R}\mathrm{Hom}_{(\mathfrak{g}, K_\mathbb{R})}^{\mathrm{top}}(\bullet, \mathrm{C}^\infty(G_\mathbb{R}))$ and $\Gamma_c(G_\mathbb{R}; \mathscr{D}ist_{G_\mathbb{R}}) \overset{\mathbf{L}}{\otimes}_{(\mathfrak{g}, K_\mathbb{R})} \bullet$ in Subsection 9.5.

$$\mathrm{MG}(M^*) \simeq \mathbf{R}\mathrm{Hom}^{\mathrm{top}}_{U(\mathfrak{g})}(M, C^\infty(G_\mathbb{R})),$$

$$\mathrm{mg}(M) \simeq \Gamma_c(G_\mathbb{R}; \mathscr{D}ist_{G_\mathbb{R}}) \overset{\mathbf{L}}{\otimes}_{U(\mathfrak{g})} M$$

for any Harish-Chandra module M.

1.6 Integral Transforms

Let Y and Z be algebraic manifolds, and consider the diagram:

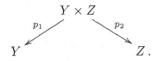

We assume that Y is projective. For $\mathscr{N} \in \mathrm{D}^\mathrm{b}(\mathscr{D}_Y)$ and $\mathscr{K} \in \mathrm{D}^\mathrm{b}(\mathscr{D}_{Y \times Z})$ we define their convolution

$$\mathscr{N} \overset{\mathbf{D}}{\circ} \mathscr{K} := \mathbf{D}p_{2*}(\mathbf{D}p_1^* \mathscr{N} \overset{\mathbf{D}}{\otimes} \mathscr{K}) \in \mathrm{D}^\mathrm{b}(\mathscr{D}_Z),$$

where $\mathbf{D}p_{2*}$, $\mathbf{D}p_1^*$, $\overset{\mathbf{D}}{\otimes}$ are the direct image, inverse image, tensor product functors for D-modules (see §3). Similarly, for $K \in \mathrm{D}^\mathrm{b}(\mathbb{C}_{Y^{\mathrm{an}} \times Z^{\mathrm{an}}})$ and $F \in \mathrm{D}^\mathrm{b}(\mathbb{C}_{Z^{\mathrm{an}}})$, we define their convolution

$$K \circ F := \mathbf{R}(p_1^{\mathrm{an}})_!(K \otimes (p_2^{\mathrm{an}})^{-1}F) \in \mathrm{D}^\mathrm{b}(\mathbb{C}_{Y^{\mathrm{an}}}).$$

Let $\mathrm{DR}_{Y \times Z} \colon \mathrm{D}^\mathrm{b}(\mathscr{D}_{Y \times Z}) \to \mathrm{D}^\mathrm{b}(\mathbb{C}_{Y^{\mathrm{an}} \times Z^{\mathrm{an}}})$ be the de Rham functor. Then we have the following integral transform formula.

Theorem 1.6.1. *For $\mathscr{K} \in \mathrm{D}^\mathrm{b}_{\mathrm{hol}}(\mathscr{D}_{Y \times Z})$, $\mathscr{N} \in \mathrm{D}^\mathrm{b}_{\mathrm{coh}}(\mathscr{D}_Y)$ and $F \in \mathrm{D}^\mathrm{b}(\mathbb{C}_{Z^{\mathrm{an}}})$, set $K = \mathrm{DR}_{Y \times Z}(\mathscr{K}) \in \mathrm{D}^\mathrm{b}_{\mathbb{C}\text{-}\mathrm{c}}(\mathbb{C}_{Y^{\mathrm{an}} \times Z^{\mathrm{an}}})$. If $\mathscr{N}$ and $\mathscr{K}$ are non-characteristic, then we have an isomorphism*

$$\mathbf{R}\mathrm{Hom}_{\mathscr{D}_Z}((\mathscr{N} \overset{\mathbf{D}}{\circ} \mathscr{K}) \otimes F, \mathscr{O}_{Z^{\mathrm{an}}}) \simeq \mathbf{R}\mathrm{Hom}_{\mathscr{D}_Y}(\mathscr{N} \otimes (K \circ F), \mathscr{O}_{Y^{\mathrm{an}}})[d_Y - 2d_Z].$$

Note that $\mathscr{N}$ and $\mathscr{K}$ are non-characteristic if $\left(\mathrm{Ch}(\mathscr{N}) \times T_Z^* Z\right) \cap \mathrm{Ch}(\mathscr{K}) \subset T_{Y \times Z}^*(Y \times Z)$, where Ch denotes the characteristic variety (see §8).

Its equivariant version also holds.

Let us apply this to the following situation. Let G, $G_\mathbb{R}$, K, $K_\mathbb{R}$, X, S be as before, and consider the diagram:

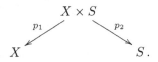

Theorem 1.6.2. *For* $\mathscr{K} \in D^b_{G,\mathrm{coh}}(\mathscr{D}_{X \times S})$, $\mathscr{N} \in D^b_{\mathrm{coh}}(\mathrm{Mod}(\mathscr{D}_X, G))$ *and* $F \in D^b_{G_\mathbb{R}, \mathbb{R}\text{-}c}(\mathbb{C}_{S^{\mathrm{an}}})$, *set* $K = DR_{X \times S}(\mathscr{K}) \in D^b_{G_\mathbb{R}, \mathbb{C}\text{-}c}(\mathbb{C}_{X^{\mathrm{an}} \times S^{\mathrm{an}}})$. *Then we have an isomorphism*

$$(1.6.1) \qquad \mathbf{R}\mathrm{Hom}^{\mathrm{top}}_{\mathscr{D}_S}((\mathscr{N} \overset{\mathbf{D}}{\circ} \mathscr{K}) \otimes F, \mathscr{O}_{S^{\mathrm{an}}})$$
$$\simeq \mathbf{R}\mathrm{Hom}^{\mathrm{top}}_{\mathscr{D}_X}(\mathscr{N} \otimes (K \circ F), \mathscr{O}_{X^{\mathrm{an}}})[d_X - 2d_S]$$

in $D^b(\mathbf{FN}_{G_\mathbb{R}})$.

Note that the non-characteristic condition in Theorem 1.6.1 is automatically satisfied in this case.

1.7 Commutativity of Fig. 1

Let us apply Theorem 1.6.2 in order to show the commutativity of Fig. 1. Let us start by taking $\mathscr{M} \in \mathrm{Mod}_{K,\mathrm{coh}}(\mathscr{D}_X)$. Then, by the Beilinson-Bernstein correspondence, $\mathscr{M}$ corresponds to the Harish-Chandra module $M := \Gamma(X; \mathscr{M})$. Let us set $\mathscr{K} = \mathrm{Ind}^G_K(\mathscr{M}) \in \mathrm{Mod}_{G,\mathrm{coh}}(\mathscr{D}_{X \times S})$. If we set $\mathscr{N} = \mathscr{D}_X \otimes \Omega^{\otimes -1}_X \in \mathrm{Mod}(\mathscr{D}_X, G)$, then $\mathscr{N} \overset{\mathbf{D}}{\circ} \mathscr{K} \in D^b(\mathrm{Mod}(\mathscr{D}_S, G))$. By the equivalence of categories $\mathrm{Mod}(\mathscr{D}_S, G) \simeq \mathrm{Mod}(\mathfrak{g}, K)$, $\mathscr{N} \overset{\mathbf{D}}{\circ} \mathscr{K}$ corresponds to $M \in \mathrm{Mod}(\mathfrak{g}, K)$. Now we take $F = \mathbb{C}_{S_\mathbb{R}}[-d_S]$. Then the left-hand side of (1.6.1) coincides with

$$\mathbf{R}\mathrm{Hom}^{\mathrm{top}}_{\mathscr{D}_S}(\mathscr{N} \overset{\mathbf{D}}{\circ} \mathscr{K}, \mathbf{R}\mathscr{H}om(\mathbb{C}_{S_\mathbb{R}}[-d_S], \mathscr{O}_{S^{\mathrm{an}}})) \simeq \mathbf{R}\mathrm{Hom}^{\mathrm{top}}_{\mathscr{D}_S}(\mathscr{N} \overset{\mathbf{D}}{\circ} \mathscr{K}, \mathscr{B}_{S_\mathbb{R}}),$$

where $\mathscr{B}_{S_\mathbb{R}}$ is the sheaf of hyperfunctions on $S_\mathbb{R}$. Since $\mathscr{N} \overset{\mathbf{D}}{\circ} \mathscr{K}$ is an elliptic $\mathscr{D}_S$-module, we have

$$\mathbf{R}\mathrm{Hom}^{\mathrm{top}}_{\mathscr{D}_S}(\mathscr{N} \overset{\mathbf{D}}{\circ} \mathscr{K}, \mathscr{B}_{S_\mathbb{R}}) \simeq \mathbf{R}\mathrm{Hom}^{\mathrm{top}}_{\mathscr{D}_S}(\mathscr{N} \overset{\mathbf{D}}{\circ} \mathscr{K}, \mathscr{C}^\infty_{S_\mathbb{R}}),$$

where $\mathscr{C}^\infty_{S_\mathbb{R}}$ is the sheaf of C^∞-functions on $S_\mathbb{R}$. The equivalence of categories $\mathrm{Mod}(\mathscr{D}_S, G) \simeq \mathrm{Mod}(\mathfrak{g}, K)$ implies

$$\mathbf{R}\mathrm{Hom}^{\mathrm{top}}_{\mathscr{D}_S}(\mathscr{N} \overset{\mathbf{D}}{\circ} \mathscr{K}, \mathscr{C}^\infty_{S_\mathbb{R}}) \simeq \mathbf{R}\mathrm{Hom}^{\mathrm{top}}_{U(\mathfrak{g})}(M, C^\infty(G_\mathbb{R})).$$

Hence we have calculated the left-hand side of (1.6.1):

$$\mathbf{R}\mathrm{Hom}^{\mathrm{top}}_{\mathscr{D}_S}((\mathscr{N} \overset{\mathbf{D}}{\circ} \mathscr{K}) \otimes F, \mathscr{O}_{S^{\mathrm{an}}}) \simeq \mathbf{R}\mathrm{Hom}^{\mathrm{top}}_{U(\mathfrak{g})}(M, C^\infty(G_\mathbb{R})).$$

Now let us calculate the right-hand side of (1.6.1). Since we have

$$K := DR_{X \times S}\mathscr{K} = DR_{X \times S}(\mathrm{Ind}^G_K(\mathscr{M}))$$
$$\simeq \mathrm{Ind}^{G^{\mathrm{an}}}_{K^{\mathrm{an}}}(DR_X(\mathscr{M})),$$

$K \circ F$ is nothing but $\Phi(\mathrm{DR}_X(\mathscr{M}))[-2d_S]$. Therefore the right-hand side of (1.6.1) is isomorphic to $\mathbf{R}\mathrm{Hom}_{\mathbb{C}}^{\mathrm{top}}(\Phi(\mathrm{DR}_X(\mathscr{M})), \Omega_{X^{\mathrm{an}}}[d_X])$.

Finally we obtain

$$
(1.7.1) \qquad
\begin{aligned}
&\mathbf{R}\mathrm{Hom}_{U(\mathfrak{g})}^{\mathrm{top}}(\Gamma(X; \mathscr{M}), \mathrm{C}^{\infty}(G_{\mathbb{R}})) \\
&\qquad \simeq \mathbf{R}\mathrm{Hom}_{\mathbb{C}}^{\mathrm{top}}(\Phi(\mathrm{DR}_X(\mathscr{M})), \Omega_{X^{\mathrm{an}}}[d_X]),
\end{aligned}
$$

or

$$(1.7.2) \qquad \mathrm{MG}(\Gamma(X; \mathscr{M})^*) \simeq \mathbf{R}\mathrm{Hom}_{\mathbb{C}}^{\mathrm{top}}(\Phi(\mathrm{DR}_X(\mathscr{M})), \Omega_{X^{\mathrm{an}}}[d_X]).$$

By duality, we have

$$(1.7.3) \qquad \mathrm{mg}(\Gamma(X; \mathscr{M})) \simeq \mathbf{R}\Gamma_c^{\mathrm{top}}(X^{\mathrm{an}}; \Phi(\mathrm{DR}_X(\mathscr{M})) \otimes \mathscr{O}_{X^{\mathrm{an}}}).$$

This is the commutativity of Fig. 1.

1.8 Example

Let us illustrate the results explained so far by taking $SL(2, \mathbb{R}) \simeq SU(1,1)$ as an example. We set

$$G_{\mathbb{R}} = SU(1,1) = \left\{ \begin{pmatrix} \alpha & \beta \\ \bar{\beta} & \bar{\alpha} \end{pmatrix}; \alpha, \beta \in \mathbb{C}, |\alpha|^2 - |\beta|^2 = 1 \right\} \subset G = SL(2, \mathbb{C}),$$

$$K_{\mathbb{R}} = \left\{ \begin{pmatrix} \alpha & 0 \\ 0 & \bar{\alpha} \end{pmatrix}; \alpha \in \mathbb{C}, |\alpha| = 1 \right\} \subset K = \left\{ \begin{pmatrix} \alpha & 0 \\ 0 & \alpha^{-1} \end{pmatrix}; \alpha \in \mathbb{C} \setminus \{0\} \right\},$$

$$X = \mathbb{P}^1.$$

Here G acts on the flag manifold $X = \mathbb{P}^1 = \mathbb{C} \sqcup \{\infty\}$ by

$$\begin{pmatrix} a & b \\ c & d \end{pmatrix} : z \longmapsto \frac{az + b}{cz + d}.$$

Its infinitesimal action $L_X : \mathfrak{g} \to \Gamma(X; \Theta_X)$ (with the sheaf Θ_X of vector fields on X) is given by

$$h := \begin{pmatrix} 1 & 0 \\ 0 & -1 \end{pmatrix} \longmapsto -2z\frac{d}{dz},$$

$$e := \begin{pmatrix} 0 & 1 \\ 0 & 0 \end{pmatrix} \longmapsto -\frac{d}{dz},$$

$$f := \begin{pmatrix} 0 & 0 \\ 1 & 0 \end{pmatrix} \longmapsto z^2\frac{d}{dz}.$$

We have

$$\Gamma(X; \mathscr{D}_X) = U(\mathfrak{g})/U(\mathfrak{g})\Delta,$$

where $\Delta = h(h-2) + 4ef = h(h+2) + 4fe \in \mathfrak{z}(\mathfrak{g})$.

The flag manifold X has three K-orbits:

$$\{0\}, \ \{\infty\} \text{ and } X \setminus \{0, \infty\}.$$

The corresponding three $G_\mathbb{R}$-orbits are

$$X_-, \ X_+ \text{ and } X_\mathbb{R},$$

where $X_\pm = \{z \in \mathbb{P}^1 \, ; \, |z| \gtrless 1\}$ and $X_\mathbb{R} = \{z \in \mathbb{C} \, ; \, |z| = 1\}$.

Let $j_0 : X \setminus \{0\} \hookrightarrow X$, $j_\infty : X \setminus \{\infty\} \hookrightarrow X$ and $j_{0,\infty} : X \setminus \{0, \infty\} \hookrightarrow X$ be the open embeddings. Then we have K-equivariant $\mathscr{D}_X$-modules $\mathscr{O}_X$, $j_{0*}j_0^{-1}\mathscr{O}_X$, $j_{\infty*}j_\infty^{-1}\mathscr{O}_X$ and $j_{0,\infty*}j_{0,\infty}^{-1}\mathscr{O}_X$. We have the inclusion relation:

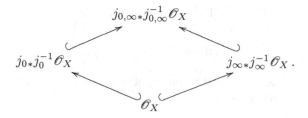

There exist four irreducible K-equivariant $\mathscr{D}_X$-modules:

$$\mathscr{M}_0 = \mathcal{H}^1_{\{0\}}(\mathscr{O}_X) \simeq j_{0*}j_0^{-1}\mathscr{O}_X/\mathscr{O}_X \simeq j_{0,\infty*}j_{0,\infty}^{-1}\mathscr{O}_X/j_{\infty*}j_\infty^{-1}\mathscr{O}_X,$$

$$\mathscr{M}_\infty = \mathcal{H}^1_{\{\infty\}}(\mathscr{O}_X) \simeq j_{\infty*}j_\infty^{-1}\mathscr{O}_X/\mathscr{O}_X \simeq j_{0,\infty*}j_{0,\infty}^{-1}\mathscr{O}_X/j_{0*}j_0^{-1}\mathscr{O}_X,$$

$$\mathscr{M}_{0,\infty} = \mathscr{O}_X,$$

$$\mathscr{M}_{1/2} = \mathscr{O}_X\sqrt{z} = \mathscr{D}_X/\mathscr{D}_X(L_X(h)+1).$$

Here, $\mathscr{M}_0$ and $\mathscr{M}_\infty$ correspond to the K-orbits $\{0\}$ and $\{\infty\}$, respectively, while both $\mathscr{M}_{0,\infty}$ and $\mathscr{M}_{1/2}$ correspond to the open K-orbit $X \setminus \{0, \infty\}$. Note that the isotropy subgroup K_z of K at $z \in X \setminus \{0, \infty\}$ is isomorphic to $\{1, -1\}$, and $\mathscr{M}_{0,\infty}$ corresponds to the trivial representation of K_z and $\mathscr{M}_{1/2}$ corresponds to the non-trivial one-dimensional representation of K_z. By the Beilinson-Bernstein correspondence, we obtain four irreducible Harish-Chandra modules with the trivial infinitesimal character:

$$M_0 = \mathscr{O}_X(X \setminus \{0\})/\mathbb{C} = \mathbb{C}[z^{-1}]/\mathbb{C} \simeq U(\mathfrak{g})/(U(\mathfrak{g})(h-2) + U(\mathfrak{g})\,f),$$

$$M_\infty = \mathscr{O}_X(X \setminus \{\infty\})/\mathbb{C} \simeq \mathbb{C}[z]/\mathbb{C} \simeq U(\mathfrak{g})/(U(\mathfrak{g})(h+2) + U(\mathfrak{g})\,e),$$

$$M_{0,\infty} = \mathscr{O}_X(X) = \mathbb{C} \simeq U(\mathfrak{g})/(U(\mathfrak{g})\,h + U(\mathfrak{g})\,e + U(\mathfrak{g})\,f),$$

$$M_{1/2} = \mathbb{C}[z, z^{-1}]\sqrt{z} \simeq U(\mathfrak{g})/(U(\mathfrak{g})(h+1) + U(\mathfrak{g})\Delta).$$

Among them, $M_{0,\infty}$ and $M_{1/2}$ are self-dual, namely they satisfy $M^* \simeq M$. We have $(M_0)^* \simeq M_\infty$.

By the de Rham functor, the irreducible K-equivariant $\mathscr{D}_X$-modules are transformed to irreducible K^{an}-equivariant perverse sheaves as follows:

$$\mathrm{DR}_X(\mathscr{M}_0) = \mathbb{C}_{\{0\}}[-1],$$
$$\mathrm{DR}_X(\mathscr{M}_\infty) = \mathbb{C}_{\{\infty\}}[-1],$$
$$\mathrm{DR}_X(\mathscr{M}_{0,\infty}) = \mathbb{C}_{X^{\mathrm{an}}},$$
$$\mathrm{DR}_X(\mathscr{M}_{1/2}) = \mathbb{C}_{X^{\mathrm{an}}}\sqrt{z}.$$

Here $\mathbb{C}_{X^{\mathrm{an}}}\sqrt{z}$ is the locally constant sheaf on $X^{\mathrm{an}} \setminus \{0, \infty\}$ of rank one (extended by zero over X^{an}) with the monodromy -1 around 0 and ∞.

Their images by the Matsuki correspondence (see Proposition 9.4.3) are

$$\Phi(\mathrm{DR}_X(\mathscr{M}_0)) \simeq \mathbb{C}_{X_-}[1],$$
$$\Phi(\mathrm{DR}_X(\mathscr{M}_\infty)) \simeq \mathbb{C}_{X_+}[1],$$
$$\Phi(\mathrm{DR}_X(\mathscr{M}_{0,\infty})) \simeq \mathbb{C}_{X^{\mathrm{an}}},$$
$$\Phi(\mathrm{DR}_X(\mathscr{M}_{1/2})) \simeq \mathbb{C}_{X_{\mathbb{R}}}\sqrt{z}.$$

Note that $\mathbb{C}_{X_{\mathbb{R}}}\sqrt{z}$ is a local system on $X_{\mathbb{R}}$ of rank one with the monodromy -1. Hence (1.7.2) reads as

$$\mathrm{MG}(M_0^*) \simeq \mathrm{MG}(M_\infty) \simeq \mathbf{R}\mathrm{Hom}_{\mathbb{C}}^{\mathrm{top}}(\mathbb{C}_{X_-}[1], \Omega_{X^{\mathrm{an}}}[1]) \simeq \Omega_{X^{\mathrm{an}}}(X_-),$$

$$\mathrm{MG}(M_\infty^*) \simeq \mathrm{MG}(M_0) \simeq \mathbf{R}\mathrm{Hom}_{\mathbb{C}}^{\mathrm{top}}(\mathbb{C}_{X_+}[1], \Omega_{X^{\mathrm{an}}}[1]) \simeq \Omega_{X^{\mathrm{an}}}(X_+),$$

$$\mathrm{MG}(M_{0,\infty}^*) \simeq \mathrm{MG}(M_{0,\infty}) \simeq \mathbf{R}\mathrm{Hom}_{\mathbb{C}}^{\mathrm{top}}(\mathbb{C}_{X^{\mathrm{an}}}, \Omega_{X^{\mathrm{an}}}[1])$$
$$\simeq H^1(X^{\mathrm{an}}; \Omega_{X^{\mathrm{an}}}) \simeq \mathbb{C},$$

$$\mathrm{MG}(M_{1/2}^*) \simeq \mathrm{MG}(M_{1/2}) \simeq \mathbf{R}\mathrm{Hom}_{\mathbb{C}}^{\mathrm{top}}(\mathbb{C}_{X_{\mathbb{R}}}\sqrt{z}, \Omega_{X^{\mathrm{an}}}[1])$$
$$\simeq \Gamma\big(X_{\mathbb{R}}; \mathscr{B}_{X_{\mathbb{R}}} \otimes \Omega_{X_{\mathbb{R}}} \otimes \mathbb{C}_{X_{\mathbb{R}}}\sqrt{z}\big).$$

Here $\mathscr{B}_{X_{\mathbb{R}}}$ is the sheaf of hyperfunctions on $X_{\mathbb{R}}$. Note that the exterior differentiation gives isomorphisms

$$\mathscr{O}_{X^{\mathrm{an}}}(X_\pm)/\mathbb{C} \xrightarrow[d]{\sim} \Omega_{X^{\mathrm{an}}}(X_\pm),$$

$$\Gamma(X_{\mathbb{R}}; \mathscr{B}_{X_{\mathbb{R}}} \otimes \mathbb{C}_{X_{\mathbb{R}}}\sqrt{z}) \xrightarrow[d]{\sim} \Gamma(X_{\mathbb{R}}; \mathscr{B}_{X_{\mathbb{R}}} \otimes \Omega_{X_{\mathbb{R}}} \otimes \mathbb{C}_{X_{\mathbb{R}}}\sqrt{z}).$$

In fact, we have

$$\mathrm{mg}(M_0) \simeq \Omega_{X^{\mathrm{an}}}(\overline{X_+}) \subset \Omega_{X^{\mathrm{an}}}(X_+) \simeq \mathrm{MG}(M_0),$$
$$\mathrm{mg}(M_\infty) \simeq \Omega_{X^{\mathrm{an}}}(\overline{X_-}) \subset \Omega_{X^{\mathrm{an}}}(X_-) \simeq \mathrm{MG}(M_\infty),$$
$$\mathrm{mg}(M_{0,\infty}) \xrightarrow{\sim} \mathrm{MG}(M_{0,\infty}) \simeq \mathbb{C},$$
$$\mathrm{mg}(M_{1/2}) \simeq \Gamma(X_{\mathbb{R}}; \mathscr{A}_{X_{\mathbb{R}}} \otimes \mathbb{C}_{X_{\mathbb{R}}}\sqrt{z}) \subset \Gamma(X_{\mathbb{R}}; \mathscr{B}_{X_{\mathbb{R}}} \otimes \mathbb{C}_{X_{\mathbb{R}}}\sqrt{z}) \simeq \mathrm{MG}(M_{1/2}).$$

Here $\mathscr{A}_{X_{\mathbb{R}}}$ is the sheaf of real analytic functions on $X_{\mathbb{R}}$.

For example, by (1.7.3), $\mathrm{mg}(M_0) \simeq \mathbf{R}\Gamma_c^{\mathrm{top}}(X^{\mathrm{an}}; \mathbb{C}_{X_-}[1] \otimes \mathscr{O}_{X^{\mathrm{an}}})$. The exact sequence

$$0 \to \mathbb{C}_{X_-} \to \mathbb{C}_{X^{\mathrm{an}}} \to \mathbb{C}_{\overline{X_+}} \to 0$$

yields the exact sequence:

$$H^0(X^{\mathrm{an}}; \mathbb{C}_{X_-} \otimes \mathscr{O}_{X^{\mathrm{an}}}) \to H^0(X^{\mathrm{an}}; \mathbb{C}_{X^{\mathrm{an}}} \otimes \mathscr{O}_{X^{\mathrm{an}}}) \to H^0(X^{\mathrm{an}}; \mathbb{C}_{\overline{X_+}} \otimes \mathscr{O}_{X^{\mathrm{an}}})$$
$$\to H^0(X^{\mathrm{an}}; \mathbb{C}_{X_-}[1] \otimes \mathscr{O}_{X^{\mathrm{an}}}) \to H^0(X^{\mathrm{an}}; \mathbb{C}_{X^{\mathrm{an}}}[1] \otimes \mathscr{O}_{X^{\mathrm{an}}}),$$

in which $H^0(X^{\mathrm{an}}; \mathbb{C}_{X_-} \otimes \mathscr{O}_{X^{\mathrm{an}}}) = \{u \in \mathscr{O}_{X^{\mathrm{an}}}(X^{\mathrm{an}}); \mathrm{supp}(u) \subset X_-\} = 0$, $H^0(X^{\mathrm{an}}; \mathbb{C}_{X^{\mathrm{an}}} \otimes \mathscr{O}_{X^{\mathrm{an}}}) = \mathscr{O}_{X^{\mathrm{an}}}(X^{\mathrm{an}}) = \mathbb{C}$ and $H^0(X^{\mathrm{an}}; \mathbb{C}_{X^{\mathrm{an}}}[1] \otimes \mathscr{O}_{X^{\mathrm{an}}}) = H^1(X^{\mathrm{an}}; \mathscr{O}_{X^{\mathrm{an}}}) = 0$.

Hence we have

$$\mathbf{R}\Gamma_c^{\mathrm{top}}(X^{\mathrm{an}}; \mathbb{C}_{X_-}[1] \otimes \mathscr{O}_{X^{\mathrm{an}}}) \simeq \mathscr{O}_{X^{\mathrm{an}}}(\overline{X_+})/\mathbb{C}.$$

The exterior differentiation gives an isomorphism

$$\mathscr{O}_{X^{\mathrm{an}}}(\overline{X_+})/\mathbb{C} \xrightarrow{\ \sim\ }_{d} \Omega_{X^{\mathrm{an}}}(\overline{X_+}).$$

Note that we have

$$\mathrm{HC}(\Omega_{X^{\mathrm{an}}}(\overline{X_+})) \simeq \mathrm{HC}(\Omega_{X^{\mathrm{an}}}(X_+))$$
$$\simeq \Omega_X(X \setminus \{0\}) \xleftarrow{\ \sim\ }_{d} \mathscr{O}_X(X \setminus \{0\})/\mathbb{C} \simeq M_0.$$

1.9 Organization of the Note

So far, we have explained Fig. 1 briefly. We shall explain more details in the subsequent sections.

The category of representations of $G_{\mathbb{R}}$ is not an abelian category, but it is a so-called quasi-abelian category and we can consider its derived category. In §2, we explain the derived category of a quasi-abelian category following J.-P. Schneiders [26].

In §3, we introduce the notion of quasi-G-equivariant D-modules, and study their derived category. We construct the pull-back and push-forward functors for $\mathrm{D}^b(\mathrm{Mod}(\mathscr{D}_X, G))$, and prove that they commute with the forgetful functor $\mathrm{D}^b(\mathrm{Mod}(\mathscr{D}_X, G)) \to \mathrm{D}^b(\mathrm{Mod}(\mathscr{D}_X))$.

In §4, we explain the equivariant derived category following Bernstein-Lunts [4].

In §5, we define $\mathbf{R}\mathrm{Hom}_{\mathscr{D}_Z}^{\mathrm{top}}(\mathscr{M} \otimes F, \mathscr{O}_{Z^{\mathrm{an}}})$ and study its functorial properties.

In §6, we prove the ellipticity theorem, which says that, for a real form $i: X_{\mathbb{R}} \hookrightarrow X$, $\mathbf{R}\mathrm{Hom}_{\mathscr{D}_X}^{\mathrm{top}}(\mathscr{M}, \mathscr{C}_{X_{\mathbb{R}}}^\infty) \longrightarrow \mathbf{R}\mathrm{Hom}_{\mathscr{D}_X}^{\mathrm{top}}(\mathscr{M} \otimes i_* i^! \mathbb{C}_{X_{\mathbb{R}}}, \mathscr{O}_{X^{\mathrm{an}}})$ is an isomorphism when $\mathscr{M}$ is an elliptic D-module. In order to construct this morphism, we use the Whitney functor introduced by Kashiwara-Schapira [20].

If we want to deal with non-trivial infinitesimal characters, we need to twist sheaves and D-modules. In §7, we explain these twistings.

In §8, we prove the integral transform formula explained in the subsection 1.6.

In §9, we apply these results to the representation theory of real semisimple Lie groups. We construct the arrows in Fig. 1.

As an application of §9, we give a proof of the cohomology vanishing theorem $H^j(\mathbf{R}\mathrm{Hom}^{\mathrm{top}}_{U(\mathfrak{g})}(M, C^\infty(G_\mathbb{R}))) = 0$ $(j \neq 0)$ and its dual statement

$$H^j(\Gamma_c(G_\mathbb{R}; \mathscr{D}ist_{G_\mathbb{R}}) \overset{\mathbf{L}}{\otimes}_{U(\mathfrak{g})} M) = 0 \text{ in } §10.$$

2 Derived Categories of Quasi-abelian Categories

2.1 Quasi-abelian Categories

The representations of real semisimple groups are realized on topological vector spaces, and they do not form an abelian category. However, they form a so-called quasi-abelian category. In this section, we shall review the results of J.-P. Schneiders on the theory of quasi-abelian categories and their derived categories. For more details, we refer the reader to [26].

Let $\mathcal{C}$ be an additive category admitting the kernels and the cokernels. Let us recall that, for a morphism $f\colon X \to Y$ in $\mathcal{C}$, $\mathrm{Im}(f)$ is the kernel of $Y \to \mathrm{Coker}(f)$, and $\mathrm{Coim}(f)$ is the cokernel of $\mathrm{Ker}(f) \to X$. Then f decomposes as $X \to \mathrm{Coim}(f) \to \mathrm{Im}(f) \to Y$. We say that f is *strict* if $\mathrm{Coim}(f) \to \mathrm{Im}(f)$ is an isomorphism. Note that a monomorphism (resp. epimorphism) $f\colon X \to Y$ is strict if and only if $X \to \mathrm{Im}(f)$ (resp. $\mathrm{Coim}(f) \to Y$) is an isomorphism. Note that, for any morphism $f\colon X \to Y$, the morphisms $\mathrm{Ker}(f) \to X$ and $\mathrm{Im}(f) \to Y$ are strict monomorphisms, and $X \to \mathrm{Coim}(f)$ and $Y \to \mathrm{Coker}(f)$ are strict epimorphisms. Note also that a morphism f is strict if and only if it factors as $i \circ s$ with a strict epimorphism s and a strict monomorphism i.

Definition 2.1.1. A *quasi-abelian* category is an additive category admitting the kernels and the cokernels which satisfies the following conditions:

  (i) the strict epimorphisms are stable by base changes,
  (ii) the strict monomorphisms are stable by co-base changes.

The condition (i) means that, for any strict epimorphism $u\colon X \to Y$ and a morphism $Y' \to Y$, setting $X' = X \times_Y Y' = \mathrm{Ker}(X \oplus Y' \to Y)$, the composition $X' \to X \oplus Y' \to Y'$ is a strict epimorphism. The condition (ii) is the similar condition obtained by reversing arrows.

Note that, for any morphism $f\colon X \to Y$ in a quasi-abelian category, $\operatorname{Coim}(f) \to \operatorname{Im}(f)$ is a monomorphism and an epimorphism.

Remark that if $\mathcal{C}$ is a quasi-abelian category, then its opposite category $\mathcal{C}^{\mathrm{op}}$ is also quasi-abelian.

We recall that an abelian category is an additive category such that it admits the kernels and the cokernels and all the morphisms are strict.

Example 2.1.2. (i) Let **Top** be the category of Hausdorff locally convex topological vector spaces. Then **Top** is a quasi-abelian category. For a morphism $f\colon X \to Y$, $\operatorname{Ker}(f)$ is $f^{-1}(0)$ with the induced topology from X, $\operatorname{Coker}(f)$ is $Y/\overline{f(X)}$ with the quotient topology of Y, $\operatorname{Coim}(f)$ is $f(X)$ with the quotient topology of X and $\operatorname{Im}(f)$ is $\overline{f(X)}$ with the induced topology from Y. Hence f is strict if and only if $f(X)$ is a closed subspace of Y and the topology on $f(X)$ induced from X coincides with the one induced from Y.

(ii) Let E be a Hausdorff locally convex topological vector space. Let us recall that a subset B of E is *bounded* if for any neighborhood U of 0 there exists $c > 0$ such that $B \subset cU$. A family $\{f_i\}$ of linear functionals on E is called *equicontinuous* if there exists a neighborhood U of $0 \in E$ such that $f_i(U) \subset \{c \in \mathbb{C}\,;\, |c| < 1\}$ for any i. For two complete locally convex topological vector spaces E and F, a continuous linear map $f\colon E \to F$ is called *nuclear* if there exist an equicontinuous sequence $\{h_n\}_{n \geqslant 1}$ of linear functionals on E, a bounded sequence $\{v_n\}_{n \geqslant 1}$ of elements of F and a sequence $\{c_n\}$ in $\mathbb{C}$ such that $\sum |c_n| < \infty$ and $f(x) = \sum_n c_n h_n(x) v_n$ for all $x \in E$.

A Fréchet nuclear space (FN space, for short) is a Fréchet space E such that any homomorphism from E to a Banach space is nuclear. It is equivalent to saying that E is isomorphic to the projective limit of a sequence of Banach spaces $F_1 \leftarrow F_2 \leftarrow \cdots$ such that $F_n \to F_{n-1}$ are nuclear for all n. We denote by **FN** the full subcategory of **Top** consisting of Fréchet nuclear spaces.

A dual Fréchet nuclear space (DFN space, for short) is the inductive limit of a sequence of Banach spaces $F_1 \to F_2 \to \cdots$ such that $F_n \to F_{n+1}$ are injective and nuclear for all n. We denote by **DFN** the full subcategory of **Top** consisting of dual Fréchet nuclear spaces.

A closed linear subspace of an FN space (resp. a DFN space), as well as the quotient of an FN space (resp. a DFN space) by a closed subspace, is also an FN space (resp. a DFN space). Hence, both **FN** and **DFN** are quasi-abelian.

A morphism $f\colon E \to F$ in **FN** or **DFN** is strict if and only if $f(E)$ is a closed subspace of F.

The category **DFN** is equivalent to the opposite category **FN**^{op} of **FN** by $E \mapsto E^*$, where E^* is the strong dual of E.

Note that if M is a C^∞-manifold (countable at infinity), then the space $C^\infty(M)$ of C^∞-functions on M is an FN space. The space $\Gamma_c(M; \mathscr{D}ist_M)$

of distributions with compact support is a DFN space. If X is a complex manifold (countable at infinity), the space $\mathscr{O}_X(X)$ of holomorphic functions is an FN space. For a compact subset K of X, the space $\mathscr{O}_X(K)$ of holomorphic functions defined on a neighborhood of K is a DFN space.

(iii) Let G be a Lie group. A Fréchet nuclear G-module is an FN space E with a continuous G-action, namely G acts on E and the action map $G \times E \to E$ is continuous. Let us denote by $\mathbf{FN}_G$ the category of Fréchet nuclear G-modules. It is also a quasi-abelian category. Similarly we define the notion of dual Fréchet nuclear G-modules and the category $\mathbf{DFN}_G$. The category $(\mathbf{FN}_G)^{\mathrm{op}}$ and $\mathbf{DFN}_G$ are equivalent.

2.2 Derived Categories

Let $\mathcal{C}$ be a quasi-abelian category. A complex X in $\mathcal{C}$ consists of objects X^n ($n \in \mathbb{Z}$) and morphisms $d_X^n \colon X^n \to X^{n+1}$ such that $d_X^{n+1} \circ d_X^n = 0$. The morphisms d_X^n are called the *differentials* of X. Morphisms between complexes are naturally defined. Then the complexes in $\mathcal{C}$ form an additive category, which will be denoted by $\mathrm{C}(\mathcal{C})$. For a complex X and $k \in \mathbb{Z}$, let $X[k]$ be the complex defined by

$$X[k]^n = X^{n+k} \quad d_{X[k]}^n = (-1)^k d_X^{n+k}.$$

Then $X \mapsto X[k]$ is an equivalence of categories, called the *translation functor*.

We say that a complex X is a *strict* complex if all the differentials d_X^n are strict. We say that a complex X is *strictly exact* if $\mathrm{Coker}(d_X^{n-1}) \to \mathrm{Ker}(d_X^{n+1})$ is an isomorphism for all n. Note that $d_X^n \colon X^n \to X^{n+1}$ decomposes into

$$X^n \twoheadrightarrow \mathrm{Coker}(d_X^{n-1}) \twoheadrightarrow \mathrm{Coim}(d_X^n) \to \mathrm{Im}(d_X^n) \rightarrowtail \mathrm{Ker}(d_X^{n+1}) \rightarrowtail X^{n+1}.$$

If X is strictly exact, then X is a strict complex and $0 \to \mathrm{Ker}(d_X^n) \to X^n \to \mathrm{Ker}(d_X^{n+1}) \to 0$ is strictly exact.

For a morphism $f \colon X \to Y$ in $\mathrm{C}(\mathcal{C})$, its mapping cone $\mathrm{Mc}(f)$ is defined by

$$\mathrm{Mc}(f)^n = X^{n+1} \oplus Y^n \text{ and } d_{\mathrm{Mc}(f)}^n = \begin{pmatrix} -d_X^{n+1} & 0 \\ f^{n+1} & d_Y^n \end{pmatrix}.$$

Then we have a sequence of canonical morphisms in $\mathrm{C}(\mathcal{C})$:

$$(2.2.1) \qquad X \xrightarrow{f} Y \xrightarrow{\alpha(f)} \mathrm{Mc}(f) \xrightarrow{\beta(f)} X[1].$$

Let $\mathrm{K}(\mathcal{C})$ be the *homotopy category*, which is defined as follows: $\mathrm{Ob}(\mathrm{K}(\mathcal{C})) = \mathrm{Ob}(\mathrm{C}(\mathcal{C}))$ and, for $X, Y \in \mathrm{K}(\mathcal{C})$, we define

$$\mathrm{Hom}_{\mathrm{K}(\mathcal{C})}(X, Y) = \mathrm{Hom}_{\mathrm{C}(\mathcal{C})}(X, Y)/\mathrm{Ht}(X, Y),$$

where

$\mathrm{Ht}(X,Y) = \{f \in \mathrm{Hom}_{C(\mathcal{C})}(X,Y)\,;$ there exist $h^n\colon X^n \to Y^{n-1}$ such that
$$f^n = d_Y^{n-1} \circ h^n + h^{n+1} \circ d_X^n \text{ for all } n\}.$$

A morphism in $\mathrm{Ht}(X,Y)$ is sometimes called a morphism *homotopic to zero*.

A *triangle* in $\mathrm{K}(\mathcal{C})$ is a sequence of morphisms

$$X \xrightarrow{f} Y \xrightarrow{g} Z \xrightarrow{h} X[1]$$

such that $g \circ f = 0$, $h \circ g = 0$, $f[1] \circ h = 0$. For example, the image of (2.2.1) in $\mathrm{K}(\mathcal{C})$ is a triangle for any morphism $f \in C(\mathcal{C})$. A triangle in $\mathrm{K}(\mathcal{C})$ is called a *distinguished triangle* if it is isomorphic to the image of the triangle (2.2.1) by the functor $C(\mathcal{C}) \to \mathrm{K}(\mathcal{C})$ for some morphism $f \in C(\mathcal{C})$. The additive category $\mathrm{K}(\mathcal{C})$ with the translation functor $\bullet[1]$ and the family of distinguished triangles is a *triangulated category* (see e.g. [19]).

Note that if two complexes X and Y are isomorphic in $\mathrm{K}(\mathcal{C})$, and if X is a strictly exact complex, then so is Y. Let $\mathcal{E}$ be the subcategory of $\mathrm{K}(\mathcal{C})$ consisting of strictly exact complexes. Then $\mathcal{E}$ is a triangulated subcategory, namely it is closed by the translation functors $[k]$ ($k \in \mathbb{Z}$), and if $X \to Y \to Z \to X[1]$ is a distinguished triangle and $X, Y \in \mathcal{E}$, then $Z \in \mathcal{E}$.

We define the derived category $\mathrm{D}(\mathcal{C})$ as the quotient category $\mathrm{K}(\mathcal{C})/\mathcal{E}$. It is defined as follows. A morphism $f\colon X \to Y$ is called a *quasi-isomorphism* (qis for short) if, embedding it in a distinguished triangle $X \xrightarrow{f} Y \to Z \to X[1]$, Z belongs to $\mathcal{E}$. For a chain of morphisms $X \xrightarrow{f} Y \xrightarrow{g} Z$ in $\mathrm{K}(\mathcal{C})$, if two of f, g and $g \circ f$ are qis, then all the three are qis.

With this terminology, $\mathrm{Ob}(\mathrm{D}(\mathcal{C})) = \mathrm{Ob}(\mathrm{K}(\mathcal{C}))$ and for $X, Y \in \mathrm{D}(\mathcal{C})$,

$$\mathrm{Hom}_{D(\mathcal{C})}(X,Y) \simeq \varinjlim_{X' \xrightarrow{\text{qis}} X} \mathrm{Hom}_{K(\mathcal{C})}(X',Y)$$

$$\xrightarrow{\sim} \varinjlim_{X' \xrightarrow{\text{qis}} X, \, Y \xrightarrow{\text{qis}} Y'} \mathrm{Hom}_{K(\mathcal{C})}(X',Y')$$

$$\xleftarrow{\sim} \varinjlim_{Y \xrightarrow{\text{qis}} Y'} \mathrm{Hom}_{K(\mathcal{C})}(X,Y').$$

The composition of morphisms $f\colon X \to Y$ and $g\colon Y \to Z$ is visualized by the following diagram:

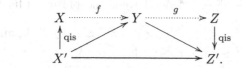

A morphism in $\mathrm{K}(\mathcal{C})$ induces an isomorphism in $\mathrm{D}(\mathcal{C})$ if and only if it is a quasi-isomorphism.

A triangle $X \to Y \to Z \to X[1]$ in $D(\mathcal{C})$ is called a distinguished triangle if it is isomorphic to the image of a distinguished triangle in $K(\mathcal{C})$. Then $D(\mathcal{C})$ is also a triangulated category.

Note that if $X \xrightarrow{f} Y \xrightarrow{g} Z$ is a sequence of morphisms in $C(\mathcal{C})$ such that $0 \to X^n \to Y^n \to Z^n \to 0$ is strictly exact for all n, then the natural morphism $\mathrm{Mc}(f) \to Z$ is a qis, and we have a distinguished triangle

$$X \to Y \to Z \to X[1]$$

in $D(\mathcal{C})$.

We denote by $C^+(\mathcal{C})$ (resp. $C^-(\mathcal{C})$, $C^b(\mathcal{C})$) the full subcategory of $C(\mathcal{C})$ consisting of objects X such that $X^n = 0$ for $n \ll 0$ (resp. $n \gg 0$, $|n| \gg 0$). Let $D^*(\mathcal{C})$ ($* = +, -, b$) be the full subcategory of $D(\mathcal{C})$ whose objects are isomorphic to the image of objects of $C^*(\mathcal{C})$. Similarly, we define the full subcategory $K^*(\mathcal{C})$ of $K(\mathcal{C})$.

We call $D^b(\mathcal{C})$ the *bounded derived category* of $\mathcal{C}$.

2.3 t-Structure

Let us define various truncation functors for $X \in C(\mathcal{C})$:

$$
\begin{array}{llllllll}
\tau^{\leqslant n}X & : \cdots \longrightarrow & X^{n-1} \longrightarrow & \mathrm{Ker}\, d_X^n & \longrightarrow & 0 & \longrightarrow & 0 & \longrightarrow \cdots \\
\tau^{\leqslant n+1/2}X & : \cdots \longrightarrow & X^{n-1} \longrightarrow & X^n & \longrightarrow & \mathrm{Im}\, d_X^n & \longrightarrow & 0 & \longrightarrow \cdots \\
\tau^{\geqslant n}X & : \cdots \longrightarrow & 0 & \longrightarrow & \mathrm{Coker}\, d_X^{n-1} & \longrightarrow & X^{n+1} & \longrightarrow & X^{n+2} \longrightarrow \cdots \\
\tau^{\geqslant n+1/2}X & : \cdots \longrightarrow & 0 & \longrightarrow & \mathrm{Coim}\, d_X^n & \longrightarrow & X^{n+1} & \longrightarrow & X^{n+2} \longrightarrow \cdots
\end{array}
$$

for $n \in \mathbb{Z}$. Then we have morphisms

$$\tau^{\leqslant s}X \longrightarrow \tau^{\leqslant t}X \longrightarrow X \longrightarrow \tau^{\geqslant s}X \longrightarrow \tau^{\geqslant t}X$$

for $s, t \in \frac{1}{2}\mathbb{Z}$ such that $s \leqslant t$. We can easily check that the functors $\tau^{\leqslant s}, \tau^{\geqslant s} \colon C(\mathcal{C}) \to C(\mathcal{C})$ send the morphisms homotopic to zero to morphisms homotopic to zero and the quasi-isomorphisms to quasi-isomorphisms. Hence, they induce the functors

$$\tau^{\leqslant s}, \tau^{\geqslant s} \colon D(\mathcal{C}) \to D(\mathcal{C})$$

and morphisms $\tau^{\leqslant s} \to \mathrm{id} \to \tau^{\geqslant s}$. We have isomorphisms of functors:

$$\tau^{\leqslant s} \circ \tau^{\leqslant t} \simeq \tau^{\leqslant \min(s,t)}, \quad \tau^{\geqslant s} \circ \tau^{\geqslant t} \simeq \tau^{\geqslant \max(s,t)}, \text{ and}$$
$$\tau^{\leqslant s} \circ \tau^{\geqslant t} \simeq \tau^{\geqslant t} \circ \tau^{\leqslant s} \qquad \text{for } s, t \in \frac{1}{2}\mathbb{Z}.$$

We set $\tau^{>s} = \tau^{\geqslant s+1/2}$ and $\tau^{<s} = \tau^{\leqslant s-1/2}$.

We have a distinguished triangle in $D(\mathcal{C})$:

$$\tau^{\leqslant s}X \longrightarrow X \longrightarrow \tau^{>s}X \overset{}{\longrightarrow} (\tau^{\leqslant s}X)[1].$$

For $s \in \frac{1}{2}\mathbb{Z}$, set

$$\mathrm{D}^{\leqslant s}(\mathcal{C}) = \left\{X \in \mathrm{D}(\mathcal{C}) \,;\, \tau^{\leqslant s}X \to X \text{ is an isomorphism}\right\}$$
$$= \left\{X \in \mathrm{D}(\mathcal{C}) \,;\, \tau^{>s}X \simeq 0\right\},$$
$$\mathrm{D}^{\geqslant s}(\mathcal{C}) = \left\{X \in \mathrm{D}(\mathcal{C}) \,;\, X \to \tau^{\geqslant s}X \text{ is an isomorphism}\right\}$$
$$= \left\{X \in \mathrm{D}(\mathcal{C}) \,;\, \tau^{<s}X \simeq 0\right\}.$$

Then $\{\mathrm{D}^{\leqslant s}(\mathcal{C})\}_{s \in \frac{1}{2}\mathbb{Z}}$ is an increasing sequence of full subcategories of $\mathrm{D}(\mathcal{C})$, and $\{\mathrm{D}^{\geqslant s}(\mathcal{C})\}_{s \in \frac{1}{2}\mathbb{Z}}$ is a decreasing sequence of full subcategories of $\mathrm{D}(\mathcal{C})$.

Note that $\mathrm{D}^{+}(\mathcal{C})$ (resp. $\mathrm{D}^{-}(\mathcal{C})$) is the union of all the $\mathrm{D}^{\geqslant s}(\mathcal{C})$'s (resp. all the $\mathrm{D}^{\leqslant s}(\mathcal{C})$'s), and $\mathrm{D}^{b}(\mathcal{C})$ is the intersection of $\mathrm{D}^{+}(\mathcal{C})$ and $\mathrm{D}^{-}(\mathcal{C})$.

The functor $\tau^{\leqslant s} \colon \mathrm{D}(\mathcal{C}) \to \mathrm{D}^{\leqslant s}(\mathcal{C})$ is a right adjoint functor of the inclusion functor $\mathrm{D}^{\leqslant s}(\mathcal{C}) \hookrightarrow \mathrm{D}(\mathcal{C})$, and $\tau^{\geqslant s} \colon \mathrm{D}(\mathcal{C}) \to \mathrm{D}^{\geqslant s}(\mathcal{C})$ is a left adjoint functor of $\mathrm{D}^{\geqslant s}(\mathcal{C}) \hookrightarrow \mathrm{D}(\mathcal{C})$.

Set $\mathrm{D}^{>s}(\mathcal{C}) = \mathrm{D}^{\geqslant s+1/2}(\mathcal{C})$ and $\mathrm{D}^{<s}(\mathcal{C}) = \mathrm{D}^{\leqslant s-1/2}(\mathcal{C})$.

The pair $(\mathrm{D}^{\leqslant s}(\mathcal{C}), \mathrm{D}^{>s-1}(\mathcal{C}))$ is a t-structure of $\mathrm{D}(\mathcal{C})$ (see [3] and also [18]) for any $s \in \frac{1}{2}\mathbb{Z}$. Hence, $\mathrm{D}^{\leqslant s}(\mathcal{C}) \cap \mathrm{D}^{>s-1}(\mathcal{C})$ is an abelian category. The triangulated category $\mathrm{D}(\mathcal{C})$ is equivalent to the derived category of $\mathrm{D}^{\leqslant s}(\mathcal{C}) \cap \mathrm{D}^{>s-1}(\mathcal{C})$. The full subcategory $\mathrm{D}^{\leqslant 0}(\mathcal{C}) \cap \mathrm{D}^{\geqslant 0}(\mathcal{C})$ is equivalent to $\mathcal{C}$.

For $X \in \mathrm{C}(\mathcal{C})$ and an integer n, the following conditions are equivalent:

(i) d_X^n is strict,

(ii) $\tau^{\leqslant n}X \to \tau^{\leqslant n+1/2}X$ is a quasi-isomorphism,

(iii) $\tau^{\geqslant n+1/2}X \to \tau^{\geqslant n+1}X$ is a quasi-isomorphism.

Hence, for an object X of $\mathrm{D}(\mathcal{C})$, X is represented by some strict complex if and only if all complexes in $\mathrm{C}(\mathcal{C})$ representing X are strict complexes. In such a case, we say that X is *strict*. Then, its cohomology group $H^n(X) :=$ $\mathrm{Coker}(X^{n-1} \to \mathrm{Ker}(d_X^n)) \simeq \mathrm{Ker}(\mathrm{Coker}(d_X^{n-1}) \to X^{n+1})$ has a sense as an object of $\mathcal{C}$. The following lemma is immediate.

Lemma 2.3.1. *Let $X \to Y \to Z \overset{+1}{\longrightarrow} X[1]$ be a distinguished triangle, and assume that X and Y are strict. If $H^n(X) \to H^n(Y)$ is a strict morphism for all n, then Z is strict. Moreover we have a strictly exact sequence:*

$$\cdots \longrightarrow H^n(X) \longrightarrow H^n(Y) \longrightarrow H^n(Z) \longrightarrow H^{n+1}(X) \longrightarrow H^{n+1}(Y) \longrightarrow \cdots.$$

Remark 2.3.2. When $\mathcal{C}$ is either **FN** or **DFN**, a complex X in $\mathcal{C}$ is strictly exact if and only if it is exact as a complex of vector spaces forgetting the topology. A complex X is strict if and only if the image of the differential d_X^n is closed in X^{n+1} for all n. Hence, denoting by $\mathscr{F}$ the functor from $\mathrm{D}(\mathbf{FN})$ (resp. $\mathrm{D}(\mathbf{DFN})$) to $\mathrm{D}(\mathrm{Mod}(\mathbb{C}))$, a morphism f in $\mathrm{D}(\mathbf{FN})$ (resp. $\mathrm{D}(\mathbf{DFN})$) is an isomorphism if and only if so is $\mathscr{F}(u)$.

3 Quasi-equivariant D-Modules

3.1 Definition

For the theory of D-modules, we refer the reader to [16].

Let us recall the definition of quasi-equivariant D-modules (cf. [15]).

Let G be an affine algebraic group over $\mathbb{C}$ and $\mathfrak{g}$ its Lie algebra. A G-*module* is by definition a vector space V endowed with an action of G such that $g \mapsto gv$ is a regular function on G for any $v \in V$, i.e., there exist finitely many vectors $\{v_i\}_i$ in V and regular functions $\{a_i(g)\}_i$ on G such that $gv = \sum_i a_i(g)v_i$ for any $g \in G$. It is equivalent to saying that there is a homomorphism $V \to \mathscr{O}_G(G) \otimes V$ (i.e., $v \mapsto \sum_i a_i(g) \otimes v_i$) such that for any $g \in G$ the action $\mu_g \in \mathrm{End}_\mathbb{C}(V)$ is given by $V \to \mathscr{O}_G(G) \otimes V \xrightarrow{i_g^*} V$, where the last arrow i_g^* is induced by the evaluation map $\mathscr{O}_G(G) \to \mathbb{C}$ at g. Hence the G-module structure is equivalent to the co-module structure over the cogebra $\mathscr{O}_G(G)$.

We denote by $\mathrm{Mod}\,(G)$ the category of G-modules, and by $\mathrm{Mod}_f(G)$ the category of finite-dimensional G-modules. It is well-known that any G-module is a union of finite-dimensional sub-G-modules.

Let us recall the definition of $(\mathfrak{g}, H)$-modules for a subgroup H of G.

Definition 3.1.1. Let H be a closed subgroup of G with a Lie algebra $\mathfrak{h}$. A $(\mathfrak{g}, H)$-module is a vector space M endowed with an H-module structure and a $\mathfrak{g}$-module structure such that

(i) the $\mathfrak{h}$-module structure on M induced by the H-module structure coincides with the one induced by the $\mathfrak{g}$-module structure,

(ii) the multiplication homomorphism $\mathfrak{g} \otimes M \to M$ is H-linear, where H acts on $\mathfrak{g}$ by the adjoint action.

Let us denote by $\mathrm{Mod}(\mathfrak{g}, H)$ the category of $(\mathfrak{g}, H)$-modules.

Let X be a smooth algebraic variety with a G-action (we call it *algebraic G-manifold*). Let $\mu \colon G \times X \to X$ denote the action morphism and $\mathrm{pr} \colon G \times X \to X$ the projection. We shall define $p_k \colon G \times G \times X \to G \times X$ $(k = 0, 1, 2)$ by

$$p_0(g_1, g_2, x) = (g_1, g_2 x),$$
$$p_1(g_1, g_2, x) = (g_1 g_2, x), \qquad \mu(g, x) = gx,$$
$$p_2(g_1, g_2, x) = (g_2, x). \qquad \mathrm{pr}(g, x) = x,$$

Then we have a simplicial diagram

$$G \times G \times X \; \underset{p_2}{\overset{p_0}{\underset{p_1}{\rightrightarrows}}} \; G \times X \; \underset{\mathrm{pr}}{\overset{\mu}{\rightrightarrows}} \; X.$$

It means that these morphisms satisfy the commutation relations:

$$\mu \circ p_0 = \mu \circ p_1,$$
$$\mathrm{pr} \circ p_1 = \mathrm{pr} \circ p_2,$$
$$\mu \circ p_2 = \mathrm{pr} \circ p_0.$$

Definition 3.1.2. A *G-equivariant* $\mathscr{O}_X$-module is an $\mathscr{O}_X$-module $\mathscr{F}$ endowed with an isomorphism of $\mathscr{O}_{G \times X}$-modules:

$$(3.1.1) \qquad\qquad \beta \colon \mu^* \mathscr{F} \xrightarrow{\sim} \mathrm{pr}^* \mathscr{F}$$

such that the following diagram commutes (*associative law*):

$$(3.1.2)$$

$$
\begin{array}{ccc}
p_1^* \mu^* \mathscr{F} & \xrightarrow{\quad p_1^* \beta \quad} & p_1^* \mathrm{pr}^* \mathscr{F} \\
\| & & \| \\
p_0^* \mu^* \mathscr{F} \xrightarrow{p_0^* \beta} p_0^* \mathrm{pr}^* \mathscr{F} & = p_2^* \mu^* \mathscr{F} \xrightarrow{p_2^* \beta} & p_2^* \mathrm{pr}^* \mathscr{F} .
\end{array}
$$

We denote by $\mathrm{Mod}\,(\mathscr{O}_X, G)$ the category of G-equivariant $\mathscr{O}_X$-modules which are *quasi-coherent* as $\mathscr{O}_X$-modules.

For a G-equivariant $\mathscr{O}_X$-module $\mathscr{F}$, we can define an action of the Lie algebra $\mathfrak{g}$ on $\mathscr{F}$, i.e., a Lie algebra homomorphism:

$$(3.1.3) \qquad\qquad L_{\mathrm{v}} \colon \mathfrak{g} \to \mathrm{End}_{\mathbb{C}}(\mathscr{F})$$

as follows. Let us denote by

$$(3.1.4) \qquad\qquad L_X \colon \mathfrak{g} \to \Theta_X(X) \to \mathscr{D}_X(X)$$

the infinitesimal action of G on X. Here, Θ_X denotes the sheaf of vector fields on X, and $\mathscr{D}_X$ denotes the sheaf of differential operators. It is a Lie algebra homomorphism. Let us denote by

$$(3.1.5) \qquad\qquad L_G \colon \mathfrak{g} \to \Gamma(G; \mathscr{D}_G)$$

the Lie algebra homomorphism derived by the left action of G on itself. Then its image is the space of right invariant vector fields on G. Denoting by $i \colon X \to G \times X$ the map $x \mapsto (e, x)$, we define

$$(3.1.6) \quad L_{\mathrm{v}}(A)s = i^* \Big((L_G(A) \boxtimes \mathrm{id})(\beta \mu^*(s)) \Big) \quad \text{for } A \in \mathfrak{g} \text{ and } s \in \mathscr{F}.$$

It is a derivation, namely

$$L_{\mathrm{v}}(A)(as) = \big(L_X(A)a \big)\, s + a \big(L_{\mathrm{v}}(A)s \big) \quad \text{for } A \in \mathfrak{g},\ a \in \mathscr{O}_X \text{ and } s \in \mathscr{F}.$$

The notion of equivariance of D-modules is defined similarly to the one of equivariant $\mathscr{O}$-modules. However, there are two options in the D-module case. Let $\mathscr{O}_G \boxtimes \mathscr{D}_X$ denote the subring $\mathscr{O}_{G \times X} \otimes_{\mathrm{pr}^{-1} \mathscr{O}_X} \mathrm{pr}^{-1} \mathscr{D}_X$ of $\mathscr{D}_{G \times X}$. There are two ring morphisms

$$\mathrm{pr}^{-1} \mathscr{D}_X \to \mathscr{O}_G \boxtimes \mathscr{D}_X \quad \text{and} \quad \mathscr{O}_{G \times X} \to \mathscr{O}_G \boxtimes \mathscr{D}_X.$$

Definition 3.1.3. A *quasi-G-equivariant* $\mathscr{D}_X$-module is a $\mathscr{D}_X$-module $\mathscr{M}$ endowed with an $\mathscr{O}_G \boxtimes \mathscr{D}_X$-linear isomorphism

$$(3.1.7) \qquad\qquad \beta : \mathbf{D}\mu^* \mathscr{M} \xrightarrow{\sim} \mathbf{D}\mathrm{pr}^* \mathscr{M}$$

such that the following diagram commutes (*associative law*):

$$
\begin{array}{ccc}
\mathbf{D}p_1^* \mathbf{D}\mu^* \mathscr{M} & \xrightarrow{\mathbf{D}p_1^* \beta} & \mathbf{D}p_1^* \mathbf{D}\mathrm{pr}^* \mathscr{M} \\
\| & & \| \\
\mathbf{D}p_0^* \mathbf{D}\mu^* \mathscr{M} \xrightarrow{\mathbf{D}p_0^* \beta} \mathbf{D}p_0^* \mathbf{D}\mathrm{pr}^* \mathscr{M} = \mathbf{D}p_2^* \mathbf{D}\mu^* \mathscr{M} & \xrightarrow{\mathbf{D}p_2^* \beta} & \mathbf{D}p_2^* \mathbf{D}\mathrm{pr}^* \mathscr{M}.
\end{array}
$$

Here $\mathbf{D}\mu^*$, $\mathbf{D}p_0^*$, etc. are the pull-back functors for D-modules (see §3.4). If moreover β is $\mathscr{D}_{G \times X}$-linear, $\mathscr{M}$ is called *G-equivariant*.

For quasi-G-equivariant $\mathscr{D}_X$-modules $\mathscr{M}$ and $\mathscr{N}$, a G-equivariant morphism $u : \mathscr{M} \to \mathscr{N}$ is a $\mathscr{D}_X$-linear homomorphism $u : \mathscr{M} \to \mathscr{N}$ such that

$$
\begin{array}{ccc}
\mathbf{D}\mu^* \mathscr{M} & \xrightarrow{\beta} & \mathbf{D}\mathrm{pr}^* \mathscr{M} \\
{\scriptstyle \mathbf{D}\mu^* u} \downarrow & & \downarrow {\scriptstyle \mathbf{D}\mathrm{pr}^* u} \\
\mathbf{D}\mu^* \mathscr{N} & \xrightarrow{\beta} & \mathbf{D}\mathrm{pr}^* \mathscr{N}
\end{array}
$$

commutes. Let us denote by $\mathrm{Mod}(\mathscr{D}_X, G)$ the category of quasi-coherent quasi-G-equivariant $\mathscr{D}_X$-modules, and by $\mathrm{Mod}_G(\mathscr{D}_X)$ the full subcategory of $\mathrm{Mod}(\mathscr{D}_X, G)$ consisting of quasi-coherent G-equivariant $\mathscr{D}_X$-modules. Then they are abelian categories, and the functor $\mathrm{Mod}_G(\mathscr{D}_X) \to \mathrm{Mod}(\mathscr{D}_X, G)$ is fully faithful and exact, and the functors $\mathrm{Mod}(\mathscr{D}_X, G) \to \mathrm{Mod}(\mathscr{D}_X) \to \mathrm{Mod}(\mathscr{O}_X)$ and $\mathrm{Mod}(\mathscr{D}_X, G) \to \mathrm{Mod}(\mathscr{O}_X, G)$ are exact.

Roughly speaking, quasi-equivariance means the following. For $g \in G$ let $\mu_g : X \to X$ denotes the multiplication map. Then a $\mathscr{D}_X$-linear isomorphism $\beta_g : \mu_g^* \mathscr{M} \xrightarrow{\sim} \mathscr{M}$ is given in such a way that it depends algebraically on g and satisfies the chain condition $\beta_{g_1 g_2} = \beta_{g_2} \circ \beta_{g_1}$ for $g_1, g_2 \in G$: the diagram

$$
\begin{array}{ccc}
\mu_{g_2}^* \mu_{g_1}^* \mathscr{M} & \xrightarrow{\beta_{g_1}} & \mu_{g_2}^* \mathscr{M} \\
\| & & \downarrow {\scriptstyle \beta_{g_2}} \\
\mu_{g_1 g_2}^* \mathscr{M} & \xrightarrow{\beta_{g_1 g_2}} & \mathscr{M}
\end{array}
$$

is commutative.

Example 3.1. (i) If $\mathscr{F}$ is a G-equivariant $\mathscr{O}_X$-module, then $\mathscr{D}_X \otimes_{\mathscr{O}_X} \mathscr{F}$ is a quasi-G-equivariant $\mathscr{D}_X$-module.

(ii) Let $P_1, \ldots, P_\ell$ be a family of G-invariant differential operators on X. Then $\mathscr{D}_X/\left(\sum_i \mathscr{D}_X P_i\right)$ is a quasi-G-equivariant $\mathscr{D}_X$-module.

Let $\mathscr{M}$ be a quasi-G-equivariant $\mathscr{D}_X$-module. Then the G-equivariant $\mathcal{O}_X$-module structure on $\mathscr{M}$ induces the Lie algebra homomorphism

$$L_{\mathrm{v}} \colon \mathfrak{g} \to \mathrm{End}_{\mathbb{C}}(\mathscr{M}).$$

On the other hand, the $\mathscr{D}_X$-module structure on $\mathscr{M}$ induces the Lie algebra homomorphism

$$\alpha_D \colon \mathfrak{g} \to \Gamma(X; \mathscr{D}_X) \to \mathrm{End}_{\mathbb{C}}(\mathscr{M}).$$

Hence we have:

$$\begin{aligned}
\alpha_D(A)s &= i^*\big((L_G(A) \boxtimes 1)(\mu^*(s))\big) \\
L_{\mathrm{v}}(A)s &= i^*\big((L_G(A) \boxtimes 1)(\beta \circ \mu^*(s))\big)
\end{aligned} \qquad \text{for } s \in \mathscr{M} \text{ and } A \in \mathfrak{g}.$$

Set

$$\gamma_{\mathscr{M}} = L_{\mathrm{v}} - \alpha_D \colon \mathfrak{g} \to \mathrm{End}_{\mathbb{C}}(\mathscr{M}).$$

Since we have

$$[L_{\mathrm{v}}(A), P] = [\alpha_D(A), P] \text{ for any } A \in \mathfrak{g} \text{ and } P \in \mathscr{D}_X,$$

the homomorphism $\gamma_{\mathscr{M}}$ sends $\mathfrak{g}$ to $\mathrm{End}_{\mathscr{D}_X}(\mathscr{M})$. The homomorphism $\gamma_{\mathscr{M}} \colon \mathfrak{g} \to \mathrm{End}_{\mathscr{D}_X}(\mathscr{M})$ vanishes if and only if $L_G(A) \boxtimes 1 \in \Theta_{G \times X}$ commutes with β for all $A \in \mathfrak{g}$. Thus we have obtained the following lemma.

Lemma 3.1.4. *Let $\mathscr{M}$ be a quasi-G-equivariant $\mathscr{D}_X$-module. Let $\gamma_{\mathscr{M}}$ be as above. Then we have*

(i) *$\gamma_{\mathscr{M}}$ is a Lie algebra homomorphism $\mathfrak{g} \to \mathrm{End}_{\mathscr{D}_X}(\mathscr{M})$,*

(ii) *$\mathscr{M}$ is G-equivariant if and only if $\gamma_{\mathscr{M}} = 0$.*

Thus $\mathscr{M}$ has a $(\mathscr{D}_X, U(\mathfrak{g}))$-bimodule structure.

When G acts transitively on X, we have the following description of quasi-equivariant D-modules.

Proposition 3.1.5 ([15]). *Let $X = G/H$ for a closed subgroup H of G, and let $i \colon \mathrm{pt} \to X$ be the map associated with $e \bmod H$. Then $\mathscr{M} \mapsto i^*\mathscr{M}$ gives equivalences of categories*

$$\begin{array}{ccc}
\mathrm{Mod}(\mathscr{D}_X, G) & \xrightarrow{\sim} & \mathrm{Mod}(\mathfrak{g}, H) \\
\cup & & \cup \\
\mathrm{Mod}_G(\mathscr{D}_X) & \xrightarrow{\sim} & \mathrm{Mod}(H/H^\circ),
\end{array}$$

where H° is the connected component of H containing the identity.

The $\mathfrak{g}$-module structure on $i^*\mathscr{M}$ is given by $\gamma_{\mathscr{M}}$. We remark that $\mathrm{Mod}(H/H^\circ)$ is embedded in $\mathrm{Mod}(\mathfrak{g}, H)$ in such a way that $\mathfrak{g}$ acts trivially on the vector spaces in $\mathrm{Mod}(H/H^\circ)$.

Remark 3.1.6. The inclusion functor $\mathrm{Mod}_G(\mathscr{D}_X) \to \mathrm{Mod}(\mathscr{D}_X, G)$ has a left adjoint functor and a right adjoint functor

$$\mathscr{M} \longmapsto \mathbb{C} \otimes_{U(\mathfrak{g})} \mathscr{M} \quad \text{and} \quad \mathscr{M} \longmapsto \mathscr{H}om_{U(\mathfrak{g})}(\mathbb{C}, \mathscr{M}).$$

Here $U(\mathfrak{g})$ acts on $\mathscr{M}$ via $\gamma_{\mathscr{M}}$.

3.2 Derived Categories

Recall that $\mathrm{Mod}(\mathscr{D}_X, G)$ denotes the abelian category of quasi-coherent quasi-G-equivariant $\mathscr{D}_X$-modules. There are the forgetful functor

$$\mathrm{Mod}(\mathscr{D}_X, G) \to \mathrm{Mod}(\mathscr{O}_X, G)$$

and

$$\mathscr{D}_X \otimes_{\mathscr{O}_X} \bullet : \mathrm{Mod}(\mathscr{O}_X, G) \to \mathrm{Mod}(\mathscr{D}_X, G).$$

They are adjoint functors to each other. Namely there is a functorial isomorphism in $\mathscr{F} \in \mathrm{Mod}(\mathscr{O}_X, G)$ and $\mathscr{M} \in \mathrm{Mod}(\mathscr{D}_X, G)$

$$(3.2.1) \quad \mathrm{Hom}_{\mathrm{Mod}(\mathscr{O}_X, G)}(\mathscr{F}, \mathscr{M}) \cong \mathrm{Hom}_{\mathrm{Mod}(\mathscr{D}_X, G)}(\mathscr{D}_X \otimes_{\mathscr{O}_X} \mathscr{F}, \mathscr{M}).$$

Note that, for $\mathscr{F} \in \mathrm{Mod}(\mathscr{O}_X, G)$, the morphism $\gamma_{\mathscr{M}} : \mathfrak{g} \to \mathrm{End}_{\mathscr{D}_X}(\mathscr{M})$ for $\mathscr{M} = \mathscr{D}_X \otimes_{\mathscr{O}_X} \mathscr{F}$ is given by $\gamma_{\mathscr{M}}(A)(P \otimes s) = -PL_X(A) \otimes s + P \otimes L_v(A)s$ for $A \in \mathfrak{g}$, $P \in \mathscr{D}_X$, $s \in \mathscr{F}$. Hence $\mathscr{D}_X \otimes_{\mathscr{O}_X} \mathscr{F}$ is not a G-equivariant $\mathscr{D}_X$-module in general.

Let $\mathrm{Mod}_{\mathrm{coh}}(\mathscr{D}_X, G)$ denote the full subcategory of $\mathrm{Mod}(\mathscr{D}_X, G)$ consisting of coherent quasi-G-equivariant $\mathscr{D}_X$-modules. Similarly let us denote by $\mathrm{Mod}_{\mathrm{coh}}(\mathscr{O}_X, G)$ the category of coherent G-equivariant $\mathscr{O}_X$-modules.

We shall introduce the following intermediate category.

Definition 3.2.1. A quasi-coherent $\mathscr{O}_X$-module (resp. $\mathscr{D}_X$-module) is called *countably coherent* if it is locally generated by countably many sections.

Note that if $\mathscr{F}$ is a countably coherent $\mathscr{O}_X$-module, then there exists locally an exact sequence $\mathscr{O}_X^{\oplus I} \to \mathscr{O}_X^{\oplus J} \to \mathscr{F} \to 0$ where I and J are countable sets.

Note also that any coherent $\mathscr{D}_X$-module is countably coherent over $\mathscr{O}_X$. Hence a quasi-coherent $\mathscr{D}_X$-module is countably coherent, if and only if so is it as an $\mathscr{O}_X$-module.

Note that countably coherent $\mathscr{O}$-modules are stable by inverse images, direct images and tensor products.

Let $\mathrm{Mod}_{\mathrm{cc}}(\mathscr{D}_X, G)$ denote the full subcategory of $\mathrm{Mod}(\mathscr{D}_X, G)$ consisting of countably coherent quasi-G-equivariant $\mathscr{D}_X$-modules.

Let us denote by $\mathrm{D}(\mathscr{D}_X, G)$ the derived category of $\mathrm{Mod}(\mathscr{D}_X, G)$. Let $\mathrm{D}_{\mathrm{cc}}(\mathscr{D}_X, G)$ (resp. $\mathrm{D}_{\mathrm{coh}}(\mathscr{D}_X, G)$) denotes the full subcategory of $\mathrm{D}(\mathscr{D}_X, G)$ consisting of objects whose cohomologies belong to $\mathrm{Mod}_{\mathrm{cc}}(\mathscr{D}_X, G)$ (resp. $\mathrm{Mod}_{\mathrm{coh}}(\mathscr{D}_X, G)$).

Let us denote by $\mathrm{D}^{\mathrm{b}}(\mathscr{D}_X, G)$ the full subcategory of $\mathrm{D}(\mathscr{D}_X, G)$ consisting of objects with bounded cohomologies. We define similarly $\mathrm{D}_{\mathrm{cc}}^{\mathrm{b}}(\mathscr{D}_X, G)$ and $\mathrm{D}_{\mathrm{coh}}^{\mathrm{b}}(\mathscr{D}_X, G)$.

Proposition 3.2.2. *The functors*

$$D^b(\mathrm{Mod}_{cc}(\mathscr{D}_X, G)) \to D^b_{cc}(\mathscr{D}_X, G) \qquad and$$
$$D^b(\mathrm{Mod}_{coh}(\mathscr{D}_X, G)) \to D^b_{coh}(\mathscr{D}_X, G)$$

are equivalences of categories.

This follows easily from the following lemma and a standard argument (e.g. cf. [19])

Lemma 3.2.3. *Any quasi-coherent G-equivariant $\mathscr{O}_X$-module is a union of coherent G-equivariant $\mathscr{O}_X$-submodules. Similarly, any quasi-coherent quasi-G-equivariant $\mathscr{D}_X$-module is a union of coherent quasi-G-equivariant $\mathscr{D}_X$-submodules.*

3.3 Sumihiro's Result

Hereafter we shall assume that X is *quasi-projective*, i.e., X is isomorphic to a subscheme of the projective space $\mathbb{P}^n$ for some n. In such a case, $\mathrm{Mod}(\mathscr{D}_X, G)$ has enough objects so that $D^b(\mathscr{D}_X, G)$ is a desired derived category, namely, the forgetful functor $D^b(\mathscr{D}_X, G) \to D^b(\mathscr{D}_X)$ commutes with various functors such as pull-back functors, push-forward functors, etc. This follows from the following result due to Sumihiro [28].

Proposition 3.3.1. *Let X be a quasi-projective G-manifold.*

(i) *There exists a G-equivariant ample invertible $\mathscr{O}_X$-module.*
(ii) *There exists a G-equivariant open embedding from X into a projective G-manifold.*

In the sequel, we assume

(3.3.1) X is a quasi-projective G-manifold.

Let $\mathscr{L}$ be a G-equivariant ample invertible $\mathscr{O}_X$-module.

Lemma 3.3.2. *Let $\mathscr{F}$ be a coherent G-equivariant $\mathscr{O}_X$-module. Then, for $n \gg 0$, there exist a finite-dimensional G-module V and a G-equivariant surjective homomorphism*

(3.3.2) $\mathscr{L}^{\otimes -n} \otimes V \twoheadrightarrow \mathscr{F}.$

Proof. For $n \gg 0$, $\mathscr{F} \otimes \mathscr{L}^{\otimes n}$ is generated by global sections. Take a finite-dimensional G-submodule V of the G-module $\Gamma(X; \mathscr{F} \otimes \mathscr{L}^{\otimes n})$ such that $V \otimes \mathscr{O}_X \to \mathscr{F} \otimes \mathscr{L}^{\otimes n}$ is surjective. Then this gives a desired homomorphism.
Q.E.D.

Lemma 3.3.2 implies the following exactitude criterion.

Lemma 3.3.3. *Let* $\mathscr{M}' \to \mathscr{M} \to \mathscr{M}''$ *be a sequence in* $\mathrm{Mod}\,(\mathscr{O}_X, G)$. *If* $\mathrm{Hom}_{\mathrm{Mod}\,(\mathscr{O}_X, G)}(\mathscr{E}, \mathscr{M}') \to \mathrm{Hom}_{\mathrm{Mod}\,(\mathscr{O}_X, G)}(\mathscr{E}, \mathscr{M}) \to \mathrm{Hom}_{\mathrm{Mod}\,(\mathscr{O}_X, G)}(\mathscr{E}, \mathscr{M}'')$ *is exact for any locally free G-equivariant $\mathscr{O}_X$-module $\mathscr{E}$ of finite rank, then* $\mathscr{M}' \to \mathscr{M} \to \mathscr{M}''$ *is exact.*

Let us denote by $\mathrm{Mod}_{lf}(\mathscr{D}_X, G)$ the full subcategory of $\mathrm{Mod}_{\mathrm{coh}}(\mathscr{D}_X, G)$ consisting of objects of the form $\mathscr{D}_X \otimes_{\mathscr{O}_X} \mathscr{E}$ for a locally free coherent G-equivariant $\mathscr{O}_X$-module $\mathscr{E}$. By Lemma 3.3.2, for any $\mathscr{M} \in \mathrm{Mod}_{\mathrm{coh}}(\mathscr{D}_X, G)$, there exists a surjective G-equivariant homomorphism $\mathscr{N} \to \mathscr{M}$ with $\mathscr{N} \in \mathrm{Mod}_{lf}(\mathscr{D}_X, G)$.

Lemma 3.3.2 together with standard arguments (see e.g. [19]), we obtain

Proposition 3.3.4. *For any* $\mathscr{M} \in \mathrm{K}^-\big(\mathrm{Mod}_{\mathrm{coh}}(\mathscr{D}_X, G)\big)$ *there exist* $\mathscr{N} \in \mathrm{K}^-\big(\mathrm{Mod}_{lf}(\mathscr{D}_X, G)\big)$ *and a quasi-isomorphism* $\mathscr{N} \to \mathscr{M}$.

The abelian category $\mathrm{Mod}\,(\mathscr{D}_X, G)$ is a Grothendieck category. By a general theory of homological algebra, we have the following proposition (see e.g. [19]).

Proposition 3.3.5. *Any object of* $\mathrm{Mod}\,(\mathscr{D}_X, G)$ *is embedded in an injective object of* $\mathrm{Mod}\,(\mathscr{D}_X, G)$.

Injective objects of $\mathrm{Mod}\,(\mathscr{D}_X, G)$ have the following properties.

Lemma 3.3.6. *The forgetful functor* $\mathrm{Mod}\,(\mathscr{D}_X, G) \to \mathrm{Mod}\,(\mathscr{O}_X, G)$ *sends the injective objects to injective objects.*

This follows from (3.2.1) and the exactitude of $\mathscr{F} \mapsto \mathscr{D}_X \otimes_{\mathscr{O}_X} \mathscr{F}$.

Lemma 3.3.7. *Let* $\mathscr{I}$ *be an injective object of* $\mathrm{Mod}\,(\mathscr{O}_X, G)$. *Then the functor* $\mathscr{F} \mapsto \mathscr{H}om_{\mathscr{O}_X}(\mathscr{F}, \mathscr{I})$ *is an exact functor from* $\mathrm{Mod}_{\mathrm{coh}}(\mathscr{O}_X, G)^{\mathrm{op}}$ *to* $\mathrm{Mod}\,(\mathscr{O}_X, G)$.

Proof. By Lemma 3.3.3, it is enough to remark that, for any locally free $\mathscr{E} \in \mathrm{Mod}_{\mathrm{coh}}(\mathscr{O}_X, G)$,

$$\mathrm{Hom}_{\mathrm{Mod}\,(\mathscr{O}_X, G)}\big(\mathscr{E}, \mathscr{H}om_{\mathscr{O}_X}(\mathscr{F}, \mathscr{I})\big) \cong \mathrm{Hom}_{\mathrm{Mod}\,(\mathscr{O}_X, G)}(\mathscr{E} \otimes_{\mathscr{O}_X} \mathscr{F}, \mathscr{I})$$

is an exact functor in $\mathscr{F}$.   Q.E.D.

Proposition 3.3.8. *Let* $\mathscr{I}$ *be an injective object of* $\mathrm{Mod}\,(\mathscr{O}_X, G)$. *Then for any* $\mathscr{F} \in \mathrm{Mod}_{\mathrm{coh}}(\mathscr{O}_X, G)$,

$$(3.3.3) \qquad \mathscr{E}xt^k_{\mathscr{O}_X}(\mathscr{F}, \mathscr{I}) = 0 \quad and \quad \mathrm{Ext}^k_{\mathscr{O}_X}(\mathscr{F}, \mathscr{I}) = 0 \quad for\ k > 0.$$

Proof. Let us prove first the global case.

(1) Projective case.   Assume first that X is projective. We have

$$\operatorname{Ext}^k_{\mathscr{O}_X}(\mathscr{F}, \mathscr{I}) = \varinjlim_{\mathscr{E}} \operatorname{Ext}^k_{\mathscr{O}_X}(\mathscr{F}, \mathscr{E})$$

where $\mathscr{E}$ ranges over the set of coherent G-equivariant $\mathscr{O}_X$-submodules of $\mathscr{I}$. Hence it is enough to show that for such an $\mathscr{E}$

$$\beta \colon \operatorname{Ext}^k_{\mathscr{O}_X}(\mathscr{F}, \mathscr{E}) \to \operatorname{Ext}^k_{\mathscr{O}_X}(\mathscr{F}, \mathscr{I})$$

vanishes. We shall prove this by the induction on $k > 0$.

For $n \gg 0$, there exists a G-equivariant surjective morphism $V \otimes \mathscr{L}^{\otimes -n} \to \mathscr{F} \to 0$ by Lemma 3.3.2, which induces an exact sequence $0 \to \mathscr{F}' \to V \otimes \mathscr{L}^{\otimes -n} \to \mathscr{F} \to 0$. We may assume that n is so large that $H^m(X; \mathscr{E} \otimes \mathscr{L}^{\otimes n}) = 0$ for any $m > 0$. Then $\operatorname{Ext}^m_{\mathscr{O}_X}(V \otimes \mathscr{L}^{\otimes -n}, \mathscr{E}) = V^* \otimes H^m(X; \mathscr{E} \otimes \mathscr{L}^{\otimes n}) = 0$ for $m > 0$, and hence we obtain a commutative diagram with exact rows:

$$\begin{array}{ccccccc}
\operatorname{Ext}^{k-1}_{\mathscr{O}_X}(V \otimes \mathscr{L}^{\otimes -n}, \mathscr{E}) & \longrightarrow & \operatorname{Ext}^{k-1}_{\mathscr{O}_X}(\mathscr{F}', \mathscr{E}) & \xrightarrow{\;\alpha\;} & \operatorname{Ext}^{k}_{\mathscr{O}_X}(\mathscr{F}, \mathscr{E}) & \longrightarrow & 0. \\
\downarrow & & \downarrow & & \beta\downarrow & & \\
\operatorname{Ext}^{k-1}_{\mathscr{O}_X}(V \otimes \mathscr{L}^{\otimes -n}, \mathscr{I}) & \xrightarrow{\;\gamma\;} & \operatorname{Ext}^{k-1}_{\mathscr{O}_X}(\mathscr{F}', \mathscr{I}) & \xrightarrow{\;\delta\;} & \operatorname{Ext}^{k}_{\mathscr{O}_X}(\mathscr{F}, \mathscr{I}). & &
\end{array}$$

The homomorphism γ is surjective, because $\mathscr{I}$ is injective for $k = 1$ and the induction hypothesis implies $\operatorname{Ext}^{k-1}_{\mathscr{O}_X}(\mathscr{F}', \mathscr{I}) = 0$ for $k > 1$. Hence we have $\delta = 0$, and the surjectivity of α implies $\beta = 0$.

(2) General case.   Let us embed X in a projective G-manifold $\bar{X}$ and let $j \colon X \hookrightarrow \bar{X}$ be the open embedding. Since

$$\operatorname{Hom}_{\mathscr{O}_{\bar{X}}}(\mathscr{N}, j_* \mathscr{I}) = \operatorname{Hom}_{\mathscr{O}_X}(j^{-1}\mathscr{N}, \mathscr{I})$$

for $\mathscr{N} \in \operatorname{Mod}(\mathscr{O}_{\bar{X}}, G)$, $j_* \mathscr{I}$ is an injective object of $\operatorname{Mod}(\mathscr{O}_{\bar{X}}, G)$. Let J be the defining ideal of $\bar{X} \setminus X$. Then J is a coherent G-equivariant ideal of $\mathscr{O}_{\bar{X}}$. Let us take a coherent G-equivariant $\mathscr{O}_{\bar{X}}$-module $\overline{\mathscr{F}}$ such that $\overline{\mathscr{F}}|_X \simeq \mathscr{F}$. Then, the isomorphism (see [6])

$$\operatorname{Ext}^k_{\mathscr{O}_X}(\mathscr{F}, \mathscr{I}) = \varinjlim_{n} \operatorname{Ext}^k_{\mathscr{O}_{\bar{X}}}(\overline{\mathscr{F}} \otimes_{\mathscr{O}_X} J^n, j_* \mathscr{I})$$

implies the desired result.

The local case can be proved similarly to the proof in (1) by using Lemma 3.3.7.   Q.E.D.

Proposition 3.3.9. *Let $f \colon X \to Y$ be a G-equivariant morphism of quasi-projective G-manifolds. Then for any injective object $\mathscr{I}$ of $\operatorname{Mod}(\mathscr{O}_X, G)$ and $\mathscr{F} \in \operatorname{Mod}_{\mathrm{coh}}(\mathscr{O}_X, G)$, we have*

$$R^k f_*\big(\mathscr{H}om_{\mathscr{O}_X}(\mathscr{F}, \mathscr{I})\big) = 0 \quad \text{for } k > 0.$$

Proof. The proof is similar to the proof of the preceding proposition. The morphism $f: X \to Y$ can be embedded in $\bar{f}: \bar{X} \to \bar{Y}$ for projective G-manifolds $\bar{X}$ and $\bar{Y}$. Let $j: X \to \bar{X}$ be the open embedding. Let J be the defining ideal of $\bar{X} \setminus X$. Then, extending $\mathscr{F}$ to a coherent G-equivariant $\mathscr{O}_{\bar{X}}$-module $\bar{\mathscr{F}}$, one has

$$R^k f_* \left(\mathscr{H}om_{\mathscr{O}_X}(\mathscr{F}, \mathscr{I}) \right) \simeq \varinjlim_n R^k \bar{f}_* \left(\mathscr{H}om_{\mathscr{O}_X}(\bar{\mathscr{F}} \otimes J^n, j_* \mathscr{I}) \right)|_Y.$$

Hence, we may assume from the beginning that X and Y are projective. Then we can argue similarly to (1) in the proof of Proposition 3.3.8, once we prove

(3.3.4) $\mathscr{F} \mapsto f_* \mathscr{H}om_{\mathscr{O}_X}(\mathscr{F}, \mathscr{I})$ is an exact functor in $\mathscr{F} \in \mathrm{Mod}_{\mathrm{coh}}(\mathscr{O}_X, G)$.

This follows from Lemma 3.3.3 and the exactitude of the functor

$$\mathrm{Hom}_{\mathrm{Mod}(\mathscr{O}_Y, G)}(\mathscr{E}, f_* \mathscr{H}om_{\mathscr{O}_X}(\mathscr{F}, \mathscr{I})) \simeq \mathrm{Hom}_{\mathrm{Mod}(\mathscr{O}_X, G)}(f^* \mathscr{E} \otimes_{\mathscr{O}_X} \mathscr{F}, \mathscr{I})$$

in $\mathscr{F}$ for any locally free G-equivariant coherent $\mathscr{O}_Y$-module $\mathscr{E}$.   Q.E.D.

By this proposition, we obtain the following corollary.

Corollary 3.3.10. *Let $\mathscr{I}$ be an injective object of $\mathrm{Mod}(\mathscr{D}_X, G)$. Then for any morphism $f: X \to Y$ of quasi-projective G-manifolds and a coherent locally free G-equivariant $\mathscr{O}_X$-module $\mathscr{E}$*

(3.3.5) $$R^k f_*(\mathscr{E} \otimes_{\mathscr{O}_X} \mathscr{I}) = 0 \quad \text{for } k > 0.$$

Lemma 3.3.11. *For any morphism $f: X \to Y$ and $\mathscr{M} \in \mathrm{Mod}_{\mathrm{cc}}(\mathscr{D}_X, G)$ and a coherent locally free G-equivariant $\mathscr{O}_X$-module $\mathscr{E}$, there exists a monomorphism $\mathscr{M} \to \mathscr{M}'$ in $\mathrm{Mod}_{\mathrm{cc}}(\mathscr{D}_X, G)$ such that $R^k f_*(\mathscr{E} \otimes_{\mathscr{O}_X} \mathscr{M}') = 0$ for any $k \neq 0$.*

Proof. Let us take a monomorphism $\mathscr{M} \to \mathscr{I}$ where $\mathscr{I}$ is an injective object of $\mathrm{Mod}(\mathscr{D}_X, G)$. Let us construct, by the induction on n, an increasing sequence $\{\mathscr{M}_n\}_{n \geqslant 0}$ of countably coherent subobjects of $\mathscr{I}$ such that $\mathscr{M}_0 = \mathscr{M}$ and

(3.3.6) $R^k f_*(\mathscr{E} \otimes \mathscr{M}_n) \to R^k f_*(\mathscr{E} \otimes \mathscr{M}_{n+1})$ vanishes for $k \neq 0$.

Assuming that $\mathscr{M}_n$ has been constructed, we shall construct $\mathscr{M}_{n+1}$. We have

$$\varinjlim_{\mathscr{N} \subset \mathscr{I}} R^k f_*(\mathscr{E} \otimes \mathscr{N}) \cong R^k f_*(\mathscr{E} \otimes \mathscr{I}) = 0 \quad \text{for } k \neq 0.$$

Here $\mathscr{N}$ ranges over the set of countably coherent subobjects of $\mathscr{I}$. Since $R^k f_*(\mathscr{E} \otimes \mathscr{M}_n)$ is countably coherent, there exists a countably coherent subobject $\mathscr{M}_{n+1}$ of $\mathscr{I}$ such that $\mathscr{M}_n \subset \mathscr{M}_{n+1}$ and the morphism $R^k f_*(\mathscr{E} \otimes \mathscr{M}_n) \to R^k f_*(\mathscr{E} \otimes \mathscr{M}_{n+1})$ vanishes for $k \neq 0$.

Then $\mathscr{M}' := \varinjlim_n \mathscr{M}_n$ satisfies the desired condition, because (3.3.6) implies

$$R^k f_*(\mathscr{E} \otimes \mathscr{M}') \simeq \varinjlim_n R^k f_*(\mathscr{E} \otimes \mathscr{M}_n) \simeq 0$$

for $k \neq 0$.   Q.E.D.

3.4 Pull-back Functors

Let $f\colon X \to Y$ be a morphism of quasi-projective algebraic manifolds. Set $\mathscr{D}_{X \to Y} = \mathscr{O}_X \otimes_{f^{-1}\mathscr{O}_Y} f^{-1}\mathscr{D}_Y$. Then $\mathscr{D}_{X \to Y}$ has a structure of a $(\mathscr{D}_X, f^{-1}\mathscr{D}_Y)$-bimodule. It is countably coherent as a $\mathscr{D}_X$-module. Then

$$f^*\colon \mathscr{N} \mapsto \mathscr{D}_{X \to Y} \otimes_{\mathscr{D}_Y} \mathscr{N} = \mathscr{O}_X \otimes_{\mathscr{O}_Y} \mathscr{N}$$

gives a right exact functor from $\mathrm{Mod}(\mathscr{D}_Y)$ to $\mathrm{Mod}(\mathscr{D}_X)$. It is left derivable, and we denote by $\mathbf{D}f^*$ its left derived functor:

$$\mathbf{D}f^*\colon \mathrm{D}^{\mathrm{b}}(\mathscr{D}_Y) \to \mathrm{D}^{\mathrm{b}}(\mathscr{D}_X).$$

Now let $f\colon X \to Y$ be a G-equivariant morphism of quasi-projective algebraic G-manifolds. Then $f^*\colon \mathscr{N} \mapsto \mathscr{D}_{X \to Y} \otimes_{\mathscr{D}_Y} \mathscr{N} = \mathscr{O}_X \otimes_{\mathscr{O}_Y} \mathscr{N}$ gives also a right exact functor:

$$f^*\colon \mathrm{Mod}(\mathscr{D}_Y, G) \to \mathrm{Mod}(\mathscr{D}_X, G).$$

Lemma 3.3.2 implies that any quasi-coherent quasi-G-equivariant $\mathscr{D}_Y$-module has a finite resolution by quasi-coherent quasi-G-equivariant $\mathscr{D}_Y$-modules flat over $\mathscr{O}_Y$. Hence the functor $f^*\colon \mathrm{Mod}(\mathscr{D}_Y, G) \to \mathrm{Mod}(\mathscr{D}_X, G)$ is left derivable. We denote its left derived functor by $\mathbf{D}f^*$:

(3.4.1) $\mathbf{D}f^*\colon \mathrm{D}^{\mathrm{b}}(\mathscr{D}_Y, G) \to \mathrm{D}^{\mathrm{b}}(\mathscr{D}_X, G).$

By the construction, the diagram

$$
\begin{array}{ccccc}
\mathrm{D}^{\mathrm{b}}(\mathscr{D}_Y, G) & \longrightarrow & \mathrm{D}^{\mathrm{b}}(\mathscr{D}_Y) & \longrightarrow & \mathrm{D}^{\mathrm{b}}(\mathscr{O}_Y) \\
\downarrow{\scriptstyle \mathbf{D}f^*} & & \downarrow{\scriptstyle \mathbf{D}f^*} & & \downarrow{\scriptstyle \mathbf{L}f^*} \\
\mathrm{D}^{\mathrm{b}}(\mathscr{D}_X, G) & \longrightarrow & \mathrm{D}^{\mathrm{b}}(\mathscr{D}_X) & \longrightarrow & \mathrm{D}^{\mathrm{b}}(\mathscr{O}_X)
\end{array}
$$

commutes. The functor $\mathbf{D}f^*$ sends $\mathrm{D}^{\mathrm{b}}_{\mathrm{cc}}(\mathscr{D}_Y, G)$ to $\mathrm{D}^{\mathrm{b}}_{\mathrm{cc}}(\mathscr{D}_X, G)$. If f is a smooth morphism, then $\mathbf{D}f^*$ sends $\mathrm{D}^{\mathrm{b}}_{\mathrm{coh}}(\mathscr{D}_Y, G)$ to $\mathrm{D}^{\mathrm{b}}_{\mathrm{coh}}(\mathscr{D}_X, G)$.

3.5 Push-forward Functors

Let $f\colon X \to Y$ be a morphism of quasi-projective algebraic manifolds. Recall that the push-forward functor

$$(3.5.1) \qquad\qquad \mathbf{D}f_*\colon \mathrm{D}^{\mathrm{b}}(\mathscr{D}_X) \to \mathrm{D}^{\mathrm{b}}(\mathscr{D}_Y)$$

is defined by $Rf_*(\mathscr{D}_{Y \leftarrow X} \overset{\mathrm{L}}{\otimes}_{\mathscr{D}_X} \mathscr{M})$. Here $\mathscr{D}_{Y \leftarrow X}$ is an $(f^{-1}\mathscr{D}_Y, \mathscr{D}_X)$-bimodule $f^{-1}\mathscr{D}_Y \otimes_{f^{-1}\mathscr{O}_Y} \Omega_{X/Y}$, where we use the notations:

$$\Omega_X := \Omega_X^{\mathrm{d}_X} \quad \text{and} \quad \Omega_{X/Y} := \Omega_X \otimes \Omega_Y^{\otimes -1}.$$

Let $f\colon X \to Y$ be a G-equivariant morphism of quasi-projective algebraic G-manifolds. Let us define the push-forward functor

$$(3.5.2) \qquad\qquad \mathbf{D}f_*\colon \mathrm{D}^{\mathrm{b}}(\mathscr{D}_X, G) \to \mathrm{D}^{\mathrm{b}}(\mathscr{D}_Y, G)$$

in the equivariant setting.

In order to calculate $\mathscr{D}_{Y \leftarrow X} \overset{\mathrm{L}}{\otimes}_{\mathscr{D}_X} \mathscr{M}$, let us take a resolution of $\mathscr{D}_{Y \leftarrow X}$ by flat $\mathscr{D}_X$-modules:

$$
\begin{aligned}
0 \leftarrow \mathscr{D}_{Y \leftarrow X} &\leftarrow f^{-1}(\mathscr{D}_Y \otimes \Omega_Y^{\otimes -1}) \otimes \Omega_X^{\mathrm{d}_X} \otimes \mathscr{D}_X \\
&\leftarrow f^{-1}(\mathscr{D}_Y \otimes \Omega_Y^{\otimes -1}) \otimes \Omega_X^{\mathrm{d}_X - 1} \otimes \mathscr{D}_X \\
&\leftarrow \cdots\cdots\cdots\cdots\cdots\cdots\cdots \leftarrow \\
&\leftarrow f^{-1}(\mathscr{D}_Y \otimes \Omega_Y^{\otimes -1}) \otimes \Omega_X^0 \otimes \mathscr{D}_X \leftarrow 0.
\end{aligned}
$$
$(3.5.3)$

It is an exact sequence of $(f^{-1}\mathscr{D}_Y, \mathscr{D}_X)$-bimodules. Thus, for a complex $\mathscr{M}$ of $\mathscr{D}_X$-modules, $\mathscr{D}_{Y \leftarrow X} \overset{\mathrm{L}}{\otimes}_{\mathscr{D}_X} \mathscr{M}$ is represented by the complex of $f^{-1}\mathscr{D}_Y$-modules

$$(3.5.4) \qquad f^{-1}(\mathscr{D}_Y \otimes_{\mathscr{O}_Y} \Omega_Y^{\otimes -1}) \otimes_{f^{-1}\mathscr{O}_Y} \Omega_X^{\bullet} \otimes_{\mathscr{O}_X} \mathscr{M}[\mathrm{d}_X].$$

The differential of the complex $(3.5.4)$ is given as follows. First note that there is a left $\mathscr{D}_Y$-linear homomorphism

$$d\colon \mathscr{D}_Y \otimes_{\mathscr{O}_Y} \Omega_Y^{\otimes -1} \to \mathscr{D}_Y \otimes_{\mathscr{O}_Y} \Omega_Y^{\otimes -1} \otimes_{\mathscr{O}_Y} \Omega_Y^1$$

given by

$$d(P \otimes dy^{\otimes -1}) = -\sum_j P \frac{\partial}{\partial y_i} \otimes dy^{\otimes -1} \otimes dy_j.$$

Here $(y_1, \ldots, y_m)$ is a local coordinate system of Y, $dy^{\otimes -1} = (dy_1 \wedge \cdots \wedge dy_m)^{\otimes -1}$ and $P \in \mathscr{D}_Y$. We define the morphism

$$\varphi\colon f^{-1}(\mathscr{D}_Y \otimes \Omega_Y^{\otimes -1} \otimes \Omega_Y^{\bullet}) \otimes \Omega_X^{\bullet} \to f^{-1}(\mathscr{D}_Y \otimes \Omega_Y^{\otimes -1}) \otimes \Omega_X^{\bullet}$$

by $a \otimes \theta \otimes \omega \mapsto a \otimes (f^* \theta \wedge \omega)$ for $a \in \mathscr{D}_Y \otimes \Omega_Y^{\otimes -1}$, $\theta \in \Omega_Y^\bullet$ and $\omega \in \Omega_X^\bullet$.
Then, taking a local coordinate system $(x_1, \ldots, x_n)$ of X, the differential d of
(3.5.4) is given by

$$d(a \otimes \omega \otimes u)$$
$$= \varphi(da \otimes \omega) \otimes u + a \otimes d\omega \otimes u$$
$$+ \sum_i a \otimes (dx_i \wedge \omega) \otimes \frac{\partial}{\partial x_i} u + (-1)^p a \otimes \omega \otimes du$$

for $a \in \mathscr{D}_Y \otimes_{\mathscr{O}_Y} \Omega_Y^{\otimes -1}$, $\omega \in \Omega_X^p$ and $u \in \mathscr{M}$.

We now define the functor

$$\mathrm{K} f_* \colon \mathrm{K}^+(\mathrm{Mod}(\mathscr{D}_X, G)) \to \mathrm{K}^+(\mathrm{Mod}(\mathscr{D}_Y, G))$$

by

$$\mathrm{K} f_*(\mathscr{M}) := f_*\big(f^{-1}(\mathscr{D}_Y \otimes_{\mathscr{O}_Y} \Omega_Y^{\otimes -1}) \otimes_{f^{-1}\mathscr{O}_Y} \Omega_X^\bullet \otimes_{\mathscr{O}_X} \mathscr{M}\big)[d_X]$$
$$\cong \mathscr{D}_Y \otimes_{\mathscr{O}_Y} \Omega_Y^{\otimes -1} \otimes_{\mathscr{O}_Y} f_*(\Omega_X^\bullet \otimes_{\mathscr{O}_X} \mathscr{M})[d_X].$$

For an injective object $\mathscr{M}$ of $\mathrm{Mod}(\mathscr{D}_X, G)$, Corollary 3.3.10 implies

(3.5.5) $R^k f_*(\Omega_X^p \otimes_{\mathscr{O}_X} \mathscr{M}) = 0$ for any p and any $k > 0$.

Hence if $\mathscr{I}^\bullet$ is an exact complex in $\mathrm{Mod}(\mathscr{D}_X, G)$ such that all $\mathscr{I}^n$ are injective, then $\mathrm{K} f_*(\mathscr{I}^\bullet)$ is exact. Hence $\mathrm{K} f_*$ is right derivable. Let $\mathbf{D}f_*$ be its right derived functor:

$$\mathbf{D}f_* \colon \mathrm{D}^+(\mathrm{Mod}(\mathscr{D}_X, G)) \to \mathrm{D}^+(\mathrm{Mod}(\mathscr{D}_Y, G)).$$

For a complex $\mathscr{M}$ in $\mathrm{Mod}(\mathscr{D}_X, G)$ bounded from below, we have

(3.5.6) $\mathrm{K} f_* \mathscr{M} \xrightarrow{\sim} \mathbf{D}f_*(\mathscr{M})$

as soon as $R^k f_*(\Omega_X^p \otimes_{\mathscr{O}_X} \mathscr{M}^n) = 0$ for all $k \neq 0$ and p, n.

By the construction, the following diagram commutes.

$$
\begin{array}{ccc}
\mathrm{D}^+(\mathrm{Mod}(\mathscr{D}_X, G)) & \xrightarrow{\mathbf{D}f_*} & \mathrm{D}^+(\mathrm{Mod}(\mathscr{D}_Y, G)) \\
\downarrow & & \downarrow \\
\mathrm{D}^+(\mathrm{Mod}(\mathscr{D}_X)) & \xrightarrow{\mathbf{D}f_*} & \mathrm{D}^+(\mathrm{Mod}(\mathscr{D}_Y)).
\end{array}
$$

Since $\mathbf{D}f_*$ sends $\mathrm{D}^b(\mathrm{Mod}(\mathscr{D}_X))$ to $\mathrm{D}^b(\mathrm{Mod}(\mathscr{D}_Y))$, we conclude that $\mathbf{D}f_*$ sends $\mathrm{D}^b(\mathscr{D}_X, G)$ to $\mathrm{D}^b(\mathscr{D}_Y, G)$, and $\mathrm{D}^b_{cc}(\mathscr{D}_X, G)$ to $\mathrm{D}^b_{cc}(\mathscr{D}_Y, G)$.

Proposition 3.5.1. *The restriction*

$$K_{cc} f_* \colon K^b(\operatorname{Mod}_{cc}(\mathscr{D}_X, G)) \to K^b(\operatorname{Mod}_{cc}(\mathscr{D}_Y, G))$$

of $K f_*$ *is right derivable and the diagram*

$$\begin{array}{ccc} D^b(\operatorname{Mod}_{cc}(\mathscr{D}_X, G)) & \xrightarrow{\sim} & D^b_{cc}(\mathscr{D}_X, G) \\ {\scriptstyle \mathbf{R}(K_{cc} f_*)} \downarrow & & \downarrow {\scriptstyle D f_*} \\ D^b(\operatorname{Mod}_{cc}(\mathscr{D}_Y, G)) & \xrightarrow{\sim} & D^b_{cc}(\mathscr{D}_Y, G) \end{array}$$

quasi-commutes.

Proof. It is enough to show that, for any $\mathscr{M} \in K^b(\operatorname{Mod}_{cc}(\mathscr{D}_X, G))$, we can find a quasi-isomorphism $\mathscr{M} \to \mathscr{M}'$ such that the morphism $K f_*(\mathscr{M}') \to D f_*(\mathscr{M})$ is an isomorphism in $D^b(\mathscr{D}_Y, G)$. In order to have such an isomorphism, it is enough to show that $\mathscr{M}'$ satisfies the condition in (3.5.6). By Lemma 3.3.11, we have a quasi-isomorphism $\mathscr{M} \to \mathscr{M}'$ such that $\mathscr{M}'$ is a complex in $\operatorname{Mod}_{cc}(\mathscr{D}_X, G)$ bounded below and satisfies the condition in (3.5.6). Since the cohomological dimension of $\mathbf{R} f_*$ is finite, by taking n sufficiently large, the truncated complex $\tau^{\leqslant n} \mathscr{M}'$ satisfies the condition in (3.5.6), and $\mathscr{M} \to \tau^{\leqslant n} \mathscr{M}'$ is a quasi-isomorphism.   Q.E.D.

Note that, if f is projective, $D f_*$ sends $D^b_{coh}(\mathscr{D}_X, G)$ to $D^b_{coh}(\mathscr{D}_Y, G)$ (see [16]).

3.6 External and Internal Tensor Products

Let X and Y be two algebraic G-manifolds. Let $q_1 \colon X \times Y \to X$ and $q_2 \colon X \times Y \to Y$ be the projections. Then for $\mathscr{M}_1 \in \operatorname{Mod}(\mathscr{D}_X, G)$ and $\mathscr{M}_2 \in \operatorname{Mod}(\mathscr{D}_Y, G)$, $\mathscr{M}_1 \boxtimes \mathscr{M}_2 = (\mathscr{O}_{X \times Y} \otimes_{q_1^{-1} \mathscr{O}_X} q_1^{-1} \mathscr{M}_1) \otimes_{q_2^{-1} \mathscr{O}_Y} q_2^{-1} \mathscr{M}_2$ has a structure of quasi-G-equivariant $\mathscr{D}_{X \times Y}$-module. Since this is an exact bi-functor, we obtain

$$\bullet \boxtimes \bullet \; \colon D^b(\mathscr{D}_X, G) \times D^b(\mathscr{D}_Y, G) \to D^b(\mathscr{D}_{X \times Y}, G).$$

Taking pt as Y, we obtain

$$\bullet \otimes \bullet \; \colon D^b(\mathscr{D}_X, G) \times D^b(\operatorname{Mod}(G)) \to D^b(\mathscr{D}_X, G).$$

Here $\operatorname{Mod}(G)$ denotes the category of G-modules.

For two quasi-G-equivariant $\mathscr{D}_X$-modules $\mathscr{M}_1$ and $\mathscr{M}_2$, the $\mathscr{O}_X$-module $\mathscr{M}_1 \otimes_{\mathscr{O}_X} \mathscr{M}_2$ has a structure of $\mathscr{D}_X$-module by

$$v(s_1 \otimes s_2) = (v s_1) \otimes s_2 + s_1 \otimes (v s_2) \quad \text{for } v \in \Theta_X \text{ and } s_\nu \in \mathscr{M}_\nu.$$

Since this is G-equivariant, we obtain the right exact bi-functor

$$\bullet \otimes \bullet \; \colon \operatorname{Mod}(\mathscr{D}_X, G) \times \operatorname{Mod}(\mathscr{D}_X, G) \to \operatorname{Mod}(\mathscr{D}_X, G).$$

Taking its left derived functor, we obtain

$$\bullet \overset{\mathrm{D}}{\otimes} \bullet \quad : \mathrm{D}^{\mathrm{b}}(\mathscr{D}_X, G) \times \mathrm{D}^{\mathrm{b}}(\mathscr{D}_X, G) \longrightarrow \mathrm{D}^{\mathrm{b}}(\mathscr{D}_X, G).$$

We have

$$\mathscr{M}_1 \overset{\mathrm{D}}{\otimes} \mathscr{M}_2 \simeq \mathscr{M}_1 \otimes_{\mathscr{O}_X} \mathscr{M}_2$$

if either $\mathscr{M}_1$ or $\mathscr{M}_2$ are complexes in $\mathrm{Mod}(\mathscr{D}_X, G)$ flat over $\mathscr{O}_X$.

The functor $\bullet \overset{\mathrm{D}}{\otimes} \bullet$ sends $\mathrm{D}^{\mathrm{b}}_{\mathrm{cc}}(\mathscr{D}_X, G) \times \mathrm{D}^{\mathrm{b}}_{\mathrm{cc}}(\mathscr{D}_X, G)$ to $\mathrm{D}^{\mathrm{b}}_{\mathrm{cc}}(\mathscr{D}_X, G)$.
Note that, denoting by $\delta \colon X \to X \times X$ the diagonal embedding, we have

$$\mathscr{M}_1 \overset{\mathrm{D}}{\otimes} \mathscr{M}_2 \simeq \mathrm{D}\delta^*(\mathscr{M}_1 \boxtimes \mathscr{M}_2).$$

Lemma 3.6.1. *For $\mathscr{F} \in \mathrm{Mod}(\mathscr{O}_X, G)$ and $\mathscr{M} \in \mathrm{Mod}(\mathscr{D}_X, G)$, there exists a canonical isomorphism in $\mathrm{Mod}(\mathscr{D}_X, G)$:*

$$(3.6.1) \qquad (\mathscr{D}_X \otimes_{\mathscr{O}_X} \mathscr{F}) \otimes \mathscr{M} \simeq \mathscr{D}_X \otimes_{\mathscr{O}_X} (\mathscr{F} \otimes_{\mathscr{O}_X} \mathscr{M}).$$

Here $\mathscr{F} \otimes_{\mathscr{O}_X} \mathscr{M}$ in the right-hand side is regarded as a G-equivariant $\mathscr{O}_X$-module.

The proof is similar to the one in [16] in the non-equivariant case.

3.7 Semi-outer Hom

Let $\mathscr{M} \in \mathrm{Mod}_{\mathrm{coh}}(\mathscr{D}_X, G)$ and $\mathscr{M}' \in \mathrm{Mod}(\mathscr{D}_X, G)$. Then the vector space $\mathrm{Hom}_{\mathscr{D}_X}(\mathscr{M}, \mathscr{M}')$ has a structure of G-modules as follows:

$$\mathrm{Hom}_{\mathscr{D}_X}(\mathscr{M}, \mathscr{M}') \to \mathrm{Hom}_{\mathscr{D}_{G \times X}}(\mu^*\mathscr{M}, \mu^*\mathscr{M}')$$
$$\to \mathrm{Hom}_{\mathscr{O}_G \boxtimes \mathscr{D}_X}(\mu^*\mathscr{M}, \mu^*\mathscr{M}') \simeq \mathrm{Hom}_{\mathscr{O}_G \boxtimes \mathscr{D}_X}(\mathrm{pr}^*\mathscr{M}, \mathrm{pr}^*\mathscr{M}')$$
$$\simeq \mathrm{Hom}_{\mathscr{D}_X}(\mathscr{M}, \mathrm{pr}_*\mathrm{pr}^*\mathscr{M}') \simeq \mathrm{Hom}_{\mathscr{D}_X}(\mathscr{M}, \mathscr{O}_G(G) \otimes_{\mathbb{C}} \mathscr{M}')$$
$$\simeq \mathscr{O}_G(G) \otimes \mathrm{Hom}_{\mathscr{D}_X}(\mathscr{M}, \mathscr{M}').$$

Here the last isomorphism follows from the fact that $\mathscr{M}$ is coherent.
We can easily see that for any $V \in \mathrm{Mod}(G)$

$$(3.7.1) \quad \mathrm{Hom}_{\mathrm{Mod}(G)}(V, \mathrm{Hom}_{\mathscr{D}_X}(\mathscr{M}, \mathscr{M}')) \cong \mathrm{Hom}_{\mathrm{Mod}(\mathscr{D}_X, G)}(V \otimes \mathscr{M}, \mathscr{M}').$$

Since $V \mapsto V \otimes \mathscr{M}$ is an exact functor, (3.7.1) implies the following lemma.

Lemma 3.7.1. *Let $\mathscr{I}$ be an injective object of $\mathrm{Mod}(\mathscr{D}_X, G)$ and $\mathscr{M} \in \mathrm{Mod}_{\mathrm{coh}}(\mathscr{D}_X, G)$. Then $\mathrm{Hom}_{\mathscr{D}_X}(\mathscr{M}, \mathscr{I})$ is an injective object of $\mathrm{Mod}(G)$.*

Let $\mathbf{R}\mathrm{Hom}_{\mathscr{D}_X}(\mathscr{M}, \bullet)$ be the right derived functor of $\mathrm{Hom}_{\mathscr{D}_X}(\mathscr{M}, \bullet)$:

$$\mathbf{R}\mathrm{Hom}_{\mathscr{D}_X}(\bullet, \bullet) \colon \mathrm{D}^{\mathrm{b}}_{\mathrm{coh}}(\mathscr{D}_X, G)^{\mathrm{op}} \times \mathrm{D}^+(\mathscr{D}_X, G) \to \mathrm{D}^+(\mathrm{Mod}(G)).$$

By (3.7.1) and Lemma 3.7.1, we have

$$(3.7.2)\,\mathrm{Hom}_{\mathrm{D}^+(\mathscr{D}_X,G)}(V \otimes \mathscr{M}, \mathscr{M}') \cong \mathrm{Hom}_{\mathrm{D}^+(\mathrm{Mod}\,(G))}(V, \mathbf{R}\mathrm{Hom}_{\mathscr{D}_X}(\mathscr{M}, \mathscr{M}'))$$

for $V \in \mathrm{D}^b(\mathrm{Mod}\,(G))$, $\mathscr{M} \in \mathrm{D}^b_{\mathrm{coh}}(\mathscr{D}_X, G)$ and $\mathscr{M}' \in \mathrm{D}^+(\mathscr{D}_X, G)$. In particular we have

$$(3.7.3)\,\,\mathrm{Hom}_{\mathrm{D}^+(\mathscr{D}_X,G)}(\mathscr{M}, \mathscr{M}') \cong \mathrm{Hom}_{\mathrm{D}^+(\mathrm{Mod}\,(G))}(\mathbb{C}, \mathbf{R}\mathrm{Hom}_{\mathscr{D}_X}(\mathscr{M}, \mathscr{M}')).$$

Lemma 3.7.2.   (i) $\mathbf{R}\mathrm{Hom}_{\mathscr{D}_X}(\bullet, \bullet)$ *sends* $\mathrm{D}^b_{\mathrm{coh}}(\mathscr{D}_X, G)^{\mathrm{op}} \times \mathrm{D}^b(\mathscr{D}_X, G)$ *to* $\mathrm{D}^b(\mathrm{Mod}\,(G))$.
(ii) *Let* $\mathscr{F}_G$ *denote the functors forgetting G-structures:*

$$\mathscr{F}_G: \begin{aligned} \mathrm{D}^b(\mathscr{D}_X, G) &\to \mathrm{D}^b(\mathscr{D}_X), \\ \mathrm{D}^b(\mathrm{Mod}\,(G)) &\to \mathrm{D}^b(\mathbb{C}). \end{aligned}$$

Then $\mathscr{F}_G \mathbf{R}\mathrm{Hom}_{\mathscr{D}_X}(\mathscr{M}, \mathscr{N}) \cong \mathbf{R}\mathrm{Hom}_{\mathscr{D}_X}(\mathscr{F}_G \mathscr{M}, \mathscr{F}_G \mathscr{N})$ *for any* $\mathscr{M} \in \mathrm{D}^b_{\mathrm{coh}}(\mathscr{D}_X, G)$ *and* $\mathscr{N} \in \mathrm{D}^b(\mathscr{D}_X, G)$.

Proof.   We may assume that $\mathscr{M} \in \mathrm{D}^b(\mathrm{Mod}_{\mathrm{coh}}(\mathscr{D}_X, G))$ by Proposition 3.2.2. Then, for an injective complex $\mathscr{N}$ in $\mathrm{Mod}\,(\mathscr{D}_X, G)$, we have

$$\mathscr{F}_G \mathbf{R}\mathrm{Hom}_{\mathscr{D}_X}(\mathscr{M}, \mathscr{N}) \simeq \mathrm{Hom}_{\mathscr{D}_X}(\mathscr{M}, \mathscr{N}) \simeq \mathbf{R}\mathrm{Hom}_{\mathscr{D}_X}(\mathscr{F}_G \mathscr{M}, \mathscr{F}_G \mathscr{N})$$

by Proposition 3.3.8. This shows (ii), and (i) follows from the fact that the global homological dimension of $\mathrm{Mod}\,(\mathscr{D}_X)$ is at most $2\dim X$ (see [16]).
$$\text{Q.E.D.}$$

Remark that this shows that the global homological dimension of $\mathrm{Mod}\,(\mathscr{D}_X, G)$ is finite. Indeed, the arguments of the preceding lemma show that for $\mathscr{M} \in \mathrm{Mod}_{\mathrm{coh}}(\mathscr{D}_X, G)$ and $\mathscr{N} \in \mathrm{Mod}_{\mathrm{coh}}(\mathscr{D}_X, G)$, $H^n(\mathbf{R}\mathrm{Hom}_{\mathscr{D}_X}(\mathscr{M}, \mathscr{N})) = 0$ for $n > 2\dim X$. On the other hand, the global homological dimension of $\mathrm{Mod}\,(G)$ is at most $\dim G$ (or more precisely the dimension of the unipotent radical of G). Thus (3.7.3) shows that

$$\mathrm{Hom}_{\mathrm{D}(\mathscr{D}_X,G)}(\mathscr{M}, \mathscr{N}) \simeq \mathrm{Hom}_{\mathrm{D}(\mathrm{Mod}(G))}(\mathbb{C}, \mathbf{R}\mathrm{Hom}_{\mathscr{D}_X}(\mathscr{M}, \mathscr{N})) = 0$$

for $n > \dim G + 2\dim X$. Therefore, the global homological dimension of $\mathrm{Mod}_{\mathrm{coh}}(\mathscr{D}_X, G)$ is at most $\dim G + 2\dim X$. Hence so is $\mathrm{Mod}\,(\mathscr{D}_X, G)$.

3.8 Relations of Push-forward and Pull-back Functors

Statements

Let $f: X \to Y$ be a G-equivariant morphism of quasi-projective G-manifolds. Then $\mathbf{D}f^*$ and $\mathbf{D}f_*$ are adjoint functors in two ways. We use the notations: $d_{X/Y} = \dim X - \dim Y$.

Theorem 3.8.1. *Let $f: X \to Y$ be a G-equivariant morphism of quasi-projective G-manifolds.*

(i) *Assume that f is smooth. Then there exists a functorial isomorphism in $\mathscr{M} \in D^b_{coh}(\mathscr{D}_X, G)$ and $\mathscr{N} \in D^b_{coh}(\mathscr{D}_Y, G)$:*

$$(3.8.1) \quad \mathrm{Hom}_{D^b(\mathscr{D}_Y, G)}(\mathscr{N}, \mathbf{D}f_*\mathscr{M}) \cong \mathrm{Hom}_{D^b(\mathscr{D}_X, G)}(\mathbf{D}f^*\mathscr{N}[-d_{X/Y}], \mathscr{M}).$$

(ii) *Assume that f is smooth and projective. Then there exists a functorial isomorphism in $\mathscr{M} \in D^b_{coh}(\mathscr{D}_X, G)$ and $\mathscr{N} \in D^b_{coh}(\mathscr{D}_Y, G)$:*

$$(3.8.2) \quad \mathrm{Hom}_{D^b(\mathscr{D}_Y, G)}(\mathbf{D}f_*\mathscr{M}, \mathscr{N}) \cong \mathrm{Hom}_{D^b(\mathscr{D}_X, G)}(\mathscr{M}, \mathbf{D}f^*\mathscr{N}[d_{X/Y}]).$$

This theorem will be proved at the end of this subsection.

By Theorem 3.8.1, we obtain the following morphisms for a smooth and projective morphism $f: X \to Y$, $\mathscr{M} \in D^b_{coh}(\mathscr{D}_X, G)$ and $\mathscr{N} \in D^b_{coh}(\mathscr{D}_Y, G)$.

$$(3.8.3) \qquad \mathbf{D}f^*\mathbf{D}f_*\mathscr{M}[-d_{X/Y}] \to \mathscr{M},$$
$$(3.8.4) \qquad \mathscr{M} \to \mathbf{D}f^*\mathbf{D}f_*\mathscr{M}[d_{X/Y}],$$
$$(3.8.5) \qquad \mathbf{D}f_*\mathbf{D}f^*\mathscr{N}[d_{X/Y}] \to \mathscr{N},$$
$$(3.8.6) \qquad \mathscr{N} \to \mathbf{D}f_*\mathbf{D}f^*\mathscr{N}[-d_{X/Y}].$$

Residue Morphism

In order to prove Theorem 3.8.1, we shall first define the morphism

$$(3.8.7) \qquad \mathbf{D}f_*\mathscr{O}_X[d_{X/Y}] \to \mathscr{O}_Y.$$

Let $f: X \to Y$ be a smooth and projective morphism. Let $\mathscr{F}_G$ be the functor from $D^b(\mathscr{D}_Y, G)$ to $D^b(\mathscr{D}_Y)$. Then we have, by the theory of D-modules

$$(3.8.8) \qquad \mathscr{F}_G(\mathbf{D}f_*\mathscr{O}_X[d_{X/Y}]) \to \mathscr{O}_Y$$

in $D^b(\mathscr{D}_Y)$. This morphism (3.8.8) gives a $\mathscr{D}_Y$-linear homomorphism

$$H^{d_{X/Y}}(\mathbf{D}f_*\mathscr{O}_X) \to \mathscr{O}_Y.$$

Since this is canonical, this commutes with the action of any element of $G(\mathbb{C})$. Hence this is a morphism in $\mathrm{Mod}(\mathscr{D}_Y, G)$. On the other hand, we have

$$H^j(\mathbf{D}f_*\mathscr{O}_X) = 0 \quad \text{for } j > d_{X/Y}.$$

We have therefore a morphism in $D^b(\mathscr{D}_Y, G)$.

$$\mathbf{D}f_*\mathscr{O}_X[d_{X/Y}] \to \tau^{\geq 0}(\mathbf{D}f_*\mathscr{O}_X[d_{X/Y}]) = H^{d_{X/Y}}(\mathbf{D}f_*\mathscr{O}_X).$$

Therefore, we obtain a morphism $\mathbf{D}f_*\mathscr{O}_X[d_{X/Y}] \to \mathscr{O}_Y$ in $D^b(\mathscr{D}_Y, G)$.

Lemma 3.8.2 (Projection formula). *There is a functorial isomorphism in* $\mathscr{M} \in D^b(\mathscr{D}_X, G)$ *and* $\mathscr{N} \in D^b(\mathscr{D}_Y, G)$

$$\mathbf{D}f_*(\mathscr{M} \overset{D}{\otimes} \mathbf{D}f^*\mathscr{N}) \simeq (\mathbf{D}f_*\mathscr{M}) \overset{D}{\otimes} \mathscr{N}.$$

Since this is proved similarly to the non-equivariant case, we omit the proof (see e.g. [16]).

By this lemma, we obtain the residue morphism:

$$(3.8.9) \qquad \mathrm{Res}_{X/Y} \colon \mathbf{D}f_*\mathbf{D}f^*\mathscr{N}[\mathrm{d}_{X/Y}] \to \mathscr{N},$$

as the compositions of a chain of morphisms

$$\begin{aligned}
\mathbf{D}f_*\mathbf{D}f^*\mathscr{N}[\mathrm{d}_{X/Y}] &\simeq \mathbf{D}f_*(\mathscr{O}_X[\mathrm{d}_{X/Y}] \overset{D}{\otimes} \mathbf{D}f^*\mathscr{N}) \\
&\simeq (\mathbf{D}f_*\mathscr{O}_X[\mathrm{d}_{X/Y}]) \overset{D}{\otimes} \mathscr{N} \\
&\to \mathscr{O}_Y \overset{D}{\otimes} \mathscr{N} \simeq \mathscr{N}.
\end{aligned}$$

Proof of Theorem 3.8.1

We shall prove first the isomorphism (3.8.1) in Theorem 3.8.1. For $\mathscr{N} \in K^+(\mathrm{Mod}(\mathscr{D}_Y, G))$, we have a quasi-isomorphism

$$\mathscr{N} \leftarrow \mathscr{D}_Y \otimes \Omega_Y^{\otimes -1} \otimes \Omega_Y^{\bullet} \otimes \mathscr{N}[\mathrm{d}_Y]$$

and a morphism

$$\begin{aligned}
\mathscr{D}_Y \otimes \Omega_Y^{\otimes -1} \otimes \Omega_Y^{\bullet} \otimes \mathscr{N} &\to \mathscr{D}_Y \otimes \Omega_Y^{\otimes -1} \otimes f_*(\Omega_X^{\bullet} \otimes f^*\mathscr{N}) \\
&\simeq K f_*(f^*\mathscr{N})[-\mathrm{d}_X].
\end{aligned}$$

Thus we obtain a morphism in $D^b(\mathscr{D}_Y, G)$:

$$(3.8.10) \qquad \mathscr{N} \to \mathbf{D}f_*\mathbf{D}f^*\mathscr{N}[-\mathrm{d}_{X/Y}],$$

even if f is not assumed to be smooth projective. This gives a chain of homomorphisms

$$\begin{aligned}
\mathrm{Hom}_{D^b(\mathscr{D}_X, G)}(\mathbf{D}f^*\mathscr{N}[-\mathrm{d}_{X/Y}], \mathscr{M}) \\
\to \mathrm{Hom}_{D^b(\mathscr{D}_Y, G)}(\mathbf{D}f_*\mathbf{D}f^*\mathscr{N}[-\mathrm{d}_{X/Y}], \mathbf{D}f_*\mathscr{M}) \\
\to \mathrm{Hom}_{D^b(\mathscr{D}_Y, G)}(\mathscr{N}, \mathbf{D}f_*\mathscr{M}).
\end{aligned}$$

Let us prove that the composition is an isomorphism when f is smooth. Similarly as above, we have a morphism in $D(\mathrm{Mod}(G))$

$$\mathbf{R}\mathrm{Hom}_{\mathscr{D}_X}(\mathbf{D}f^*\mathscr{N}[-\mathrm{d}_{X/Y}], \mathscr{M}) \to \mathbf{R}\mathrm{Hom}_{\mathscr{D}_Y}(\mathscr{N}, \mathbf{D}f_*\mathscr{M}).$$

By the theory of D-modules, forgetting the equivariance, this is an isomorphism in $D^b(\mathbb{C})$, assuming that f is smooth (see [16]). Hence this is an isomorphism in $D^b(\mathrm{Mod}(G))$. Finally we obtain by (3.7.3)

$$\mathrm{Hom}_{D^b(\mathscr{D}_X, G)}(\mathbf{D}f^*\mathscr{N}[-d_{X/Y}], \mathscr{M})$$

$$\simeq \mathrm{Hom}_{D^b(\mathrm{Mod}(G))}\big(\mathbb{C}, \mathbf{R}\mathrm{Hom}_{\mathscr{D}_X}(\mathbf{D}f^*\mathscr{N}[-d_{X/Y}], \mathscr{M})\big)$$

$$\xrightarrow{\sim} \mathrm{Hom}_{D^b(\mathrm{Mod}(G))}\big(\mathbb{C}, \mathbf{R}\mathrm{Hom}_{\mathscr{D}_Y}(\mathscr{N}, \mathbf{D}f_*\mathscr{M})\big)$$

$$\simeq \mathrm{Hom}_{D^b(\mathscr{D}_Y, G)}(\mathscr{N}, \mathbf{D}f_*\mathscr{M}).$$

The proof of (3.8.2) is similar using $\mathrm{Res}_{X/Y}\colon \mathbf{D}f_*\mathbf{D}f^*\mathscr{N}[d_{X/Y}] \to \mathscr{N}$ given in (3.8.9) instead of (3.8.10).

3.9 Flag Manifold Case

We shall apply Theorem 3.8.1 when $X = G/P$ and $Y = \{\mathrm{pt}\}$, where P is a parabolic subgroup of a reductive group G. Note that X is a projective G-manifold. Then, we obtain the following duality isomorphism.

Lemma 3.9.1. *For any finite-dimensional G-module E and a $(\mathfrak{g}, P)$-module M finitely generated over $U(\mathfrak{g})$, we have an isomorphism*

$$(3.9.1) \qquad \mathrm{Ext}^{2\dim(G/P)-j}_{(\mathfrak{g},P)}(M, E) \cong \mathrm{Hom}_{\mathbb{C}}\big(\mathrm{Ext}^j_{(\mathfrak{g},P)}(E, M), \mathbb{C}\big).$$

Proof.
The category $\mathrm{Mod}_{\mathrm{coh}}(\mathscr{D}_X, G)$ is equivalent to the category $\mathrm{Mod}_f(\mathfrak{g}, P)$ of $(\mathfrak{g}, P)$-modules finitely generated over $U(\mathfrak{g})$, and $\mathrm{Mod}_{\mathrm{coh}}(\mathscr{D}_Y, G)$ is equivalent to the category $\mathrm{Mod}_f(G)$ of finite-dimensional G-modules (see Proposition 3.1.5). The functor $\mathbf{D}f^*$ is induced by the functor $V \mapsto V$ from $\mathrm{Mod}(G)$ to $\mathrm{Mod}(\mathfrak{g}, P)$. The right adjoint functor to the last functor is given by

$$(3.9.2) \qquad\qquad M \longmapsto \bigoplus_V V \otimes \mathrm{Hom}_{(\mathfrak{g},P)}(V, M).$$

Here V ranges over the isomorphism classes of irreducible G-modules. Hence the functor $\mathbf{D}f_*[-d_{X/Y}]$, the right adjoint functor of $\mathbf{D}f^*$, is the right derived functor of the functor (3.9.2). Hence (3.8.2) implies that

$$\prod_V \mathrm{Hom}_{D(\mathrm{Mod}(G))}(V \otimes \mathbf{R}\mathrm{Hom}_{(\mathfrak{g},P)}(V, M)[d_X], E[j])$$

$$\simeq \mathrm{Hom}_{D(\mathrm{Mod}(\mathfrak{g},P))}(M, E[j][d_X]).$$

The last term is isomorphic to $\mathrm{Ext}^{d_X+j}_{(\mathfrak{g},P)}(M, E)$, and the first term is isomorphic to $\mathrm{Hom}_{\mathbb{C}}\big(\mathrm{Ext}^{d_X-j}_{(\mathfrak{g},P)}(E, M), \mathbb{C}\big)$ because, when E and V are irreducible, we have

$$\mathrm{Hom}_{D(\mathrm{Mod}(G))}(V, E[j]) = \begin{cases} \mathbb{C} & \text{if } V \simeq E \text{ and } j = 0, \\ 0 & \text{otherwise.} \end{cases}$$

Q.E.D.

4 Equivariant Derived Category

4.1 Introduction

In the case of quasi-equivariant D-modules, the category has enough objects, and it is enough to consider the derived category of the abelian category of quasi-equivariant D-modules. However the categories of equivariant sheaves have not enough objects, and the derived category of the abelian category of equivariant sheaves is not an appropriate category. In order to avoid this difficulty, we have to enrich spaces themselves. In this paper, we follow a definition of the equivariant derived categories due to Bernstein-Lunts [4].

4.2 Sheaf Case

Let G be a real Lie group and X a (separated) locally compact space with G-action. We assume that X has a finite soft dimension (e.g. a finite-dimensional topological manifold). We call such an X a G-space. If X is a manifold, we call it a G-manifold.

In this paper, we say that G acts *freely* if the morphism $\tilde{\mu} \colon G \times X \to X \times X$ $((g,x) \mapsto (gx, x))$ is a closed embedding. Therefore, if X is a G-manifold with a free action of G, then X/G exists as a (separated) topological manifold.

Let $\mathrm{Mod}\,(\mathbb{C}_X)$ be the category of sheaves of $\mathbb{C}$-vector spaces on X. We denote by $\mathrm{D}^{\mathrm{b}}(\mathbb{C}_X)$ the bounded derived category of $\mathrm{Mod}\,(\mathbb{C}_X)$.

Let $\mu \colon G \times X \to X$ be the action map and $\mathrm{pr} \colon G \times X \to X$ the projection.

Definition 4.2.1. A sheaf F of $\mathbb{C}$-vector spaces is called G-*equivariant* if it is endowed with an isomorphism $\mu^{-1}F \xrightarrow{\sim} \mathrm{pr}^{-1}F$ satisfying the associative law as in (3.1.2).

Let us denote by $\mathrm{Mod}_G(\mathbb{C}_X)$ the abelian categories of G-equivariant sheaves.

If G acts freely, then we have the equivalence of categories:

$$\mathrm{Mod}\,(\mathbb{C}_{X/G}) \xrightarrow{\sim} \mathrm{Mod}_G(\mathbb{C}_X).$$

We will construct the equivariant derived category $\mathrm{D}^{\mathrm{b}}_G(\mathbb{C}_X)$ which has suitable functorial properties and satisfies the condition:

$$\text{if } G \text{ acts freely on } X, \text{ then } \mathrm{D}^{\mathrm{b}}(\mathbb{C}_{X/G}) \simeq \mathrm{D}^{\mathrm{b}}_G(\mathbb{C}_X).$$

Assume that there is a sequence of G-equivariant morphisms

$$V_1 \longrightarrow V_2 \longrightarrow V_3 \longrightarrow \cdots$$

where V_k is a connected G-manifold with a free action and

(4.2.1)
    (i) $H^n(V_k; \mathbb{C})$ is finite-dimensional for any n,k,

    (ii) for each $n > 0$, $H^n(V_k; \mathbb{C}) = 0$ for $k \gg 0$.

Any real semisimple Lie group with finite center has such a sequence $\{V_k\}$. If G is embedded in some $GL_N(\mathbb{C})$ as a closed subgroup, we can take $V_k = \{f \in \mathrm{Hom}_\mathbb{C}(\mathbb{C}^N, \mathbb{C}^{N+k}); f \text{ is injective}\}$. If G is a connected real semi-simple group with finite center, then we can take $(G \times V_k)/K$ as V_k, where K is a maximal compact subgroup of G and V_k is the one for K. Note that G/K is contractible.

The condition (4.2.1) implies

$$\mathbb{C} \xrightarrow{\sim} \text{``}\varprojlim_k\text{''}\, R\Gamma(V_k; \mathbb{C}).$$

This follows from the following lemma (see e.g. [19, Exercise 15.1]).

Lemma 4.2.2. *Let $\mathcal{C}$ be an abelian category. Let $\{X_n\}_{n \in \mathbb{Z}_{\geq 0}}$ be a projective system in $\mathrm{D}^b(\mathcal{C})$. Assume that it satisfies the conditions:*

(i) *for any $k \in \mathbb{Z}$, "$\varprojlim_n$" $H^k(X_n)$ is representable by an object of $\mathcal{C}$,*

(ii) *one of the following conditions holds:*

   (a) *there exist $a \leq b$ such that $H^k(X_n) \simeq 0$ for $k > b$ and "$\varprojlim_n$" $H^k(X_n) \simeq 0$ for $k < a$,*

   (b) *$\mathcal{C}$ has finite homological dimension, and there exist $a \leq b$ such that "$\varprojlim_n$" $H^k(X_n) \simeq 0$ unless $a \leq k \leq b$,*

Then "$\varprojlim_n$" X_n is representable by an object of $\mathrm{D}^b(\mathcal{C})$.

For example, we say that "$\varprojlim_n$" X_n is representable by $X \in \mathrm{D}^b(\mathcal{C})$ if there exists an isomorphism $\varinjlim_n \mathrm{Hom}_{\mathrm{D}^b(\mathcal{C})}(X_n, Y) \simeq \mathrm{Hom}_{\mathrm{D}^b(\mathcal{C})}(X, Y)$ functorially in $Y \in \mathrm{D}^b(\mathcal{C})$. In such a case, X is unique up to an isomorphism, and we write $X = \text{``}\varprojlim_n\text{''}\, X_n$.

Let us denote by $p_k: V_k \times X \to X$ the second projection and by $\pi_k: V_k \times X \to (V_k \times X)/G$ the quotient map. Here the action of G on $V_k \times X$ is the diagonal action. We denote by the same letter i_k the maps $V_k \times X \to V_{k+1} \times X$ and $(V_k \times X)/G \to (V_{k+1} \times X)/G$.

Definition 4.2.3. Let $\mathrm{D}^b_G(\mathbb{C}_X)$ be the category whose objects are $F = (F_\infty, F_k, j_k, \varphi_k\ (k = 1, 2, \ldots))$ where $F_\infty \in \mathrm{D}^b(\mathbb{C}_X)$, $F_k \in \mathrm{D}^b\left(\mathbb{C}_{(V_k \times X)/G}\right)$ and $j_k: i_k^{-1} F_{k+1} \xrightarrow{\sim} F_k$ and $\varphi_k: p_k^{-1} F_\infty \xrightarrow{\sim} \pi_k^{-1} F_k$ are such that the diagram

$$
\begin{array}{ccc}
i_k^{-1} p_{k+1}^{-1} F_\infty & \xrightarrow{\ \sim\ } & p_k^{-1} F_\infty \\
\downarrow{\scriptstyle \varphi_{k+1}} & & \downarrow{\scriptstyle \varphi_k} \\
i_k^{-1} \pi_{k+1}^{-1} F_{k+1} & \xrightarrow[\ j_k\ ]{\sim} & \pi_k^{-1} F_k
\end{array}
$$

commutes. The morphisms in $D^b_G(\mathbb{C}_X)$ are defined in an evident way.

The category $D^b_G(\mathbb{C}_X)$ is a triangulated category in an obvious way, and the triangulated category $D^b_G(\mathbb{C}_X)$ does not depend on the choice of a sequence $\{V_k\}_k$ (see [4]). We call $D^b_G(\mathbb{C}_X)$ the *equivariant derived category*.

By the condition (4.2.1), we have

$$(4.2.2) \qquad \text{“}\varprojlim_k\text{”} \, \mathbf{R}p_{k*}\pi_k^{-1}F_k \cong F_\infty.$$

Indeed, we have

$$\text{“}\varprojlim_k\text{”} \, \mathbf{R}p_{k*}\pi_k^{-1}F_k \cong \text{“}\varprojlim_k\text{”} \, \mathbf{R}p_{k*}p_k^{-1}F_\infty \cong \text{“}\varprojlim_k\text{”} \left(F_\infty \otimes \mathbf{R}\Gamma(V_k;\mathbb{C}) \right)$$

and $\text{“}\varprojlim_k\text{”} \, \mathbf{R}\Gamma(V_k;\mathbb{C}) \simeq \mathbb{C}$ by (4.2.1).

There exists a functor of triangulated categories (called the *forgetful functor*):

$$\mathscr{F}_G : D^b_G(\mathbb{C}_X) \to D^b(\mathbb{C}_X).$$

Note that a morphism u in $D^b_G(\mathbb{C}_X)$ is an isomorphism if and only if $\mathscr{F}_G(u)$ is an isomorphism in $D^b(\mathbb{C}_X)$.

By taking the cohomology groups, we obtain cohomological functors:

$$H^n : D^b_G(\mathbb{C}_X) \to \mathrm{Mod}_G(\mathbb{C}_X).$$

Lemma 4.2.4. *Assume that G acts freely on X. Then $D^b_G(\mathbb{C}_X)$ is equivalent to $D^b(\mathbb{C}_{X/G})$.*

Proof. The functor $D^b(\mathbb{C}_{X/G}) \to D^b_G(\mathbb{C}_X)$ is obviously defined, and its quasi-inverse $D^b_G(\mathbb{C}_X) \to D^b(\mathbb{C}_{X/G})$ is given by $F \mapsto \text{“}\varprojlim_k\text{”} \, \mathbf{R}q_{k*}(F_k)$, where q_k is the map $(V_k \times X)/G \to X/G$. Note that $\text{“}\varprojlim_k\text{”} \, \mathbf{R}q_{k*}(F_k) \cong \tau^{\leqslant a}\mathbf{R}q_{l*}(F_l)$ for $l \gg a \gg 0$.   Q.E.D.

Since $\mathrm{Mod}(\mathbb{C}_{X/G})$ is equivalent to $\mathrm{Mod}_G(\mathbb{C}_X)$ in such a case, we have

$$(4.2.3) \qquad \text{if } G \text{ acts freely on } X, \text{ then } D^b(\mathrm{Mod}_G(\mathbb{C}_X)) \xrightarrow{\sim} D^b_G(\mathbb{C}_X).$$

For a G-equivariant map $f : X \to Y$, we can define the functors

$$f^{-1}, \, f^! : D^b_G(\mathbb{C}_Y) \to D^b_G(\mathbb{C}_X)$$

and

$$\mathbf{R}f_!, \, \mathbf{R}f_* : D^b_G(\mathbb{C}_X) \to D^b_G(\mathbb{C}_Y).$$

The functors $\mathbf{R}f_!$ and f^{-1} are left adjoint functors of $f^!$ and $\mathbf{R}f_*$, respectively. Moreover they commute with the forgetful functor $D^b_G(\mathbb{C}_X) \to D^b(\mathbb{C}_X)$.

4.3 Induction Functor

The following properties are easily checked.

(4.3.1)
> For a group morphism $H \to G$ and a G-manifold X, there exists a canonical functor (*restriction functor*)
> $$\operatorname{Res}_H^G \colon D_G^b(\mathbb{C}_X) \to D_H^b(\mathbb{C}_X).$$

(4.3.2)
> If H is a closed normal subgroup of G and if H acts freely on a G-manifold X, then
> $$D_G^b(\mathbb{C}_X) \simeq D_{G/H}^b(\mathbb{C}_{X/H}).$$

For $F \in D_G^b(\mathbb{C}_X)$, we denote by F/H the corresponding object of $D_{G/H}^b(\mathbb{C}_{X/H})$.

Let H be a closed subgroup of G and X an H-manifold. Then we have a chain of equivalences of triangulated categories

$$D_H^b(\mathbb{C}_X) \simeq D_{H \times G}^b(\mathbb{C}_{X \times G}) \simeq D_G^b(\mathbb{C}_{(X \times G)/H})$$

by (4.3.2). Here $H \times G$ acts on $X \times G$ by $(h, g)(x, g') = (hx, gg'h^{-1})$. Let us denote the composition by

(4.3.3)
$$\operatorname{Ind}_H^G \colon D_H^b(\mathbb{C}_X) \xrightarrow{\sim} D_G^b(\mathbb{C}_{(X \times G)/H}).$$

When X is a G-manifold, we have $(X \times G)/H \simeq X \times (G/H)$, and we obtain an equivalence of categories

(4.3.4) $\quad \operatorname{Ind}_H^G \colon D_H^b(\mathbb{C}_X) \xrightarrow{\sim} D_G^b(\mathbb{C}_{X \times (G/H)})$ when X is a G-manifold.

Note that the action of G on $X \times (G/H)$ is the diagonal action.

4.4 Constructible Sheaves

Assume that X is a complex algebraic variety and a real Lie group G acts real analytically on the associated complex manifold X^{an}. We denote by $D_{G,\mathbb{R}\text{-c}}^b(\mathbb{C}_{X^{\mathrm{an}}})$ the full subcategory of $D_G^b(\mathbb{C}_{X^{\mathrm{an}}})$ consisting of $\mathbb{R}$-constructible objects. Here $F \in D_G^b(\mathbb{C}_{X^{\mathrm{an}}})$ is called $\mathbb{R}$-*constructible* if it satisfies the following two conditions:

(i) $\dim H^j(F)_x < \infty$ for any $x \in X^{\mathrm{an}}$.
(ii) there exists a finite family $\{Z_\alpha\}$ of locally closed subsets of X^{an} such that
$\quad$ (a) $X^{\mathrm{an}} = \bigcup_\alpha Z_\alpha$,

(b) each Z_α is subanalytic in $(\overline{X})^{\mathrm{an}}$ for any (or equivalently, some) compactification $X \hookrightarrow \overline{X}$ of X,

(c) $H^j(F)|_{Z_\alpha}$ is locally constant .

For subanalyticity and $\mathbb{R}$-constructibility, see e.g. [18].

We say that F is $\mathbb{C}$-*constructible* (or *constructible*, for short) if we assume further that each Z_α is the associated topological set of a subscheme of X.

We denote by $\mathrm{D}^{\mathrm{b}}_{G,\,\mathbb{C}\text{-c}}(\mathbb{C}_{X^{\mathrm{an}}})$ the full subcategory of $\mathrm{D}^{\mathrm{b}}_G(\mathbb{C}_{X^{\mathrm{an}}})$ consisting of constructible objects.

4.5 D-module Case

The construction of the equivariant derived category for sheaves can be applied similarly to the equivariant derived categories of D-modules.

Let G be an affine algebraic group. Let us take a sequence of connected algebraic G-manifolds

$$(4.5.1) \qquad\qquad V_1 \longrightarrow V_2 \longrightarrow V_3 \longrightarrow \cdots$$

such that

$$(4.5.2) \qquad \begin{array}{l} G \text{ acts freely on } V_k, \text{ and} \\ \text{for any } n > 0,\ \mathrm{Ext}^n_{\mathscr{D}_{V_k}}(\mathscr{O}_{V_k}, \mathscr{O}_{V_k}) \cong H^n(V_k^{\mathrm{an}}; \mathbb{C}) = 0 \text{ for } k \gg 0. \end{array}$$

Such a sequence $\{V_k\}_k$ exists. With the aid of $\{V_k\}_k$, we can define the equivariant derived category of D-modules similarly to the sheaf case. Let X be a quasi-projective algebraic G-manifold. Let us denote by $p_k \colon V_k \times X \to X$ the second projection and by $\pi_k \colon V_k \times X \to (V_k \times X)/G$ the quotient morphism.[3] We denote by the same letter i_k the maps $V_k \times X \to V_{k+1} \times X$ and $(V_k \times X)/G \to (V_{k+1} \times X)/G$.

Definition 4.5.1. Let $\mathrm{D}^{\mathrm{b}}_G(\mathscr{D}_X)$ be the category whose objects are $\mathscr{M} = (\mathscr{M}_\infty,\ \mathscr{M}_k,\ j_k,\ \varphi_k\ (k \in \mathbb{Z}_{\geqslant 1}))$ where $\mathscr{M}_\infty \in \mathrm{D}^{\mathrm{b}}(\mathscr{D}_X)$, $\mathscr{M}_k \in \mathrm{D}^{\mathrm{b}}(\mathscr{D}_{(V_k \times X)/G})$ and $j_k \colon \mathrm{D}i_k^* \mathscr{M}_{k+1} \xrightarrow{\sim} \mathscr{M}_k$ and $\varphi_k \colon \mathrm{D}p_k^* \mathscr{M}_\infty \xrightarrow{\sim} \mathrm{D}\pi_k^* \mathscr{M}_k$ are such that the diagram

$$
\begin{array}{ccc}
\mathrm{D}i_k^* \mathrm{D}p_{k+1}^* \mathscr{M}_\infty & \overset{\sim}{=\!=\!=} & \mathrm{D}p_k^* \mathscr{M}_\infty \\
\Big\downarrow{\scriptstyle \varphi_{k+1}} & & \Big\downarrow{\scriptstyle \varphi_k} \\
\mathrm{D}i_k^* \mathrm{D}\pi_{k+1}^* \mathscr{M}_{k+1} & \xrightarrow{\ j_k\ } & \mathrm{D}\pi_k^* \mathscr{M}_k
\end{array}
$$

commutes.

---

[3] The quotient $(V_k \times X)/G$ may not exist as a scheme, but it exists as an algebraic space. Although we do not develop here, we have the theory of D-modules on algebraic spaces. Alternatively, we can use $\mathrm{Mod}_G(\mathscr{D}_{V_k \times X})$ instead of $\mathrm{Mod}(\mathscr{D}_{(V_k \times X)/G})$.

Note that we have a canonical functor

$$D_G^b(\mathscr{D}_X) \to D^b(\mathscr{D}_X, G).$$

We denote by $D_{G,\mathrm{coh}}^b(\mathscr{D}_X)$ the full triangulated subcategory of $D_G^b(\mathscr{D}_X)$ consisting of objects $\mathscr{M}$ with coherent cohomologies.

Similarly to the sheaf case, we have the following properties.

(4.5.3) For a morphism $f\colon X \to Y$ of quasi-projective G-manifolds, we can define the pull-back functor $\mathbf{D}f^*\colon D_G^b(\mathscr{D}_Y) \to D_G^b(\mathscr{D}_X)$ and the push-forward functor $\mathbf{D}f_*\colon D_G^b(\mathscr{D}_X) \to D_G^b(\mathscr{D}_Y)$.

(4.5.4) The canonical functor $D_G^b(\mathscr{D}_X) \to D^b(\mathscr{D}_X, G)$ commutes with the pull-back and push-forward functors.

(4.5.5) For a closed algebraic subgroup H of G and an algebraic G-manifold X, there exists a canonical functor $\mathrm{Res}_H^G\colon D_G^b(\mathscr{D}_X) \to D_H^b(\mathscr{D}_X)$.

(4.5.6) If H is a normal subgroup of G and if H acts freely on X and if X/H exists, then $D_G^b(\mathscr{D}_X) \simeq D_{G/H}^b(\mathscr{D}_{X/H})$.

(4.5.7) If H is a closed algebraic subgroup of G and X is an algebraic G-manifold, then we have
$$\mathrm{Ind}_H^G\colon D_H^b(\mathscr{D}_X) \xrightarrow{\sim} D_G^b(\mathscr{D}_{X \times (G/H)}).$$

4.6 Equivariant Riemann-Hilbert Correspondence

Let X be a quasi-projective manifold. Let us denote by X^{an} the associated complex manifold. Accordingly, $\mathscr{O}_{X^{\mathrm{an}}}$ is the sheaf of holomorphic functions on X^{an}. Then there exists a morphism of ringed spaces $\pi\colon X^{\mathrm{an}} \to X$. We denote by $\mathscr{D}_{X^{\mathrm{an}}}$ the sheaf of differential operators with holomorphic coefficients on X^{an}. For a $\mathscr{D}_X$-module $\mathscr{M}$, we denote by $\mathscr{M}^{\mathrm{an}}$ the associated $\mathscr{D}_{X^{\mathrm{an}}}$-module $\mathscr{D}_{X^{\mathrm{an}}} \otimes_{\pi^{-1}\mathscr{D}_X} \pi^{-1}\mathscr{M} \simeq \mathscr{O}_{X^{\mathrm{an}}} \otimes_{\pi^{-1}\mathscr{O}_X} \pi^{-1}\mathscr{M}$.

Let us denote by $D_{\mathrm{hol}}^b(\mathscr{D}_X)$ (resp. $D_{\mathrm{rh}}^b(\mathscr{D}_X)$) the full subcategory of $D^b(\mathscr{D}_X)$ consisting of objects with holonomic cohomologies (resp. regular holonomic cohomologies) (see [16]). Then the de Rham functor

$$\mathrm{DR}_X := \mathbf{R}\mathscr{H}om_{\mathscr{D}_{X^{\mathrm{an}}}}(\mathscr{O}_{X^{\mathrm{an}}}, \bullet^{\mathrm{an}})\colon D^b(\mathscr{D}_X) \to D^b(\mathbb{C}_X)$$

sends $D_{\mathrm{hol}}^b(\mathscr{D}_X)$ to $D_{\mathbb{C}\text{-}c}^b(\mathbb{C}_{X^{\mathrm{an}}})$.

Then we have the following Riemann-Hilbert correspondence.

Theorem 4.6.1 ([12]). *The functor DR_X gives an equivalence of categories:*

$$(4.6.1) \qquad \mathrm{DR}_X : D_{\mathrm{rh}}^b(\mathscr{D}_X) \xrightarrow{\sim} D_{\mathbb{C}\text{-}c}^b(\mathbb{C}_{X^{\mathrm{an}}}).$$

Now, let G be an affine algebraic group and X a quasi-projective G-manifold. Then we define similarly $\mathrm{D}^{\mathrm{b}}_{G,\mathrm{hol}}(\mathscr{D}_X)$ and $\mathrm{D}^{\mathrm{b}}_{G,\mathrm{rh}}(\mathscr{D}_X)$ as full subcategories of $\mathrm{D}^{\mathrm{b}}_G(\mathscr{D}_X)$. Then we can define the equivariant de Rham functor:

$$\mathrm{DR}_X\colon \mathrm{D}^{\mathrm{b}}_{G,\mathrm{hol}}(\mathscr{D}_X) \to \mathrm{D}^{\mathrm{b}}_{G^{\mathrm{an}},\mathbb{C}\text{-c}}(\mathbb{C}_{X^{\mathrm{an}}}).$$

Theorem 4.6.1 implies the following theorem.

Theorem 4.6.2. *The functor* DR_X *gives an equivalence of categories:*

$$(4.6.2) \qquad \mathrm{DR}_X\colon \mathrm{D}^{\mathrm{b}}_{G,\mathrm{rh}}(\mathscr{D}_X) \xrightarrow{\sim} \mathrm{D}^{\mathrm{b}}_{G^{\mathrm{an}},\mathbb{C}\text{-c}}(\mathbb{C}_{X^{\mathrm{an}}}).$$

5 Holomorphic Solution Spaces

5.1 Introduction

Let G be an affine complex algebraic group and let X be a quasi-projective G-manifold. Recall that we denote by X^{an} the associated complex manifold and, for a $\mathscr{D}_X$-module $\mathscr{M}$, we denote by $\mathscr{M}^{\mathrm{an}}$ the associated $\mathscr{D}_{X^{\mathrm{an}}}$-module $\mathscr{D}_{X^{\mathrm{an}}} \otimes_{\pi^{-1}\mathscr{D}_X} \pi^{-1}\mathscr{M} \simeq \mathscr{O}_{X^{\mathrm{an}}} \otimes_{\pi^{-1}\mathscr{O}_X} \pi^{-1}\mathscr{M}$. Here $\pi\colon X^{\mathrm{an}} \to X$ is the canonical morphism of ringed spaces.

Let $G_{\mathbb{R}}$ be a real Lie group and let $G_{\mathbb{R}} \to G^{\mathrm{an}}$ be a morphism of Lie groups. Hence $G_{\mathbb{R}}$ acts on X^{an}.

Recall that $\mathbf{FN}_{G_{\mathbb{R}}}$ is the category of Fréchet nuclear $G_{\mathbb{R}}$-modules (see Example 2.1.2 (iii)). We denote by $\mathrm{D}^{\mathrm{b}}_{\mathrm{cc}}(\mathscr{D}_X,G)$ the full subcategory of $\mathrm{D}^{\mathrm{b}}(\mathscr{D}_X,G)$ consisting of objects with countably coherent cohomologies, by $\mathrm{D}^{\mathrm{b}}_{G_{\mathbb{R}},\mathrm{ctb}}(\mathbb{C}_{X^{\mathrm{an}}})$ the full subcategory of $\mathrm{D}^{\mathrm{b}}_{G_{\mathbb{R}}}(\mathbb{C}_{X^{\mathrm{an}}})$ consisting of objects with countable sheaves as cohomology groups (see § 5.2), and by $\mathrm{D}^{\mathrm{b}}(\mathbf{FN}_{G_{\mathbb{R}}})$ the bounded derived category of $\mathbf{FN}_{G_{\mathbb{R}}}$.

In this section, we shall define

$$\mathbf{R}\mathrm{Hom}_{\mathscr{D}_X}(\mathscr{M} \otimes K, \mathscr{O}_{X^{\mathrm{an}}})$$

as an object of $\mathrm{D}^{\mathrm{b}}(\mathbf{FN}_{G_{\mathbb{R}}})$ for $\mathscr{M} \in \mathrm{D}^{\mathrm{b}}_{\mathrm{cc}}(\mathscr{D}_X,G)$ and $K \in \mathrm{D}^{\mathrm{b}}_{G_{\mathbb{R}},\mathrm{ctb}}(\mathbb{C}_{X^{\mathrm{an}}})$. Here, we write $\mathbf{R}\mathrm{Hom}_{\mathscr{D}_X}(\mathscr{M} \otimes K, \mathscr{O}_{X^{\mathrm{an}}})$ instead of $\mathbf{R}\mathrm{Hom}_{\pi^{-1}\mathscr{D}_X}(\pi^{-1}\mathscr{M} \otimes K, \mathscr{O}_{X^{\mathrm{an}}}) \simeq \mathbf{R}\mathrm{Hom}_{\mathscr{D}_{X^{\mathrm{an}}}}(\mathscr{M}^{\mathrm{an}} \otimes K, \mathscr{O}_{X^{\mathrm{an}}})$ for short.

We also prove the dual statement. Let $\mathbf{DFN}_{G_{\mathbb{R}}}$ be the category of dual Fréchet nuclear $G_{\mathbb{R}}$-modules. We will define

$$\mathbf{R}\Gamma_{\mathrm{c}}(X^{\mathrm{an}}; K \otimes \Omega_{X^{\mathrm{an}}} \overset{\mathbf{L}}{\otimes}_{\mathscr{D}_X} \mathscr{M})$$

as an object of $\mathrm{D}^{\mathrm{b}}(\mathbf{DFN}_{G_{\mathbb{R}}})$ for $\mathscr{M}$ and K as above. We then prove that $\mathbf{R}\mathrm{Hom}_{\mathscr{D}_X}(\mathscr{M} \otimes K, \mathscr{O}_{X^{\mathrm{an}}})$ and $\mathbf{R}\Gamma_{\mathrm{c}}(X^{\mathrm{an}}; K \otimes \Omega_{X^{\mathrm{an}}} \overset{\mathbf{L}}{\otimes}_{\mathscr{D}_X} \mathscr{M})[d_X]$ are dual to each other.

5.2 Countable Sheaves

Let X be a topological manifold (countable at infinity).

Proposition 5.2.1. *Let F be a sheaf of $\mathbb{C}$-vector spaces on X. Then the following conditions are equivalent.*

(i) *for any compact subset K of X, $\Gamma(K; F)$ is countable-dimensional,*

(ii) *for any compact subset K of X, $H^n(K; F)$ is countable-dimensional for all n,*

(iii) *for any x and an open neighborhood U of x, there exists an open neighborhood V of x such that $V \subset U$ and $\mathrm{Im}(\Gamma(U; F) \to \Gamma(V; F))$ is countable-dimensional,*

(iv) *there exist a countable family of open subsets $\{U_i\}_i$ of X and an epimorphism $\oplus_i \mathbb{C}_{U_i} \twoheadrightarrow F$.*

If X is a real analytic manifold, then the above conditions are also equivalent to

(a) *there exist a countable family of subanalytic open subsets $\{U_i\}_i$ of X and an epimorphism $\oplus_i \mathbb{C}_{U_i} \twoheadrightarrow F$.*

Proof. For compact subsets K_1 and K_2, we have an exact sequence

$$H^{n-1}(K_1 \cap K_2; F) \longrightarrow H^n(K_1 \cup K_2; F) \longrightarrow H^n(K_1; F) \oplus H^n(K_2; F).$$

Hence, if K_1, K_2 and $K_1 \cap K_2$, satisfy the condition (i) or (ii), then so does $K_1 \cup K_2$. Hence the conditions (i) and (ii) are local properties. Since the other conditions are also local, we may assume from the beginning that X is real analytic.

(ii)$\Rightarrow$(i)$\Rightarrow$ (iii) are obvious.

(iii)$\Rightarrow$(iv) Let us take a countable base of open subsets $\{U_s\}_{s \in S}$ of X. Then, for each $s \in S$, there exists a countable open covering $\{V_i\}_{i \in I(s)}$ of U_s such that $\mathrm{Im}(\Gamma(U_s; F) \to \Gamma(V_i; F))$ is countable-dimensional. Then the natural morphism

$$\bigoplus_{s \in S, \, i \in I(s)} \mathrm{Im}(\Gamma(U_s; F) \to \Gamma(V_i; F)) \otimes \mathbb{C}_{V_i} \to F$$

is an epimorphism.

(iv)$\Rightarrow$(a) follows from the fact that each $\mathbb{C}_{U_i}$ is a quotient of a countable direct sum of sheaves of the form $\mathbb{C}_V$ with a subanalytic open subset V.

(a)$\Rightarrow$(ii) We shall prove it by the descending induction on n. Assume that F satisfies the condition (a). Let us take an exact sequence

$$0 \to F' \to L \to F \to 0,$$

such that $L \simeq \oplus_i \mathbb{C}_{U_i}$ for a countable family $\{U_i\}_i$ of subanalytic open subsets of X. Then, for any relatively compact subanalytic open subset

W, $H^k(W; \mathbb{C}_{U_i})$ is finite-dimensional (see e.g. [18]). Hence, the cohomology group $H^k(K; \mathbb{C}_{U_i}) \cong \varinjlim_{K \subset W} H^k(W; \mathbb{C}_{U_i})$ is countable-dimensional, and so is $H^k(K; L) \simeq \oplus_i H^k(K; \mathbb{C}_{U_i})$. Therefore L satisfies (i), which implies that F' also satisfies the condition (i) and hence the condition (a). By the induction hypothesis, $H^{n+1}(K; F')$ is countable-dimensional. By the exact sequence

$$H^n(K; L) \to H^n(K; F) \to H^{n+1}(K; F'),$$

$H^n(K; F)$ is countable-dimensional.   Q.E.D.

Definition 5.2.2. A sheaf F of complex vector spaces on X is called a *countable sheaf* if F satisfies the equivalent conditions in Proposition 5.2.1.

Let us denote by $\mathrm{Mod}_{\mathrm{ctb}}(\mathbb{C}_X)$ the full subcategory of $\mathrm{Mod}(\mathbb{C}_X)$ consisting of countable sheaves. Then, $\mathrm{Mod}_{\mathrm{ctb}}(\mathbb{C}_X)$ is closed by subobjects, quotients and extensions. Moreover it is closed by a countable inductive limits. Let us denote by $\mathrm{D}^b_{\mathrm{ctb}}(\mathbb{C}_X)$ the full subcategory of $\mathrm{D}^b(\mathbb{C}_X)$ consisting of objects whose cohomology groups are countable sheaves. It is a triangulated subcategory of $\mathrm{D}^b(\mathbb{C}_X)$.

Lemma 5.2.3.   (i) *If* F, $F' \in \mathrm{D}^b_{\mathrm{ctb}}(\mathbb{C}_X)$, *then* $F \otimes F'' \in \mathrm{D}^b_{\mathrm{ctb}}(\mathbb{C}_X)$.
  (ii) *For* $F \in \mathrm{D}^b(\mathbb{C}_X)$, *the following conditions are equivalent.*
    (a) $F \in \mathrm{D}^b_{\mathrm{ctb}}(\mathbb{C}_X)$,
    (b) $H^n(K; F)$ *is countable-dimensional for any compact subset* K *and any integer* n,
    (c) $H^n_c(U; F)$ *is countable-dimensional for any open subset* U *and any integer* n.
  (iii) *Let* $f : X \to Y$ *be a continuous map of topological manifolds. Then* $\mathbf{R}f_! F \in \mathrm{D}^b_{\mathrm{ctb}}(\mathbb{C}_Y)$ *for any* $F \in \mathrm{D}^b_{\mathrm{ctb}}(\mathbb{C}_X)$.

Proof. (i) follows from (iv) in Proposition 5.2.1.
(ii) (b)$\Rightarrow$(c) If U is relatively compact, it follows from the exact sequence $H^{n-1}(K \setminus U; F) \to H^n_c(U; F) \to H^n(K; F)$ for a compact set $K \supset U$, and if U is arbitrary, it follows from $H^n_c(U; F) = \varinjlim_{V \subset\subset U} H^n_c(V; F)$.

(c)$\Rightarrow$(b) follows from the exact sequence

$$H^n_c(X; F) \to H^n(K; F) \to H^{n+1}_c(X \setminus K; F).$$

(a)$\Rightarrow$(b) Let us show that $H^n(K; \tau^{\leqslant k} F)$ is countable-dimensional by the induction on k. If $H^n(K; \tau^{\leqslant k-1} F)$ is countable-dimensional, the exact sequence

$$H^n(K; \tau^{\leqslant k-1} F) \to H^n(K; \tau^{\leqslant k} F) \to H^{n-k}(K; H^k(F))$$

shows that $H^n(K; \tau^{\leqslant k} F)$ is countable-dimensional.

(b)$\Rightarrow$(a) We shall show that $H^k(F)$ is a countable sheaf by the induction on k. Assume that $\tau^{<k}F \in D^b_{ctb}(\mathbb{C}_X)$. Then, for any compact subset K, we have the exact sequence

$$H^n(K;F) \to H^n(K;\tau^{\geq k}F)) \to H^{n+1}(K;\tau^{<k}F).$$

Since $H^{n+1}(K;\tau^{<k}F)$ is countable-dimensional by (a)$\Rightarrow$(b), $H^n(K;\tau^{\geq k}F))$ is also countable-dimensional. In particular, $\Gamma(K;H^k(F)) = H^k(K;\tau^{\geq k}F)$ is countable-dimensional.

(iii)   For any open subset V of Y, $H^n_c(V;\mathbf{R}f_!F) \simeq H^n_c(f^{-1}(V);F)$ is countable-dimensional.   Q.E.D.

The following lemma is immediate.

Lemma 5.2.4. *Let F be a countable sheaf and let $H \twoheadrightarrow F$ be an epimorphism. Then there exist a countable sheaf F' and a morphism $F' \to H$ such that the composition $F' \to H \to F$ is an epimorphism.*

By Lemma 5.2.4, we have the following lemma.

Lemma 5.2.5. *The functor $D^b(\mathrm{Mod}_{ctb}(\mathbb{C}_X)) \to D^b_{ctb}(\mathbb{C}_X)$ is an equivalence of triangulated categories.*

More precisely, we have the following.

Lemma 5.2.6. *Let F be a bounded complex of sheaves such that all the co-homology groups are countable. Then we can find a bounded complex F' of countable sheaves and a quasi-isomorphism $F' \to F$.*

If a Lie group G acts on a real analytic manifold X, we denote by $\mathrm{Mod}_{G,ctb}(\mathbb{C}_X)$ the category of G-equivariant sheaves of $\mathbb{C}$-vector spaces which are countable.

Remark 5.2.7. A sheaf F of $\mathbb{C}$-vector spaces on X is not necessarily countable even if F_x is finite-dimensional for all $x \in X$. Indeed, the sheaf $\oplus_{x \in X}\mathbb{C}_{\{x\}}$ on X is such an example.

5.3 C^∞-Solutions

Let X, G and $G_{\mathbb{R}}$ be as in §5.1. Let $X_{\mathbb{R}}$ be a real analytic submanifold of X^{an} invariant by the $G_{\mathbb{R}}$-action such that $T_xX \cong \mathbb{C} \otimes_{\mathbb{R}} T_xX_{\mathbb{R}}$ for any $x \in X_{\mathbb{R}}$. Let M be a differentiable $G_{\mathbb{R}}$-manifold. Let us denote by $\mathscr{C}^\infty_{X_{\mathbb{R}} \times M}$ the sheaf of C^∞-functions on $X_{\mathbb{R}} \times M$. Then, $\mathscr{C}^\infty_{X_{\mathbb{R}} \times M}$ is an $f^{-1}\mathscr{D}_X$-module, where $f\colon X^{an} \times M \to X$ is a canonical map. For $\mathscr{M} \in \mathrm{Mod}(\mathscr{D}_X)$ and $K \in \mathrm{Mod}(\mathbb{C}_{X_{\mathbb{R}} \times M})$, we write $\mathrm{Hom}_{\mathscr{D}_X}\left(\mathscr{M} \otimes K, \mathscr{C}^\infty_{X_{\mathbb{R}} \times M}\right)$ instead of $\mathrm{Hom}_{f^{-1}\mathscr{D}_X}\left(f^{-1}\mathscr{M} \otimes K, \mathscr{C}^\infty_{X_{\mathbb{R}} \times M}\right)$ for short.

Lemma 5.3.1. *For any countable sheaf K on $X_{\mathbb{R}} \times M$ and a countably co-herent $\mathscr{D}_X$-module $\mathscr{M}$, $\mathrm{Hom}_{\mathscr{D}_X}\left(\mathscr{M} \otimes K, \mathscr{C}^\infty_{X_{\mathbb{R}} \times M}\right)$ has a structure of Fréchet nuclear space.*

Proof. The topology is the weakest topology such that, for any open subset U of X, any open subset V of $(U^{\text{an}} \cap X_{\mathbb{R}}) \times M$ and $s \in \Gamma(U; \mathscr{M})$, $t \in \Gamma(V; K)$, the homomorphism

$$(5.3.1) \qquad \text{Hom}_{\mathscr{D}_X} \left(\mathscr{M} \otimes K, \mathscr{C}^\infty_{X_{\mathbb{R}} \times M} \right) \ni \varphi \mapsto \varphi(s \otimes t) \in \text{C}^\infty(V)$$

is a continuous map. Here, $\text{C}^\infty(V)$ is the space of C^∞-functions on V.

There exist a countable index set A and a family of open subsets $\{U_a\}_{a \in A}$ of X, open subsets $\{V_a\}_{a \in A}$ of $X_{\mathbb{R}} \times M$ and $s_a \in \Gamma(U_a; \mathscr{M})$, $t_a \in \Gamma(V_a; K)$ satisfying the following properties:

(i) $V_a \subset U_a^{\text{an}} \times M$,
(ii) $\{s_a\}_{a \in A}$ generates $\mathscr{M}$, namely, $\mathscr{M}_x = \sum_{x \in U_a} (\mathscr{D}_X)_x (s_a)_x$ for any $x \in X$,
(iii) $\{t_a\}_{a \in A}$ generates K, namely, $K_x \simeq \sum_{x \in V_a} \mathbb{C}(t_a)_x$ for any $x \in X_{\mathbb{R}} \times M$.

Then by the morphisms (5.3.1), $\{s_a\}$ and $\{t_a\}$ induce an injection

$$\text{Hom}_{\mathscr{D}_X} \left(\mathscr{M} \otimes K, \mathscr{C}^\infty_{X_{\mathbb{R}} \times M} \right) \rightarrowtail \prod_{a \in A} \text{C}^\infty(V_a).$$

We can easily see that its image is a closed subspace of $\prod_{a \in A} \text{C}^\infty(V_a)$, and the induced topology coincides with the weakest topology introduced in the beginning. Since $\text{C}^\infty(V_a)$ is a Fréchet nuclear space and a countable product of Fréchet nuclear spaces is also a Fréchet nuclear space, $\prod_{a \in A} \text{C}^\infty(V_a)$ is a Fréchet nuclear space. Hence, its closed subspace $\text{Hom}_{\mathscr{D}_X} \left(\mathscr{M} \otimes K, \mathscr{C}^\infty_{X_{\mathbb{R}} \times M} \right)$ is also a Fréchet nuclear space.   Q.E.D.

Let $\mathscr{E}^{(p,q,r)}_{X^{\text{an}} \times M}$ denote the sheaf of differential forms on $X^{\text{an}} \times M$ with C^∞-coefficients which are (p,q)-forms with respect to X^{an}, and r-forms with respect to M. We set $\mathscr{E}^{(0,n)}_{X^{\text{an}} \times M} = \oplus_{n=q+r} \mathscr{E}^{(0,q,r)}_{X^{\text{an}} \times M}$. Then $\mathscr{E}^{(0,\bullet)}_{X^{\text{an}} \times M}$ is a complex of $p^{-1}\mathscr{D}_{X^{\text{an}}}$-modules, and it is quasi-isomorphic to $p^{-1}\mathscr{O}_{X^{\text{an}}}$, where $p: X^{\text{an}} \times M \to X^{\text{an}}$ is the projection.

Lemma 5.3.2. *For any* $K \in \text{Mod}_{G_{\mathbb{R}}, \text{ctb}}(\mathbb{C}_{X^{\text{an}} \times M})$ *and* $\mathscr{M} \in \text{Mod}_{\text{cc}}(\mathscr{D}_X, G)$, $\text{Hom}_{\mathscr{D}_X} \left(\mathscr{M} \otimes K, \mathscr{E}^{(0,q)}_{X^{\text{an}} \times M} \right)$ *has a Fréchet nuclear* $G_{\mathbb{R}}$-*module structure.*

The proof is similar to the previous lemma.

We denote by $\text{Hom}^{\text{top}}_{\mathscr{D}_X} \left(\mathscr{M} \otimes K, \mathscr{C}^\infty_{X_{\mathbb{R}} \times M} \right)$ and $\text{Hom}^{\text{top}}_{\mathscr{D}_X} \left(\mathscr{M} \otimes K, \mathscr{E}^{(0,n)}_{X^{\text{an}} \times M} \right)$ the corresponding space endowed with the Fréchet nuclear $G_{\mathbb{R}}$-module structure.

5.4 Definition of $\text{R}\,\text{Hom}^{\text{top}}$

Let us take a differentiable $G_{\mathbb{R}}$-manifold M with a free $G_{\mathbb{R}}$-action. Then we have an equivalence of categories:

$$(5.4.1) \qquad \begin{array}{ccc} \text{Mod}_{G_{\mathbb{R}}}(\mathbb{C}_{X^{\text{an}} \times M}) & \simeq & \text{Mod}(\mathbb{C}_{(X^{\text{an}} \times M)/G_{\mathbb{R}}}) \\ \cup & & \cup \\ \text{Mod}_{G_{\mathbb{R}}, \text{ctb}}(\mathbb{C}_{X^{\text{an}} \times M}) & \simeq & \text{Mod}_{\text{ctb}}(\mathbb{C}_{(X^{\text{an}} \times M)/G_{\mathbb{R}}}). \end{array}$$

Definition 5.4.1. A countable $G_{\mathbb{R}}$-equivariant sheaf K on $X^{\mathrm{an}} \times M$ is called *standard* if K is isomorphic to $\bigoplus_{j \in J} (E_j)_{U_j}$, where $\{U_j\}_{j \in J}$ is a countable family of $G_{\mathbb{R}}$-invariant open subsets of $X^{\mathrm{an}} \times M$ and E_j is a $G_{\mathbb{R}}$-equivariant local system on U_j of finite rank. Note the $(E_j)_{U_j}$ is the extension of E_j to the sheaf on $X^{\mathrm{an}} \times M$ such that $(E_j)_{U_j}|_{(X^{\mathrm{an}} \times M) \setminus U_j} = 0$.

Let us denote by $\mathrm{Mod}_{G_{\mathbb{R}}, \mathrm{stand}}(\mathbb{C}_{X^{\mathrm{an}} \times M})$ the full additive subcategory of $\mathrm{Mod}_{G_{\mathbb{R}}}(\mathbb{C}_{X^{\mathrm{an}} \times M})$ consisting of standard sheaves. With this terminology, we obtain the following lemma by (5.4.1) and Proposition 5.2.1.

Lemma 5.4.2. *For any $K \in \mathrm{C}^-\big(\mathrm{Mod}_{G_{\mathbb{R}}}(\mathbb{C}_{X^{\mathrm{an}} \times M})\big)$ with countable sheaves as cohomologies, there exist $K' \in \mathrm{C}^-\big(\mathrm{Mod}_{G_{\mathbb{R}}, \mathrm{stand}}(\mathbb{C}_{X^{\mathrm{an}} \times M})\big)$ and a quasi-isomorphism $K' \to K$ in $\mathrm{C}^-\big(\mathrm{Mod}_{G_{\mathbb{R}}}(\mathbb{C}_{X^{\mathrm{an}} \times M})\big)$.*

Similarly we introduce the following notion.

Definition 5.4.3. A countably coherent quasi-G-equivariant $\mathscr{D}_X$-module $\mathscr{M}$ is called *standard* if $\mathscr{M}$ is isomorphic to $\mathscr{D}_X \otimes_{\mathscr{O}_X} \mathscr{E}$ where $\mathscr{E}$ is a countably coherent locally free G-equivariant $\mathscr{O}_X$-module.

Let $\mathrm{Mod}_{\mathrm{stand}}(\mathscr{D}_X, G)$ denote the full additive subcategory of $\mathrm{Mod}(\mathscr{D}_X, G)$ consisting of standard modules.

For $K \in \mathrm{K}^{\mathrm{b}}\big(\mathrm{Mod}_{G_{\mathbb{R}}, \mathrm{ctb}}(\mathbb{C}_{X^{\mathrm{an}} \times M})\big)$ and $\mathscr{M} \in \mathrm{K}^{\mathrm{b}}\big(\mathrm{Mod}_{\mathrm{cc}}(\mathscr{D}_X, G)\big)$, we defined the complex $\mathrm{Hom}^{\mathrm{top}}_{\mathscr{D}_X}\big(\mathscr{M} \otimes K, \mathscr{E}^{(0, \bullet)}_{X^{\mathrm{an}} \times M}\big)$ of Fréchet nuclear $G_{\mathbb{R}}$-modules in Lemma 5.3.2.

Lemma 5.4.4.   (i) *Let $\mathscr{N} \in \mathrm{Mod}_{\mathrm{stand}}(\mathscr{D}_X)$ and $L \in \mathrm{Mod}_{\mathrm{stand}}(\mathbb{C}_{X^{\mathrm{an}} \times M})$. Then, we have*

$$\mathrm{Ext}^j_{\mathscr{D}_X}\big(\mathscr{N} \otimes L, \mathscr{E}^{(0,q)}_{X^{\mathrm{an}} \times M}\big) = 0$$

for any $j \neq 0$ and any q.
(ii) *We have isomorphisms in $\mathrm{D}^{\mathrm{b}}(\mathbb{C})$:*

$$\mathrm{Hom}_{\mathscr{D}_X}\big(\mathscr{M} \otimes K, \mathscr{E}^{(0, \bullet)}_{X^{\mathrm{an}} \times M}\big) \xrightarrow{\sim} \mathbf{R}\mathrm{Hom}_{\mathscr{D}_X}\big(\mathscr{M} \otimes K, \mathscr{E}^{(0, \bullet)}_{X^{\mathrm{an}} \times M}\big)$$

$$\simeq \mathbf{R}\mathrm{Hom}_{\mathscr{D}_X}\big(\mathscr{M} \otimes K, p^{-1}\mathscr{O}_{X^{\mathrm{an}}}\big)$$

for $\mathscr{M} \in \mathrm{K}^-(\mathrm{Mod}_{\mathrm{stand}}(\mathscr{D}_X))$ and $K \in \mathrm{K}^-(\mathrm{Mod}_{\mathrm{stand}}(\mathbb{C}_{X^{\mathrm{an}} \times M}))$. Here $p \colon X^{\mathrm{an}} \times M \to X^{\mathrm{an}}$ is the projection.

Proof. (i)   Since $\mathscr{N}$ is a locally free $\mathscr{D}_X$-module and $\mathscr{E}^{(0,q)}_{X^{\mathrm{an}} \times M}$ is a soft sheaf, we have $\mathscr{E}xt^j_{\mathscr{D}_X}\big(\mathscr{N}, \mathscr{E}^{(0,q)}_{X^{\mathrm{an}} \times M}\big) = 0$ for $j \neq 0$. Hence, $\mathbf{R}\mathscr{H}om_{\mathscr{D}_X}\big(\mathscr{N}, \mathscr{E}^{(0,q)}_{X^{\mathrm{an}} \times M}\big)$ is represented by $\mathscr{H}om_{\mathscr{D}_X}\big(\mathscr{N}, \mathscr{E}^{(0,q)}_{X^{\mathrm{an}} \times M}\big)$. Since $\mathscr{H}om_{\mathscr{D}_X}\big(\mathscr{N}, \mathscr{E}^{(0,q)}_{X^{\mathrm{an}} \times M}\big)$ has locally a $\mathscr{C}^{\infty}_{X^{\mathrm{an}} \times M}$-module structure, it is a soft sheaf. Hence, we obtain $\mathrm{Ext}^j_{\mathscr{D}_X}\big(L, \mathscr{H}om_{\mathscr{D}_X}(\mathscr{N}, \mathscr{E}^{(0,q)}_{X^{\mathrm{an}} \times M})\big) = 0$ for $j \neq 0$. Finally, we conclude that $\mathscr{H}om_{\mathbb{C}}\big(L, \mathscr{H}om_{\mathscr{D}_X}(\mathscr{N}, \mathscr{E}^{(0,q)}_{X^{\mathrm{an}} \times M})\big)$ represents

$$\mathbf{R}\mathscr{H}om_{\mathbb{C}}(L, \mathbf{R}\mathscr{H}om_{\mathscr{D}_X}(\mathscr{N}, \mathscr{E}^{(0,q)}_{X^{\mathrm{an}} \times M})) \simeq \mathbf{R}\mathscr{H}om_{\mathscr{D}_X}(\mathscr{N} \otimes L, \mathscr{E}^{(0,q)}_{X^{\mathrm{an}} \times M}).$$

(ii) follows immediately from (i).   Q.E.D.

Proposition 5.4.5. *Let us assume that* $K \in \mathrm{K}^{\mathrm{b}}\big(\mathrm{Mod}_{G_{\mathbb{R}},\mathrm{ctb}}(\mathbb{C}_{X^{\mathrm{an}} \times M})\big)$ *and* $\mathscr{M} \in \mathrm{K}^{\mathrm{b}}\big(\mathrm{Mod}_{\mathrm{cc}}(\mathscr{D}_X, G)\big)$. *Then,*

$$\underset{\mathscr{M}', K'}{\text{``}\varinjlim\text{''}} \mathrm{Hom}^{\mathrm{top}}_{\mathscr{D}_X}\big(\mathscr{M}' \otimes K', \mathscr{E}^{(0,\bullet)}_{X^{\mathrm{an}} \times M}\big)$$

is representable in $\mathrm{D}^{\mathrm{b}}(\mathbf{FN}_{G_{\mathbb{R}}})$. *Here,* $\mathscr{M}' \to \mathscr{M}$ *ranges over the quasi-isomorphisms in* $\mathrm{K}^{-}\big(\mathrm{Mod}_{\mathrm{cc}}(\mathscr{D}_X, G)\big)$ *and* $K' \to K$ *ranges over the quasi-isomorphisms in* $\mathrm{K}^{-}\big(\mathrm{Mod}_{G_{\mathbb{R}},\mathrm{ctb}}(\mathbb{C}_{X^{\mathrm{an}} \times M})\big)$. *Moreover, forgetting the topology and the equivariance, it is isomorphic to* $\mathbf{R}\mathrm{Hom}_{\mathscr{D}_X}(\mathscr{M} \otimes K, p^{-1}\mathscr{O}_{X^{\mathrm{an}}})$. *Here* $p: X^{\mathrm{an}} \times M \to X^{\mathrm{an}}$ *is the projection.*

Proof. There exist $\mathscr{M}' \in \mathrm{K}^{-}(\mathrm{Mod}_{\mathrm{stand}}(\mathscr{D}_X, G))$ and a quasi-isomorphism $\mathscr{M}' \to \mathscr{M}$. Similarly by Lemma 5.4.2, there exist $K' \in \mathrm{K}^{-}(\mathrm{Mod}_{G_{\mathbb{R}},\mathrm{stand}}(\mathbb{C}_X))$ and a quasi-isomorphism $K' \to K$.

Then

$$(5.4.2) \qquad \mathrm{Hom}_{\mathscr{D}_X}\big(\mathscr{M}' \otimes K', \mathscr{E}^{(0,\bullet)}_{X^{\mathrm{an}} \times M}\big) \to \mathbf{R}\mathrm{Hom}_{\mathscr{D}_X}(\mathscr{M} \otimes K, p^{-1}\mathscr{O}_{X^{\mathrm{an}}})$$

is an isomorphism in $\mathrm{D}(\mathbb{C})$ by the preceding lemma.

To complete the proof, it is enough to remark that, if a morphism in $\mathrm{K}(\mathbf{FN}_{G_{\mathbb{R}}})$ is a quasi-isomorphism in $\mathrm{K}(\mathrm{Mod}(\mathbb{C}))$ forgetting the topology and the equivariance, then it is a quasi-isomorphism in $\mathrm{K}(\mathbf{FN}_{G_{\mathbb{R}}})$.   Q.E.D.

Definition 5.4.6. Assume that $G_{\mathbb{R}}$ acts freely on M. For $\mathscr{M} \in \mathrm{D}^{\mathrm{b}}_{\mathrm{cc}}(\mathscr{D}_X, G)$ and $K \in \mathrm{D}^{\mathrm{b}}_{G_{\mathbb{R}},\mathrm{ctb}}(\mathbb{C}_{X^{\mathrm{an}} \times M})$, we define $\mathbf{R}\mathrm{Hom}^{\mathrm{top}}_{\mathscr{D}_X}(\mathscr{M} \otimes K, \mathscr{E}^{(0,\bullet)}_{X^{\mathrm{an}} \times M})$ as the object

$$\underset{\mathscr{M}', K'}{\text{``}\varinjlim\text{''}} \mathrm{Hom}^{\mathrm{top}}_{\mathscr{D}_X}\big(\mathscr{M}' \otimes K', \mathscr{E}^{(0,\bullet)}_{X^{\mathrm{an}} \times M}\big)$$

of $\mathrm{D}^{\mathrm{b}}(\mathbf{FN}_{G_{\mathbb{R}}})$. Here, $\mathscr{M}'$ ranges over the set of objects of $\mathrm{K}^{-}\big(\mathrm{Mod}_{\mathrm{cc}}(\mathscr{D}_X, G)\big)$ isomorphic to $\mathscr{M}$ in $\mathrm{D}_{\mathrm{cc}}(\mathscr{D}_X, G)$, and K' ranges over the set of objects of $\mathrm{K}^{-}\big(\mathrm{Mod}_{G_{\mathbb{R}},\mathrm{ctb}}(\mathbb{C}_{X^{\mathrm{an}} \times M})\big)$ isomorphic to K in $\mathrm{D}\big(\mathrm{Mod}_{G_{\mathbb{R}}}(\mathbb{C}_{X^{\mathrm{an}} \times M})\big)$.

Let us take a sequence of $G_{\mathbb{R}}$-manifolds with a free $G_{\mathbb{R}}$-action:

$$(5.4.3) \qquad\qquad V_1 \longrightarrow V_2 \longrightarrow V_3 \longrightarrow \cdots$$

as in (4.2.1).

Lemma 5.4.7. *For* $\mathscr{M} \in \mathrm{D}^{\mathrm{b}}_{\mathrm{cc}}(\mathscr{D}_X, G)$ *and* $K \in \mathrm{D}^{\mathrm{b}}_{G_{\mathbb{R}},\mathrm{ctb}}(\mathbb{C}_{X^{\mathrm{an}}})$, *Then*

$$\tau^{\leq a} \mathbf{R}\mathrm{Hom}^{\mathrm{top}}_{\mathscr{D}_X}\big(\mathscr{M} \otimes p_k^{-1}K, \mathscr{E}^{(0,\bullet)}_{X^{\mathrm{an}} \times V_k}\big)$$

does not depend on $k \gg a \gg 0$ *as an object of* $\mathrm{D}^{\mathrm{b}}(\mathbf{FN}_{G_{\mathbb{R}}})$. *Here* $p_k: X^{\mathrm{an}} \times V_k \to X^{\mathrm{an}}$ *is the projection.*

Proof. Forgetting the topology and the equivariance, we have

$$\mathbf{R}\mathrm{Hom}_{\mathscr{D}_X}^{\mathrm{top}}(\mathscr{M}\otimes p_k^{-1}K,\mathscr{E}_{X^{\mathrm{an}}\times V_k}^{(0,\,\bullet)})\simeq \mathbf{R}\mathrm{Hom}_{\mathscr{D}_X}(\mathscr{M}\otimes p_k^{-1}K,p_k^{-1}\mathscr{O}_{X^{\mathrm{an}}})$$
$$\simeq \mathbf{R}\mathrm{Hom}_{\mathscr{D}_X}(\mathscr{M}\otimes K,\mathscr{O}_{X^{\mathrm{an}}})\otimes \mathbf{R}\,\Gamma(V_k;\mathbb{C}),$$

and

$$\tau^{\leqslant a}\Big(\mathbf{R}\mathrm{Hom}_{\mathscr{D}_X}(\mathscr{M}\otimes K,\mathscr{O}_{X^{\mathrm{an}}})\otimes \mathbf{R}\,\Gamma(V_k;\mathbb{C})\Big)\simeq \mathbf{R}\mathrm{Hom}_{\mathscr{D}_X}(\mathscr{M}\otimes K,\mathscr{O}_{X^{\mathrm{an}}})$$

for $k\gg a\gg 0$.   Q.E.D.

Definition 5.4.8. We define

$$\mathbf{R}\mathrm{Hom}_{\mathscr{D}_X}^{\mathrm{top}}(\mathscr{M}\otimes K,\mathscr{O}_{X^{\mathrm{an}}})$$

as $\tau^{\leqslant a}\,\mathbf{R}\mathrm{Hom}_{\mathscr{D}_X}^{\mathrm{top}}(\mathscr{M}\otimes p_k^{-1}K,\mathscr{E}_{X^{\mathrm{an}}\times V_k}^{(0,\,\bullet)})$ for $k\gg a\gg 0$.

Note that

$$\mathbf{R}\mathrm{Hom}_{\mathscr{D}_X}^{\mathrm{top}}(\mathscr{M}\otimes K,\mathscr{O}_{X^{\mathrm{an}}})$$
$$\simeq \text{``}\varprojlim_k\text{''}\,\mathbf{R}\mathrm{Hom}_{\mathscr{D}_X}^{\mathrm{top}}(\mathscr{M}\otimes p_k^{-1}K,\mathscr{E}_{X^{\mathrm{an}}\times V_k}^{(0,\,\bullet)}).$$

Note that, forgetting the topology and the equivariance, $\mathbf{R}\mathrm{Hom}_{\mathscr{D}_X}^{\mathrm{top}}(\mathscr{M}\otimes K,\mathscr{O}_{X^{\mathrm{an}}})$ is isomorphic to $\mathbf{R}\mathrm{Hom}_{\mathscr{D}_X}(\mathscr{M}\otimes K,\mathscr{O}_{X^{\mathrm{an}}})\in\mathrm{D}(\mathbb{C})$.

5.5 DFN Version

In this subsection, let us define $\mathbf{R}\,\Gamma_c^{\mathrm{top}}(X^{\mathrm{an}};K\otimes \Omega_{X^{\mathrm{an}}}\overset{\mathbf{L}}{\otimes}_{\mathscr{D}_X}\mathscr{M})$, which is the dual of $\mathbf{R}\mathrm{Hom}_{\mathscr{D}_X}^{\mathrm{top}}(\mathscr{M}\otimes K,\mathscr{O}_{X^{\mathrm{an}}})$. Since the construction is similar to the one of $\mathbf{R}\mathrm{Hom}_{\mathscr{D}_X}^{\mathrm{top}}(\mathscr{M}\otimes K,\mathscr{O}_{X^{\mathrm{an}}})$, we shall be brief.

Let us denote by $\mathscr{D}ist_{X^{\mathrm{an}}}^{(p,q)}$ the sheaf of (p,q)-forms on X^{an} with distributions as coefficients. Then for any open subset U of X^{an}, $\Gamma_c(U;\mathscr{D}ist_{X^{\mathrm{an}}}^{(p,q)})$ is endowed with a DFN-topology and it is the dual topological space of the FN-space $\mathscr{E}_{X^{\mathrm{an}}}^{(\mathrm{d}_X-p,\mathrm{d}_X-q)}(U)$. Hence for $\mathscr{M}\in\mathrm{K}^-(\mathrm{Mod}_{\mathrm{stand}}(\mathscr{D}_X))$ and $F\in\mathrm{K}^-(\mathrm{Mod}_{\mathrm{stand}}(\mathbb{C}_{X^{\mathrm{an}}}))$, $\Gamma_c(X^{\mathrm{an}};K\otimes \mathscr{D}ist_{X^{\mathrm{an}}}^{(\mathrm{d}_X,\,\bullet)}\otimes_{\mathscr{D}_X}\mathscr{M})[\mathrm{d}_X]$ is a complex of DFN-spaces, and it is the dual of $\mathrm{Hom}_{\mathscr{D}_X}^{\mathrm{top}}(\mathscr{M}\otimes K,\mathscr{E}_{X^{\mathrm{an}}}^{(0,\,\bullet)})$. We denote by $\Gamma_c^{\mathrm{top}}(X^{\mathrm{an}};K\otimes \mathscr{D}ist_{X^{\mathrm{an}}}^{(\mathrm{d}_X,\,\bullet)}\otimes_{\mathscr{D}_X}\mathscr{M})$ the complex of DFN-spaces $\Gamma_c(X^{\mathrm{an}};K\otimes \mathscr{D}ist_{X^{\mathrm{an}}}^{(\mathrm{d}_X,\,\bullet)}\otimes_{\mathscr{D}_X}\mathscr{M})$. If we forget the topology, it is isomorphic to $\mathbf{R}\,\Gamma_c(X^{\mathrm{an}};K\otimes \Omega_{X^{\mathrm{an}}}\otimes_{\mathscr{D}_X}\mathscr{M})\in\mathrm{D}^{\mathrm{b}}(\mathbb{C})$. Thus we have defined a functor:

$$\mathbf{R}\,\Gamma_c^{\mathrm{top}}(X^{\mathrm{an}};\,\bullet\,\otimes \Omega_{X^{\mathrm{an}}}\overset{\mathbf{L}}{\otimes}_{\mathscr{D}_X}\,\bullet):\ \mathrm{D}_{\mathrm{ctb}}^{\mathrm{b}}(\mathbb{C}_{X^{\mathrm{an}}})\times \mathrm{D}_{\mathrm{cc}}^{\mathrm{b}}(\mathscr{D}_X)\to \mathrm{D}^{\mathrm{b}}(\mathbf{DFN}).$$

When X is a quasi-projective G-manifold, we can define its equivariant version

$$\mathbf{R}\,\Gamma_{\mathrm{c}}^{\mathrm{top}}(X^{\mathrm{an}};\bullet\otimes\Omega_{X^{\mathrm{an}}}\overset{\mathbf{L}}{\otimes}_{\mathscr{D}_X}\bullet)\colon \mathrm{D}_{G_{\mathbb{R}},\mathrm{ctb}}^{\mathrm{b}}(\mathbb{C}_{X^{\mathrm{an}}})\times \mathrm{D}_{\mathrm{cc}}^{\mathrm{b}}(\mathscr{D}_X,G)\to \mathrm{D}^{\mathrm{b}}(\mathbf{DFN}_{G_{\mathbb{R}}}).$$

We have

$$(5.5.1)\quad \mathbf{R}\,\Gamma_{\mathrm{c}}^{\mathrm{top}}(X^{\mathrm{an}};K\otimes\Omega_{X^{\mathrm{an}}}\overset{\mathbf{L}}{\otimes}_{\mathscr{D}_X}\mathscr{M})[\mathrm{d}_X]\cong \big(\mathbf{R}\mathrm{Hom}_{\mathscr{D}_X}^{\mathrm{top}}(\mathscr{M}\otimes K,\mathscr{O}_{X^{\mathrm{an}}})\big)^{*}.$$

Here $(\bullet)^{*}\colon \mathrm{D}^{\mathrm{b}}(\mathbf{FN}_{G_{\mathbb{R}}})^{\mathrm{op}}\overset{\sim}{\longrightarrow}\mathrm{D}^{\mathrm{b}}(\mathbf{DFN}_{G_{\mathbb{R}}})$ is the functor induced by the duality.

If we forget the topology and the equivariance, $\mathbf{R}\,\Gamma_{\mathrm{c}}^{\mathrm{top}}(X^{\mathrm{an}};K\otimes\Omega_{X^{\mathrm{an}}}\otimes_{\mathscr{D}_X}\mathscr{M})$ is isomorphic to $\mathbf{R}\,\Gamma_{\mathrm{c}}(X^{\mathrm{an}};K\otimes\Omega_{X^{\mathrm{an}}}\otimes_{\mathscr{D}_X}\mathscr{M})\in \mathrm{D}^{\mathrm{b}}(\mathbb{C})$.

5.6 Functorial Properties of $\mathbf{R}\mathrm{Hom}^{\mathrm{top}}$

Statements

We shall study how $\mathbf{R}\mathrm{Hom}^{\mathrm{top}}$ behaves under G-equivariant morphisms of G-manifolds. We shall keep the notations $G, G_{\mathbb{R}}$ as in §5.1.

Let $f\colon X\to Y$ be a G-equivariant morphism of quasi-projective algebraic G-manifolds. Let $f^{\mathrm{an}}\colon X^{\mathrm{an}}\to Y^{\mathrm{an}}$ be the associated holomorphic map.

Theorem 5.6.1. (i) *Assume that f is smooth and projective. Then, there exists a canonical isomorphism in* $\mathrm{D}^{\mathrm{b}}(\mathbf{FN}_{G_{\mathbb{R}}})$:

$$\mathbf{R}\mathrm{Hom}_{\mathscr{D}_X}^{\mathrm{top}}(\mathscr{M}\otimes(f^{\mathrm{an}})^{-1}L,\mathscr{O}_{X^{\mathrm{an}}})\simeq \mathbf{R}\mathrm{Hom}_{\mathscr{D}_Y}^{\mathrm{top}}(\mathbf{D}f_*\mathscr{M}\otimes L,\mathscr{O}_{Y^{\mathrm{an}}})[-\mathrm{d}_{X/Y}]$$

for $\mathscr{M}\in \mathrm{D}_{\mathrm{coh}}^{\mathrm{b}}(\mathscr{D}_X,G)$ and $L\in \mathrm{D}_{G_{\mathbb{R}},\mathrm{ctb}}^{\mathrm{b}}(\mathbb{C}_Y)$.

(ii) *Assume that f is smooth. Then, there exists a canonical isomorphism in* $\mathrm{D}^{\mathrm{b}}(\mathbf{FN}_{G_{\mathbb{R}}})$:

$$\mathbf{R}\mathrm{Hom}_{\mathscr{D}_Y}^{\mathrm{top}}(\mathscr{N}\otimes\mathbf{R}(f^{\mathrm{an}})_!K,\mathscr{O}_{Y^{\mathrm{an}}})\simeq \mathbf{R}\mathrm{Hom}_{\mathscr{D}_X}^{\mathrm{top}}(\mathbf{D}f^*\mathscr{N}\otimes K,\mathscr{O}_{X^{\mathrm{an}}})[2\mathrm{d}_{X/Y}]$$

for $\mathscr{N}\in \mathrm{D}_{\mathrm{coh}}^{\mathrm{b}}(\mathscr{D}_Y,G)$ and $K\in \mathrm{D}_{G_{\mathbb{R}},\mathrm{ctb}}^{\mathrm{b}}(\mathbb{C}_X)$.

Preparation

Let us take a sequence $\{V_k\}$ as in (4.2.1).

Let $\mathscr{N}\in \mathrm{D}_{\mathrm{cc}}^{\mathrm{b}}(\mathscr{D}_Y,G)$ and $L\in \mathrm{D}_{G_{\mathbb{R}},\mathrm{ctb}}^{\mathrm{b}}(\mathbb{C}_{Y^{\mathrm{an}}})$. Then, by the definition, we have

$$\mathbf{R}\mathrm{Hom}_{\mathscr{D}_Y}^{\mathrm{top}}(\mathscr{N}\otimes(L\boxtimes\mathbb{C}_{V_k}),\mathscr{E}_{Y^{\mathrm{an}}\times V_k}^{(0,\bullet)})=\underset{\mathscr{N}',L'}{\text{``}\varinjlim\text{''}}\mathrm{Hom}_{\mathscr{D}_Y}^{\mathrm{top}}(\mathscr{N}'\otimes L',\mathscr{E}_{Y^{\mathrm{an}}\times V_k}^{(0,\bullet)}).$$

Here, $\mathscr{N}'$ ranges over the objects of $\mathrm{K}^{-}\big(\mathrm{Mod}_{\mathrm{cc}}(\mathscr{D}_Y,G)\big)$ isomorphic to $\mathscr{N}$ in $\mathrm{D}(\mathrm{Mod}(\mathscr{D}_Y,G))$, and L' ranges over the objects of $\mathrm{K}^{-}\big(\mathrm{Mod}_{G_{\mathbb{R}},\mathrm{ctb}}(\mathbb{C}_{Y^{\mathrm{an}}\times V_k})\big)$ isomorphic to $L\boxtimes\mathbb{C}_{V_k}$ in $\mathrm{D}_{G_{\mathbb{R}},\mathrm{ctb}}^{\mathrm{b}}(\mathbb{C}_{Y^{\mathrm{an}}\times V_k})$. Then the morphism

$$\mathscr{D}_{X \to Y} \underset{\mathscr{D}_Y}{\otimes} \mathscr{E}^{(0,\bullet)}_{Y^{\mathrm{an}} \times V_k} \to \mathscr{E}^{(0,\bullet)}_{X^{\mathrm{an}} \times V_k}$$

induces morphisms in $\mathrm{D}^{\mathrm{b}}(\mathbf{FN}_{G_{\mathbb{R}}})$:

$$\mathrm{Hom}^{\mathrm{top}}_{\mathscr{D}_Y}(\mathscr{N}' \otimes L', \mathscr{E}^{(0,\bullet)}_{Y^{\mathrm{an}} \times V_k}) \to \mathrm{Hom}^{\mathrm{top}}_{\mathscr{D}_X}(f^* \mathscr{N}' \otimes (f^{\mathrm{an}} \times \mathrm{id}_{V_k})^{-1} L', \mathscr{E}^{(0,\bullet)}_{X^{\mathrm{an}} \times V_k})$$
$$\to \mathbf{R}\mathrm{Hom}^{\mathrm{top}}_{\mathscr{D}_X}(\mathbf{D}f^* \mathscr{N} \otimes ((f^{\mathrm{an}})^{-1} L \boxtimes \mathbb{C}_{V_k}), \mathscr{E}^{(0,\bullet)}_{X^{\mathrm{an}} \times V_k}).$$

Thus we obtain a morphism

$$\mathbf{R}\mathrm{Hom}^{\mathrm{top}}_{\mathscr{D}_Y}(\mathscr{N} \otimes (L \boxtimes \mathbb{C}_{V_k}), \mathscr{E}^{(0,\bullet)}_{Y^{\mathrm{an}} \times V_k})$$
$$\to \mathbf{R}\mathrm{Hom}^{\mathrm{top}}_{\mathscr{D}_X}(\mathbf{D}f^* \mathscr{N} \otimes ((f^{\mathrm{an}})^{-1} L \boxtimes \mathbb{C}_{V_k}), \mathscr{E}^{(0,\bullet)}_{X^{\mathrm{an}} \times V_k})$$

for $\mathscr{N} \in \mathrm{D}^{\mathrm{b}}_{\mathrm{cc}}(\mathscr{D}_Y, G)$ and $L \in \mathrm{D}^{\mathrm{b}}_{G_{\mathbb{R}}, \mathrm{ctb}}(\mathbb{C}_{Y_{\mathrm{an}}})$. Taking the projective limit with respect to k, we obtain

$$(5.6.1) \quad \mathbf{R}\mathrm{Hom}^{\mathrm{top}}_{\mathscr{D}_Y}(\mathscr{N} \otimes L, \mathscr{O}_{Y^{\mathrm{an}}}) \to \mathbf{R}\mathrm{Hom}^{\mathrm{top}}_{\mathscr{D}_X}(\mathbf{D}f^* \mathscr{N} \otimes (f^{\mathrm{an}})^{-1} L, \mathscr{O}_{X^{\mathrm{an}}}).$$

Here, f is arbitrary.

Proof of Theorem 5.6.1

Let us first prove (i). For $\mathscr{M}$ and L as in (i), we have morphisms

$$\mathbf{R}\mathrm{Hom}^{\mathrm{top}}_{\mathscr{D}_Y}(\mathbf{D}f_* \mathscr{M} \otimes L[\mathrm{d}_{X/Y}], \mathscr{O}_{Y^{\mathrm{an}}})$$
$$\to \mathbf{R}\mathrm{Hom}^{\mathrm{top}}_{\mathscr{D}_X}(\mathbf{D}f^* \mathbf{D}f_* \mathscr{M}[\mathrm{d}_{X/Y}] \otimes (f^{\mathrm{an}})^{-1} L, \mathscr{O}_{X^{\mathrm{an}}})$$
$$\to \mathbf{R}\mathrm{Hom}^{\mathrm{top}}_{\mathscr{D}_X}(\mathscr{M} \otimes (f^{\mathrm{an}})^{-1} L, \mathscr{O}_{X^{\mathrm{an}}}).$$

Here, the first arrow is given by (5.6.1) and the last arrow is given by $\mathscr{M} \to \mathbf{D}f^* \mathbf{D}f_* \mathscr{M}[\mathrm{d}_{X/Y}]$ (see (3.8.4)).

We shall prove that the composition

$$\mathbf{R}\mathrm{Hom}^{\mathrm{top}}_{\mathscr{D}_Y}(\mathbf{D}f_* \mathscr{M}[\mathrm{d}_{X/Y}] \otimes L, \mathscr{O}_{Y^{\mathrm{an}}}) \to \mathbf{R}\mathrm{Hom}^{\mathrm{top}}_{\mathscr{D}_X}(\mathscr{M} \otimes (f^{\mathrm{an}})^{-1} L, \mathscr{O}_{X^{\mathrm{an}}})$$

is an isomorphism in $\mathrm{D}^{\mathrm{b}}(\mathbf{FN}_{G_{\mathbb{R}}})$.

In order to see this, it is enough to show that it is an isomorphism in $\mathrm{D}^{\mathrm{b}}(\mathbb{C})$. Then the result follows from the result of D-modules:

$$\mathbf{R}\mathscr{H}om_{\mathscr{D}_Y}(\mathbf{D}f_* \mathscr{M}[\mathrm{d}_{X/Y}], \mathscr{O}_{Y^{\mathrm{an}}}) \cong \mathbf{R}(f^{\mathrm{an}})_* \mathbf{R}\mathscr{H}om_{\mathscr{D}_X}(\mathscr{M}, \mathscr{O}_{X^{\mathrm{an}}}).$$

The proof of (ii) is similar. Let $\mathscr{N}$ and K be as in (ii), then we have a sequence of morphisms

$$\mathbf{R}\mathrm{Hom}^{\mathrm{top}}_{\mathscr{D}_Y}(\mathscr{N} \otimes \mathbf{R}f^{\mathrm{an}}_! K, \mathscr{O}_{Y^{\mathrm{an}}})$$
$$\to \mathbf{R}\mathrm{Hom}^{\mathrm{top}}_{\mathscr{D}_X}(\mathbf{D}f^* \mathscr{N} \otimes (f^{\mathrm{an}})^{-1} \mathbf{R}f^{\mathrm{an}}_! K, \mathscr{O}_{X^{\mathrm{an}}})$$

$$\to \mathbf{R}\mathrm{Hom}_{\mathscr{D}_X}^{\mathrm{top}}(\mathbf{D}f^*\mathscr{N} \otimes K, \mathscr{O}_{X^{\mathrm{an}}})[2\mathrm{d}_{X/Y}].$$

Here the last arrow is obtained by

$$K \to (f^{\mathrm{an}})^!\mathbf{R}(f^{\mathrm{an}})_!K \cong (f^{\mathrm{an}})^{-1}\mathbf{R}(f^{\mathrm{an}})_!K[2\mathrm{d}_{X/Y}].$$

The rest of arguments is similar to the proof of (i) by reducing it to the corresponding result in the D-module theory:

$$\mathbf{R}\mathscr{H}om_{\mathscr{D}_X}(\mathbf{D}f^*\mathscr{N}, \mathscr{O}_{X^{\mathrm{an}}}) \simeq (f^{\mathrm{an}})^{-1}\mathbf{R}\mathscr{H}om_{\mathscr{D}_Y}(\mathscr{N}, \mathscr{O}_{Y^{\mathrm{an}}}).$$

5.7 Relation with the de Rham Functor

Let X be an algebraic G-manifold. First assume that G acts freely on X. Let $p\colon X \to X/G$ be the projection. Let $\mathscr{M} \in \mathrm{D}_{\mathrm{cc}}^{\mathrm{b}}(\mathscr{D}_X, G)$ and $K \in \mathrm{D}_{G_{\mathbb{R}},\mathrm{ctb}}^{\mathrm{b}}(\mathbb{C}_{X^{\mathrm{an}}})$. Let $\mathscr{L}$ be an object of $\mathrm{D}_{G,\mathrm{hol}}^{\mathrm{b}}(\mathscr{D}_X)$. Let $\mathscr{L}/G$ be the object of $\mathrm{D}_{\mathrm{hol}}^{\mathrm{b}}(\mathscr{D}_{X/G})$ corresponding to $\mathscr{L}$. Set $L = \mathrm{DR}_X(\mathscr{L}) \in \mathrm{D}_{G^{\mathrm{an}}, \mathbb{C}\text{-c}}^{\mathrm{b}}(\mathbb{C}_{X^{\mathrm{an}}})$ (see Subsection 4.6). Then the corresponding object $L/G^{\mathrm{an}} \in \mathrm{D}_{\mathbb{C}\text{-c}}^{\mathrm{b}}(\mathbb{C}_{(X/G)^{\mathrm{an}}})$ is isomorphic to $\mathrm{DR}_{X/G}(\mathscr{L}/G)$.

Let us represent $\mathscr{M}$ by an object of $\mathrm{K}_{\mathrm{stand}}^-(\mathscr{D}_X, G)$ and $\mathscr{L}/G$ by an object $\widetilde{\mathscr{L}} \in \mathrm{K}^-(\mathrm{Mod}_{\mathrm{stand}}(\mathscr{D}_{X/G}))$. Then $\mathscr{L}$ is represented by $p^*\widetilde{\mathscr{L}}$. Since $L/G^{\mathrm{an}} \simeq \mathscr{H}om_{\mathscr{D}_{(X/G)^{\mathrm{an}}}}(\mathscr{D}_{(X/G)^{\mathrm{an}}} \otimes \overset{\bullet}{\bigwedge}\Theta_{(X/G)^{\mathrm{an}}}, \widetilde{\mathscr{L}}^{\mathrm{an}})$ belongs to $\mathrm{D}_{\mathrm{ctb}}^{\mathrm{b}}(\mathbb{C}_{(X/G)^{\mathrm{an}}})$, there exist $F \in \mathrm{K}^-(\mathrm{Mod}_{\mathrm{stand}}(\mathbb{C}_{(X/G)^{\mathrm{an}}}))$ and a quasi-isomorphism

$$F \to \mathscr{H}om_{\mathscr{D}_{(X/G)^{\mathrm{an}}}}(\mathscr{D}_{(X/G)^{\mathrm{an}}} \otimes \overset{\bullet}{\bigwedge}\Theta_{(X/G)^{\mathrm{an}}}, \widetilde{\mathscr{L}}^{\mathrm{an}})$$

by Lemma 5.2.6. Thus we obtain a morphism of complexes of $\mathscr{D}_{(X/G)^{\mathrm{an}}}$-modules:

$$(5.7.1) \qquad \mathscr{D}_{(X/G)^{\mathrm{an}}} \otimes \overset{\bullet}{\bigwedge}\Theta_{(X/G)^{\mathrm{an}}} \otimes F \to \widetilde{\mathscr{L}}^{\mathrm{an}}.$$

Let M be a differentiable manifold with a free $G_{\mathbb{R}}$-action. Then for any $E \in \mathrm{K}_{G_{\mathbb{R}},\mathrm{stand}}^-(\mathbb{C}_{X^{\mathrm{an}} \times M})$, the morphism (5.7.1) induces morphisms

$$\mathrm{Hom}_{\mathscr{D}_X}^{\mathrm{top}}(\mathscr{M} \otimes_{\mathscr{O}_X} \mathscr{L} \otimes E, \mathscr{E}_{X^{\mathrm{an}} \times M}^{(0,\bullet)})$$

$$\simeq \mathrm{Hom}_{\mathscr{D}_{X^{\mathrm{an}}}}^{\mathrm{top}}(\mathscr{M}^{\mathrm{an}} \otimes_{(p^{\mathrm{an}})^{-1}\mathscr{O}_{(X/G)^{\mathrm{an}}}} (p^{\mathrm{an}})^{-1}\widetilde{\mathscr{L}}^{\mathrm{an}} \otimes_{\mathbb{C}} E, \mathscr{E}_{X^{\mathrm{an}} \times M}^{(0,\bullet)})$$

$$\to \mathrm{Hom}_{\mathscr{D}_{X^{\mathrm{an}}}}^{\mathrm{top}}(\mathscr{M}^{\mathrm{an}} \underset{(p^{\mathrm{an}})^{-1}\mathscr{O}_{(X/G)^{\mathrm{an}}}}{\otimes} (p^{\mathrm{an}})^{-1}(\mathscr{D}_{(X/G)^{\mathrm{an}}} \otimes \overset{\bullet}{\bigwedge}\Theta_{(X/G)^{\mathrm{an}}} \otimes F) \otimes_{\mathbb{C}} E, \mathscr{E}_{X^{\mathrm{an}} \times M}^{(0,\bullet)})$$

$$\simeq \mathrm{Hom}_{\mathscr{D}_X}^{\mathrm{top}}\left((\mathscr{M} \underset{p^{-1}\mathscr{O}_{X/G}}{\otimes} p^{-1}(\mathscr{D}_{X/G} \otimes \overset{\bullet}{\bigwedge}\Theta_{X/G})) \otimes ((p^{\mathrm{an}})^{-1}F \otimes E), \mathscr{E}_{X^{\mathrm{an}} \times M}^{(0,\bullet)}\right).$$

On the other hand, we have an isomorphism in $\mathrm{D}^{\mathrm{b}}(\mathbf{FN}_{G_{\mathbb{R}}})$:

$$\mathbf{R}\mathrm{Hom}^{\mathrm{top}}_{\mathscr{D}_X}(\mathscr{M}\otimes_{\mathbb{C}}(p^{-1}F\otimes_{\mathbb{C}}E),\mathscr{E}^{(0,\bullet)}_{X^{\mathrm{an}}\times M})$$

$$\simeq \mathrm{Hom}^{\mathrm{top}}_{\mathscr{D}_X}\Big((\mathscr{M}\underset{p^{-1}\mathscr{O}_{X/G}}{\otimes}p^{-1}(\mathscr{D}_{X/G}\otimes\bigwedge\Theta_{X/G}))\otimes((p^{\mathrm{an}})^{-1}F\otimes E),\mathscr{E}^{(0,\bullet)}_{X^{\mathrm{an}}\times M}\Big),$$

because $\mathscr{M}\otimes_{p^{-1}\mathscr{O}_{X/G}}p^{-1}(\mathscr{D}_{X/G}\otimes\bigwedge^{\bullet}\Theta_{X/G})\to\mathscr{M}$ is a quasi-isomorphism, and $\mathscr{M}\otimes_{p^{-1}\mathscr{O}_{X/G}}p^{-1}(\mathscr{D}_{X/G}\otimes\bigwedge^{\bullet}\Theta_{X/G})$ and $(p^{\mathrm{an}})^{-1}F\otimes E$ are standard complexes. Thus we obtain a morphism in $\mathrm{D}^{\mathrm{b}}(\mathbf{FN}_{G_{\mathbb{R}}})$

$$(5.7.2)\qquad \begin{aligned}&\mathbf{R}\mathrm{Hom}^{\mathrm{top}}_{\mathscr{D}_X}((\mathscr{M}\overset{\mathbf{D}}{\otimes}\mathscr{L})\otimes E,\mathscr{E}^{(0,\bullet)}_{X^{\mathrm{an}}\times M})\\ &\qquad\to\mathbf{R}\mathrm{Hom}^{\mathrm{top}}_{\mathscr{D}_X}(\mathscr{M}\otimes(\mathrm{DR}_X(\mathscr{L})\otimes E),\mathscr{E}^{(0,\bullet)}_{X^{\mathrm{an}}\times M})\end{aligned}$$

for $\mathscr{M}\in\mathrm{D}^{\mathrm{b}}_{\mathrm{cc}}(\mathscr{D}_X,G)$, $\mathscr{L}\in\mathrm{D}^{\mathrm{b}}_{G,\mathrm{hol}}(\mathscr{D}_X)$ and $E\in\mathrm{D}^{\mathrm{b}}_{G_{\mathbb{R}},\mathrm{ctb}}(\mathbb{C}_{X^{\mathrm{an}}\times M})$.

Let us take a sequence $\{V_k\}$ as in (4.2.1). Let $K\in\mathrm{D}^{\mathrm{b}}_{G_{\mathbb{R}},\mathrm{ctb}}(\mathbb{C}_{X^{\mathrm{an}}})$. Setting $M=V_k$, $E=K\boxtimes\mathbb{C}_{V_k}$ in (5.7.2), and then taking the projective limit with respect to k, we obtain

$$(5.7.3)\qquad \begin{aligned}&\mathbf{R}\mathrm{Hom}^{\mathrm{top}}_{\mathscr{D}_X}((\mathscr{M}\overset{\mathbf{D}}{\otimes}\mathscr{L})\otimes K,\mathscr{O}_{X^{\mathrm{an}}})\\ &\qquad\to\mathbf{R}\mathrm{Hom}^{\mathrm{top}}_{\mathscr{D}_X}(\mathscr{M}\otimes(\mathrm{DR}_X(\mathscr{L})\otimes K),\mathscr{O}_{X^{\mathrm{an}}}).\end{aligned}$$

When the action of G is not free, we can also define the morphism (5.7.3) replacing X with $V_k\times X$, and then taking the projective limit with respect to k. Here $\{V_k\}$ is as in (4.5.1). Thus we obtain the following lemma.

Lemma 5.7.1. *Let $\mathscr{M}\in\mathrm{D}^{\mathrm{b}}_{\mathrm{cc}}(\mathscr{D}_X,G)$ and $K\in\mathrm{D}^{\mathrm{b}}_{G_{\mathbb{R}},\mathrm{ctb}}(\mathbb{C}_{X^{\mathrm{an}}})$. Then for any $\mathscr{L}\in\mathrm{D}^{\mathrm{b}}_{G,\mathrm{hol}}(\mathscr{D}_X)$, there exists a canonical morphism in $\mathrm{D}^{\mathrm{b}}(\mathbf{FN}_{G_{\mathbb{R}}})$:*

$$\begin{aligned}&\mathbf{R}\mathrm{Hom}^{\mathrm{top}}_{\mathscr{D}_X}((\mathscr{M}\overset{\mathbf{D}}{\otimes}\mathscr{L})\otimes K,\mathscr{O}_{X^{\mathrm{an}}})\\ &\qquad\to\mathbf{R}\mathrm{Hom}^{\mathrm{top}}_{\mathscr{D}_X}(\mathscr{M}\otimes(\mathrm{DR}_X(\mathscr{L})\otimes K),\mathscr{O}_{X^{\mathrm{an}}}).\end{aligned}$$

For a coherent $\mathscr{D}_X$-module $\mathscr{N}$, let us denote by $\mathrm{Ch}(\mathscr{N})\subset T^*X$ the characteristic variety of $\mathscr{N}$ (see [16]). For a submanifold Y of X, we denote by $T^*_Y X$ the conormal bundle to Y. In particular, $T^*_X X$ is nothing but the zero section of the cotangent bundle T^*X.

Theorem 5.7.2. *Let $\mathscr{M}\in\mathrm{D}^{\mathrm{b}}_{\mathrm{coh}}(\mathscr{D}_X,G)$, $\mathscr{L}\in\mathrm{D}^{\mathrm{b}}_{G,\mathrm{hol}}(\mathscr{D}_X)$. Assume that $\mathscr{M}$ and $\mathscr{L}$ are non-characteristic, i.e.*

$$(5.7.4)\qquad\qquad \mathrm{Ch}(\mathscr{M})\cap\mathrm{Ch}(\mathscr{L})\subset T^*_X X.$$

Then, for any $K\in\mathrm{D}^{\mathrm{b}}_{G_{\mathbb{R}},\mathrm{ctb}}(\mathbb{C}_{X^{\mathrm{an}}})$, we have an isomorphism in $\mathrm{D}^{\mathrm{b}}(\mathbf{FN}_{G_{\mathbb{R}}})$:

$$\mathbf{R}\mathrm{Hom}^{\mathrm{top}}_{\mathscr{D}_X}((\mathscr{M}\overset{\mathbf{D}}{\otimes}\mathscr{L})\otimes K,\mathscr{O}_{X^{\mathrm{an}}})\overset{\sim}{\to}\mathbf{R}\mathrm{Hom}^{\mathrm{top}}_{\mathscr{D}_X}(\mathscr{M}\otimes(\mathrm{DR}_X(\mathscr{L})\otimes K),\mathscr{O}_{X^{\mathrm{an}}}).$$

Proof. It is enough to show the result forgetting the topology and the equivariance. Then this follows from the well-known result

$$\mathbf{R}\mathcal{H}om_{\mathcal{D}_X}(\mathcal{M} \overset{\mathrm{D}}{\otimes} \mathcal{L}, \mathcal{O}_{X^{\mathrm{an}}})$$

$$\overset{\sim}{\longleftarrow} \mathbf{R}\mathcal{H}om_{\mathcal{D}_X}(\mathcal{M}, \mathcal{O}_{X^{\mathrm{an}}}) \otimes_{\mathbb{C}} \mathbf{R}\mathcal{H}om_{\mathcal{D}_X}(\mathcal{L}, \mathcal{O}_{X^{\mathrm{an}}})$$

$$\overset{\sim}{\longrightarrow} \mathbf{R}\mathcal{H}om_{\mathcal{D}_X}(\mathcal{M}, \mathcal{O}_{X^{\mathrm{an}}}) \otimes_{\mathbb{C}} \mathbf{R}\mathcal{H}om_{\mathbb{C}_{X^{\mathrm{an}}}}(\mathrm{DR}_X(\mathcal{L}), \mathbb{C}_{X^{\mathrm{an}}})$$

$$\overset{\sim}{\longrightarrow} \mathbf{R}\mathcal{H}om_{\mathbb{C}}(\mathrm{DR}_X(\mathcal{L}), \mathbf{R}\mathcal{H}om_{\mathcal{D}_X}(\mathcal{M}, \mathcal{O}_{X^{\mathrm{an}}}))$$

$$\overset{\sim}{\longrightarrow} \mathbf{R}\mathcal{H}om_{\mathcal{D}_X}(\mathcal{M} \otimes \mathrm{DR}_X(\mathcal{L}), \mathcal{O}_{X^{\mathrm{an}}}).$$

Here, the first and the third isomorphisms need the non-characteristic condition (see [18]).   Q.E.D.

6 Whitney Functor

6.1 Whitney Functor

In §5, we defined $\mathbf{R}\mathrm{Hom}^{\mathrm{top}}_{\mathcal{D}_X}(\mathcal{M} \otimes K, \mathcal{O}_{X^{\mathrm{an}}})$ as an object of $\mathrm{D}^{\mathrm{b}}(\mathbf{FN}_{G_{\mathbb{R}}})$. In this section, we introduce its C^{∞}-version. We use the Whitney functor developed in Kashiwara-Schapira [20].

Theorem 6.1.1 ([20]). *Let M be a real analytic manifold. Then there exists an exact functor*

$$\bullet \overset{\mathrm{w}}{\otimes} \mathscr{C}^{\infty}_M : \mathrm{Mod}_{\mathbb{R}\text{-}c}(\mathbb{C}_M) \to \mathrm{Mod}(\mathcal{D}_M).$$

Moreover, for any $F \in \mathrm{Mod}_{\mathbb{R}\text{-}c}(\mathbb{C}_M)$, $\Gamma(M; F\overset{\mathrm{w}}{\otimes}\mathscr{C}^{\infty}_M)$ is endowed with a Fréchet nuclear topology, and

$$\Gamma(M; \bullet \overset{\mathrm{w}}{\otimes} \mathscr{C}^{\infty}_M) : \mathrm{Mod}_{\mathbb{R}\text{-}c}(\mathbb{C}_M) \to \mathbf{FN}$$

is an exact functor.

Remark 6.1.2.   (i) For a subanalytic open subset U, $\Gamma(M; \mathbb{C}_U \overset{\mathrm{w}}{\otimes} \mathscr{C}^{\infty}_M)$ is the set of C^{∞}-functions f defined on M such that all the derivatives of f vanish at any point outside U. Its topology is the induced topology of $C^{\infty}(M)$.

(ii) For a closed real analytic submanifold N of M, the sheaf $\mathbb{C}_N \overset{\mathrm{w}}{\otimes} \mathscr{C}^{\infty}_M$ is isomorphic to the completion $\varprojlim_n \mathscr{C}^{\infty}_M/I^n$, where I is the ideal of $\mathscr{C}^{\infty}_M$ consisting of C^{∞}-functions vanishing on N.

(iii) In this paper, the Whitney functor is used only for the purpose of the construction of the morphism in Proposition 6.3.2. However, with this functor and $\mathcal{T}hom$ (see [20]), we can construct the C^{∞}-globalization and the distribution globalization of Harish-Chandra modules.

Hence we can define the functor

$$\bullet \overset{\mathrm{w}}{\otimes} \mathscr{C}_M^\infty : \mathrm{D}_{\mathbb{R}\text{-c}}^{\mathrm{b}}(\mathbb{C}_M) \to \mathrm{D}^{\mathrm{b}}(\mathscr{D}_M),$$

$$\mathbf{R}\Gamma^{\mathrm{top}}(M; \bullet \overset{\mathrm{w}}{\otimes} \mathscr{C}_M^\infty) : \mathrm{D}_{\mathbb{R}\text{-c}}^{\mathrm{b}}(\mathbb{C}_M) \to \mathrm{D}^{\mathrm{b}}(\mathbf{FN}).$$

For any $F \in \mathrm{Mod}_{\mathbb{R}\text{-c}}(\mathbb{C}_M)$, we have a morphism

$$F \overset{\mathrm{w}}{\otimes} \mathscr{C}_M^\infty \to \mathscr{H}om_\mathbb{C}(\mathscr{H}om_\mathbb{C}(F, \mathbb{C}_M), \mathscr{C}_M^\infty),$$

which induces a morphism in $\mathrm{D}^{\mathrm{b}}(\mathbf{FN})$

$$(6.1.1) \qquad \mathbf{R}\Gamma^{\mathrm{top}}(M; F \overset{\mathrm{w}}{\otimes} \mathscr{C}_M^\infty) \longrightarrow \mathbf{R}\mathrm{Hom}_\mathbb{C}^{\mathrm{top}}(F^*, \mathscr{C}_M^\infty)$$

for $F \in \mathrm{D}_{\mathbb{R}\text{-c}}^{\mathrm{b}}(\mathbb{C}_M)$, where $F^* := \mathbf{R}\mathscr{H}om(F, \mathbb{C}_M)$.

If a real Lie group H acts on M, we can define

$$\Gamma^{\mathrm{top}}(M; \bullet \overset{\mathrm{w}}{\otimes} \mathscr{C}_M^\infty) : \mathrm{Mod}_{H, \mathbb{R}\text{-c}}(\mathbb{C}_M) \to \mathbf{FN}_H.$$

Note that, for a complex manifold X and $F \in \mathrm{D}_{\mathbb{R}\text{-c}}^{\mathrm{b}}(\mathbb{C}_X)$, $F \overset{\mathrm{w}}{\otimes} \mathscr{O}_X \in \mathrm{D}^{\mathrm{b}}(\mathscr{D}_X)$ is defined as $F \overset{\mathrm{w}}{\otimes} \mathscr{E}_{X^{\mathrm{an}}}^{(0, \bullet)}$.

6.2 The Functor $\mathbf{R}\mathrm{Hom}_{\mathscr{D}_X}^{\mathrm{top}}(\bullet, \bullet \overset{\mathrm{w}}{\otimes} \mathscr{O}_{X^{\mathrm{an}}})$

Let X, G, $G_\mathbb{R}$ be as in §5.1.

For $\mathscr{M} \in \mathrm{D}_{\mathrm{cc}}^{\mathrm{b}}(\mathscr{D}_X, G)$, $F \in \mathrm{D}_{G_\mathbb{R}, \mathbb{R}\text{-c}}^{\mathrm{b}}(\mathbb{C}_X)$, let us define $\mathbf{R}\mathrm{Hom}_{\mathscr{D}_X}^{\mathrm{top}}(\mathscr{M}, F \overset{\mathrm{w}}{\otimes} \mathscr{O}_{X^{\mathrm{an}}})$ as an object of $\mathrm{D}^{\mathrm{b}}(\mathbf{FN}_{G_\mathbb{R}})$, which is isomorphic to $\mathbf{R}\mathrm{Hom}_{\mathscr{D}_X}(\mathscr{M}, F \overset{\mathrm{w}}{\otimes} \mathscr{O}_{X^{\mathrm{an}}})$ forgetting the topology and the equivariance. The construction is similar to the one in §5.

Let M be a $G_\mathbb{R}$-manifold with a free $G_\mathbb{R}$-action. For $\mathscr{M} \in \mathrm{Mod}_{\mathrm{cc}}(\mathscr{D}_X, G)$ and $F \in \mathrm{Mod}_{G_\mathbb{R}, \mathbb{R}\text{-c}}(\mathbb{C}_{X^{\mathrm{an}} \times M})$, we endow $\mathrm{Hom}_{\mathscr{D}_X}(\mathscr{M}, F \overset{\mathrm{w}}{\otimes} \mathscr{E}_{X^{\mathrm{an}} \times M}^{(0, p)})$ with a Fréchet nuclear $G_\mathbb{R}$-module structure as in Lemma 5.3.1. Hence, for $\mathscr{M} \in \mathrm{K}^-(\mathrm{Mod}_{\mathrm{cc}}(\mathscr{D}_X, G))$ and $F \in \mathrm{K}^-(\mathrm{Mod}_{G_\mathbb{R}, \mathbb{R}\text{-c}}(\mathbb{C}_{X^{\mathrm{an}} \times M}))$, we can regard the complex $\mathrm{Hom}_{\mathscr{D}_X}(\mathscr{M}, F \overset{\mathrm{w}}{\otimes} \mathscr{E}_{X^{\mathrm{an}} \times M}^{(0, \bullet)})$ as an object of $\mathrm{D}^{\mathrm{b}}(\mathbf{FN}_{G_\mathbb{R}})$. Taking the inductive limit with respect to $\mathscr{M}$, we obtain $\mathbf{R}\mathrm{Hom}_{\mathscr{D}_X}^{\mathrm{top}}(\mathscr{M}, F \overset{\mathrm{w}}{\otimes} \mathscr{E}_{X^{\mathrm{an}} \times M}^{(0, \bullet)}) \in \mathrm{D}^{\mathrm{b}}(\mathbf{FN}_{G_\mathbb{R}})$ for $\mathscr{M} \in \mathrm{D}_{\mathrm{cc}}^{\mathrm{b}}(\mathscr{D}_X, G)$ and $F \in \mathrm{D}_{G_\mathbb{R}, \mathbb{R}\text{-c}}^{\mathrm{b}}(\mathbb{C}_{X^{\mathrm{an}} \times M})$.

Let us take a sequence $\{V_k\}$ as in (4.2.1). Let $\mathscr{M} \in \mathrm{D}_{\mathrm{cc}}^{\mathrm{b}}(\mathscr{D}_X, G)$ and $F \in \mathrm{D}_{G_\mathbb{R}, \mathbb{R}\text{-c}}^{\mathrm{b}}(\mathbb{C}_{X^{\mathrm{an}}})$. Forgetting the topology and the equivariance, we have

$$\mathbf{R}\mathrm{Hom}_{\mathscr{D}_X}^{\mathrm{top}}(\mathscr{M}, (F \boxtimes \mathbb{C}_{V_k}) \overset{\mathrm{w}}{\otimes} \mathscr{E}_{X^{\mathrm{an}} \times V_k}^{(0, \bullet)})$$

$$\simeq \mathbf{R}\mathrm{Hom}_{\mathscr{D}_X}(\mathscr{M}, F \overset{\mathrm{w}}{\otimes} \mathscr{O}_{X^{\mathrm{an}}}) \otimes \mathbf{R}\Gamma(V_k; \mathbb{C}) \quad \text{in } \mathrm{D}^{\mathrm{b}}(\mathbb{C}).$$

As in Definition 5.4.8, we define

$$\mathbf{R}\mathrm{Hom}^{\mathrm{top}}_{\mathscr{D}_X}(\mathscr{M}, F \overset{\mathrm{w}}{\otimes} \mathscr{O}_{X^{\mathrm{an}}}) = \tau^{\leqslant a}\,\mathbf{R}\mathrm{Hom}^{\mathrm{top}}_{\mathscr{D}_X}(\mathscr{M}, (F \boxtimes \mathbb{C}_{V_k}) \overset{\mathrm{w}}{\otimes} \mathscr{E}^{(0,\bullet)}_{X^{\mathrm{an}} \times V_k})$$

for $k \gg a \gg 0$.

Thus we have defined the functor

$$(6.2.1) \qquad \begin{aligned} &\mathbf{R}\mathrm{Hom}^{\mathrm{top}}_{\mathscr{D}_X}(\bullet, \bullet \overset{\mathrm{w}}{\otimes} \mathscr{O}_{X^{\mathrm{an}}}) \\ &\qquad : \mathrm{D}^{\mathrm{b}}_{\mathrm{cc}}(\mathscr{D}_X, G)^{\mathrm{op}} \times \mathrm{D}^{\mathrm{b}}_{G_{\mathbb{R}}, \mathbb{R}\text{-}\mathrm{c}}(\mathbb{C}_{X^{\mathrm{an}}}) \to \mathrm{D}^{\mathrm{b}}(\mathbf{FN}_{G_{\mathbb{R}}}). \end{aligned}$$

By (6.1.1), we have a morphism

$$(6.2.2) \qquad \mathbf{R}\mathrm{Hom}^{\mathrm{top}}_{\mathscr{D}_X}(\mathscr{M}, F \overset{\mathrm{w}}{\otimes} \mathscr{O}_{X^{\mathrm{an}}}) \to \mathbf{R}\mathrm{Hom}^{\mathrm{top}}_{\mathscr{D}_X}(\mathscr{M} \otimes F^*, \mathscr{O}_{X^{\mathrm{an}}})$$

in $\mathrm{D}^{\mathrm{b}}(\mathbf{FN}_{G_{\mathbb{R}}})$.

6.3 Elliptic Case

Let $X_{\mathbb{R}}$ be a closed real analytic submanifold of X^{an} invariant by $G_{\mathbb{R}}$. Let $i\colon X_{\mathbb{R}} \hookrightarrow X^{\mathrm{an}}$ be the inclusion.

Assume that $T_x X \cong \mathbb{C} \otimes_{\mathbb{R}} T_x X_{\mathbb{R}}$ for any $x \in X_{\mathbb{R}}$.

In Lemma 5.3.2, we define $\mathrm{Hom}^{\mathrm{top}}_{\mathscr{D}_X}(\mathscr{M}, \mathscr{C}^{\infty}_{X_{\mathbb{R}}}) \in \mathbf{FN}_{G_{\mathbb{R}}}$ for a countably coherent quasi-G-equivariant $\mathscr{D}_X$-module $\mathscr{M}$. It is right derivable and we can define the functor

$$\mathbf{R}\mathrm{Hom}^{\mathrm{top}}_{\mathscr{D}_X}(\bullet, \mathscr{C}^{\infty}_{X_{\mathbb{R}}}) : \mathrm{D}^{\mathrm{b}}_{\mathrm{cc}}(\mathscr{D}_X, G)^{\mathrm{op}} \to \mathrm{D}^{\mathrm{b}}(\mathbf{FN}_{G_{\mathbb{R}}}).$$

Proposition 6.3.1. *For* $\mathscr{M} \in \mathrm{D}^{\mathrm{b}}_{\mathrm{cc}}(\mathscr{D}_X, G)$, *we have*

$$\mathbf{R}\mathrm{Hom}^{\mathrm{top}}_{\mathscr{D}_X}(\mathscr{M}, \mathscr{C}^{\infty}_{X_{\mathbb{R}}}) \simeq \mathbf{R}\mathrm{Hom}^{\mathrm{top}}_{\mathscr{D}_X}(\mathscr{M}, i_* \mathbb{C}_{X_{\mathbb{R}}} \overset{\mathrm{w}}{\otimes} \mathscr{O}_{X^{\mathrm{an}}}).$$

Proof. Let $\{V_k\}$ be as in the preceding section. The restriction map

$$(i_* \mathbb{C}_{X_{\mathbb{R}}} \boxtimes \mathbb{C}_{V_k}) \overset{\mathrm{w}}{\otimes} \mathscr{E}^{(0,\bullet)}_{X^{\mathrm{an}} \times V_k} \to \mathscr{E}^{(0,\bullet)}_{X_{\mathbb{R}} \times V_k}$$

induces $\mathrm{Hom}^{\mathrm{top}}_{\mathscr{D}_X}(\mathscr{M}, (i_* \mathbb{C}_{X_{\mathbb{R}}} \boxtimes \mathbb{C}_{V_k}) \overset{\mathrm{w}}{\otimes} \mathscr{E}^{(0,\bullet)}_{X^{\mathrm{an}} \times V_k}) \to \mathrm{Hom}^{\mathrm{top}}_{\mathscr{D}_X}(\mathscr{M}, \mathscr{E}^{(0,\bullet)}_{X_{\mathbb{R}} \times V_k})$ in $\mathrm{D}^{\mathrm{b}}(\mathbf{FN}_{G_{\mathbb{R}}})$. It induces a morphism

$$\mathbf{R}\mathrm{Hom}^{\mathrm{top}}_{\mathscr{D}_X}(\mathscr{M}, (i_* \mathbb{C}_{X_{\mathbb{R}}} \boxtimes \mathbb{C}_{V_k}) \overset{\mathrm{w}}{\otimes} \mathscr{E}^{(0,\bullet)}_{X^{\mathrm{an}} \times V_k}) \to \mathbf{R}\mathrm{Hom}^{\mathrm{top}}_{\mathscr{D}_X}(\mathscr{M}, \mathscr{E}^{(0,\bullet)}_{X_{\mathbb{R}} \times V_k}).$$

Taking the projective limit with respect to k, we obtain

$$\mathbf{R}\mathrm{Hom}^{\mathrm{top}}_{\mathscr{D}_X}(\mathscr{M}, i_* \mathbb{C}_{X_{\mathbb{R}}} \overset{\mathrm{w}}{\otimes} \mathscr{O}_{X^{\mathrm{an}}}) \to \mathbf{R}\mathrm{Hom}^{\mathrm{top}}_{\mathscr{D}_X}(\mathscr{M}, \mathscr{C}^{\infty}_{X_{\mathbb{R}}}).$$

Forgetting the topology and the equivariance, it is an isomorphism since $i_* \mathbb{C}_{X_{\mathbb{R}}} \overset{\mathrm{w}}{\otimes} \mathscr{O}_{X^{\mathrm{an}}} \simeq \mathscr{C}^{\infty}_{X_{\mathbb{R}}}$ (see [20]).   Q.E.D.

Proposition 6.3.2. *There exists a canonical morphism in* $\mathrm{D}^b(\mathbf{FN}_{G_\mathbb{R}})$:

$$\mathbf{R}\mathrm{Hom}^{\mathrm{top}}_{\mathscr{D}_X}(\mathscr{M},\mathscr{C}^\infty_{X_\mathbb{R}}) \longrightarrow \mathbf{R}\mathrm{Hom}^{\mathrm{top}}_{\mathscr{D}_X}(\mathscr{M}\otimes i_*i^!\mathbb{C}_{X^{\mathrm{an}}},\mathscr{O}_{X^{\mathrm{an}}})$$

for $\mathscr{M}\in \mathrm{D}^b_{cc}(\mathscr{D}_X,G)$.

Proof. This follows from the preceding proposition, $(i_*\mathbb{C}_{X_\mathbb{R}})^* \simeq i_*i^!\mathbb{C}_{X^{\mathrm{an}}}$ and (6.2.2).   Q.E.D.

Proposition 6.3.3. *Let us assume that* $\mathscr{M}\in \mathrm{D}^b_{\mathrm{coh}}(\mathscr{D}_X,G)$ *is elliptic i.e.* $\mathrm{Ch}(\mathscr{M})\cap T^*_{X_\mathbb{R}}X \subset T^*_XX$ *(cf. e.g. [16]). Then we have*

$$\mathbf{R}\mathrm{Hom}^{\mathrm{top}}_{\mathscr{D}_X}(\mathscr{M},\mathscr{C}^\infty_{X_\mathbb{R}}) \xrightarrow{\ \sim\ } \mathbf{R}\mathrm{Hom}^{\mathrm{top}}_{\mathscr{D}_X}(\mathscr{M}\otimes i_*i^!\mathbb{C}_{X^{\mathrm{an}}},\mathscr{O}_{X^{\mathrm{an}}})$$

in $\mathrm{D}^b(\mathbf{FN}_{G_\mathbb{R}})$.

Proof. Let $\mathscr{B}_{X_\mathbb{R}} = \mathbf{R}\mathscr{H}om_\mathbb{C}(i_*i^!\mathbb{C}_{X^{\mathrm{an}}},\mathscr{O}_{X^{\mathrm{an}}})$ be the sheaf of hyperfunctions on $X_\mathbb{R}$. Forgetting the topology and the equivariance, we have isomorphisms:

$$\mathbf{R}\mathrm{Hom}_{\mathscr{D}_X}(\mathscr{M}\otimes i_*i^!\mathbb{C}_{X^{\mathrm{an}}},\mathscr{O}_{X^{\mathrm{an}}}) \simeq \mathbf{R}\mathrm{Hom}_{\mathscr{D}_X}(\mathscr{M},\mathscr{B}_{X_\mathbb{R}})$$
$$\simeq \mathbf{R}\mathrm{Hom}_{\mathscr{D}_X}(\mathscr{M},\mathscr{C}^\infty_{X_\mathbb{R}}).$$

Here, the last isomorphism follows from the ellipticity of $\mathscr{M}$. Hence we obtain the desired result.   Q.E.D.

7 Twisted Sheaves

7.1 Twisting Data

If we deal with the non-integral infinitesimal character case in the representation theory by a geometric method, we need to *twist* sheaves. In this note, we shall not go into a systematic study of twisted sheaves, but introduce it here in an ad hoc manner by using the notion of twisting data. (See [15] for more details.)

A *twisting data* τ (for twisting sheaves) over a topological space X is a triple $(X_0 \xrightarrow{\pi} X, L, m)$. Here $\pi\colon X_0 \to X$ is a continuous map admitting a section locally on X, L is an invertible $\mathbb{C}_{X_0\times_X X_0}$-module and m is an isomorphism

$$m\colon p_{12}^{-1}L \otimes p_{23}^{-1}L \xrightarrow{\ \sim\ } p_{23}^{-1}L \quad \text{on } X_2.$$

Here and hereafter, we denote by X_n the fiber product of (n+1) copies of X_0 over X, by p_i ($i = 1,2$) the i-th projection from X_1 to X_0, by p_{ij} ($i,j = 1,2,3$)

the (i,j)-th projection from X_2 to X_1, and so on. We assume that the isomorphism m satisfies the associative law: the following diagram of morphisms of sheaves on X_3 is commutative.

(7.1.1)

$$
\begin{array}{ccc}
p_{12}^{-1}L \otimes p_{23}^{-1}L \otimes p_{34}^{-1}L & =\!=\!=\!= & p_{12}^{-1}L \otimes p_{234}^{-1}\big(p_{12}^{-1}L \otimes p_{23}^{-1}L\big) \\
\| & & \downarrow m \\
p_{123}^{-1}\big(p_{12}^{-1}L \otimes p_{23}^{-1}L\big) \otimes p_{34}^{-1}L & & p_{12}^{-1}L \otimes p_{234}^{-1}p_{13}^{-1}L \\
\downarrow m & & \| \\
p_{123}^{-1}p_{13}^{-1}L \otimes p_{34}^{-1}L & & p_{12}^{-1}L \otimes p_{24}^{-1}L \\
\| & & \| \\
p_{13}^{-1}L \otimes p_{34}^{-1}L & & p_{124}^{-1}\big(p_{12}^{-1}L \otimes p_{23}^{-1}L\big) \\
\| & & \downarrow m \\
p_{134}^{-1}\big(p_{12}^{-1}L \otimes p_{23}^{-1}L\big) & & p_{124}^{-1}p_{13}^{-1}L \\
\downarrow m & & \| \\
p_{134}^{-1}p_{13}^{-1}L & =\!=\!=\!= & p_{14}^{-1}L .
\end{array}
$$

In other words, for $(x_1, x_2, x_3) \in X_0 \times_X X_0 \times_X X_0$, an isomorphism

$$
m(x_1, x_2, x_3)\colon L_{(x_1, x_2)} \otimes L_{(x_2, x_3)} \xrightarrow{\;\sim\;} L_{(x_1, x_3)}
$$

is given in a locally constant manner in (x_1, x_2, x_3) such that the diagram

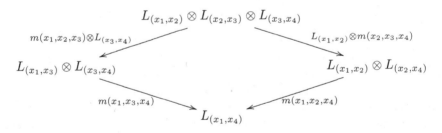

is commutative for $(x_1, x_2, x_3, x_4) \in X_0 \times_X X_0 \times_X X_0 \times_X X_0$.

In particular, we have $i^{-1}L \cong \mathbb{C}_{X_0}$, where $i\colon X_0 \hookrightarrow X_1$ is the diagonal embedding. Indeed, for $x' \in X_0$, $m(x', x', x')$ gives $L_{(x', x')} \otimes L_{(x', x')} \xrightarrow{\;\sim\;} L_{(x', x')}$ and hence an isomorphism $L_{(x', x')} \xrightarrow{\;\sim\;} \mathbb{C}$.

7.2 Twisted Sheaf

Let $\tau = (X_0 \xrightarrow{\pi} X, L, m)$ be a twisting data on X. A *twisted sheaf* F on X with twist τ (or simply τ-twisted sheaf) is a sheaf F on X_0 equipped with an isomorphism $\beta\colon L\otimes p_2^{-1}F \xrightarrow{\sim} p_1^{-1}F$ such that we have a commutative diagram on X_2

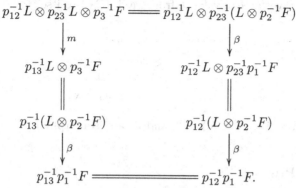

$$p_{12}^{-1}L \otimes p_{23}^{-1}L \otimes p_3^{-1}F \;=\!=\!=\; p_{12}^{-1}L \otimes p_{23}^{-1}(L \otimes p_2^{-1}F)$$

In particular, F is locally constant on each fiber of π. We can similarly define a twisted sheaf on an open subset U of X. Let $\mathrm{Mod}_\tau(\mathbb{C}_U)$ denote the category of τ-twisted sheaves on U. Then $\mathfrak{Mod}_\tau(\mathbb{C}_X)\colon U \mapsto \mathrm{Mod}_\tau(\mathbb{C}_U)$ is a stack (a sheaf of categories) on X (see e.g. [19]).

If $\pi\colon X_0 \to X$ admits a section $s\colon X \to X_0$, then the category $\mathrm{Mod}_\tau(\mathbb{C}_X)$ is equivalent to the category $\mathrm{Mod}(\mathbb{C}_X)$ of sheaves on X. Indeed, the functor $\mathrm{Mod}_\tau(\mathbb{C}_X) \to \mathrm{Mod}(\mathbb{C}_X)$ is given by $F \mapsto s^{-1}F$ and the quasi-inverse is given by $G \mapsto \tilde{s}^{-1}L \otimes \pi^{-1}G$, where $\tilde{s}$ is the map $x' \mapsto \big(x', s\pi(x')\big)$ from X_0 to X_1. Hence the stack $\mathfrak{Mod}_\tau(\mathbb{C}_X)$ is locally equivalent to the stack $\mathfrak{Mod}(\mathbb{C}_X)$ of sheaves on X. Conversely, a stack locally equivalent to the stack $\mathfrak{Mod}(\mathbb{C}_X)$ is equivalent to $\mathfrak{Mod}_\tau(\mathbb{C}_X)$ for some twisting data τ (see [15]).

Let τ_{triv} be the twisting data $(X \xrightarrow{\mathrm{id}} X, \mathbb{C}_X)$. Then $\mathrm{Mod}_{\tau_{\mathrm{triv}}}(\mathbb{C}_X)$ is equivalent to $\mathrm{Mod}(\mathbb{C}_X)$.

For a twisting data τ on X, we denote by $\mathrm{D}_\tau^{\mathrm{b}}(\mathbb{C}_X)$ the bounded derived category $\mathrm{D}^{\mathrm{b}}(\mathrm{Mod}_\tau(\mathbb{C}_X))$.

7.3 Morphism of Twisting Data

Let $\tau = (X_0 \xrightarrow{\pi} X, L, m)$ and $\tau' = (X_0' \xrightarrow{\pi'} X, L', m')$ be two twisting data. A morphism from τ to τ' is a pair $u = (f, \varphi)$ of a map $f\colon X_0 \to X_0'$ over X and an isomorphism $\varphi\colon L \xrightarrow{\sim} f_1^{-1}L'$ compatible with m and m'. Here f_1 is the map $f \underset{X}{\times} f\colon X_0 \underset{X}{\times} X_0 \to X_0' \underset{X}{\times} X_0'$. One can easily see that a morphism $u\colon \tau \to \tau'$ gives an equivalence of categories $u^*\colon \mathrm{Mod}_{\tau'}(\mathbb{C}_X) \xrightarrow{\sim} \mathrm{Mod}_\tau(\mathbb{C}_X)$ by $F \mapsto f^{-1}F$. Hence we say that twisting data τ and τ' are equivalent in this case.

Let us discuss briefly what happens if there are two morphisms $u = (f, \varphi)$ and $u' = (f', \varphi')$ from τ to τ'. Let $g\colon X_0 \to X_0' \underset{X}{\times} X_0'$ be the map $x' \mapsto \big(f(x'), f'(x')\big)$. Then the invertible sheaf $K' = g^{-1}L'$ on X_0 satisfies $p_1^{-1}K' \cong p_2^{-1}K'$, and there exists an invertible sheaf K on X such that

$$\pi^{-1}K \cong g^{-1}L'.$$

Then, $\bullet \otimes K$ gives an equivalence from $\mathrm{Mod}_\tau(\mathbb{C}_X)$ to itself, and the diagram

$$\mathrm{Mod}_{\tau'}(\mathbb{C}_X) \xrightarrow{u'^*} \mathrm{Mod}_\tau(\mathbb{C}_X)$$

quasi-commutes (i.e. $(\bullet \otimes K) \circ u'^*$ and u^* are isomorphic).

7.4 Tensor Product

Let $\tau' = (X_0' \to X, L', m')$ and $\tau'' = (X_0'' \to X, L'', m'')$ be two twisting data on X. Then their tensor product $\tau' \otimes \tau''$ is defined as follows: $\tau' \otimes \tau'' = (X_0 \to X, L, m)$, where $X_0 = X_0' \underset{X}{\times} X_0''$, $L = q_1^{-1}L' \otimes q_2^{-1}L''$ with the projections $q_1 \colon X_1 \simeq X_1' \times X_1'' \to X_1'$ and $q_2 \colon X_1 \to X_1''$, and $m = m' \otimes m''$. Then we can define the bi-functor

$$(7.4.1) \qquad \bullet \otimes \bullet \colon \mathfrak{Mod}_{\tau'}(\mathbb{C}_X) \times \mathfrak{Mod}_{\tau''}(\mathbb{C}_X) \to \mathfrak{Mod}_{\tau' \otimes \tau''}(\mathbb{C}_X)$$

by $(F', F'') \mapsto r_1^{-1}F' \otimes r_2^{-1}F''$, where $r_1 \colon X_0 \to X_0'$ and $r_2 \colon X_0 \to X_0''$ are the projections.

For a twisting data $\tau = (X_0 \to X, L, m)$, let $\tau^{\otimes -1}$ be the twisting data $\tau^{\otimes -1} := (X_0 \to X, L^{\otimes -1}, m^{\otimes -1})$. Note that $L^{\otimes -1} \simeq r^{-1}L$, where $r \colon X_1 \to X_1$ is the map $(x', x'') \mapsto (x'', x')$. Then we can easily see that $\tau \otimes \tau^{\otimes -1}$ is canonically equivalent to the trivial twisting data. Hence we obtain

$$\bullet \otimes \bullet \colon \mathfrak{Mod}_\tau(\mathbb{C}_X) \times \mathfrak{Mod}_{\tau^{\otimes -1}}(\mathbb{C}_X) \to \mathfrak{Mod}(\mathbb{C}_X).$$

For twisting data τ and τ', we have a functor

$$(7.4.2) \quad \mathscr{H}om(\bullet, \bullet) \colon \mathfrak{Mod}_\tau(\mathbb{C}_X)^{\mathrm{op}} \times \mathfrak{Mod}_{\tau'}(\mathbb{C}_X) \to \mathfrak{Mod}_{\tau^{\otimes -1} \otimes \tau'}(\mathbb{C}_X).$$

They induce functors:

$$(7.4.3) \quad \begin{aligned} \bullet \otimes \bullet \colon & \ \mathrm{D}^{\mathrm{b}}_\tau(\mathbb{C}_Y) \otimes \mathrm{D}^{\mathrm{b}}_{\tau'}(\mathbb{C}_Y) \to \mathrm{D}^{\mathrm{b}}_{\tau \otimes \tau'}(\mathbb{C}_Y) \quad \text{and} \\ \mathbf{R}\mathscr{H}om(\bullet, \bullet) \colon & \ \mathrm{D}^{\mathrm{b}}_\tau(\mathbb{C}_Y)^{\mathrm{op}} \times \mathrm{D}^{\mathrm{b}}_{\tau'}(\mathbb{C}_Y) \to \mathrm{D}^{\mathrm{b}}_{\tau^{\otimes -1} \otimes \tau'}(\mathbb{C}_Y). \end{aligned}$$

7.5 Inverse and Direct Images

Let $f \colon X \to Y$ be a continuous map and let $\tau = (Y_0 \xrightarrow{\pi} Y, L_Y, m_Y)$ be a twisting data on Y. Then one can define naturally the pull-back $f^*\tau$. This is the twisting data $(X_0 \to X, L_X, m_X)$ on X, where X_0 is the fiber product

$X \times_Y Y_0$, L_X is the inverse image of L_Y by the map $X_1 \to Y_1$ and m_X is the isomorphism induced by m_Y.

Then, similarly to the non-twisted case, we can define

(7.5.1)
$$f^{-1} : \quad \mathrm{Mod}_\tau(\mathbb{C}_Y) \to \mathrm{Mod}_{f^*\tau}(\mathbb{C}_X),$$
$$f_*, f_! : \mathrm{Mod}_{f^*\tau}(\mathbb{C}_X) \to \mathrm{Mod}_\tau(\mathbb{C}_Y).$$

They have right derived functors:

(7.5.2)
$$f^{-1} : \mathrm{D}^b_\tau(\mathbb{C}_Y) \to \mathrm{D}^b_{f^*\tau}(\mathbb{C}_X),$$
$$\mathbf{R}f_*, \mathbf{R}f_! : \mathrm{D}^b_{f^*\tau}(\mathbb{C}_X) \to \mathrm{D}^b_\tau(\mathbb{C}_Y).$$

The functor $Rf_!$ has a right adjoint functor

(7.5.3)
$$f^! : \mathrm{D}^b_\tau(\mathbb{C}_Y) \to \mathrm{D}^b_{f^*\tau}(\mathbb{C}_X).$$

7.6 Twisted Modules

Let $\tau = (X_0 \xrightarrow{\pi} X, L, m)$ be a twisting data on X. Let $\mathscr{A}$ be a sheaf of $\mathbb{C}$-algebras on X. Then we can define the category $\mathrm{Mod}_\tau(\mathscr{A})$ of τ-twisted $\mathscr{A}$-modules. A τ-twisted $\mathscr{A}$-module is a pair (F, β) of a $\pi^{-1}\mathscr{A}$-module F on X_0 and a $p^{-1}\mathscr{A}$-linear isomorphism $\beta \colon L \otimes p_2^{-1}F \xrightarrow{\sim} p_1^{-1}F$ satisfying the chain condition (7.2.1). Here $p \colon X_1 \to X$ is the projection. The stack $\mathfrak{Mod}_\tau(\mathscr{A})$ of τ-twisted $\mathscr{A}$-modules is locally equivalent to the stack $\mathfrak{Mod}(\mathscr{A})$ of $\mathscr{A}$-modules.

7.7 Equivariant Twisting Data

Let G be a Lie group, and let X be a topological G-manifold. A G-equivariant twisting data on X is a twisting data $\tau = (X_0 \xrightarrow{\pi} X, L, m)$ such that X_0 is a G-manifold, π is G-equivariant and L is G-equivariant, as well as m. Let $\mu \colon G \times X \to X$ be the multiplication map and $\mathrm{pr} \colon G \times X \to X$ the projection. Then the two twisting data $\mu^*\tau$ and $\mathrm{pr}^*\tau$ on $G \times X$ are canonically isomorphic. We can then define the G-equivariant derived category $\mathrm{D}^b_{G,\tau}(\mathbb{C}_X)$, similarly to the non-twisted case.

If G acts freely on X, then denoting by $p \colon X \to X/G$ the projection, we can construct the quotient twisting data τ/G on X/G such that $\tau \cong p^*(\tau/G)$, and we have an equivalence

$$\mathrm{D}^b_{G,\tau}(\mathbb{C}_X) \simeq \mathrm{D}^b_{\tau/G}(\mathbb{C}_{X/G}).$$

7.8 Character Local System

In order to construct twisting data, the following notion is sometimes useful.

Let H be a real Lie group. Let $\mu\colon H \times H \to H$ be the multiplication map and $q_j\colon H \times H \to H$ be the j-th projection ($j = 1, 2$). A *character local system* on H is by definition an invertible $\mathbb{C}_H$-module L equipped with an isomorphism $m\colon q_1^{-1}L \otimes q_2^{-1}L \xrightarrow{\sim} \mu^{-1}L$ satisfying the associativity law: denoting by $m(h_1, h_2)\colon L_{h_1} \otimes L_{h_2} \to L_{h_1 h_2}$ the morphism given by m, the following diagram commutes for $h_1, h_2, h_3 \in H$

$$(7.8.1) \quad \begin{array}{ccc} L_{h_1} \otimes L_{h_2} \otimes L_{h_3} & \xrightarrow{\ m(h_1,h_2)\ } & L_{h_1 h_2} \otimes L_{h_3} \\[2pt] {\scriptstyle m(h_2,h_3)} \downarrow & & \downarrow {\scriptstyle m(h_1 h_2, h_3)} \\[2pt] L_{h_1} \otimes L_{h_2 h_3} & \xrightarrow{\ m(h_1, h_2 h_3)\ } & L_{h_1 h_2 h_3}. \end{array}$$

Let $\mathfrak{h}$ be the Lie algebra of H. For $A \in \mathfrak{h}$, let $L_H(A)$ and $R_H(A)$ denote the vector fields on H defined by

$$(7.8.2)\; \big(L_H(A)f\big)(h) = \left.\frac{d}{dt}f(e^{-tA}h)\right|_{t=0} \text{ and } \big(R_H(A)f\big)(h) = \left.\frac{d}{dt}f(he^{tA})\right|_{t=0}.$$

Let us take an H-invariant element λ of $\operatorname{Hom}_{\mathbb{R}}(\mathfrak{h}, \mathbb{C}) \simeq \operatorname{Hom}_{\mathbb{C}}(\mathbb{C} \otimes_{\mathbb{R}} \mathfrak{h}, \mathbb{C})$. Hence λ satisfies $\lambda([\mathfrak{h}, \mathfrak{h}]) = 0$. Let L_λ be the sheaf of functions f on H satisfying $R_H(A)f = \lambda(A)f$ for all $A \in \mathfrak{h}$, or equivalently $L_H(A)f = -\lambda(A)f$ for all $A \in \mathfrak{h}$. Then L_λ is a local system on H of rank one. Regarding $q_1^{-1}L_\lambda$, $q_2^{-1}L_\lambda$ and $\mu^{-1}L_\lambda$ as subsheaves of the sheaf $\mathscr{O}_{H \times H}$ of functions on $H \times H$, the multiplication morphism $\mathscr{O}_{H \times H} \otimes \mathscr{O}_{H \times H} \to \mathscr{O}_{H \times H}$ induces an isomorphism

$$(7.8.3) \qquad m\colon q_1^{-1}L_\lambda \otimes q_2^{-1}L_\lambda \xrightarrow{\sim} \mu^{-1}L_\lambda.$$

With this data, L_λ has a structure of a character local system.

If λ lifts to a character $\chi\colon H \to \mathbb{C}^*$, then L_λ is isomorphic to the trivial character local system $\mathbb{C}_H = L_0$ by $\mathbb{C}_H \xrightarrow{\sim} L_\lambda \subset \mathscr{O}_H$ given by χ.

For $\lambda, \lambda' \in \operatorname{Hom}_H(\mathfrak{h}, \mathbb{C})$, we have

$$(7.8.4) \qquad L_\lambda \otimes L_{\lambda'} \cong L_{\lambda + \lambda'}$$

compatible with m.

7.9 Twisted Equivariance

Let H, λ, L_λ be as in the preceding subsection. Let X be an H-manifold. Let $\mathrm{pr}\colon H \times X \to X$ and $q\colon H \times X \to H$ be the projections and $\mu\colon H \times X \to X$ the multiplication map.

Definition 7.9.1. *An (H, λ)-equivariant sheaf on X is a pair (F, β) where F is a $\mathbb{C}_X$-module and β is an isomorphism*

$$(7.9.1) \qquad \beta \colon q^{-1}L_\lambda \otimes \mathrm{pr}^{-1}F \xrightarrow{\sim} \mu^{-1}F$$

satisfying the following associativity law: letting $\beta(h, x) \colon (L_\lambda)_h \otimes F_x \xrightarrow{\sim} F_{hx}$ be the induced morphism for $(h, x) \in H \times X$, the following diagram commutes for $(h_1, h_2, x) \in H \times H \times X$:

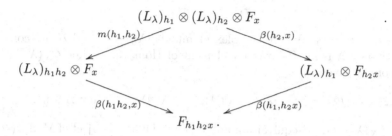

Let us denote by $\mathrm{Mod}_{(H,\lambda)}(\mathbb{C}_X)$ the category of (H, λ)-equivariant sheaves on X. It is an abelian category.

If $\lambda = 0$, then $\mathrm{Mod}_{(H,\lambda)}(\mathbb{C}_X) \simeq \mathrm{Mod}_H(\mathbb{C}_X)$.

For $x \in X$ and $h \in H$, we have a chain of isomorphisms

$$(7.9.2) \qquad F_x \xleftarrow[\beta]{\sim} (L_\lambda)_{h^{-1}} \otimes F_{hx} \xrightarrow{\sim} \mathbb{C} \otimes F_{hx} \simeq F_{hx}.$$

Here $(L_\lambda)_{h^{-1}} \xrightarrow{\sim} \mathbb{C}$ is induced by the evaluation map $(\mathscr{O}_H)_{h^{-1}} \to \mathbb{C}$. Let H_x be the isotropy subgroup at $x \in X$ and $\mathfrak{h}_x$ its Lie algebra. Then, $(7.9.2)$ gives a group homomorphism

$$H_x \to \mathrm{Aut}(F_x).$$

Its infinitesimal representation coincides with $\mathfrak{h}_x \xrightarrow{-\lambda} \mathbb{C} \to \mathrm{End}_\mathbb{C}(F_x)$.

Lemma 7.9.2 ([15]). *Let X be a homogeneous space of H and $x \in X$. Then $\mathrm{Mod}_{(H,\lambda)}(\mathbb{C}_X)$ is equivalent to the category of H_x-modules M such that its infinitesimal representation coincides with $\mathfrak{h}_x \xrightarrow{-\lambda} \mathbb{C} \to \mathrm{End}_\mathbb{C}(M)$.*

7.10 Twisting Data Associated with Principal Bundles

Let $\pi \colon X_0 \to X$ be a principal bundle with a real Lie group H as a structure group. We use the convention that H acts from the left on X_0. Let $\mathfrak{h}$ be the Lie algebra of H and λ an H-invariant element of $\mathrm{Hom}_\mathbb{R}(\mathfrak{h}, \mathbb{C})$. Let L_λ be the corresponding character local character system. Let us identify $X_0 \times_X X_0$ with $H \times X_0$ by the isomorphism $H \times X_0 \xrightarrow{\sim} X_0 \times_X X_0$ given by $(h, x') \mapsto (hx', x')$. Then the projection map $H \times X_0 \to H$ gives $q \colon X_0 \times_X X_0 \to H ((hx', x') \mapsto h)$, and the multiplication isomorphism $(7.8.3)$ induces

$$p_{12}^{-1}(q^{-1}L_\lambda) \otimes p_{23}^{-1}(q^{-1}L_\lambda) \xrightarrow{\sim} p_{13}^{-1}(q^{-1}L_\lambda).$$

Thus $(X_0 \to X, q^{-1}L_\lambda)$ is a twisting data on X. We denote it by τ_λ. By the definition, we have an equivalence of categories:

$$(7.10.1) \qquad\qquad \operatorname{Mod}_{\tau_\lambda}(\mathbb{C}_X) \cong \operatorname{Mod}_{(H,\lambda)}(\mathbb{C}_{X_0}).$$

For $\lambda, \lambda' \in \operatorname{Hom}_H(\mathfrak{h}, \mathbb{C})$, we have

$$\tau_\lambda \otimes \tau_{\lambda'} \cong \tau_{\lambda+\lambda'}.$$

Assume that X, X_0 are complex manifolds, $X_0 \to X$ and H are complex analytic and λ is an H-invariant element of $\operatorname{Hom}_\mathbb{C}(\mathfrak{h}, \mathbb{C})$. Let $\mathscr{O}_X(\lambda)$ be the sheaf on X_0 given by

$$(7.10.2) \quad \mathscr{O}_X(\lambda) = \{\varphi \in \mathscr{O}_{X_0}\,;\, L_X(A)\varphi = -\lambda(A)\varphi \quad \text{for any } A \in \mathfrak{h}\}.$$

Then $\mathscr{O}_X(\lambda)$ is (H,λ)-equivariant and we regard it as an object of $\operatorname{Mod}_{\tau_\lambda}(\mathscr{O}_X)$.

7.11 Twisting (D-module Case)

So far, we discussed the twisting in the topological framework. Now let us investigate the twisting in the D-module framework. This is similar to the topological case. Referring the reader to [15] for treatments in a more general situation, we restrict ourselves to the twisting arising from a principal bundle as in § 7.10.

Let H be a complex affine algebraic group, $\mathfrak{h}$ its Lie algebra and let $R_H, L_H \colon \mathfrak{h} \to \Theta_H$ be the Lie algebra homomorphisms defined by (7.8.2). For $\lambda \in \operatorname{Hom}_H(\mathfrak{h}, \mathbb{C})$, let us define the $\mathscr{D}_H$-module $\mathscr{L}_\lambda = \mathscr{D}_H u_\lambda$ by the defining relation $R_H(A)u_\lambda = \lambda(A)u_\lambda$ for any $A \in \mathfrak{h}$ (which is equivalent to the relation: $L_H(A)u_\lambda = -\lambda(A)u_\lambda$ for any $A \in \mathfrak{h}$). Hence we have $L_\lambda \cong \mathscr{H}om_{\mathscr{D}_H}(\mathscr{L}_\lambda, \mathscr{O}_{H^{\mathrm{an}}})$. Let $\mu \colon H \times H \to H$ be the multiplication morphism. Then we have $\mathscr{D}_{H \times H}$-linear isomorphism

$$(7.11.1) \qquad\qquad m \colon \mathscr{L}_\lambda \overset{\mathbf{D}}{\boxtimes} \mathscr{L}_\lambda \xrightarrow{\sim} \mathbf{D}\mu^* \mathscr{L}_\lambda$$

by $m(u_\lambda \boxtimes u_\lambda) = \mu^*(u_\lambda)$. It satisfies the associative law similar to (7.8.1) (i.e., (7.11.2) with $\mathscr{M} = \mathscr{L}_\lambda$ and $\beta = m$). For $\lambda, \lambda' \in \operatorname{Hom}_H(\mathfrak{h}, \mathbb{C})$, there is an isomorphism

$$\mathscr{L}_\lambda \overset{\mathbf{D}}{\otimes} \mathscr{L}_{\lambda'} \cong \mathscr{L}_{\lambda+\lambda'}$$

that is compatible with m.

Let X be a complex algebraic H-manifold. Then we can define the notion of (H,λ)-equivariant $\mathscr{D}_X$-module as in §7.9. Let us denote by $\mu \colon H \times X \to X$ the multiplication morphism.

Definition 7.11.1. An (H, λ)-equivariant $\mathscr{D}_X$-module is a pair $(\mathscr{M}, \beta)$ where $\mathscr{M}$ is a $\mathscr{D}_X$-module and β is a $\mathscr{D}_{H \times X}$-linear isomorphism

$$\beta \colon \mathscr{L}_\lambda \overset{\mathbf{D}}{\boxtimes} \mathscr{M} \overset{\sim}{\longrightarrow} \mathbf{D}\mu^* \mathscr{M}$$

satisfying the associativity law: the following diagram on $H \times H \times X$ commutes.

(7.11.2)

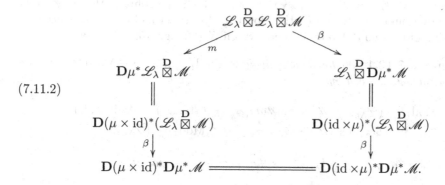

Then the quasi-coherent (H, λ)-equivariant $\mathscr{D}_X$-modules form an abelian category. We denote it by $\operatorname{Mod}_{(H,\lambda)}(\mathscr{D}_X)$.

Note that any (H, λ)-equivariant $\mathscr{D}_X$-module may be regarded as a quasi-H-equivariant $\mathscr{D}_X$-module since $\mathscr{L}_\lambda = \mathscr{O}_H u_\lambda \cong \mathscr{O}_H$ as an $\mathscr{O}_H$-module, and $m(u_\lambda \boxtimes u_\lambda) = \mu^* u_\lambda$. Thus we have a fully faithful exact functor

$$\operatorname{Mod}_{(H,\lambda)}(\mathscr{D}_X) \to \operatorname{Mod}(\mathscr{D}_X, H).$$

Similarly to Lemma 3.1.4, we can prove the following lemma (see [15]).

Lemma 7.11.2. An object $\mathscr{M}$ of $\operatorname{Mod}(\mathscr{D}_X, H)$ is isomorphic to the image of an object of $\operatorname{Mod}_{(H,\lambda)}(\mathscr{D}_X)$ if and only if $\gamma_\mathscr{M} \colon \mathfrak{h} \to \operatorname{End}_{\mathscr{D}_X}(\mathscr{M})$ coincides with the composition $\mathfrak{h} \overset{\lambda}{\to} \mathbb{C} \to \operatorname{End}_{\mathscr{D}_X}(\mathscr{M})$.

Note that for $\lambda, \lambda' \in \operatorname{Hom}_H(\mathfrak{h}, \mathbb{C})$, $\mathscr{L}_\lambda \otimes \mathscr{L}_{\lambda'} \cong \mathscr{L}_{\lambda+\lambda'}$ gives the right exact functor

$$\bullet \overset{\mathbf{D}}{\otimes} \bullet = \bullet \underset{\mathscr{O}_X}{\otimes} \bullet \colon \operatorname{Mod}_{(H,\lambda)}(\mathscr{D}_X) \times \operatorname{Mod}_{(H,\lambda')}(\mathscr{D}_X) \to \operatorname{Mod}_{(H, \lambda+\lambda')}(\mathscr{D}_X).$$

Note that for $\mathscr{M} \in \operatorname{Mod}_{(H,\lambda)}(\mathscr{D}_X)$, the sheaf $\mathscr{H}om_{\mathscr{D}_X}(\mathscr{M}, \mathscr{O}_{X^{\mathrm{an}}})$ is an $(H^{\mathrm{an}}, \lambda)$-equivariant sheaf on X^{an}.

7.12 Ring of Twisted Differential Operators

Let $\pi \colon X_0 \to X$ be a principal H-bundle over X, and $\lambda \in \operatorname{Hom}_H(\mathfrak{h}, \mathbb{C})$. Let $\mathscr{N}_\lambda = \mathscr{D}_{X_0} v_\lambda$ be the $\mathscr{D}_{X_0}$-module defined by the defining relation $L_{X_0}(A) v_\lambda =$

$-\lambda(A)v_\lambda$. Then $\mathscr{N}_\lambda$ is an (H,λ)-equivariant $\mathscr{D}_{X_0}$-module in an evident way. We set

$$\mathscr{D}_{X,\lambda} = \left\{f \in \pi_* \mathscr{E}nd_{\mathscr{D}_{X_0}}(\mathscr{N}_\lambda)\,;\, f \text{ is } H\text{-equivariant}\right\}^{\mathrm{op}}.$$

Here op means the opposite ring. Then $\mathscr{D}_{X,\lambda}$ is a ring on X, and $\mathscr{N}_\lambda$ is a right $\pi^{-1}\mathscr{D}_{X,\lambda}$-module.

If there is a section s of $\pi\colon X_0 \to X$, then the composition $\mathscr{D}_{X,\lambda} \to \mathscr{E}nd_{\mathscr{D}_X^{\mathrm{op}}}(s^*\mathscr{N}_\lambda) = \mathscr{E}nd_{\mathscr{D}_X^{\mathrm{op}}}(\mathscr{D}_X) = \mathscr{D}_X$ is an isomorphism. Hence $\mathscr{D}_{X,\lambda}$ is locally isomorphic to $\mathscr{D}_X$ (with respect to the étale topology), and hence it is a ring of twisted differential operators on X (cf. e.g. [15]). We have

Lemma 7.12.1. *We have an equivalence* $\mathrm{Mod}_{(H,\lambda)}(\mathscr{D}_{X_0}) \cong \mathrm{Mod}(\mathscr{D}_{X,\lambda})$. *The equivalence is given by:*

$$\mathrm{Mod}_{(H,\lambda)}(\mathscr{D}_{X_0}) \ni \widetilde{\mathscr{M}} \mapsto \pi_* \mathscr{H}om_{(\mathscr{D}_{X_0},H)}(\mathscr{N}_\lambda, \widetilde{\mathscr{M}}) \in \mathrm{Mod}(\mathscr{D}_{X,\lambda}) \quad and$$
$$\mathrm{Mod}(\mathscr{D}_{X,\lambda}) \ni \mathscr{M} \mapsto \mathscr{N}_\lambda \otimes_{\mathscr{D}_{X,\lambda}} \mathscr{M} \in \mathrm{Mod}_{(H,\lambda)}(\mathscr{D}_{X_0}).$$

Here $\pi_* \mathscr{H}om_{(\mathscr{D}_{X_0},H)}(\mathscr{N}_\lambda, \widetilde{\mathscr{M}})$ *is the sheaf which associates*

$$\mathrm{Hom}_{\mathrm{Mod}_{(H,\lambda)}(\mathscr{D}_{\pi^{-1}U})}(\mathscr{N}_\lambda|_{\pi^{-1}U}, \widetilde{\mathscr{M}}|_{\pi^{-1}U})$$

to an open set U of X.

Note that $\mathscr{O}_{X^{\mathrm{an}}}(\lambda) \cong \mathscr{H}om_{\mathscr{D}_{X_0}}(\mathscr{N}_\lambda, \mathscr{O}_{X_0^{\mathrm{an}}})$ is an $(H^{\mathrm{an}},\lambda)$-equivariant sheaf and it may be regarded as a τ_λ-twisted $\mathscr{D}_{X^{\mathrm{an}},\lambda}$-module:

$$\mathscr{O}_{X^{\mathrm{an}}}(\lambda) \in \mathrm{Mod}_{\tau_\lambda}(\mathscr{D}_{X^{\mathrm{an}},\lambda}).$$

The twisted module $\mathscr{O}_{X^{\mathrm{an}}}(\lambda)$ plays the role of $\mathscr{O}_{X^{\mathrm{an}}}$ for $\mathscr{D}_X$-modules. For example, defining by

$$\mathrm{DR}_X(\mathscr{M}) := \mathbf{R}\mathscr{H}om_{\mathscr{D}_{X^{\mathrm{an}},\lambda}}(\mathscr{O}_{X^{\mathrm{an}}}(\lambda), \mathscr{M}^{\mathrm{an}}) \quad and$$
$$\mathrm{Sol}_X(\mathscr{M}) := \mathbf{R}\mathscr{H}om_{\mathscr{D}_{X,\lambda}}(\mathscr{M}, \mathscr{O}_{X^{\mathrm{an}}}(\lambda)),$$

we obtain the functors

(7.12.1)
$$\mathrm{DR}_X\colon \mathrm{D}^{\mathrm{b}}(\mathscr{D}_{X,\lambda}) \to \mathrm{D}^{\mathrm{b}}_{\tau_{-\lambda}}(\mathbb{C}_{X^{\mathrm{an}}}),$$
$$\mathrm{Sol}_X\colon \mathrm{D}^{\mathrm{b}}(\mathscr{D}_{X,\lambda})^{\mathrm{op}} \to \mathrm{D}^{\mathrm{b}}_{\tau_\lambda}(\mathbb{C}_{X^{\mathrm{an}}}).$$

Note that we have

$$\mathrm{Mod}(\mathscr{D}_{X^{\mathrm{an}},\lambda}) \simeq \mathrm{Mod}_{\tau_{-\lambda}}(\mathscr{D}_{X^{\mathrm{an}}})$$

by $\mathscr{M} \mapsto \mathscr{O}_{X^{\mathrm{an}}}(-\lambda) \otimes \mathscr{M}$.

7.13 Equivariance of Twisted Sheaves and Twisted D-modules

Let $\pi\colon X_0 \to X$ be a principal bundle with an affine group H as a structure group, and let $\lambda \in \operatorname{Hom}_H(\mathfrak{h}, \mathbb{C})$. Assume that an affine group G acts on X_0 and X such that π is G-equivariant and the action of G commutes with the action of H. Then, as we saw in §7.6, we can define the notion of G^{an}-equivariant τ_λ-twisted $\mathbb{C}_{X^{\mathrm{an}}}$-modules, and the equivariant derived category $\mathrm{D}^{\mathrm{b}}_{G^{\mathrm{an}}, \tau_\lambda}(\mathbb{C}_{X^{\mathrm{an}}})$.

Let $\mathfrak{g}$ be the Lie algebra of G. Then, for any $A \in \mathfrak{g}$, $\mathscr{N}_\lambda \ni v_\lambda \mapsto L_{X_0}(A)v_\lambda \in \mathscr{N}_\lambda$ extends to a $\mathscr{D}_{X_0}$-linear endomorphism of $\mathscr{N}_\lambda$ and it gives an element of $\mathscr{D}_{X,\lambda}$. Hence we obtain a Lie algebra homomorphism

$$\mathfrak{g} \to \Gamma(X; \mathscr{D}_{X,\lambda}).$$

We can define the notion of quasi-G-equivariant $\mathscr{D}_{X,\lambda}$-modules and G-equivariant $\mathscr{D}_{X,\lambda}$-modules. Moreover the results in the preceding sections for non-twisted case hold with a suitable modification, and we shall not repeat them. For example, for λ, $\mu \in \operatorname{Hom}_H(\mathfrak{h}, \mathbb{C})$, we have a functor

$$\bullet \overset{\mathrm{D}}{\otimes} \bullet \colon \mathrm{D}^{\mathrm{b}}(\mathscr{D}_{X,\lambda}, G) \times \mathrm{D}^{\mathrm{b}}(\mathscr{D}_{X,\mu}, G) \to \mathrm{D}^{\mathrm{b}}(\mathscr{D}_{X,\lambda+\mu}, G).$$

If $G_{\mathbb{R}}$ is a real Lie group with a Lie group morphism $G_{\mathbb{R}} \to G^{\mathrm{an}}$,

$$\mathbf{R}\operatorname{Hom}^{\mathrm{top}}_{\mathscr{D}_{X,\lambda}}(\mathscr{M} \otimes F, \mathscr{O}_{X^{\mathrm{an}}}(\lambda)) \in \mathrm{D}^{\mathrm{b}}(\mathbf{FN}_{G_{\mathbb{R}}})$$

is well-defined for $\mathscr{M} \in \mathrm{D}^{\mathrm{b}}_{\mathrm{cc}}(\mathscr{D}_{X,\lambda}, G)$ and $F \in \mathrm{D}^{\mathrm{b}}_{G_{\mathbb{R}}, \tau_\lambda, \mathrm{ctb}}(\mathbb{C}_{X^{\mathrm{an}}})$. Note that $\mathscr{H}om_{\mathbb{C}}(F, \mathscr{O}_{X^{\mathrm{an}}}(\lambda)) \in \operatorname{Mod}(\mathscr{D}_{X^{\mathrm{an}}, \lambda})$ because $\mathscr{O}_{X^{\mathrm{an}}}(\lambda) \in \operatorname{Mod}_{\tau_\lambda}(\mathscr{D}_{X^{\mathrm{an}}, \lambda})$.

7.14 Riemann-Hilbert Correspondence

Let $\pi\colon X_0 \to X$, H, G and $\lambda \in \operatorname{Hom}_H(\mathfrak{h}, \mathbb{C})$ be as in the preceding subsection. Assume that λ vanishes on the Lie algebra of the unipotent radical of H. Then $\mathscr{L}_\lambda$ is a regular holonomic $\mathscr{D}_H$-module. Hence we can define the notion of regular holonomic $\mathscr{D}_{X,\lambda}$-module (i.e. a $\mathscr{D}_{X,\lambda}$-module $\mathscr{M}$ is regular holonomic if $\mathscr{N}_\lambda \otimes_{\mathscr{D}_{X,\lambda}} \mathscr{M}$ is a regular holonomic $\mathscr{D}_{X_0}$-module).

Assume that there are finitely many G-orbits in X. Then any coherent holonomic G-equivariant $\mathscr{D}_{X,\lambda}$-module is regular holonomic (see [15]). Hence the Riemann-Hilbert correspondence (see Subsection 4.6) implies the following result.

Theorem 7.14.1. *Assume that λ vanishes on the Lie algebra of the unipotent radical of H. If there are only finitely many G-orbits in X, then the functor*

$$\mathrm{DR}_X := \mathbf{R}\mathscr{H}om_{\mathscr{D}_{X^{\mathrm{an}}, \lambda}}(\mathscr{O}_{X^{\mathrm{an}}}(\lambda), \bullet^{\mathrm{an}})\colon \mathrm{D}^{\mathrm{b}}_{G, \mathrm{coh}}(\mathscr{D}_{X,\lambda}) \to \mathrm{D}^{\mathrm{b}}_{G^{\mathrm{an}}, \tau_{-\lambda}, \mathbb{C}\text{-}c}(\mathbb{C}_{X^{\mathrm{an}}})$$

is an equivalence of triangulated categories.

8 Integral Transforms

8.1 Convolutions

Let X, Y and Z be topological manifolds.

Let us consider a diagram

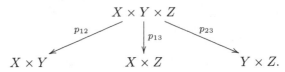

For $F \in D^b(\mathbb{C}_{X \times Y})$ and $G \in D^b(\mathbb{C}_{Y \times Z})$, we define the object $F \circ G$ of $D^b(\mathbb{C}_{X \times Z})$ by

$$(8.1.1) \qquad F \circ G := \mathbf{R}p_{13!}(p_{12}^{-1}F \otimes p_{23}^{-1}G).$$

We call it the *convolution* of F and G.

Hence we obtain the functor

$$\bullet \circ \bullet : D^b(\mathbb{C}_{X \times Y}) \times D^b(\mathbb{C}_{Y \times Z}) \longrightarrow D^b(\mathbb{C}_{X \times Z}).$$

In particular, letting X or Z be $\{pt\}$, we obtain

$$\bullet \circ \bullet : \begin{matrix} D^b(\mathbb{C}_Y) \times D^b(\mathbb{C}_{Y \times Z}) \longrightarrow D^b(\mathbb{C}_Z) \\ D^b(\mathbb{C}_{X \times Y}) \times D^b(\mathbb{C}_Y) \longrightarrow D^b(\mathbb{C}_X). \end{matrix}$$

This functor satisfies the associative law

$$(F \circ G) \circ H \simeq F \circ (G \circ H)$$

for $F \in D^b(\mathbb{C}_{X \times Y})$, $G \in D^b(\mathbb{C}_{Y \times Z})$ and $H \in D^b(\mathbb{C}_{Z \times W})$.

This can be generalized to the twisted case. Let τ_X (resp. τ_Y, τ_Z) be a twisting data on X (resp. Y, Z). Then we have a functor

$$\bullet \circ \bullet : D^b_{\tau_X \boxtimes (\tau_Y)^{\otimes -1}}(\mathbb{C}_{X \times Y}) \times D^b_{\tau_Y \boxtimes (\tau_Z)^{\otimes -1}}(\mathbb{C}_{Y \times Z}) \longrightarrow D^b_{\tau_X \boxtimes (\tau_Z)^{\otimes -1}}(\mathbb{C}_{X \times Z}).$$

Similarly, we can define the convolutions of D-modules. Let X, Y and Z be algebraic manifolds. Then we can define, for $\mathcal{M} \in D^b(\mathcal{D}_{X \times Y})$ and $\mathcal{N} \in D^b(\mathcal{D}_{Y \times Z})$, the object $\mathcal{M} \overset{D}{\circ} \mathcal{N}$ of $D^b(\mathcal{D}_{X \times Z})$ by

$$(8.1.2) \qquad \mathcal{M} \overset{D}{\circ} \mathcal{N} := \mathbf{D}p_{13*}(\mathbf{D}p_{12}^* \mathcal{M} \overset{D}{\otimes} \mathbf{D}p_{23}^* \mathcal{N}).$$

We call it the *convolution* of $\mathcal{M}$ and $\mathcal{N}$.

Hence we obtain the functor

$$\bullet \overset{D}{\circ} \bullet : D^b(\mathcal{D}_{X \times Y}) \times D^b(\mathcal{D}_{Y \times Z}) \longrightarrow D^b(\mathcal{D}_{X \times Z}).$$

If X, Y and Z are quasi-projective G-manifolds, we can define

$$\bullet \overset{D}{\circ} \bullet : D^b(\mathcal{D}_{X \times Y}, G) \times D^b(\mathcal{D}_{Y \times Z}, G) \longrightarrow D^b(\mathcal{D}_{X \times Z}, G).$$

These definitions also extend to the twisted case.

8.2 Integral Transform Formula

Let G be an affine algebraic group, and let $G_\mathbb{R}$ be a real Lie group with a Lie group morphism $G_\mathbb{R} \to G^{\mathrm{an}}$.

Let X be a projective algebraic G-manifold and Y a quasi-projective G-manifold. Let us consider the diagram

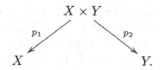

For $\mathscr{M} \in \mathrm{D}^{\mathrm{b}}_{\mathrm{coh}}(\mathscr{D}_X, G)$, $\mathscr{K} \in \mathrm{D}^{\mathrm{b}}_{G,\mathrm{hol}}(\mathscr{D}_{X \times Y})$ and $F \in \mathrm{D}^{\mathrm{b}}_{G_\mathbb{R},\mathrm{ctb}}(\mathbb{C}_{Y^{\mathrm{an}}})$, let us calculate $\mathbf{R}\mathrm{Hom}^{\mathrm{top}}_{\mathscr{D}_Y}((\mathscr{M} \overset{\mathbf{D}}{\circ} \mathscr{K}) \otimes F, \mathscr{O}_{Y^{\mathrm{an}}})$. Note that $\mathscr{M} \overset{\mathbf{D}}{\circ} \mathscr{K} \in \mathrm{D}^{\mathrm{b}}_{\mathrm{coh}}(\mathscr{D}_Y, G)$. We have by Theorem 5.6.1

$$
\begin{aligned}
&\mathbf{R}\mathrm{Hom}^{\mathrm{top}}_{\mathscr{D}_Y}((\mathscr{M} \overset{\mathbf{D}}{\circ} \mathscr{K}) \otimes F, \mathscr{O}_{Y^{\mathrm{an}}}) \\
(8.2.1)\quad &= \mathbf{R}\mathrm{Hom}^{\mathrm{top}}_{\mathscr{D}_Y}(\mathbf{D}p_{2*}(\mathbf{D}p_1^*\mathscr{M} \overset{\mathbf{D}}{\otimes} \mathscr{K}) \otimes F, \mathscr{O}_{Y^{\mathrm{an}}}) \\
&\simeq \mathbf{R}\mathrm{Hom}^{\mathrm{top}}_{\mathscr{D}_{X \times Y}}((\mathbf{D}p_1^*\mathscr{M} \overset{\mathbf{D}}{\otimes} \mathscr{K}) \otimes (p_2^{\mathrm{an}})^{-1}F, \mathscr{O}_{(X \times Y)^{\mathrm{an}}})[\mathrm{d}_X].
\end{aligned}
$$

If we assume the non-characteristic condition:

$$
(\mathrm{Ch}(\mathscr{M}) \times T_Y^*Y) \cap \mathrm{Ch}(\mathscr{K}) \subset T_{X \times Y}^*(X \times Y),
$$

Theorem 5.7.2 implies that

$$
\begin{aligned}
(8.2.2)\quad &\mathbf{R}\mathrm{Hom}^{\mathrm{top}}_{\mathscr{D}_{X \times Y}}((\mathbf{D}p_1^*\mathscr{M} \overset{\mathbf{D}}{\otimes} \mathscr{K}) \otimes (p_2^{\mathrm{an}})^{-1}F, \mathscr{O}_{(X \times Y)^{\mathrm{an}}}) \\
&\simeq \mathbf{R}\mathrm{Hom}^{\mathrm{top}}_{\mathscr{D}_{X \times Y}}(\mathbf{D}p_1^*\mathscr{M} \otimes (K \otimes (p_2^{\mathrm{an}})^{-1}F), \mathscr{O}_{(X \times Y)^{\mathrm{an}}}).
\end{aligned}
$$

Here, $K := \mathrm{DR}_{X \times Y}(\mathscr{K}) \in \mathrm{D}^{\mathrm{b}}_{G^{\mathrm{an}}, \mathbb{C}\text{-c}}(\mathbb{C}_{(X \times Y)^{\mathrm{an}}})$. Then, again by Theorem 5.6.1, we have

$$
\begin{aligned}
&\mathbf{R}\mathrm{Hom}^{\mathrm{top}}_{\mathscr{D}_{X \times Y}}(\mathbf{D}p_1^*\mathscr{M} \otimes (K \otimes (p_2^{\mathrm{an}})^{-1}F), \mathscr{O}_{(X \times Y)^{\mathrm{an}}}) \\
&\simeq \mathbf{R}\mathrm{Hom}^{\mathrm{top}}_{\mathscr{D}_X}(\mathscr{M} \otimes \mathbf{R}(p_1^{\mathrm{an}})_!(K \otimes (p_2^{\mathrm{an}})^{-1}F), \mathscr{O}_{X^{\mathrm{an}}}))[-2\mathrm{d}_Y] \\
&= \mathbf{R}\mathrm{Hom}^{\mathrm{top}}_{\mathscr{D}_X}(\mathscr{M} \otimes (K \circ F), \mathscr{O}_{X^{\mathrm{an}}})[-2\mathrm{d}_Y].
\end{aligned}
$$

Combining this with (8.2.1) and (8.2.2), we obtain the following theorem.

Theorem 8.2.1 (Integral transform formula). *Let G be an affine algebraic group, and let $G_\mathbb{R}$ be a Lie group with a Lie group morphism $G_\mathbb{R} \to G^{\mathrm{an}}$. Let X be a projective G-manifold and Y a quasi-projective G-manifold. Let $\mathscr{M} \in \mathrm{D}^{\mathrm{b}}_{\mathrm{coh}}(\mathscr{D}_X, G)$, $\mathscr{K} \in \mathrm{D}^{\mathrm{b}}_{G,\mathrm{hol}}(\mathscr{D}_{X \times Y})$ and $F \in \mathrm{D}^{\mathrm{b}}_{G_\mathbb{R},\mathrm{ctb}}(\mathbb{C}_Y)$, If the non-characteristic condition*

(8.2.3) $$(\mathrm{Ch}(\mathcal{M}) \times T_Y^*Y) \cap \mathrm{Ch}(\mathcal{K}) \subset T_{X \times Y}^*(X \times Y)$$

is satisfied, then we have an isomorphism in $\mathrm{D}^{\mathrm{b}}(\mathbf{FN}_{G_{\mathbb{R}}})$

(8.2.4) $$\begin{aligned} & \mathrm{R\,Hom}_{\mathscr{D}_Y}^{\mathrm{top}}((\mathcal{M} \overset{\mathrm{D}}{\circ} \mathcal{K}) \otimes F, \mathscr{O}_{Y^{\mathrm{an}}}) \\ & \simeq \mathrm{R\,Hom}_{\mathscr{D}_X}^{\mathrm{top}}(\mathcal{M} \otimes (\mathrm{DR}_{X \times Y}(\mathcal{K}) \circ F), \mathscr{O}_{X^{\mathrm{an}}})[\mathrm{d}_X - 2\mathrm{d}_Y]. \end{aligned}$$

Remark 8.2.2. If G acts transitively on X, then the non-characteristic condition (8.2.3) is always satisfied. Indeed, let $\mu_X : T^*X \to \mathfrak{g}^*$, $\mu_Y : T^*Y \to \mathfrak{g}^*$ and $\mu_{X \times Y} : T^*(X \times Y) \to \mathfrak{g}^*$ be the moment maps. Then we have $\mu_{X \times Y}(\xi, \eta) = \mu_X(\xi) + \mu_Y(\eta)$ for $\xi \in T^*X$ and $\eta \in T^*Y$. Since $\mathcal{K} \in \mathrm{D}_{G,\mathrm{hol}}^{\mathrm{b}}(\mathscr{D}_X)$, we have $\mathrm{Ch}(\mathcal{K}) \subset \mu_{X \times Y}^{-1}(0)$ (see [15]). Hence we have $(T^*X \times T_Y^*Y) \cap \mathrm{Ch}(\mathcal{K}) \subset \mu_X^{-1}(0) \times T_Y^*Y$. Since G acts transitively on X, we have $\mu_X^{-1}(0) = T_X^*X$.

Remark 8.2.3. Although we don't repeat here, there is a twisted version of Theorem 8.2.1.

9 Application to the Representation Theory

9.1 Notations

In this section, we shall apply the machinery developed in the earlier sections to the representation theory of real semisimple Lie groups.

Let $G_{\mathbb{R}}$ be a connected real semisimple Lie group with a finite center, and let $K_{\mathbb{R}}$ be a maximal compact subgroup of $G_{\mathbb{R}}$. Let $\mathfrak{g}_{\mathbb{R}}$ and $\mathfrak{k}_{\mathbb{R}}$ be the Lie algebra of $G_{\mathbb{R}}$ and $K_{\mathbb{R}}$, respectively. Let $\mathfrak{g}$ and $\mathfrak{k}$ be their complexifications. Let K be the complexification of $K_{\mathbb{R}}$. Let G be a connected semisimple algebraic group with the Lie algebra $\mathfrak{g}$, and assume that there is an *injective* morphism $G_{\mathbb{R}} \to G^{\mathrm{an}}$ of real Lie groups which induces the embedding $\mathfrak{g}_{\mathbb{R}} \hookrightarrow \mathfrak{g}$.[4]

Thus we obtain the diagrams:

$$\begin{array}{ccc} K_{\mathbb{R}} \longhookrightarrow K & & \mathfrak{k}_{\mathbb{R}} \longhookrightarrow \mathfrak{k} \\ \Big\uparrow \qquad \Big\uparrow & \text{and} & \Big\downarrow \qquad \Big\downarrow \\ G_{\mathbb{R}} \longhookrightarrow G & & \mathfrak{g}_{\mathbb{R}} \longhookrightarrow \mathfrak{g}. \end{array}$$

Let us take an Iwasawa decomposition

(9.1.1) $$\begin{aligned} G_{\mathbb{R}} &= K_{\mathbb{R}} A_{\mathbb{R}} N_{\mathbb{R}}, \\ \mathfrak{g}_{\mathbb{R}} &= \mathfrak{k}_{\mathbb{R}} \oplus \mathfrak{a}_{\mathbb{R}} \oplus \mathfrak{n}_{\mathbb{R}}. \end{aligned}$$

Let $\mathfrak{a}$, $\mathfrak{n}$ be the complexification of $\mathfrak{a}_{\mathbb{R}}$ and $\mathfrak{n}_{\mathbb{R}}$. Let A and N be the connected closed subgroups of G with Lie algebras $\mathfrak{a}$ and $\mathfrak{n}$, respectively.

---

[4] In this note, we assume that $G_{\mathbb{R}} \to G$ is injective. However, we can remove this condition, by regarding G/K as an orbifold.

Let $M_{\mathbb{R}} = Z_{K_{\mathbb{R}}}(\mathfrak{a}_{\mathbb{R}})$ and $\mathfrak{m}_{\mathbb{R}} = Z_{\mathfrak{k}_{\mathbb{R}}}(\mathfrak{a}_{\mathbb{R}})$. Let M and $\mathfrak{m}$ be the complexification of $M_{\mathbb{R}}$ and $\mathfrak{m}_{\mathbb{R}}$. Then we have $M = Z_K(A)$. Let P be the parabolic subgroup of G with $\mathfrak{m} \oplus \mathfrak{a} \oplus \mathfrak{n}$ as its Lie algebra, and $P_{\mathbb{R}} = M_{\mathbb{R}} A_{\mathbb{R}} N_{\mathbb{R}} \subset G_{\mathbb{R}}$.

Let us fix a Cartan subalgebra $\mathfrak{t}$ of $\mathfrak{g}$ such that

(9.1.2) $$\mathfrak{t} = \mathbb{C} \otimes_{\mathbb{R}} \mathfrak{t}_{\mathbb{R}} \quad \text{where} \quad \mathfrak{t}_{\mathbb{R}} = (\mathfrak{t} \cap \mathfrak{m}_{\mathbb{R}}) \oplus \mathfrak{a}_{\mathbb{R}}.$$

Let T be the maximal torus of G with $\mathfrak{t}$ as its Lie algebra.

We take a Borel subalgebra $\mathfrak{b}$ of $\mathfrak{g}$ containing $\mathfrak{t}$ and $\mathfrak{n}$, and let B be the Borel subgroup with $\mathfrak{b}$ as its Lie algebra.

We have

$$K \cap P = M \quad \text{and} \quad K \cap B = M \cap B, \quad K \cap T = M \cap T,$$

and $M/(M \cap B) \simeq P/B$ is the flag manifold for M.

Let Δ be the root system of $(\mathfrak{g}, \mathfrak{t})$, and take the positive root system $\Delta^+ = \{\alpha \in \Delta \,;\, \mathfrak{g}_\alpha \subset \mathfrak{b}\}$. Let $\Delta_k = \{\alpha \in \Delta \,;\, \mathfrak{g}_\alpha \subset \mathfrak{k}\} = \{\alpha \in \Delta \,;\, \mathfrak{g}_\alpha \subset \mathfrak{m}\} = \{\alpha \in \Delta \,;\, \alpha|_\mathfrak{a} = 0\}$ be the set of compact roots, and set $\Delta_k^+ = \Delta_k \cap \Delta^+$. Let ρ be the half sum of positive roots.

An element λ of $\mathfrak{t}^*$ is called *integral* if it can be lifted to a character of T. We say that $\lambda|_{\mathfrak{t} \cap \mathfrak{k}}$ is integral if it can be lifted to a character of $K \cap T = M \cap T$.

Let $\mathfrak{z}(\mathfrak{g})$ denote the center of the universal enveloping algebra $U(\mathfrak{g})$ of $\mathfrak{g}$. Let $\chi \colon \mathfrak{z}(\mathfrak{g}) \to \mathbb{C}[\mathfrak{t}^*] = S(\mathfrak{t})$ be the ring morphism given by:

$$a - (\chi(a))(\lambda) \in \mathrm{Ker}(\mathfrak{b} \xrightarrow{\lambda} \mathbb{C}) U(\mathfrak{g}) \quad \text{for any } \lambda \in \mathfrak{t}^* \text{ and } a \in \mathfrak{z}(\mathfrak{g}).$$

It means that $a \in \mathfrak{z}(\mathfrak{g})$ acts on the lowest weight module with lowest weight λ through the multiplication by the scalar $(\chi(a))(\lambda)$. For $\lambda \in \mathfrak{t}^*$, let

$$\chi_\lambda \colon \mathfrak{z}(\mathfrak{g}) \to \mathbb{C}$$

be the ring homomorphism given by $\chi_\lambda(a) := (\chi(a))(\lambda)$. Note that

(9.1.3) for $\lambda, \mu \in \mathfrak{t}^*$, $\chi_\lambda = \chi_\mu$ if and only if $w \circ \lambda = \mu$ for some $w \in W$.

Here $w \circ \lambda = w(\lambda - \rho) + \rho$ is the shifted action of the Weyl group W. We set

$$U_\lambda(\mathfrak{g}) = U(\mathfrak{g})/(U(\mathfrak{g}) \,\mathrm{Ker}(\chi_\lambda)).$$

Then $U_\lambda(\mathfrak{g})$-modules are nothing but $\mathfrak{g}$-modules with infinitesimal character χ_λ.

Let X be the flag manifold of G (the set of Borel subgroups of G). Then X is a projective G-manifold and $X \simeq G/B$. For $x \in X$, we set $B(x) = \{g \in G \,;\, gx = x\}$, $\mathfrak{b}(x) = \mathrm{Lie}(B(x))$ the Lie algebra of $B(x)$, and $\mathfrak{n}(x) = [\mathfrak{b}(x), \mathfrak{b}(x)]$ the nilpotent radical of $\mathfrak{b}(x)$. Let $x_0 \in X$ be the point of X such that $\mathfrak{b}(x_0) = \mathfrak{b}$. Then, for any $x \in X$, there exists a unique Lie algebra homomorphism $\mathfrak{b}(x) \to \mathfrak{t}$ which is equal to the composition $\mathfrak{b}(x) \xrightarrow{\mathrm{Ad}(g)} \mathfrak{b} \to \mathfrak{t}$ for any $g \in G$ such that $gx = x_0$.

Let $X_{\min} = G/P$. Let

$$\pi\colon X \to X_{\min}$$

be the canonical projection. We set $x_0^{\min} = \pi(x_0) = e \bmod P$.

Let $\tilde{p}\colon G \to X$ be the G-equivariant projection such that $\tilde{p}(e) = x_0$. Then this is a principal B-bundle. For $\lambda \in \mathfrak{t}^* = (\mathfrak{b}/\mathfrak{n})^* = \operatorname{Hom}_B(\mathfrak{b}, \mathbb{C})$, let $\mathscr{D}_{X,\lambda}$ be the ring of twisted differential operators on X with twist λ. Let τ_λ denote the G^{an}-equivariant twisting data on X^{an} corresponding to λ (see §7.10, 7.12).

Note the following lemma (see [15] and Lemma 7.9.2).

Lemma 9.1.1.   (i) *Let H be a closed algebraic subgroup of G with a Lie algebra $\mathfrak{h} \subset \mathfrak{g}$, Z an H-orbit in X and $x \in Z$. Then the category $\operatorname{Mod}_H(\mathscr{D}_{Z,\lambda})$ of H-equivariant $\mathscr{D}_{Z,\lambda}$-modules is equivalent to the category of $H \cap B(x)$-modules V whose infinitesimal representation coincides with*

$$\mathfrak{h} \cap \mathfrak{b}(x) \to \mathfrak{b}(x) \to \mathfrak{t} \xrightarrow{\lambda} \mathbb{C} \to \operatorname{End}_{\mathbb{C}}(V).$$

(ii) *Let H be a closed real Lie subgroup of G^{an} with a Lie algebra $\mathfrak{h} \subset \mathfrak{g}$, Z an H-orbit in X^{an} and $x \in Z$. Then the category $\operatorname{Mod}_{H,\tau_\lambda}(\mathbb{C}_Z)$ of H-equivariant τ_λ-twisted sheaves on Z is equivalent to the category of $H \cap B(x)$-modules V whose infinitesimal representation coincides with*

$$\mathfrak{h} \cap \mathfrak{b}(x) \to \mathfrak{b}(x) \to \mathfrak{t} \xrightarrow{-\lambda} \mathbb{C} \to \operatorname{End}_{\mathbb{C}}(V).$$

Note that, in the situation of (i), the de Rham functor gives an equivalence

$$\operatorname{Mod}_H(\mathscr{D}_{Z,\lambda}) \xrightarrow{\sim} \operatorname{Mod}_{H^{\mathrm{an}},\tau_{-\lambda}}(\mathbb{C}_{Z^{\mathrm{an}}}).$$

9.2 Beilinson-Bernstein Correspondence

Let us recall a result of Beilinson-Bernstein [1] on the correspondence of $U(\mathfrak{g})$-modules and D-modules on the flag manifold.

For $\alpha \in \Delta$, let $\alpha^\vee \in \mathfrak{t}$ be the corresponding co-root.

Definition 9.2.1. Let $\lambda \in \mathfrak{t}^*$.

(i) We say that λ is *regular* if $\langle \alpha^\vee, \lambda \rangle$ does not vanish for any $\alpha \in \Delta^+$.

(ii) We say that a weight $\lambda \in \mathfrak{t}^*$ is *integrally anti-dominant* if $\langle \alpha^\vee, \lambda \rangle \neq 1, 2, 3, \ldots$ for any $\alpha \in \Delta^+$.

Recall that $\tilde{p}\colon G \to X = G/B$ is the projection. For $\lambda \in \mathfrak{t}^* = \operatorname{Hom}_B(\mathfrak{b}, \mathbb{C})$, we have defined the twisting data τ_λ on X^{an} and the ring of twisted differential operators $\mathscr{D}_{X,\lambda}$. We defined also $\mathscr{O}_{X^{\mathrm{an}}}(\lambda)$. Recall that $\mathscr{O}_{X^{\mathrm{an}}}(\lambda)$ is a twisted $\mathscr{D}_{X^{\mathrm{an}},\lambda}$-module, and it is an object of $\operatorname{Mod}_{\tau_\lambda}(\mathscr{D}_{X^{\mathrm{an}},\lambda})$.

If λ is an integral weight, then the twisting data τ_λ is trivial and $\mathscr{O}_{X^{\mathrm{an}}}(\lambda)$ is the invertible $\mathscr{O}_{X^{\mathrm{an}}}$-module associated with the invertible $\mathscr{O}_X$-module $\mathscr{O}_X(\lambda)$:

$$(9.2.1) \qquad \mathscr{O}_X(\lambda) = \left\{ u \in \tilde{p}_* \mathscr{O}_G \,;\, u(gb^{-1}) = b^\lambda u(g) \text{ for any } b \in B \right\}.$$

Here $B \ni b \mapsto b^\lambda \in \mathbb{C}^*$ is the character of B corresponding to $\lambda \in \operatorname{Hom}_B(\mathfrak{b}, \mathbb{C})$. Note that we have

$$(9.2.2) \qquad \mathscr{D}_{X,\lambda+\mu} = \mathscr{O}_X(\mu) \otimes \mathscr{D}_{X,\lambda} \otimes \mathscr{O}_X(-\mu)$$

for any $\lambda \in \mathfrak{t}$ and any integral $\mu \in \mathfrak{t}$.

If λ is an anti-dominant integral weight (i.e., $\langle \alpha^\vee, \lambda \rangle \in \mathbb{Z}_{\leq 0}$ for any $\alpha \in \Delta^+$), let $V(\lambda)$ be the irreducible G-module with lowest weight λ. Then we have

$$(9.2.3) \qquad \Gamma(X; \mathscr{O}_X(\lambda)) \simeq V(\lambda).$$

Here the isomorphism $V(\lambda) \xrightarrow{\sim} \Gamma(X; \mathscr{O}_X(\lambda))$ is given as follows. Let us fix a highest weight vector $u_{-\lambda}$ of $V(-\lambda) = V(\lambda)^*$. Then, for any $v \in V(\lambda)$, the function $\langle v, gu_{-\lambda} \rangle$ in $g \in G$ is the corresponding global section of $\mathscr{O}_X(\lambda)$.

The following theorem is due to Beilinson-Bernstein ([1]).

Theorem 9.2.2. *Let λ be an element of $\mathfrak{t}^* \cong \mathrm{Hom}_B(\mathfrak{b}, \mathbb{C})$.*

(i) *We have*

$$H^k(X; \mathscr{D}_{X,\lambda}) \simeq \begin{cases} U_\lambda(\mathfrak{g}) & \text{for } k = 0, \\ 0 & \text{otherwise.} \end{cases}$$

(ii) *Assume that $\lambda - \rho$ is integrally anti-dominant. Then we have*
  (a) *for any quasi-coherent $\mathscr{D}_{X,\lambda}$-module $\mathscr{M}$, we have*

$$H^n(X; \mathscr{M}) = 0 \quad \text{for any } n \neq 0,$$

  (b) *for any $U_\lambda(\mathfrak{g})$-module M, we have an isomorphism*

$$M \xrightarrow{\sim} \Gamma(X; \mathscr{D}_{X,\lambda} \otimes_{U(\mathfrak{g})} M),$$

  namely, the diagram

$$(9.2.4) \qquad \mathrm{Mod}\,(U_\lambda(\mathfrak{g})) \xrightarrow{\;\mathscr{D}_{X,\lambda} \otimes_{U(\mathfrak{g})} \bullet\;} \mathrm{Mod}\,(\mathscr{D}_{X,\lambda})$$

with id and $\Gamma(X; \bullet)$ arrows to $\mathrm{Mod}\,(U_\lambda(\mathfrak{g}))$

  quasi-commutes.

(iii) *Assume that $\lambda - \rho$ is regular and integrally anti-dominant. Then*

$$\mathscr{M} \simeq \mathscr{D}_{X,\lambda} \otimes_{U(\mathfrak{g})} \Gamma(X; \mathscr{M}),$$

and we have an equivalence of categories

$$\mathrm{Mod}\,(U_\lambda(\mathfrak{g})) \simeq \mathrm{Mod}\,(\mathscr{D}_{X,\lambda}).$$

9.3 Quasi-equivariant D-modules on the Symmetric Space

We set $S = G/K$. Let $j \colon \mathrm{pt} \hookrightarrow S$ be the morphism given by the origin $s_0 \in S$. By Proposition 3.1.5, $j^* \colon \mathrm{Mod}(\mathscr{D}_S, G) \xrightarrow{\sim} \mathrm{Mod}(\mathfrak{g}, K)$ is an equivalence of categories. Since $\mathrm{D}^{\mathrm{b}}(\mathscr{D}_S, G) = \mathrm{D}^{\mathrm{b}}(\mathrm{Mod}(\mathscr{D}_S, G))$ by the definition, j^* induces an equivalence

(9.3.1) $$\mathbf{L}j^* \colon \mathrm{D}^{\mathrm{b}}(\mathscr{D}_S, G) \xrightarrow{\sim} \mathrm{D}^{\mathrm{b}}(\mathrm{Mod}(\mathfrak{g}, K)).$$

Let

$$\Psi \colon \mathrm{D}^{\mathrm{b}}(\mathrm{Mod}(\mathfrak{g}, K)) \xrightarrow{\sim} \mathrm{D}^{\mathrm{b}}(\mathscr{D}_S, G)$$

be its quasi-inverse.

Consider the diagram:

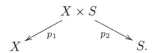

Set

$$\mathscr{M}_0 := \mathscr{D}_{X,-\lambda} \otimes_{\mathscr{O}_X} \Omega_X^{\otimes -1}.$$

It is an object of $\mathrm{Mod}_{\mathrm{coh}}(\mathscr{D}_{X,-\lambda}, G)$.

For $\mathscr{L} \in \mathrm{D}^{\mathrm{b}}_{K,\mathrm{coh}}(\mathscr{D}_{X,\lambda})$, set $\mathscr{L}_0 = \mathrm{Ind}^G_K(\mathscr{L}) \in \mathrm{D}^{\mathrm{b}}_{G,\mathrm{coh}}(\mathscr{D}_{X \times S,\lambda})$. Let us calculate $\mathscr{M}_0 \overset{\mathbf{D}}{\circ} \mathscr{L}_0 \in \mathrm{D}^{\mathrm{b}}_{\mathrm{coh}}(\mathscr{D}_S, G)$.

We have

$$\mathscr{M}_0 \overset{\mathbf{D}}{\circ} \mathscr{L}_0 = \mathbf{D}p_{2*}(\mathbf{D}p_1^* \mathscr{M}_0 \overset{\mathbf{D}}{\otimes} \mathscr{L}_0)$$

$$\simeq \mathbf{R}p_{2*}(\mathscr{D}_{S \leftarrow X \times S} \overset{\mathbf{L}}{\otimes}_{\mathscr{D}_{X \times S}} (p_1^*(\mathscr{D}_{X,-\lambda} \otimes \Omega_X^{\otimes -1}) \overset{\mathbf{L}}{\otimes}_{\mathscr{O}_{X \times S}} \mathscr{L}_0)).$$

On the other hand, we have

$$\mathscr{D}_{S \leftarrow X \times S} \overset{\mathbf{L}}{\otimes}_{\mathscr{D}_{X \times S}} (p_1^*(\mathscr{D}_{X,-\lambda} \otimes \Omega_X^{\otimes -1}) \overset{\mathbf{L}}{\otimes}_{\mathscr{O}_{X \times S}} \mathscr{L}_0)$$

$$\simeq \Omega_X \overset{\mathbf{L}}{\otimes}_{\mathscr{D}_X} (p_1^*(\mathscr{D}_{X,-\lambda} \otimes \Omega_X^{\otimes -1}) \overset{\mathbf{L}}{\otimes}_{\mathscr{O}_{X \times S}} \mathscr{L}_0)$$

$$\simeq \mathscr{L}_0.$$

Hence we obtain

(9.3.2) $$\mathscr{M}_0 \overset{\mathbf{D}}{\circ} \mathscr{L}_0 \cong \mathbf{R}p_{2*}\mathscr{L}_0.$$

It is an object of $\mathrm{D}^{\mathrm{b}}_{\mathrm{coh}}(\mathscr{D}_S, G)$.

Let $\tilde{j} \colon X \to X \times S$ be the induced morphism $(x \mapsto (x, s_0))$. Then, we have

$$\mathbf{L}j^* \mathbf{R}p_{2*}\mathscr{L}_0 \simeq \mathbf{R}\Gamma(X; \mathbf{D}\tilde{j}^* \mathscr{L}_0) \simeq \mathbf{R}\Gamma(X; \mathscr{L}).$$

Thus we obtain the following proposition.

Proposition 9.3.1. *The diagram*

$$
\begin{array}{ccc}
D^b_{K,\,\mathrm{coh}}(\mathscr{D}_{X,\lambda}) & \xrightarrow{\;\mathbf{R}\Gamma(X;\bullet)\;} & D^b(\mathrm{Mod}_f(\mathfrak{g},K)) \\
& & \downarrow{\scriptstyle\sim}\;\Psi \\
(\mathscr{D}_{X,-\lambda}\otimes\Omega_X^{\otimes-1})\overset{\mathbf{D}}{\circ}\mathrm{Ind}_K^G(\bullet) & \searrow & D^b_{\mathrm{coh}}(\mathscr{D}_S,G)
\end{array}
$$

quasi-commutes.

Proposition 9.3.2. *For any $\lambda \in \mathfrak{t}^*$, any $\mathscr{M} \in D^b_{\mathrm{coh}}(\mathscr{D}_{X,-\lambda},G)$, any $\mathscr{L} \in D^b_{K,\,\mathrm{coh}}(\mathscr{D}_{X,\lambda})$ and any integer n,*

$$
H^n(\mathscr{M}\overset{\mathbf{D}}{\circ}\mathrm{Ind}_K^G\mathscr{L}) \simeq \Psi(M)
$$

for some Harish-Chandra module M.

Proof. We know that $H^n(\mathscr{M}\overset{\mathbf{D}}{\circ}\mathrm{Ind}_K^G\mathscr{L}) \simeq \Psi(M)$ for some $M \in \mathrm{Mod}_f(\mathfrak{g},K)$. Hence we need to show that M is $\mathfrak{z}(\mathfrak{g})$-finite. Since $\mathscr{M}$ has a resolution whose components are of the form $\mathscr{D}_{X,-\lambda}\otimes(\Omega_X^{\otimes-1}\otimes\mathscr{O}_X(\mu)\otimes V)$ for an integrable $\mu \in \mathfrak{t}^*$ and a finite-dimensional G-module V, we may assume from the beginning that $\mathscr{M} = \mathscr{D}_{X,-\lambda}\otimes(\Omega_X^{\otimes-1}\otimes\mathscr{O}_X(\mu))$. In this case, we have by (9.2.2)

$$
\begin{aligned}
\mathscr{M} &\simeq \big(\mathscr{O}_X(\mu)\otimes\mathscr{D}_{X,-\lambda-\mu}\otimes\mathscr{O}_X(-\mu)\big)\otimes\big(\Omega_X^{\otimes-1}\otimes\mathscr{O}_X(\mu)\big) \\
&\simeq \mathscr{O}_X(\mu)\overset{\mathbf{D}}{\otimes}(\mathscr{D}_{X,-\lambda-\mu}\otimes\Omega_X^{\otimes-1}),
\end{aligned}
$$

which implies that

$$
\begin{aligned}
\mathbf{D}p_1^*\mathscr{M}\overset{\mathbf{D}}{\otimes}\mathrm{Ind}_K^G(\mathscr{L}) &\simeq \mathbf{D}p_1^*(\mathscr{D}_{X,-\lambda-\mu}\otimes\Omega_X^{\otimes-1})\overset{\mathbf{D}}{\otimes}\mathbf{D}p_1^*\mathscr{O}_X(\mu)\overset{\mathbf{D}}{\otimes}\mathrm{Ind}_K^G(\mathscr{L}) \\
&\simeq \mathbf{D}p_1^*(\mathscr{D}_{X,-\lambda-\mu}\otimes\Omega_X^{\otimes-1})\overset{\mathbf{D}}{\otimes}\mathrm{Ind}_K^G(\mathscr{O}_X(\mu)\overset{\mathbf{D}}{\otimes}\mathscr{L}).
\end{aligned}
$$

Hence, Proposition 9.3.1 implies

$$
\begin{aligned}
\mathscr{M}\overset{\mathbf{D}}{\circ}\mathrm{Ind}_K^G(\mathscr{L}) &\simeq (\mathscr{D}_{X,-\lambda-\mu}\otimes\Omega_X^{\otimes-1})\overset{\mathbf{D}}{\circ}\mathrm{Ind}_K^G(\mathscr{O}_X(\mu)\overset{\mathbf{D}}{\otimes}\mathscr{L}) \\
&\simeq \Psi\big(\mathbf{R}\Gamma(X;\mathscr{O}_X(\mu)\overset{\mathbf{D}}{\otimes}\mathscr{L})\big).
\end{aligned}
$$

Since $\mathscr{O}_X(\mu)\overset{\mathbf{D}}{\otimes}\mathscr{L} \in D^b_{K,\,\mathrm{coh}}(\mathscr{D}_{X,\lambda+\mu})$, its cohomology $H^n(X;\mathscr{O}_X(\mu)\overset{\mathbf{D}}{\otimes}\mathscr{L})$ is a Harish-Chandra module. Q.E.D.

9.4 Matsuki Correspondence

The following theorem is due to Matsuki ([22]).

Theorem 9.4.1. (i) *There are only finitely many K-orbits in X and also finitely many $G_{\mathbb{R}}$-orbits in X^{an}.*
(ii) *There is a one-to-one correspondence between the set of K-orbits in X and the one of $G_{\mathbb{R}}$-orbits.*

More precisely, a K-orbit E and a $G_{\mathbb{R}}$-orbit F correspond by the correspondence above if and only if one of the following equivalent conditions is satisfied:

(1) $E^{\mathrm{an}} \cap F$ is a $K_{\mathbb{R}}$-orbit,
(2) $E^{\mathrm{an}} \cap F$ is non-empty and compact.

Its sheaf-theoretic version was conjectured by the author [14] and proved by Mirković-Uzawa-Vilonen [23]. Let $S_{\mathbb{R}} = G_{\mathbb{R}}/K_{\mathbb{R}}$ be the Riemannian symmetric space and set $S = G/K$. Then S is an affine algebraic manifold. The canonical map $i\colon S_{\mathbb{R}} \hookrightarrow S^{\mathrm{an}}$ is a closed embedding.

We have the functor

$$(9.4.1) \qquad \mathrm{Ind}_{K^{\mathrm{an}}}^{G^{\mathrm{an}}} \colon \mathrm{D}_{K^{\mathrm{an}},\tau_\lambda}^{\mathrm{b}}(\mathbb{C}_{X^{\mathrm{an}}}) \to \mathrm{D}_{G^{\mathrm{an}},\tau_\lambda}^{\mathrm{b}}(\mathbb{C}_{(X \times S)^{\mathrm{an}}}).$$

We define the functor

$$\Phi \colon \mathrm{D}_{K^{\mathrm{an}},\tau_\lambda}^{\mathrm{b}}(\mathbb{C}_{X^{\mathrm{an}}}) \to \mathrm{D}_{G_{\mathbb{R}},\tau_\lambda}^{\mathrm{b}}(\mathbb{C}_{X^{\mathrm{an}}})$$

by

$$\begin{aligned} \Phi(F) &= \mathrm{Ind}_{K^{\mathrm{an}}}^{G^{\mathrm{an}}}(F) \circ i_* i^! \mathbb{C}_{S^{\mathrm{an}}}[2d_S] \\ &= \mathbf{R}(p_1^{\mathrm{an}})_!(\mathrm{Ind}_{K^{\mathrm{an}}}^{G^{\mathrm{an}}}(F) \otimes (p_2^{\mathrm{an}})^{-1} i_* i^! \mathbb{C}_{S^{\mathrm{an}}})[2d_S]. \end{aligned}$$

Here, p_1^{an} and p_2^{an} are the projections from $X^{\mathrm{an}} \times S^{\mathrm{an}}$ to X^{an} and S^{an}, respectively. Note that $i_* i^! \mathbb{C}_{S^{\mathrm{an}}}$ is isomorphic to $i_* \mathbb{C}_{S_{\mathbb{R}}}[-d_S]$ (once we give an orientation of $S_{\mathbb{R}}$). Hence we have

$$(9.4.2) \qquad \Phi(F) \simeq \mathbf{R}p_{1\mathbb{R}!}(\mathrm{Ind}_{K^{\mathrm{an}}}^{G^{\mathrm{an}}}(F)|_{X \times S_{\mathbb{R}}})[d_S] \simeq \mathbf{R}p_{1\mathbb{R}!}(\mathrm{Ind}_{K_{\mathbb{R}}}^{G_{\mathbb{R}}}(F))[d_S],$$

where $p_{1\mathbb{R}}\colon X^{\mathrm{an}} \times S_{\mathbb{R}} \to X^{\mathrm{an}}$ is the projection.

Theorem 9.4.2 ([23]). *The functor Φ induces equivalences of triangulated categories:*

$$\Phi : \quad \begin{array}{ccc} \mathrm{D}_{K^{\mathrm{an}},\tau_\lambda}^{\mathrm{b}}(\mathbb{C}_{X^{\mathrm{an}}}) & \xrightarrow{\sim} & \mathrm{D}_{G_{\mathbb{R}},\tau_\lambda}^{\mathrm{b}}(\mathbb{C}_{X^{\mathrm{an}}}) \\ \cup & & \cup \\ \mathrm{D}_{K^{\mathrm{an}},\tau_\lambda,\mathbb{C}\text{-}c}^{\mathrm{b}}(\mathbb{C}_{X^{\mathrm{an}}}) & \xrightarrow{\sim} & \mathrm{D}_{G_{\mathbb{R}},\tau_\lambda,\mathbb{R}\text{-}c}^{\mathrm{b}}(\mathbb{C}_{X^{\mathrm{an}}}). \end{array}$$

We call Φ the *Matsuki correspondence*.

For some equivariant sheaves, the Matsuki correspondence is given as follows.

Proposition 9.4.3 ([23]). *Let $i_Z \colon Z \hookrightarrow X$ be a K-orbit in X and let $i_{Z^a} \colon Z^a \hookrightarrow X^{\mathrm{an}}$ be the $G_{\mathbb{R}}$-orbit corresponding to Z.*

(i) *The restriction functors induce equivalences of categories:*

$$\mathrm{Mod}_{K^{\mathrm{an}}, \tau_\lambda}(\mathbb{C}_{Z^{\mathrm{an}}}) \xrightarrow{\sim} \mathrm{Mod}_{K_{\mathbb{R}}, \tau_\lambda}(\mathbb{C}_{Z^{\mathrm{an}} \cap Z^a}) \xleftarrow{\sim} \mathrm{Mod}_{G_{\mathbb{R}}, \tau_\lambda}(\mathbb{C}_{Z^a}).$$

(ii) *Assume that $F \in \mathrm{Mod}_{K^{\mathrm{an}}, \tau_\lambda}(\mathbb{C}_{Z^{\mathrm{an}}})$ and $F^a \in \mathrm{Mod}_{G_{\mathbb{R}}, \tau_\lambda}(\mathbb{C}_{Z^a})$ correspond by the equivalence above. Then we have*

$$\Phi(\mathbf{R}(i_Z^{\mathrm{an}})_* F) \simeq \mathbf{R}(i_{Z^a})_! F^a [2 \operatorname{codim}_X Z].$$

The K-orbit $K x_0 \subset X$ is a unique open K-orbit in X and $G_{\mathbb{R}} x_0 \subset X^{\mathrm{an}}$ is a unique closed $G_{\mathbb{R}}$-orbit in X^{an}. Set $X_{\min}^{\mathbb{R}} = G_{\mathbb{R}}/P_{\mathbb{R}}$. Then $X_{\min}^{\mathbb{R}} = G_{\mathbb{R}} x_0^{\min} = K_{\mathbb{R}} x_0^{\min}$ and it is a unique closed $G_{\mathbb{R}}$-orbit in $X_{\min}^{\mathrm{an}}$. We have

$$(9.4.3) \quad (K x_0)^{\mathrm{an}} = (\pi^{-1}(K x_0^{\min}))^{\mathrm{an}} \supset G_{\mathbb{R}} x_0 = K_{\mathbb{R}} x_0 = (\pi^{\mathrm{an}})^{-1}(X_{\min}^{\mathbb{R}}).$$

Let $j \colon K x_0 \hookrightarrow X$ be the open embedding and $j^a \colon G_{\mathbb{R}} x_0 \hookrightarrow X^{\mathrm{an}}$ the closed embedding. Then as a particular case of Proposition 9.4.3, we have an isomorphism:

$$(9.4.4) \qquad \Phi(\mathbf{R} j_*^{\mathrm{an}} F) \simeq j_*^a (F|_{G_{\mathbb{R}} x_0})$$

for any K^{an}-equivariant local system F on $K^{\mathrm{an}} x_0$.

9.5 Construction of Representations

For $M \in \mathrm{Mod}_f(\mathfrak{g}, K)$, let $\mathrm{Hom}_{(\mathfrak{g}, K_{\mathbb{R}})}(M, C^\infty(G_{\mathbb{R}}))$ be the set of homomorphisms from M to $C^\infty(G_{\mathbb{R}})$ which commute with the actions of $\mathfrak{g}$ and $K_{\mathbb{R}}$. Here, $\mathfrak{g}$ and $K_{\mathbb{R}}$ act on $C^\infty(G_{\mathbb{R}})$ through the right $G_{\mathbb{R}}$-action on $G_{\mathbb{R}}$. Then $G_{\mathbb{R}}$ acts on $\mathrm{Hom}_{(\mathfrak{g}, K_{\mathbb{R}})}(M, C^\infty(G_{\mathbb{R}}))$ through the left $G_{\mathbb{R}}$-action on $G_{\mathbb{R}}$.

Let us write by $C^\infty(G_{\mathbb{R}})^{K_{\mathbb{R}}\text{-fini}}$ the set of $K_{\mathbb{R}}$-finite vectors of $C^\infty(G_{\mathbb{R}})$. Then $C^\infty(G_{\mathbb{R}})^{K_{\mathbb{R}}\text{-fini}}$ is a $(\mathfrak{g}, K)$-module and

$$\mathrm{Hom}_{(\mathfrak{g}, K_{\mathbb{R}})}(M, C^\infty(G_{\mathbb{R}})) \simeq \mathrm{Hom}_{(\mathfrak{g}, K)}(M, C^\infty(G_{\mathbb{R}})^{K_{\mathbb{R}}\text{-fini}}).$$

Note that, in our context, $K_{\mathbb{R}}$ is connected and hence the $\mathfrak{g}$-invariance implies the $K_{\mathbb{R}}$-invariance. Therefore, we have

$$\mathrm{Hom}_{(\mathfrak{g}, K_{\mathbb{R}})}(M, C^\infty(G_{\mathbb{R}})) \simeq \mathrm{Hom}_{\mathfrak{g}}(M, C^\infty(G_{\mathbb{R}})).$$

We endow $\mathrm{Hom}_{(\mathfrak{g}, K_{\mathbb{R}})}(M, C^\infty(G_{\mathbb{R}}))$ with the Fréchet nuclear topology as in Lemma 5.3.1.

In any way, $\mathrm{Hom}_{(\mathfrak{g}, K_{\mathbb{R}})}(M, C^{\infty}(G_{\mathbb{R}}))$ has a Fréchet nuclear $G_{\mathbb{R}}$-module structure. We denote it by $\mathrm{Hom}^{\mathrm{top}}_{(\mathfrak{g}, K_{\mathbb{R}})}(M, C^{\infty}(G_{\mathbb{R}}))$. Let us denote by

$$\mathbf{R}\mathrm{Hom}^{\mathrm{top}}_{(\mathfrak{g}, K_{\mathbb{R}})}(\,\bullet\,, C^{\infty}(G_{\mathbb{R}})) : \mathrm{D}^{\mathrm{b}}(\mathrm{Mod}_f(\mathfrak{g}, K)) \to \mathrm{D}^{\mathrm{b}}(\mathbf{FN}_{G_{\mathbb{R}}})$$

its right derived functor.[5] Note that $\mathrm{Mod}_f(\mathfrak{g}, K)$ has enough projectives, and any M can be represented by a complex P of projective objects in $\mathrm{Mod}_f(\mathfrak{g}, K)$, and then $\mathbf{R}\mathrm{Hom}^{\mathrm{top}}_{(\mathfrak{g}, K_{\mathbb{R}})}(M, C^{\infty}(G_{\mathbb{R}}))$ is represented by a complex $\mathrm{Hom}^{\mathrm{top}}_{(\mathfrak{g}, K_{\mathbb{R}})}(P, C^{\infty}(G_{\mathbb{R}}))$ of Fréchet nuclear $G_{\mathbb{R}}$-modules. Note that for a finite-dimensional K-module V, $U(\mathfrak{g}) \otimes_{\mathfrak{k}} V$ is a projective object of $\mathrm{Mod}_f(\mathfrak{g}, K)$, and $\mathrm{Hom}^{\mathrm{top}}_{(\mathfrak{g}, K_{\mathbb{R}})}(U(\mathfrak{g}) \otimes_{\mathfrak{k}} V, C^{\infty}(G_{\mathbb{R}})) \simeq \mathrm{Hom}_{K_{\mathbb{R}}}(V, C^{\infty}(G_{\mathbb{R}}))$.

Since we have $\mathrm{Hom}^{\mathrm{top}}_{(\mathfrak{g}, K_{\mathbb{R}})}(M, C^{\infty}(G_{\mathbb{R}})) \simeq \mathrm{Hom}^{\mathrm{top}}_{\mathscr{D}_S}(\Psi(M), \mathscr{C}^{\infty}_{S_{\mathbb{R}}})$ for any $M \in \mathrm{Mod}_f(\mathfrak{g}, K)$, their right derived functors are isomorphic:

$$\mathbf{R}\mathrm{Hom}^{\mathrm{top}}_{(\mathfrak{g}, K_{\mathbb{R}})}(M, C^{\infty}(G_{\mathbb{R}})) \simeq \mathbf{R}\mathrm{Hom}^{\mathrm{top}}_{\mathscr{D}_S}(\Psi(M), \mathscr{C}^{\infty}_{S_{\mathbb{R}}})$$

for any $M \in \mathrm{D}^{\mathrm{b}}(\mathrm{Mod}_f(\mathfrak{g}, K))$.

Lemma 9.5.1. *Let M be a Harish-Chandra module. Then $\Psi(M)$ is an elliptic $\mathscr{D}_S$-module.*

Proof.   Let Δ be a Casimir element of $U(\mathfrak{g})$. Then there exists a non-zero polynomial $a(t)$ such that $a(\Delta)M = 0$. Hence the characteristic variety $\mathrm{Ch}(\Psi(M)) \subset T^*S$ of $\Psi(M)$ is contained in the zero locus of the principal symbol of $L_S(\Delta)$. Then the result follows from the well-known fact that the Laplacian $L_S(\Delta)|_{S_{\mathbb{R}}}$ is an elliptic differential operator on $S_{\mathbb{R}}$.   Q.E.D.

If the cohomologies of $M \in \mathrm{D}^{\mathrm{b}}(\mathrm{Mod}_f(\mathfrak{g}, K))$ are Harish-Chandra modules, then $\Psi(M)$ is elliptic, and Proposition 6.3.3 implies

$$(9.5.1)\quad \mathbf{R}\mathrm{Hom}^{\mathrm{top}}_{(\mathfrak{g}, K_{\mathbb{R}})}(M, C^{\infty}(G_{\mathbb{R}})) \xrightarrow{\sim} \mathbf{R}\mathrm{Hom}^{\mathrm{top}}_{\mathscr{D}_S}(\Psi(M) \otimes i_* i^! \mathbb{C}_{S^{\mathrm{an}}}, \mathscr{O}_{S^{\mathrm{an}}}).$$

There is a dual notion. For $M \in \mathrm{Mod}_f(\mathfrak{g}, K)$, let $\Gamma_c(G_{\mathbb{R}}; \mathscr{D}ist_{G_{\mathbb{R}}}) \otimes_{(\mathfrak{g}, K_{\mathbb{R}})} M$ be the quotient of $\Gamma_c(G_{\mathbb{R}}; \mathscr{D}ist_{G_{\mathbb{R}}}) \otimes_{\mathbb{C}} M$ by the linear subspace spanned by vectors $(R_{G_{\mathbb{R}}}(A)u) \otimes v + u \otimes (Av)$ and $(ku) \otimes (kv) - u \otimes v$ ($u \in \Gamma_c(G_{\mathbb{R}}; \mathscr{D}ist_{G_{\mathbb{R}}})$, $v \in M$, $A \in \mathfrak{g}$, $k \in K_{\mathbb{R}}$). Here, we consider it as a vector space (not considering the topology). In our case, $K_{\mathbb{R}}$ is connected, and $K_{\mathbb{R}}$ acts trivially on $\Gamma_c(G_{\mathbb{R}}; \mathscr{D}ist_{G_{\mathbb{R}}}) \otimes_{U(\mathfrak{k})} M$. Therefore, we have

$$\Gamma_c(G_{\mathbb{R}}; \mathscr{D}ist_{G_{\mathbb{R}}}) \otimes_{(\mathfrak{g}, K_{\mathbb{R}})} M \simeq \Gamma_c(G_{\mathbb{R}}; \mathscr{D}ist_{G_{\mathbb{R}}}) \otimes_{U(\mathfrak{g})} M.$$

It is a right exact functor from $\mathrm{Mod}_f(\mathfrak{g}, K)$ to the category $\mathrm{Mod}(\mathbb{C})$ of $\mathbb{C}$-vector spaces. Let

---

[5] We may write here $\mathrm{Hom}^{\mathrm{top}}_{U(\mathfrak{g})}(M, C^{\infty}(G_{\mathbb{R}}))$, but we use this notation in order to emphasize that it is calculated not on $\mathrm{Mod}(U(\mathfrak{g}))$ but on $\mathrm{Mod}_f(\mathfrak{g}, K)$.

(9.5.2) $\qquad \Gamma_c(G_{\mathbb{R}}; \mathscr{D}ist_{G_{\mathbb{R}}}) \overset{L}{\otimes}_{(\mathfrak{g}, K_{\mathbb{R}})} \bullet : D^b(\mathrm{Mod}_f(\mathfrak{g}, K)) \to D^b(\mathbb{C})$

be its left derived functor.

For any $M \in D^b(\mathrm{Mod}_f(\mathfrak{g}, K))$, we can take a quasi-isomorphism $P^\bullet \to M$ such that each P^n has a form $U(\mathfrak{g}) \otimes_{\mathfrak{k}} V^n$ for a finite-dimensional K-module V^n. Then, we have

$$\Gamma_c(G_{\mathbb{R}}; \mathscr{D}ist_{G_{\mathbb{R}}}) \otimes_{(\mathfrak{g}, K_{\mathbb{R}})} P^n \simeq \Gamma_c(G_{\mathbb{R}}; \mathscr{D}ist_{G_{\mathbb{R}}}) \otimes_{U(\mathfrak{k})} V^n$$
$$\simeq (\Gamma_c(G_{\mathbb{R}}; \mathscr{D}ist_{G_{\mathbb{R}}}) \otimes_{\mathbb{C}} V^n)^{K_{\mathbb{R}}},$$

where the superscript $K_{\mathbb{R}}$ means the set of $K_{\mathbb{R}}$-invariant vectors. The $K_{\mathbb{R}}$-module structure on $\Gamma_c(G_{\mathbb{R}}; \mathscr{D}ist_{G_{\mathbb{R}}})$ is by the right action of $K_{\mathbb{R}}$ on $G_{\mathbb{R}}$. By the left action of $G_{\mathbb{R}}$ on $G_{\mathbb{R}}$, $G_{\mathbb{R}}$ acts on $\Gamma_c(G_{\mathbb{R}}; \mathscr{D}ist_{G_{\mathbb{R}}}) \otimes_{(\mathfrak{g}, K_{\mathbb{R}})} P^n$. Hence it belongs to $\mathbf{DFN}_{G_{\mathbb{R}}}$. The object $\Gamma_c(G_{\mathbb{R}}; \mathscr{D}ist_{G_{\mathbb{R}}}) \otimes_{(\mathfrak{g}, K_{\mathbb{R}})} P^\bullet \in D^b(\mathbf{DFN}_{G_{\mathbb{R}}})$ does not depend on the choice of a quasi-isomorphism $P^\bullet \to M$, and we denote it by $\Gamma_c(G_{\mathbb{R}}; \mathscr{D}ist_{G_{\mathbb{R}}}) \overset{L}{\otimes}_{(\mathfrak{g}, K_{\mathbb{R}})} M$. Thus we have constructed a functor:

(9.5.3) $\quad \Gamma_c(G_{\mathbb{R}}; \mathscr{D}ist_{G_{\mathbb{R}}}) \overset{L}{\otimes}_{(\mathfrak{g}, K_{\mathbb{R}})} \bullet : D^b(\mathrm{Mod}_f(\mathfrak{g}, K)) \to D^b(\mathbf{DFN}_{G_{\mathbb{R}}})$.

If we forget the topology and the equivariance, (9.5.3) reduces to (9.5.2).

We have

(9.5.4) $\quad \Gamma_c(G_{\mathbb{R}}; \mathscr{D}ist_{G_{\mathbb{R}}}) \overset{L}{\otimes}_{(\mathfrak{g}, K_{\mathbb{R}})} M \simeq \left(\mathbf{R}\mathrm{Hom}^{\mathrm{top}}_{(\mathfrak{g}, K_{\mathbb{R}})} (M, C^\infty(G_{\mathbb{R}})) \right)^*$

in $D^b(\mathbf{DFN}_{G_{\mathbb{R}}})$. (Here, we fix an invariant measure on $G_{\mathbb{R}}$.)

In general, $\mathbf{R}\mathrm{Hom}^{\mathrm{top}}_{(\mathfrak{g}, K_{\mathbb{R}})}(M, C^\infty(G_{\mathbb{R}}))$ and $\Gamma_c(G_{\mathbb{R}}; \mathscr{D}ist_{G_{\mathbb{R}}}) \overset{L}{\otimes}_{(\mathfrak{g}, K_{\mathbb{R}})} M$ are not strict (see Theorems 10.4.1 and 10.4.2).

9.6 Integral Transformation Formula

Since X has finitely many K-orbits, the Riemann-Hilbert correspondence (Theorem 7.14.1) implies the following theorem.

Theorem 9.6.1. *The de Rham functor gives an equivalence of categories:*

(9.6.1) $\qquad \mathrm{DR}_X : D^b_{K, \mathrm{coh}}(\mathscr{D}_{X, \lambda}) \xrightarrow{\sim} D^b_{K^{\mathrm{an}}, \tau_{-\lambda}, \mathbb{C}\text{-}c}(\mathbb{C}_{X^{\mathrm{an}}})$.

Recall that the de Rham functor is defined by

$$\mathrm{DR}_X : \mathscr{M} \mapsto \mathbf{R}\mathscr{H}om_{\mathscr{D}_{X^{\mathrm{an}}, \lambda}}(\mathscr{O}_{X^{\mathrm{an}}}(\lambda), \mathscr{M}^{\mathrm{an}}),$$

where $\mathscr{M}^{\mathrm{an}} = \mathscr{D}_{X^{\mathrm{an}}, \lambda} \otimes_{\mathscr{D}_{X, \lambda}} \mathscr{M}$. Similarly to (9.4.1), we have the equivalence of categories:

(9.6.2) $\qquad \mathrm{Ind}^G_K : D^b_{K, \mathrm{coh}}(\mathscr{D}_{X, \lambda}) \xrightarrow{\sim} D^b_{G, \mathrm{coh}}(\mathscr{D}_{X \times S, \lambda})$

and a quasi-commutative diagram

$$
\begin{array}{ccc}
\mathrm{D}^{\mathrm{b}}_{K,\mathrm{coh}}(\mathscr{D}_{X,\lambda}) & \xrightarrow{\ \mathrm{Ind}^G_K\ } & \mathrm{D}^{\mathrm{b}}_{G,\mathrm{coh}}(\mathscr{D}_{X\times S,\lambda}) \\
{\scriptstyle \mathrm{DR}_X}\Big\downarrow & & \Big\downarrow{\scriptstyle \mathrm{DR}_{X\times S}} \\
\mathrm{D}^{\mathrm{b}}_{K^{\mathrm{an}},\tau_{-\lambda},\mathbb{C}\text{-c}}(\mathbb{C}_{X^{\mathrm{an}}}) & \xrightarrow{\ \mathrm{Ind}^{G^{\mathrm{an}}}_{K^{\mathrm{an}}}\ } & \mathrm{D}^{\mathrm{b}}_{G^{\mathrm{an}},\tau_{-\lambda},\mathbb{C}\text{-c}}(\mathbb{C}_{(X\times S)^{\mathrm{an}}}).
\end{array}
$$

Consider the diagram:

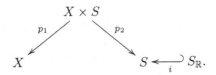

Let us take $\mathscr{L} \in \mathrm{D}^{\mathrm{b}}_{K,\mathrm{coh}}(\mathscr{D}_{X,\lambda})$ and set $\mathscr{L}_0 = \mathrm{Ind}^G_K \mathscr{L} \in \mathrm{D}^{\mathrm{b}}_G(\mathscr{D}_{X\times S,\lambda})$. Set $L = \mathrm{DR}_X(\mathscr{L}) := \mathbf{R}\mathscr{H}om_{\mathscr{D}_{X^{\mathrm{an}},\lambda}}(\mathscr{O}_{X^{\mathrm{an}}}(\lambda),\mathscr{L}^{\mathrm{an}}_0) \in \mathrm{D}^{\mathrm{b}}_{K^{\mathrm{an}},\tau_{-\lambda},\mathbb{C}\text{-c}}(\mathbb{C}_{X^{\mathrm{an}}})$ and $L_0 = \mathrm{Ind}^{G^{\mathrm{an}}}_{K^{\mathrm{an}}} L = \mathrm{DR}_{X\times S}(\mathscr{L}_0) \in \mathrm{D}^{\mathrm{b}}_{G^{\mathrm{an}},\tau_{-\lambda},\mathbb{C}\text{-c}}(\mathbb{C}_{(X\times S)^{\mathrm{an}}})$. Let $\mathscr{M} \in \mathrm{D}^{\mathrm{b}}_{\mathrm{coh}}(\mathscr{D}_{X,-\lambda},G)$.

Then Theorem 8.2.1 (see Remark 8.2.2) immediately implies the following result.

Proposition 9.6.2. *For $\mathscr{M} \in \mathrm{D}^{\mathrm{b}}_{\mathrm{coh}}(\mathscr{D}_{X,-\lambda},G)$ and $\mathscr{L} \in \mathrm{D}^{\mathrm{b}}_{K,\mathrm{coh}}(\mathscr{D}_{X,\lambda})$ and $L = \mathrm{DR}_X(\mathscr{L}) \in \mathrm{D}^{\mathrm{b}}_{K^{\mathrm{an}},\tau_{-\lambda},\mathbb{C}\text{-c}}(\mathbb{C}_{X^{\mathrm{an}}})$, we have*

$$
(9.6.3) \qquad
\begin{aligned}
&\mathbf{R}\mathrm{Hom}^{\mathrm{top}}_{\mathscr{D}_S}\big((\mathscr{M} \overset{\mathrm{D}}{\circ} \mathrm{Ind}^G_K(\mathscr{L})) \otimes i_* i^! \mathbb{C}_{S^{\mathrm{an}}}, \mathscr{O}_{S^{\mathrm{an}}}\big) \\
&\simeq \mathbf{R}\mathrm{Hom}^{\mathrm{top}}_{\mathscr{D}_{X,-\lambda}}(\mathscr{M} \otimes \Phi(L), \mathscr{O}_{X^{\mathrm{an}}}(-\lambda))[d_X]
\end{aligned}
$$

in $\mathrm{D}^{\mathrm{b}}(\mathbf{FN}_{G_{\mathbb{R}}})$.

Let us recall the equivalence $\mathbf{L}j^* \colon \mathrm{D}^{\mathrm{b}}_{\mathrm{coh}}(\mathscr{D}_S, G) \xrightarrow{\sim} \mathrm{D}^{\mathrm{b}}(\mathrm{Mod}_f(\mathfrak{g},K))$ in (9.3.1). Since $\mathbf{L}j^*(\mathscr{M} \overset{\mathrm{D}}{\circ} \mathrm{Ind}^G_K(\mathscr{L}))$ has Harish-Chandra modules as cohomologies by Proposition 9.3.2, the isomorphism (9.5.1) reads as

$$
(9.6.4) \qquad
\begin{aligned}
&\mathbf{R}\mathrm{Hom}^{\mathrm{top}}_{\mathscr{D}_S}(\mathscr{M} \overset{\mathrm{D}}{\circ} \mathrm{Ind}^G_K(\mathscr{L}) \otimes i_* i^! \mathbb{C}_{S^{\mathrm{an}}}, \mathscr{O}_{S^{\mathrm{an}}}) \\
&\simeq \mathbf{R}\mathrm{Hom}^{\mathrm{top}}_{(\mathfrak{g},K_{\mathbb{R}})}(\mathbf{L}j^*(\mathscr{M} \overset{\mathrm{D}}{\circ} \mathrm{Ind}^G_K(\mathscr{L})), \mathrm{C}^\infty(G_{\mathbb{R}}))
\end{aligned}
$$

in $\mathrm{D}^{\mathrm{b}}(\mathbf{FN}_{G_{\mathbb{R}}})$. Thus we obtain the following proposition.

Proposition 9.6.3. *For $\mathscr{M} \in \mathrm{D}^{\mathrm{b}}_{\mathrm{coh}}(\mathscr{D}_{X,-\lambda},G)$ and $\mathscr{L} \in \mathrm{D}^{\mathrm{b}}_{K,\mathrm{coh}}(\mathscr{D}_{X,\lambda})$, we have*

$$
(9.6.5) \qquad
\begin{aligned}
&\mathbf{R}\mathrm{Hom}^{\mathrm{top}}_{(\mathfrak{g},K_{\mathbb{R}})}(\mathbf{L}j^*(\mathscr{M} \overset{\mathrm{D}}{\circ} \mathrm{Ind}^G_K \mathscr{L}), \mathrm{C}^\infty(G_{\mathbb{R}})) \\
&\simeq \mathbf{R}\mathrm{Hom}^{\mathrm{top}}_{\mathscr{D}_{X,-\lambda}}(\mathscr{M} \otimes \Phi(\mathrm{DR}_X(\mathscr{L})), \mathscr{O}_{X^{\mathrm{an}}}(-\lambda))[d_X].
\end{aligned}
$$

Now let us take as $\mathscr{M}$ the quasi-G-equivariant $\mathscr{D}_{X,-\lambda}$-module

$$\mathscr{M}_0 := \mathscr{D}_{X,-\lambda} \otimes \Omega_X^{\otimes -1}.$$

Then we have by Proposition 9.3.1

$$\mathbf{L}j^*(\mathscr{M}_0 \overset{\mathbf{D}}{\circ} \mathrm{Ind}_K^G(\mathscr{L})) \simeq \mathbf{R}\,\Gamma(X; \mathscr{L}).$$

On the other hand, we have

$$\mathbf{R}\,\mathrm{Hom}^{\mathrm{top}}_{\mathscr{D}_{X,-\lambda}}(\mathscr{M}_0 \otimes \Phi(L), \mathscr{O}_{X^{\mathrm{an}}}(-\lambda))$$
$$\simeq \mathbf{R}\,\mathrm{Hom}^{\mathrm{top}}_{\mathbb{C}}(\Phi(L), \Omega_{X^{\mathrm{an}}} \otimes_{\mathscr{O}_{X^{\mathrm{an}}}} \mathscr{O}_{X^{\mathrm{an}}}(-\lambda))$$
$$\simeq \mathbf{R}\,\mathrm{Hom}^{\mathrm{top}}_{\mathbb{C}}(\Phi(L), \mathscr{O}_{X^{\mathrm{an}}}(-\lambda + 2\rho)).$$

Here the last isomorphism follows from $\Omega_X \simeq \mathscr{O}_X(2\rho)$.

Thus we obtain the following theorem.

Theorem 9.6.4. *For* $\mathscr{L} \in \mathrm{D}_K^{\mathrm{b}}(\mathscr{D}_{X,\lambda})$, *we have an isomorphism*

$$\begin{aligned} &\mathbf{R}\,\mathrm{Hom}^{\mathrm{top}}_{(\mathfrak{g},K_{\mathbb{R}})}(\mathbf{R}\,\Gamma(X; \mathscr{L}), \mathrm{C}^{\infty}(G_{\mathbb{R}})) \\ &\qquad \simeq \mathbf{R}\,\mathrm{Hom}^{\mathrm{top}}_{\mathbb{C}}(\Phi(\mathrm{DR}_X(\mathscr{L})), \mathscr{O}_{X^{\mathrm{an}}}(-\lambda + 2\rho))[\mathrm{d}_X] \end{aligned}$$

(9.6.6)

in $\mathrm{D}^{\mathrm{b}}(\mathbf{FN}_{G_{\mathbb{R}}})$.

Taking their dual, we obtain the following theorem.

Theorem 9.6.5. *For* $\mathscr{L} \in \mathrm{D}_K^{\mathrm{b}}(\mathscr{D}_{X,\lambda})$, *we have an isomorphism*

$$\begin{aligned} &\Gamma_c(G_{\mathbb{R}}; \mathscr{D}ist_{G_{\mathbb{R}}}) \overset{\mathbf{L}}{\otimes}_{(\mathfrak{g},K_{\mathbb{R}})} \mathbf{R}\,\Gamma(X; \mathscr{L}) \\ &\qquad \simeq \mathbf{R}\,\Gamma_c^{\mathrm{top}}(X^{\mathrm{an}}; \Phi(\mathrm{DR}_X(\mathscr{L})) \otimes \mathscr{O}_{X^{\mathrm{an}}}(\lambda)) \end{aligned}$$

(9.6.7)

in $\mathrm{D}^{\mathrm{b}}(\mathbf{DFN}_{G_{\mathbb{R}}})$.

These results were conjectured in [14, Conjecture 3].

10 Vanishing Theorems

10.1 Preliminary

In this section, let us show that, for any Harish-Chandra module M, the object $\mathbf{R}\,\mathrm{Hom}^{\mathrm{top}}_{(\mathfrak{g},K_{\mathbb{R}})}(M, \mathrm{C}^{\infty}(G_{\mathbb{R}}))$ of $\mathrm{D}^{\mathrm{b}}(\mathbf{FN}_{G_{\mathbb{R}}})$ is strict and

$$\mathrm{Ext}^n_{(\mathfrak{g},K_{\mathbb{R}})}(M, \mathrm{C}^{\infty}(G_{\mathbb{R}})) := H^n\big(\mathbf{R}\,\mathrm{Hom}^{\mathrm{top}}_{(\mathfrak{g},K_{\mathbb{R}})}(M, \mathrm{C}^{\infty}(G_{\mathbb{R}}))\big) = 0 \quad \text{for } n \neq 0.$$

In order to prove this, we start by the calculation of the both sides of (9.6.6) for a K-equivariant holonomic $\mathscr{D}_{X,\lambda}$-module $\mathscr{L}$ such that

$$(10.1.1) \qquad \mathscr{L} \xrightarrow{\sim} j_* j^* \mathscr{L},$$

where $j\colon Kx_0 \hookrightarrow X$ is the open embedding of the open K-orbit Kx_0 into X. There exists a cartesian product

$$
\begin{array}{ccc}
Kx_0 & \overset{j}{\lhook\joinrel\longrightarrow} & X \\
\big\downarrow & \square & \big\downarrow{\scriptstyle\pi} \\
Kx_0^{\min} & \lhook\joinrel\longrightarrow & X_{\min}.
\end{array}
$$

Since $Kx_0^{\min} \cong K/M$ is an affine variety, $Kx_0^{\min} \to X_{\min}$ is an affine morphism, and hence $j\colon Kx_0 \hookrightarrow X$ is an affine morphism. Therefore

$$(10.1.2) \qquad \mathbf{D}^n j_* j^* \mathscr{M} = 0 \quad \text{for } n \neq 0 \text{ and an arbitrary } \mathscr{M} \in \mathrm{Mod}\,(\mathscr{D}_X).$$

Hence by the hypothesis (10.1.1), we have

$$(10.1.3) \qquad \mathscr{L} \xrightarrow{\sim} \mathbf{D} j_* j^* \mathscr{L}.$$

Let V be the stalk $\mathscr{L}(x_0)$ regarded as a $(K \cap B)$-module. Then its infinitesimal action coincides with $\mathfrak{k} \cap \mathfrak{b} \to \mathfrak{b} \xrightarrow{\lambda} \mathbb{C} \to \mathrm{End}_{\mathbb{C}}(V)$ by Lemma 9.1.1.

Hence, if $\mathscr{L} \neq 0$, then we have

$$(10.1.4) \quad \lambda|_{\mathfrak{t} \cap \mathfrak{k}} \text{ is integral, in particular } \langle \alpha^\vee, \lambda \rangle \in \mathbb{Z} \text{ for any } \alpha \in \Delta_k^+.$$

Recall that we say $\lambda|_{\mathfrak{t} \cap \mathfrak{k}}$ is integral if $\lambda|_{\mathfrak{t} \cap \mathfrak{k}}$ is the differential of a character of $K \cap T = M \cap T$.

Conversely, for a $(K \cap B)$-module V whose infinitesimal action coincides with $\mathfrak{k} \cap \mathfrak{b} \to \mathfrak{b} \xrightarrow{\lambda} \mathbb{C} \to \mathrm{End}_{\mathbb{C}}(V)$, there exists a K-equivariant $\mathscr{D}_{X,\lambda}$-module $\mathscr{L}$ such that it satisfies (10.1.1) and $\mathscr{L}(x_0) \simeq V$ (see Lemma 9.1.1).

10.2 Calculation (I)

Let $\mathscr{L}$ be a K-equivariant coherent $\mathscr{D}_{X,\lambda}$-module satisfying (10.1.1).

Recall that $\pi\colon X \simeq G/B \to X_{\min} = G/P$ is a canonical morphism. Let $s\colon X_0 := \pi^{-1}(x_0^{\min}) \to X$ be the embedding. Then $X_0 \simeq P/B \simeq M/(M \cap B)$ is the flag manifold of M. Note that $\mathscr{L}|_{Kx_0}$ is a locally free $\mathscr{O}_{Kx_0}$-module ($Kx_0 = \pi^{-1}(Kx_0^{\min})$ is an open subset of X). Hence we have $\mathbf{D} s^* \mathscr{L} \simeq s^* \mathscr{L}$. Since X_0 is the flag manifold of M and $s^* \mathscr{L}$ is a $\mathscr{D}_{X_0,\lambda}$-module, we have by Theorem 9.2.2

$$(10.2.1) \qquad H^n(X_0; s^* \mathscr{L}) = 0 \quad \text{for } n \neq 0$$

under the condition:

(10.2.2) $\lambda|_{\mathfrak{t}\cap\mathfrak{k}}$ is integral and $\langle\alpha^{\vee},\lambda\rangle\in\mathbb{Z}_{\leqslant 0}$ for $\alpha\in\Delta_k^+$.

Hence $R^n\pi_*\mathscr{L}|_{Kx_0^{\min}}=0$ for $n\neq 0$, and we have

(10.2.3) $H^n(X;\mathscr{L})=H^n(Kx_0;\mathscr{L})=H^n(Kx_0^{\min};\pi_*(\mathscr{L}))$.

Since $Kx_0^{\min}\cong K/M$ is an affine variety, we obtain

(10.2.4) $\begin{aligned}&H^n(X;\mathscr{L})=0 \quad \text{for } n\neq 0\\ &\qquad\text{under the conditions (10.1.1) and (10.2.2).}\end{aligned}$

Now let us calculate $\Gamma(Kx_0;\mathscr{L})$. The sheaf $\mathscr{L}$ is a K-equivariant vector bundle on Kx_0. We have $Kx_0=K/(K\cap B)$. Hence $\mathscr{L}$ is determined by the isotropy representation of $K\cap B$ on the stalk $V:=\mathscr{L}(x_0)$ of $\mathscr{L}$ at x_0. We have as a K-module

(10.2.5) $\Gamma(X;\mathscr{L})=\Gamma(Kx_0;\mathscr{L})\cong(\mathscr{O}_K(K)\otimes V)^{K\cap B}$.

Here the action of $K\cap B$ on $\mathscr{O}_K(K)\otimes V$ is the diagonal action where the action on $\mathscr{O}_K(K)$ is through the right multiplication of $K\cap B$ on K. The superscript $K\cap B$ means the space of $(K\cap B)$-invariant vectors. The K-module structure on $(\mathscr{O}_K(K)\otimes V)^{K\cap B}$ is through the left K-action on K.

Thus we obtain the following proposition.

Proposition 10.2.1. *Assume that λ satisfies (10.2.2) and a K-equivariant holonomic $\mathscr{D}_{X,\lambda}$-module $\mathscr{L}$ satisfies (10.1.1), and set $V=\mathscr{L}(x_0)$. Then we have*

(10.2.6) $H^n(X;\mathscr{L})\cong\begin{cases}(\mathscr{O}_K(K)\otimes V)^{K\cap B} & \text{for } n=0,\\ 0 & \text{for } n\neq 0\end{cases}$

as a K-module.

For a $(\mathfrak{g},K)$-module M, we shall calculate $\mathrm{Hom}_{(\mathfrak{g},K)}(M,\Gamma(X;\mathscr{L}))$. We have the isomorphism $\mathrm{Hom}_K(M,\Gamma(X;\mathscr{L}))\xrightarrow{\sim}\mathrm{Hom}_{K\cap B}(M,V)$ by the evaluation map $\psi\colon\Gamma(X;\mathscr{L})\to\mathscr{L}(x_0)=V$. Since $\mathscr{L}$ is a $\mathscr{D}_{X,\lambda}$-module, we have

(10.2.7) $\psi(At)=\langle\lambda,A\rangle\psi(t)$ for any $A\in\mathfrak{b}$ and $t\in\Gamma(X;\mathscr{L})$.

Indeed, $L_X(A)-\langle\lambda,A\rangle\in\mathfrak{m}_{x_0}\mathscr{D}_{X,\lambda}$ for any $A\in\mathfrak{b}$, where $\mathfrak{m}_{x_0}$ is the maximal ideal of $(\mathscr{O}_X)_{x_0}$.

Lemma 10.2.2. *For any $(\mathfrak{g},K)$-module M, and $\mathscr{L}\in\mathrm{Mod}_K(\mathscr{D}_{X,\lambda})$ satisfying (10.1.1), we have*

(10.2.8) $\begin{aligned}&\mathrm{Hom}_{(\mathfrak{g},K)}(M,\Gamma(X;\mathscr{L}))\\ &\qquad\cong\{f\in\mathrm{Hom}_{K\cap B}(M,\mathscr{L}(x_0));\\ &\qquad\qquad f(As)=\langle\lambda,A\rangle f(s)\ \text{for any }A\in\mathfrak{b}\ \text{and }s\in M\}.\end{aligned}$

Proof. Set $V = \mathscr{L}(x_0)$.

For $h \in \mathrm{Hom}_{(\mathfrak{g},K)}(M, \Gamma(X; \mathscr{L}))$, let $f \in \mathrm{Hom}_{K \cap B}(M, V)$ be the element $\psi \circ h$. Since h is $\mathfrak{g}$-linear, (10.2.7) implies that f satisfies the condition: $f(As) = \psi(h(As)) = \psi(Ah(s)) = \langle \lambda, A \rangle \psi(h(s)) = \langle \lambda, A \rangle f(s)$ for any $A \in \mathfrak{b}$ and $s \in M$.

Conversely, for $f \in \mathrm{Hom}_{K \cap B}(M, V)$ such that $f(As) = \langle \lambda, A \rangle f(s)$ for $A \in \mathfrak{b}$ and $s \in M$, let $h \in \mathrm{Hom}_K(M, \Gamma(X; \mathscr{L}))$ be the corresponding element: $\psi(h(s)) = f(s)$.

Then, we obtain

(10.2.9) $h(As) = Ah(s)$ at $x = x_0$ for any $A \in \mathfrak{g}$.

Indeed, we have $\mathfrak{g} = \mathfrak{k} + \mathfrak{b}$. The equation (10.2.9) holds for $A \in \mathfrak{k}$ by the K-equivariance of h, and also for $A \in \mathfrak{b}$ because

$$f(As) = \langle \lambda, A \rangle f(s) = \langle \lambda, A \rangle \psi(h(s)) = \psi(Ah(s)).$$

Since h is K-equivariant, $h(As) = Ah(s)$ holds at any point of Kx_0. Therefore we have $h(As) = Ah(s)$.   Q.E.D.

10.3 Calculation (II)

Let $\mathscr{L} \in \mathrm{Mod}_{K, \mathrm{coh}}(\mathscr{D}_{X, \lambda})$, and set $L = \mathrm{DR}_X(\mathscr{L}) \in \mathrm{D}^b_{K^{\mathrm{an}}, \tau_{-\lambda}}(\mathbb{C}_{X^{\mathrm{an}}})$. Now, we shall calculate $\mathbf{R}\mathrm{Hom}^{\mathrm{top}}_{\mathbb{C}}(\Phi(L), \mathscr{O}_{X^{\mathrm{an}}}(-\lambda + 2\rho))[d_X]$, the right-hand side of (9.6.6), under the conditions (10.1.1) and (10.2.2). We do it forgetting the topology and the equivariance.

By the assumption (10.2.2), we can decompose $\lambda = \lambda_1 + \lambda_0$ where λ_1 is integral and $\lambda_0|_{\mathfrak{k} \cap \mathfrak{t}} = 0$. Then λ_0 may be regarded as a P-invariant map $\mathrm{Lie}(P) = \mathfrak{m} \oplus \mathfrak{a} \oplus \mathfrak{n} \to \mathbb{C}$. Hence, we can consider the twisting data $\tau_{\lambda_0, X^{\mathrm{an}}_{\mathrm{min}}}$ on $X^{\mathrm{an}}_{\mathrm{min}}$. Then, the twisting data τ_{λ_0} on X^{an} is isomorphic to $\pi^* \tau_{\lambda_0, X^{\mathrm{an}}_{\mathrm{min}}}$. Since the twisting data τ_{λ_1} is trivial, we have $\tau_\lambda \cong \pi^* \tau_{\lambda_0, X^{\mathrm{an}}_{\mathrm{min}}}$.

Since $\mathscr{L} \simeq \mathbf{D}j_* j^* \mathscr{L}$, we have $L \simeq \mathbf{R}j_* j^* L$. Hence, (9.4.4) implies that

(10.3.1) $$\Phi(L) = j^a_*(L|_{G_{\mathbb{R}} x_0}).$$

Here, $j^a \colon G_{\mathbb{R}} x_0 \hookrightarrow X$ is the closed embedding. We can regard $L|_{G_{\mathbb{R}} x_0}$ as a $G_{\mathbb{R}}$-equivariant $(\pi^* \tau_{-\lambda_0, X^{\mathrm{an}}_{\mathrm{min}}})$-twisted local system

Then there exists a $G_{\mathbb{R}}$-equivariant $\tau_{-\lambda_0, X^{\mathrm{an}}_{\mathrm{min}}}$-twisted local system $\widetilde{L}$ on $X^{\mathbb{R}}_{\mathrm{min}}$ such that $L|_{G_{\mathbb{R}} x_0} \simeq (\pi^{\mathrm{an}})^{-1} \widetilde{L}$, because the fiber of π^{an} is simply connected.

Hence, we have

$$\mathbf{R}\mathrm{Hom}_{\mathbb{C}}(\Phi(L), \mathscr{O}_{X^{\mathrm{an}}}(-\lambda + 2\rho))[d_X]$$
$$\simeq \mathbf{R}\mathrm{Hom}_{\mathbb{C}}\left((\pi^{\mathrm{an}})^{-1} \widetilde{L}, \mathscr{O}_{X^{\mathrm{an}}}(-\lambda + 2\rho)\right)[d_X]$$
$$\simeq \mathbf{R}\mathrm{Hom}_{\mathbb{C}}\left(\widetilde{L}, \mathbf{R}(\pi^{\mathrm{an}})_* \mathscr{O}_{X^{\mathrm{an}}}(-\lambda + 2\rho)\right)[d_X].$$

On the other hand, we have

$$\mathbf{R}(\pi^{\mathrm{an}})_* \mathscr{O}_{X^{\mathrm{an}}}(-\lambda + 2\rho) \simeq \mathscr{O}_{X^{\mathrm{an}}_{\min}}(-\lambda_0) \otimes \mathbf{R}(\pi^{\mathrm{an}})_* \mathscr{O}_{X^{\mathrm{an}}}(-\lambda_1 + 2\rho),$$

and we have, by the Serre-Grothendieck duality,

$$\mathbf{R}\pi_* \mathscr{O}_X(-\lambda_1 + 2\rho)[\mathrm{d}_X] \simeq \mathbf{R}\pi_* \mathbf{R}\mathscr{H}om_{\mathscr{O}_X}(\mathscr{O}_X(\lambda_1), \Omega_X)[\mathrm{d}_X]$$
$$\simeq \mathbf{R}\mathscr{H}om_{\mathscr{O}_{X_{\min}}}(\mathbf{R}\pi_* \mathscr{O}_X(\lambda_1), \Omega_{X_{\min}})[\mathrm{d}_{X_{\min}}].$$

Since $\lambda_1|_{\mathfrak{k}\cap\mathfrak{t}} = \lambda|_{\mathfrak{k}\cap\mathfrak{t}}$ is anti-dominant, $\mathbf{R}\pi_* \mathscr{O}_X(\lambda_1)$ is concentrated at degree 0 by Theorem 9.2.2 (ii), and $\mathcal{V} = \pi_* \mathscr{O}_X(\lambda_1)$ is the G-equivariant locally free $\mathscr{O}_{X_{\min}}$-module associated with the representation

$$P \to MA \to \mathrm{Aut}(V_{\lambda_1}),$$

where V_{λ_1} is the irreducible (MA)-module with lowest weight λ_1 (see (9.2.3)).
Thus we obtain

$$\mathbf{R}\mathrm{Hom}_{\mathbb{C}}(\Phi(L), \mathscr{O}_{X^{\mathrm{an}}}(-\lambda + 2\rho))[\mathrm{d}_X]$$
$$\simeq \mathbf{R}\mathrm{Hom}_{\mathscr{O}_{X_{\min}}}(\mathbf{R}\pi_* \mathscr{O}_X(\lambda_1) \otimes \mathscr{O}_{X^{\mathrm{an}}_{\min}}(\lambda_0) \otimes \widetilde{L}, \Omega_{X^{\mathrm{an}}_{\min}})[\mathrm{d}_{X_{\min}}]$$
$$\simeq \mathbf{R}\mathrm{Hom}_{\mathscr{O}_{X^{\mathrm{an}}_{\min}}}(\mathcal{V}^{\mathrm{an}} \otimes \mathscr{O}_{X^{\mathrm{an}}_{\min}}(\lambda_0) \otimes \widetilde{L}, \Omega_{X^{\mathrm{an}}_{\min}})[\mathrm{d}_{X_{\min}}].$$

On the other hand, since $\widetilde{L}$ is supported on $X^{\mathbb{R}}_{\min}$,

$$\mathbf{R}\mathscr{H}om_{\mathbb{C}}(\widetilde{L}, \mathscr{O}_{X^{\mathrm{an}}_{\min}})[\mathrm{d}_{X_{\min}}] \simeq \mathbf{R}\mathscr{H}om_{\mathbb{C}}(\widetilde{L}, \mathbf{R}\Gamma_{X^{\mathbb{R}}_{\min}}(\mathscr{O}_{X^{\mathrm{an}}_{\min}}))[\mathrm{d}_{X_{\min}}]$$
$$\simeq \mathbf{R}\mathscr{H}om_{\mathbb{C}}(\widetilde{L}, \mathscr{B}_{X^{\mathbb{R}}_{\min}} \otimes or_{X^{\mathbb{R}}_{\min}}).$$

Here, $or_{X^{\mathbb{R}}_{\min}}$ is the orientation sheaf of $X^{\mathbb{R}}_{\min}$, and $\mathscr{B}_{X^{\mathbb{R}}_{\min}} = or_{X^{\mathbb{R}}_{\min}} \otimes \mathbf{R}\mathscr{H}om_{\mathbb{C}}(\mathbb{C}_{X^{\mathbb{R}}_{\min}}, \mathscr{O}_{X^{\mathrm{an}}_{\min}})[\mathrm{d}_{X_{\min}}]$ is the sheaf of hyperfunctions. Thus we obtain

$$\mathbf{R}\mathrm{Hom}_{\mathbb{C}}(\Phi(L), \mathscr{O}_{X^{\mathrm{an}}}(-\lambda + 2\rho))[\mathrm{d}_X]$$
$$\simeq \mathbf{R}\mathrm{Hom}_{\mathscr{O}_{X_{\min}}}(\mathcal{V} \otimes \Omega^{\otimes -1}_{X_{\min}} \otimes \mathscr{O}_{X^{\mathrm{an}}_{\min}}(\lambda_0) \otimes \widetilde{L} \otimes or_{X^{\mathbb{R}}_{\min}}, \mathscr{B}_{X^{\mathbb{R}}_{\min}}).$$

Note that $\mathscr{O}_{X^{\mathrm{an}}_{\min}}(\lambda_0)$ is a $\tau_{\lambda_0, X^{\mathrm{an}}_{\min}}$-twisted sheaf and $\widetilde{L}$ is a $\tau_{-\lambda_0, X^{\mathrm{an}}_{\min}}$-twisted sheaf. Hence $\mathscr{O}_{X^{\mathrm{an}}_{\min}}(\lambda_0) \otimes \widetilde{L}$ is a (non-twisted) locally free $\mathscr{O}_{X^{\mathrm{an}}_{\min}}|_{X^{\mathbb{R}}_{\min}}$-module. Hence, so is $\mathcal{V}^{\mathrm{an}} \otimes \Omega^{\otimes -1}_{X_{\min}} \otimes \mathscr{O}_{X^{\mathrm{an}}_{\min}}(\lambda_0) \otimes \widetilde{L} \otimes or_{X^{\mathbb{R}}_{\min}}$. Since $\mathscr{B}_{X^{\mathbb{R}}_{\min}}$ is a flabby sheaf, we have

$$H^n\left(\mathbf{R}\mathrm{Hom}_{\mathscr{O}_{X^{\mathrm{an}}_{\min}}}(\mathcal{V}^{\mathrm{an}} \otimes \Omega^{\otimes -1}_{X_{\min}} \otimes \mathscr{O}_{X^{\mathrm{an}}_{\min}}(\lambda_0) \otimes \widetilde{L} \otimes or_{X^{\mathbb{R}}_{\min}}, \mathscr{B}_{X^{\mathbb{R}}_{\min}})\right) = 0 \text{ for } n \neq 0.$$

Hence, we obtain

$$H^n\left(\mathbf{R}\mathrm{Hom}_{\mathbb{C}}(\Phi(L), \mathscr{O}_{X^{\mathrm{an}}}(-\lambda + 2\rho)[\mathrm{d}_X])\right) = 0 \text{ for } n \neq 0.$$

Proposition 10.3.1. *Assume that* $\lambda \in \mathfrak{t}^*$ *satisfies* (10.2.2), *and let* $\mathscr{L}$ *be a* K-*equivariant* $\mathscr{D}_{X,\lambda}$-*module satisfying* (10.1.1). *Then we have*

(i) $\mathbf{R}\mathrm{Hom}^{\mathrm{top}}_{(\mathfrak{g},K_{\mathbb{R}})}(\Gamma(X;\mathscr{L}),\mathrm{C}^\infty(G_{\mathbb{R}})) \in \mathrm{D}^{\mathrm{b}}(\mathbf{FN}_{G_{\mathbb{R}}})$ *is strict, and*

(ii) $H^n\big(\mathbf{R}\mathrm{Hom}^{\mathrm{top}}_{(\mathfrak{g},K_{\mathbb{R}})}(\Gamma(X;\mathscr{L}),\mathrm{C}^\infty(G_{\mathbb{R}}))\big) = 0$ *for* $n \neq 0$.

Proof. Set $M = \Gamma(X;\mathscr{L})$. By (9.6.6), we have

$$\mathbf{R}\mathrm{Hom}^{\mathrm{top}}_{(\mathfrak{g},K_{\mathbb{R}})}(M,\mathrm{C}^\infty(G_{\mathbb{R}})) \simeq \mathbf{R}\mathrm{Hom}^{\mathrm{top}}_{\mathbb{C}}\big(\Phi(L),\mathscr{O}_{X^{\mathrm{an}}}(-\lambda+2\rho)[d_X]\big).$$

Hence, forgetting the topology and the equivariance, the cohomology groups of $\mathbf{R}\mathrm{Hom}^{\mathrm{top}}_{(\mathfrak{g},K_{\mathbb{R}})}(M,\mathrm{C}^\infty(G_{\mathbb{R}}))$ are concentrated at degree 0. On the other hand, $\mathbf{R}\mathrm{Hom}^{\mathrm{top}}_{(\mathfrak{g},K_{\mathbb{R}})}(M,\mathrm{C}^\infty(G_{\mathbb{R}}))$ is represented by a complex in $\mathbf{FN}_{G_{\mathbb{R}}}$ whose negative components vanish. Hence it is a strict complex.   Q.E.D.

10.4 Vanishing Theorem

By using the result of the preceding paragraph, we shall prove the following statement.

Theorem 10.4.1. *Let* N *be a Harish-Chandra module. Then we have*

(i) $\mathbf{R}\mathrm{Hom}^{\mathrm{top}}_{(\mathfrak{g},K_{\mathbb{R}})}(N,\mathrm{C}^\infty(G_{\mathbb{R}})) \in \mathrm{D}^{\mathrm{b}}(\mathbf{FN}_{G_{\mathbb{R}}})$ *is strict,*

(ii) $H^n(\mathbf{R}\mathrm{Hom}^{\mathrm{top}}_{(\mathfrak{g},K_{\mathbb{R}})}(N,\mathrm{C}^\infty(G_{\mathbb{R}}))) = 0$ *for* $n \neq 0$.

Proof. Since $\mathbf{R}\mathrm{Hom}^{\mathrm{top}}_{(\mathfrak{g},K_{\mathbb{R}})}(N,\mathrm{C}^\infty(G_{\mathbb{R}}))$ is represented by a complex in $\mathbf{FN}_{G_{\mathbb{R}}}$ whose negative components vanish, it is enough to show that, forgetting topology,

(10.4.1) $\mathrm{Ext}^n_{(\mathfrak{g},K_{\mathbb{R}})}(N,\mathrm{C}^\infty(G_{\mathbb{R}})) = 0$ for $n \neq 0$.

We shall prove this by the descending induction on n. If $n \gg 0$, this is obvious because the global dimension of $\mathrm{Mod}\,(\mathfrak{g},K)$ is finite.

We may assume that N is simple without the loss of generality.

By [2, 5], $N/\tilde{\mathfrak{n}}N \neq 0$, where $\tilde{\mathfrak{n}} = [\mathfrak{b},\mathfrak{b}]$ is the nilpotent radical of $\mathfrak{b}$. Since the center $\mathfrak{z}(\mathfrak{g})$ acts by scalar on N, $N/\tilde{\mathfrak{n}}N$ is $U(\mathfrak{t})$-finite. Hence there exists a surjective $(\mathfrak{t},T\cap K)$-linear homomorphism $N/\tilde{\mathfrak{n}}N \twoheadrightarrow V$ for some one-dimensional $(\mathfrak{t},T\cap K)$-module V. Let $\lambda \in \mathfrak{t}^*$ be the character of V. Since $S/(\mathfrak{t}\cap\tilde{\mathfrak{n}})S \to V$ is a surjective homomorphism for some irreducible M-submodule S of N, $\lambda|_{\mathfrak{t}\cap\mathfrak{t}}$ is the lowest weight of S, and hence λ satisfies (10.2.2).

Let us take a K-equivariant $(\mathscr{D}_{X,\lambda})|_{K_{x_0}}$- module $\mathscr{L}'$ such that $\mathscr{L}'(x_0) \cong V$ as $(B \cap K)$-modules, and set $\mathscr{L} = \mathbf{D}j_*\mathscr{L}'$.

Then by Lemma 10.2.2, $\mathrm{Hom}_{(\mathfrak{g},K)}(N,\Gamma(X;\mathscr{L}))$ contains a non-zero element. Thus we obtain an exact sequence of $(\mathfrak{g},K)$-modules

$$0 \to N \to M \to M' \to 0 \quad \text{with } M = \Gamma(X;\mathscr{L}).$$

This gives an exact sequence

$$\mathrm{Ext}^n_{(\mathfrak{g},K_{\mathbb{R}})}(M, \mathrm{C}^\infty(G_{\mathbb{R}})) \to \mathrm{Ext}^n_{(\mathfrak{g},K_{\mathbb{R}})}(N, \mathrm{C}^\infty(G_{\mathbb{R}}))$$
$$\to \mathrm{Ext}^{n+1}_{(\mathfrak{g},K_{\mathbb{R}})}(M', \mathrm{C}^\infty(G_{\mathbb{R}})),$$

in which the first term vanishes for $n > 0$ by Proposition 10.3.1 and the last term vanishes by the induction hypothesis. Thus we obtain the desired result.

Q.E.D.

By duality, we obtain the following proposition.

Theorem 10.4.2. *Let N be a Harish-Chandra module. Then we have*

(i) $\Gamma_c(G_{\mathbb{R}}; \mathscr{D}ist_{G_{\mathbb{R}}}) \overset{\mathrm{L}}{\otimes}_{(\mathfrak{g},K_{\mathbb{R}})} N \in \mathrm{D}^b(\mathbf{DFN}_{G_{\mathbb{R}}})$ *is strict,*

(ii) $H^n(\Gamma_c(G_{\mathbb{R}}; \mathscr{D}ist_{G_{\mathbb{R}}}) \overset{\mathrm{L}}{\otimes}_{(\mathfrak{g},K_{\mathbb{R}})} N) = 0$ *for $n \neq 0$.*

Recall that the maximal globalization functor $\mathrm{MG}\colon \mathrm{HC}(\mathfrak{g}, K) \to \mathbf{FN}_{G_{\mathbb{R}}}$ is given by

$$\mathrm{MG}(M) = H^0\big(\mathbf{R}\mathrm{Hom}^{\mathrm{top}}_{(\mathfrak{g},K_{\mathbb{R}})}(M^*, \mathrm{C}^\infty(G_{\mathbb{R}}))\big)$$

and the minimal globalization functor $\mathrm{mg}\colon \mathrm{HC}(\mathfrak{g}, K) \to \mathbf{DFN}_{G_{\mathbb{R}}}$ is given by

$$\mathrm{mg}(M) = H^0\big(\Gamma_c(G_{\mathbb{R}}; \mathscr{D}ist_{G_{\mathbb{R}}}) \overset{\mathrm{L}}{\otimes}_{(\mathfrak{g},K_{\mathbb{R}})} M\big).$$

We denote by $\mathrm{MG}_{G_{\mathbb{R}}}$ (resp. $\mathrm{mg}_{G_{\mathbb{R}}}$) the subcategory of $\mathbf{FN}_{G_{\mathbb{R}}}$ (resp. $\mathbf{DFN}_{G_{\mathbb{R}}}$) consisting of objects isomorphic to $\mathrm{MG}(M)$ (resp. $\mathrm{mg}(M)$) for a Harish-Chandra module M (see § 1.1). Then both $\mathrm{MG}_{G_{\mathbb{R}}}$ and $\mathrm{mg}_{G_{\mathbb{R}}}$ are equivalent to the category $\mathrm{HC}(\mathfrak{g}, K)$ of Harish-Chandra modules.

The above theorem together with Theorem 10.4.1 shows the following result.

Theorem 10.4.3.   (i) *The functor $M \mapsto \mathrm{MG}(M)$ (resp. $M \mapsto \mathrm{mg}(M)$) is an exact functor from the category $\mathrm{HC}(\mathfrak{g}, K)$ of Harish-Chandra modules to $\mathbf{FN}_{G_{\mathbb{R}}}$ (resp. $\mathbf{DFN}_{G_{\mathbb{R}}}$).*

(ii) *Any morphism in $\mathrm{MG}_{G_{\mathbb{R}}}$ or $\mathrm{mg}_{G_{\mathbb{R}}}$ is strict in $\mathbf{FN}_{G_{\mathbb{R}}}$ or $\mathbf{DFN}_{G_{\mathbb{R}}}$ (i.e., with a closed range).*

(iii) *Any $G_{\mathbb{R}}$-invariant closed subspace of E in $\mathrm{MG}_{G_{\mathbb{R}}}$ (resp. $\mathrm{mg}_{G_{\mathbb{R}}}$) belongs to $\mathrm{MG}_{G_{\mathbb{R}}}$ (resp. $\mathrm{mg}_{G_{\mathbb{R}}}$).*

(iv) *$\mathrm{MG}_{G_{\mathbb{R}}}$ is closed by extensions in $\mathbf{FN}_{G_{\mathbb{R}}}$, namely, if $0 \to E' \to E \to E'' \to 0$ is a strict exact sequence in $\mathbf{FN}_{G_{\mathbb{R}}}$, and E' and E'' belong to $\mathrm{MG}_{G_{\mathbb{R}}}$, then so does E. A similar statement holds for $\mathrm{mg}_{G_{\mathbb{R}}}$.*

Here the exactness in (i) means that they send the short exact sequences to strictly exact sequences.

Proof. Let us only show the statements on the maximal globalization. (i) follows immediately from Theorem 10.4.1.

(ii)   Let M, M' be Harish-Chandra modules, and let $u\colon \mathrm{MG}(M) \to \mathrm{MG}(M')$ be a morphism in $\mathbf{FN}_{G_{\mathbb{R}}}$. Then

$$\psi := \mathrm{HC}(u)\colon M \simeq \mathrm{HC}(\mathrm{MG}(M)) \to \mathrm{HC}(\mathrm{MG}(M')) \simeq M'$$

is a morphism in $\mathrm{HC}(\mathfrak{g}, K)$ and $\mathrm{MG}(\psi) = u$. Let I be the image of ψ, Then $\mathrm{MG}(M) \to \mathrm{MG}(I)$ is surjective and $\mathrm{MG}(I)$ is a closed subspace of $\mathrm{MG}(M')$ by (i).

(iii)   Let M be a Harish-Chandra module and E a $G_{\mathbb{R}}$-invariant closed subspace of $\mathrm{MG}(M)$. Then $N := \mathrm{HC}(E) \subset M$ is a Harish-Chandra module and $\mathrm{MG}(N)$ is a closed subspace of $\mathrm{MG}(M)$ by (ii), and it contains N as a dense subspace. Since E is also the closure of N, $E = \mathrm{MG}(N)$.

(iv)   We have an exact sequence $0 \to \mathrm{HC}(E') \to \mathrm{HC}(E) \to \mathrm{HC}(E'') \to 0$. Since $\mathrm{HC}(E')$ and $\mathrm{HC}(E'')$ are Harish-Chandra modules, so is $\mathrm{HC}(E)$. Hence we have a commutative diagram with strictly exact rows:

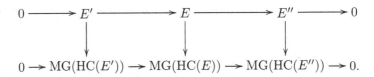

Since the left and right vertical arrows are isomorphisms, the middle vertical arrow is also an isomorphism.   Q.E.D.

Let us denote by $\mathrm{D}^{\mathrm{b}}_{\mathrm{MG}}(\mathbf{FN}_{G_{\mathbb{R}}})$ the full subcategory of $\mathrm{D}^{\mathrm{b}}(\mathbf{FN}_{G_{\mathbb{R}}})$ consisting of E such that E is strict and the cohomologies of E belong to $\mathrm{MG}_{G_{\mathbb{R}}}$. Similarly, we define $\mathrm{D}^{\mathrm{b}}_{\mathrm{mg}}(\mathbf{DFN}_{G_{\mathbb{R}}})$. Then the following result follows immediately from the preceding theorem and Lemma 2.3.1.

Corollary 10.4.4. *The category* $\mathrm{D}^{\mathrm{b}}_{\mathrm{MG}}(\mathbf{FN}_{G_{\mathbb{R}}})$ *is a triangulated full subcategory of* $\mathrm{D}^{\mathrm{b}}(\mathbf{FN}_{G_{\mathbb{R}}})$, *namely, it is closed by the translation functors, and closed by distinguished triangles (if* $E' \to E \to E'' \to E'[1]$ *is a distinguished triangle in* $\mathrm{D}^{\mathrm{b}}(\mathbf{FN}_{G_{\mathbb{R}}})$ *and* E' *and* E *belong to* $\mathrm{D}^{\mathrm{b}}_{\mathrm{MG}}(\mathbf{FN}_{G_{\mathbb{R}}})$, *then so does* E'').

This corollary together with Theorem 10.4.1 implies the following corollary.

Corollary 10.4.5. *If* $M \in \mathrm{D}^{\mathrm{b}}(\mathrm{Mod}_f(\mathfrak{g}, K))$ *has Harish-Chandra modules as cohomologies, then* $\mathbf{R}\mathrm{Hom}^{\mathrm{top}}_{(\mathfrak{g}, K_{\mathbb{R}})}(M, \mathrm{C}^{\infty}(G_{\mathbb{R}}))$ *belongs to* $\mathrm{D}^{\mathrm{b}}_{\mathrm{MG}}(\mathbf{FN}_{G_{\mathbb{R}}})$.

Hence we obtain the following theorem.

Theorem 10.4.6. *Let* $\lambda \in \mathfrak{t}^*$, $K \in \mathrm{D}^{\mathrm{b}}_{G_{\mathbb{R}}, \tau_{-\lambda}, \mathbb{R}\text{-}c}(\mathbb{C}_X)$ *and* $\mathcal{M} \in \mathrm{D}^{\mathrm{b}}_{\mathrm{coh}}(\mathscr{D}_{X,\lambda}, G)$. *Then we have*

(i) $\mathbf{R}\mathrm{Hom}^{\mathrm{top}}_{\mathscr{D}_{X,\lambda}}(\mathcal{M} \otimes K, \mathscr{O}_{X^{\mathrm{an}}}(\lambda))$ *belongs to* $\mathrm{D}^{\mathrm{b}}_{\mathrm{MG}}(\mathbf{FN}_{G_{\mathbb{R}}})$.

(ii) $\mathbf{R}\Gamma^{\mathrm{top}}_{\mathrm{c}}\big(X^{\mathrm{an}}; K \otimes \Omega_{X^{\mathrm{an}}}(-\lambda) \overset{\mathbf{L}}{\otimes}_{\mathscr{D}_{X,\lambda}} \mathcal{M}\big)$ *belongs to* $\mathrm{D}^{\mathrm{b}}_{\mathrm{mg}}(\mathbf{DFN}_{G_{\mathbb{R}}})$.

Proof. Since (ii) is the dual statement of (i), it is enough to prove (i). By Matsuki correspondence (Theorem 9.4.2), there exists $L \in D^b_{K^{\mathrm{an}}, \tau_{-\lambda}, \mathbb{C}\text{-c}}(\mathbb{C}_{X^{\mathrm{an}}})$ such that $K \simeq \Phi(L)$. By Theorem 9.6.1, there exists $\mathscr{L} \in D^b_{K, \mathrm{coh}}(\mathscr{D}_{X,\lambda})$ such that $\mathrm{DR}_X(\mathscr{L}) \simeq L$. Then Proposition 9.6.3 implies

$$\mathbf{R}\mathrm{Hom}^{\mathrm{top}}_{\mathscr{D}_{X,\lambda}}(\mathscr{M} \otimes K, \mathscr{O}_{X^{\mathrm{an}}}(\lambda))$$

$$\simeq \mathbf{R}\mathrm{Hom}^{\mathrm{top}}_{(\mathfrak{g}, K_{\mathbb{R}})}(\mathbf{L}j^*(\mathscr{M} \overset{\mathbf{D}}{\circ} \mathrm{Ind}^G_K \mathscr{L}), C^{\infty}(S_{\mathbb{R}}))[-d_X].$$

Then the result follows from Corollary 10.4.5 and Proposition 9.3.2.   Q.E.D.

Let us illustrate Theorem 10.4.6 in the case $\mathscr{M} = \mathscr{D}_{X,\lambda}$ and K is a twisted $G_{\mathbb{R}}$-equivariant sheaf supported on a $G_{\mathbb{R}}$-orbit Z of X^{an}.

Let us take a point $x \in Z$. Let V be a finite-dimensional $G_{\mathbb{R}} \cap B(x)$-module whose differential coincides with $\mathfrak{g}_{\mathbb{R}} \cap \mathfrak{b}(x) \to \mathfrak{b}(x) \overset{\lambda}{\to} \mathbb{C} \to \mathrm{End}_{\mathbb{C}}(V)$.

Then the Cauchy-Riemann equations give a complex

$$(10.4.2) \qquad \left(\mathscr{B}(G_{\mathbb{R}}) \otimes V \otimes \overset{\bullet}{\bigwedge} \mathfrak{n}(x) \right)^{G_{\mathbb{R}} \cap B(x)}.$$

Then its cohomology groups belong to $\mathrm{MG}_{G_{\mathbb{R}}}$.

Indeed, if F is the τ_{λ}-twisted local system on Z associated with V^* (see Lemma 9.1.1), then (10.4.2) is isomorphic to $\mathbf{R}\mathrm{Hom}^{\mathrm{top}}_{\mathscr{D}_{X,\lambda}}(\mathscr{M} \otimes i_! F, \mathscr{O}_{X^{\mathrm{an}}}(\lambda))$ (up to a shift). Here $i \colon Z \to X^{\mathrm{an}}$ is the embedding,

References

1. A. Beilinson and J. Bernstein, *Localization de $\mathfrak{g}$-modules*, C. R. Acad. Sci. Paris, **292** (1981), 15–18.

2. ———, ———, *A generalization of Casselman's submodule theorem*, Representation theory of reductive groups (Park City, Utah, 1982), Progr. Math., **40**, Birkhäuser Boston, Boston, MA, (1983) 35–52.

3. A. Beilinson, J. Bernstein and P. Deligne, *Faisceaux Pervers*, Astérisque, **100** (1982).

4. J. Bernstein and V. Lunts, *Equivariant sheaves and functors*, Lecture Notes in Mathematics, **1578**, Springer-Verlag, Berlin, (1994).

5. W. Casselman, *Jacquet modules for real reductive groups*, Proceedings of the International Congress of Mathematicians (Helsinki, 1978), Acad. Sci. Fennica, Helsinki, (1980) 557–563.

6. A. Grothendieck, *Cohomologie locale des faisceaux cohérents et théorèmes de Lefschetz locaux et globaux* (SGA 2), Advanced Studies in Pure Mathematics, **2**, North-Holland Publishing Co. (1968).

7. Harish-Chandra, *Representation of a semisimple Lie group on a Banach space I*, Trans. Amer. Math., Soc., **75** (1953), 185–243.

8. ———, *The characters of semisimple Lie groups*, ibid., **83** (1956), 98–163.

9. ———, *Invariant eigendistributions on semisimple Lie groups*, ibid., **119** (1965), 457–508.

10. ____, *Discrete series for semisimple Lie groups I*, Acta Math. **113** (1965), 241–318.

11. H. Hecht and J. L. Taylor, *Analytic localization of group representations*. Adv. Math. **79** no. 2 (1990), 139–212.

12. M. Kashiwara, *The Riemann-Hilbert problem for holonomic systems*, Publ. Res. Inst. Math. Sci. **20** (1984), no. 2, 319–365.

13. ____, *Character, character cycle, fixed point theorem and group representations*, Advanced Studies in Pure Math., **14** (1988) 369–378.

14. ____, *Open problems in group representation theory*, Proceeding of Taniguchi Symposium held in 1986, RIMS preprint 569.

15. ____, *Representation theory and $\mathscr{D}$-modules on flag manifolds*, Astérisque **173–174** (1989) 55–109.

16. ____, *D-modules and microlocal calculus*, Translated from the 2000 Japanese original by Mutsumi Saito, Translations of Mathematical Monographs **217** Iwanami Series in Modern Mathematics, American Mathematical Society, Providence, RI (2003).

17. M. Kashiwara, A. Kowata, K. Minemura, K. Okamoto, T. Oshima, M. Tanaka, *Eigenfunctions of invariant differential operators on a symmetric space*, Ann. of Math. (2) **107** (1978), no. 1, 1–39.

18. M. Kashiwara and P. Schapira, *Sheaves on Manifold*, Grundlehren der Mathematischen Wissenschaften **292**, Springer (1990).

19. ____, *Categories and sheaves*, Grundlehren der Mathematischen Wissenschaften **332**, Springer-Verlag, Berlin, (2006).

20. ____, *Moderate and formal cohomology associated with constructible sheaves*, Mém. Soc. Math. France (N.S.) **64**, (1996).

21. M. Kashiwara and W. Schmid, *Quasi-equivariant D-modules, equivariant derived category, and representations of reductive Lie groups*, Lie theory and geometry, Progr. Math., **123**, Birkhäuser, (1994) 457–488.

22. T. Matsuki, *Orbits on affine symmetric spaces under the action of parabolic subgroups*, Hiroshima Math., J., **12** (1982), 307–320.

23. I. Mirković, T. Uzawa and K. Vilonen, *Matsuki correspondence for sheaves*, Invent. Math., **109** (1992), 231–245.

24. S. J. Prichepionok, *A natural topology for linear representations of semisimple Lie algebras*, Soviet Math. Doklady **17** (1976) 1564–1566.

25. W. Schmid, *Boundary value problems for group invariant differential equations*, Élie Cartan et les Mathématiques d'aujourd'hui, Astérisque, (1985), 311–322.

26. J.-P. Schneiders, *Quasi-abelian categories and sheaves*, Mém. Soc. Math. Fr. (N.S.) **76** (1999).

27. L. Smithies and J. L. Taylor, *An analytic Riemann-Hilbert correspondence for semi-simple Lie groups*, Represent. Theory **4** (2000), 398–445 (electronic)

28. H. Sumihiro, *Equivariant completion*, J. Math. Kyoto Univ. **14** (1974), 1–28.

29. W. Schmid and J. A. Wolf, *Globalizations of Harish-Chandra modules*, Bulletin of AMS, **17** (1987), 117–120.

List of Notations

Index

Amenability and Margulis Super-Rigidity

Alain Valette

Institut de Mathématiques, University of Neuchâtel, 11, Rue Emile Argand,
BP 158, CH-2009 Neuchâtel, Switzerland
`alain.valette@unine.ch`

1 Introduction

Amenability was introduced in 1929 by J. von Neumann [vN29] for discrete groups, and in 1950 by M. Day [Day50] for general locally compact groups. Originating from harmonic analysis and representation theory, amenability extended to a well-established body of mathematics, with applications in: dynamical systems, operator algebras, graph theory, metric geometry,... One definite advantage of amenability for groups is the equivalence of various, apparently remote, characterizations. So in Chapter 1 we survey the classical theory of amenability for a locally compact group G, (our basic reference being Appendix G in [BHV]) and we establish the equivalence between:

- G is amenable, in the sense that every action of G by homeomorphisms on a compact space X, fixes some probability measure on X;
- any affine G-action on a compact convex set (in a locally convex Hausdorff space) has a fixed point;
- G admits an invariant mean;

- (Reiter's property (P_1)) For every compact subset $Q \subset G$ and $\epsilon > 0$, there exists $f \in L^1(G)_{1,+}$ such that

$$\max_{x \in Q} \|\lambda_G(x)f - f\| \leq \epsilon;$$

- (Reiter's property (P_2)) The left regular representation λ_G on $L^2(G)$ almost has invariant vectors;
- the representation $\infty \lambda_G$ almost has invariant vectors.

In Chapter 2, we digress on ergodic theory for group actions on measure spaces. The goal of the chapter is to establish Moore's ergodicity theorem, stating that if Γ is a lattice in a non-compact simple Lie group G, and H is a non-compact closed subgroup of G, then Γ acts ergodically on G/H. We deduce it from the Howe-Moore vanishing theorem, stating that coefficients of unitary representations of G having no non-zero fixed vector, go to zero at infinity of G. Our basic reference for that chapter is [BM00].

In Chapter 3, we explain how amenability is used in the proof of Margulis' super-rigidity theorem. Although semisimple Lie groups are very far from being amenable, they contain a co-compact amenable subgroup (namely a minimal parabolic subgroup) and this fact, together with Moore's ergodicity theorem, plays a crucial role in super-rigidity. References for this Chapter are [Mar91] and [Zim84].

The presentation follows rather closely the CIME course taught at San Servolo in June 2004. I thank heartily Enrico Casadio Tarabusi, Andrea D'Agnolo and Massimo Picardello for bringing me to that magical place.

2 Amenability for Locally Compact Groups

2.1 Definition, Examples, and First Characterizations

For a compact space X, we denote by $M(X)$ the space of probability measures on X.

Definition 2.1. *A locally compact group G is **amenable** if, for every compact space X endowed with a G-action, there exists a G-fixed point in $M(X)$ (i.e. G fixes a probability measure on X)*

Example 2.1. : Compact groups are amenable.

Indeed, let dg be normalized Haar measure on the compact group G. Let G act on the compact space X. Pick any $\mu \in M(X)$. Then $\nu = \int_G (g_* \mu) \, dg$ is a fixed point in $M(X)$.

Example 2.2. : $SL_2(\mathbf{R})$ is not amenable.

To see this, let $SL_2(\mathbf{R})$ act by fractional linear transformations on $P^1(\mathbf{R}) = \mathbf{R} \cup \{\infty\}$. Then $SL_2(\mathbf{R})$ fixes no measure at all on $P^1(\mathbf{R})$. Indeed look at subgroups

$$N = \left\{ \begin{pmatrix} 1 & * \\ 0 & 1 \end{pmatrix} \right\} \quad (= \text{ translations on } \mathbf{R});$$

$$A = \left\{ \begin{pmatrix} a & 0 \\ 0 & a^{-1} \end{pmatrix} : a > 0 \right\} \quad (= \text{ dilations on } \mathbf{R});$$

The only N-invariant measures on $P^1(\mathbf{R})$ are of the form $s\, dx + t\, \delta_\infty$ (where dx is Lebesgue measure on $\mathbf{R}$). Among these, the only A-invariant measures are the $t\, \delta_\infty$'s. But those are not w-invariant, where $w = \begin{pmatrix} 0 & 1 \\ -1 & 0 \end{pmatrix}$ (so that $w(x) = \frac{-1}{x}$).

Proposition 2.1. *The following are equivalent:*

  i) G *is amenable;*
 ii) *Any affine action of* G *on a (non-empty) convex, compact subset of a Hausdorff, locally convex topological vector space, has a fixed point.*

Proof: $(ii) \Rightarrow (i)$: If X is a compact space, then $M(X)$ is a convex subset of $C(X)^*$ (space of all Borel measures on X), and $M(X)$ is compact in the weak-$*$ topology.

$(i) \Rightarrow (ii)$: Let C be a compact convex subset in E. To each $\mu \in M(C)$, we associate its barycentre $b(\mu) \in C$: this is the unique point in C such that, for every $f \in E^*$:

$$f(b(\mu)) = \int_C f(c)\, d\mu(c)$$

(formally: $b(\mu) = \int_C c\, d\mu(c)$). If $\mu = \sum_i \lambda_i \delta_{c_i}$ is an atomic measure, i.e. a convex combination of Dirac masses, then $b(\mu) = \sum_i \lambda_i c_i$, and this can be extended to $M(C)$ using density of atomic measures in $M(C)$ (see Theorem 3.27 in [Rud73] for details). Clearly b commutes with affine maps of C: $b(T_*\mu) = T(b(\mu))$. In particular, if μ is a G-fixed probability measure on C, then $b(\mu)$ is a G-fixed point in C. □

Here is the famous Markov-Kakutani theorem.

Theorem 2.1. *Every abelian group is amenable.*

Proof: Let G be an abelian group, acting on a compact convex subset C in E. For $g \in G$, define $A_n(g) : C \to C$ by

$$A_n(g)x = \frac{1}{n+1} \sum_{i=0}^{n} g^i x.$$

Let $\mathcal{G}$ be the semi-group generated by the $A_n(g)$'s ($n \geq 0$, $g \in G$). For every $\gamma \in \mathcal{G}$, the set $\gamma(C)$ is convex compact.

Claim: $\bigcap_{\gamma \in \mathcal{G}} \gamma(C) \neq \emptyset$

It is enough to see that $\gamma_1(C) \cap \ldots \cap \gamma_m(C) \neq \emptyset$, for $\gamma_1, \ldots, \gamma_m \in \mathcal{G}$. Set $\gamma = \gamma_1 \gamma_2 \ldots \gamma_m \in \mathcal{G}$. Since $\mathcal{G}$ is abelian: $\gamma(C) \subset \gamma_i(C)$ for $i = 1, \ldots, m$, proving the claim.

Take $x_0 \in \bigcap_{\gamma \in \mathcal{G}} \gamma(C)$. We claim that x_0 is $\mathcal{G}$-fixed.

For every $n \geq 0, g \in G$, there exists $x \in C$ such that $A_n(g)x = x_0$. For $f \in E^*$:

$$|f(x_0 - gx_0)| = \left| f\left(\frac{1}{n+1} \left(\sum_{i=0}^{n} g^i x - \sum_{i=0}^{n} g^{i+1} x \right) \right) \right|$$

$$= \frac{1}{n+1} |f(x - g^{n+1}x)| \leq \frac{2K}{n+1}$$

where $K = \max\{|f(c)| : c \in C\}$. So $f(x_0) = f(gx_0)$ for every $f \in E^*$, therefore $x_0 = gx_0$. $\qquad\Box$

Definition 2.2. A **mean** on G is a linear form m on $L^\infty(G)$, such that:

i) $m(1) = 1$;
ii) $m(f) \geq 0$ for every $f \in L^\infty(G)$, $f \geq 0$

Example 2.3. If μ is a Borel probability measure on G, absolutely continuous with respect to Haar measure, then $m(f) = \int_G f \, d\mu$ defines a mean on G.

There are some important differences between probability measures and means:

1. means make up a convex compact subset in $L^\infty(G)^*$ (for the weak-* topology);
2. for $A \in \mathcal{B}$ (the Borel subsets of G), let χ_A be the characteristic function on G; let m be a mean on G, set $m(A) = m(\chi_A)$. The map $m : \mathcal{B} \to [0, 1] : A \mapsto m(A)$ satisfies:
   (i) $m(G) = 1$;
   (ii) If $A_1, \ldots, A_n$ are pairwise disjoint, then $m(A_1 \cup \ldots \cup A_n) = m(A_1) + \ldots + m(A_n)$.
   This second property is **finite additivity** (as opposed to σ-additivity).

In other words, we may think of a mean as a probability measure which is only finitely additive.

Proposition 2.2. The following are equivalent:

i) G is amenable;
ii) G admits an invariant mean.

Proof: $(i) \Rightarrow (ii)$ Follows from compactness and convexity of the set of means.

$(ii) \Rightarrow (i)$ Let G act on a compact space X. Fix $x_0 \in X$. For $f \in C(X)$, set $\phi_f(g) = f(gx_0)$. The map

$$\phi : C(X) \to L^\infty(G) : f \mapsto \phi_f$$

is G-equivariant. So if m is an invariant mean on G, then $\mu(f) = m(\phi_f)$ defines a G-invariant linear functional on $C(X)$. This μ is positive, unital, so by the Riesz representation theorem it is a probability measure on X. $\qquad\square$

Example 2.4. The free group $\mathbb{F}_2$ on two generators a, b is *not* amenable.

To see it, assume by contradiction that $\mathbb{F}_2$ is amenable. Let m be an invariant mean. Set

$$A = \{w \in \mathbb{F}_2 : w \text{ starts with a non-zero power (positive or negative)} of a\}.$$

Then $A \cup aA = \mathbb{F}_2$, so $m(A) + m(aA) \geq 1$ and $m(A) = m(aA)$, so $m(A) \geq \frac{1}{2}$. On the other hand A, bA, b^2A are pairwise disjoint, so $m(A) + m(bA) + m(b^2A) \leq 1$; with $m(A) = m(bA) = m(b^2A)$, this gives $m(A) \leq \frac{1}{3}$, a contradiction.

2.2 Stability Properties

Proposition 2.3. *Every closed subgroup of an amenable group is amenable.*

We postpone the proof until the end of·section 2.5.

Proposition 2.4. *:*

i) *Every quotient of an amenable group is amenable.*
ii) *Let $1 \to N \to G \to G/N \to 1$ be a short exact sequence, with N closed, amenable in G. The following are equivalent:*
  - *G is amenable;*
  - *G/N is amenable.*

Proof: (i) Every action of G/N can be seen as an action of G.

(ii) Assume that G/N is amenable. Let G act affinely on a non-empty, compact convex subset C. Since N is amenable, the set C^N of N-fixed points is convex, compact and *non-empty*. Since N is normal, the set C^N is G-invariant, and the G-action factors through G/N. We conclude by amenability of G/N. $\qquad\square$

Example 2.5. Solvable groups are amenable.

This is proved by induction on the length of the derived series.

Example 2.6. Borel subgroups are amenable. More precisely, if $G = KAN$ is a semisimple Lie group, and $P = MAN$ is a minimal parabolic subgroup, then P is amenable.

Example 2.7. Non-compact semisimple Lie groups are not amenable.

Indeed, upon replacing G by $G/Z(G)$, we may assume that G has trivial centre. By root theory, G has a closed subgroup isomorphic to $(P)SL_2(\mathbf{R})$, so G is not amenable.

Proposition 2.5. *A connected Lie group G is amenable if and only if G is an extension of a solvable group by a compact group.*

Proof: Let $G = RS$ be a Levi decomposition (with R closed, normal, solvable, and S semisimple). Then G is amenable if and only if $S/(R \cap S)$ is amenable, if and only if S is compact. □

2.3 Lattices in Locally Compact Groups

Definition 2.3. *A discrete subgroup $\Gamma \subset G$ is a **lattice** if G/Γ carries a G-invariant probability measure. A lattice Γ is **uniform**, or **co-compact**, if G/Γ is compact.*

Example 2.8. 1) $\Gamma = \mathbb{Z}^n$ is a uniform lattice in $G = \mathbf{R}^n$;
2) The discrete Heisenberg group

$$\Gamma = H(\mathbb{Z}) = \left\{ \begin{pmatrix} 1 & m & p \\ 0 & 1 & n \\ 0 & 0 & 1 \end{pmatrix} : m, n, p \in \mathbb{Z} \right\}$$

is a uniform lattice in the Heisenberg group $G = H(\mathbf{R})$;
3) $\Gamma = \mathbb{Z}^2 \rtimes_{\begin{pmatrix} 2 & 1 \\ 1 & 1 \end{pmatrix}^n} \mathbb{Z}$ is a uniform lattice in

$$SOL = \mathbf{R}^2 \rtimes_{\begin{pmatrix} e^t & 0 \\ 0 & e^{-t} \end{pmatrix}} \mathbf{R};$$

4) $\Gamma = SL_n(\mathbb{Z})$ ($n \geq 2$) is a non-uniform lattice in $G = SL_n(\mathbf{R})$;
5) $\Gamma = Sp_n(\mathbb{Z})$ ($n \geq 1$) is a non-uniform lattice in $G = Sp_n(\mathbf{R})$;
6) The free group $\mathbb{F}_2$ can be embedded as a non-uniform lattice in $SL_2(\mathbf{R})$. E.g.,

$$\mathbb{F}_2 \simeq \langle \begin{pmatrix} 1 & 2 \\ 0 & 1 \end{pmatrix}, \begin{pmatrix} 1 & 0 \\ 2 & 1 \end{pmatrix} \rangle$$

is of index 12 in $SL_2(\mathbb{Z})$;
7) Let Γ_g be the fundamental group of a closed Riemann surface of genus $g \geq 2$. Then Γ_g embeds as a uniform lattice in $G = PSL_2(\mathbf{R})$.

More examples of lattices will be given in section 4.5.

Proposition 2.6. *Let Γ be a lattice in G. The following are equivalent:*

i) G is amenable;

ii) Γ is amenable.

Proof of $(ii) \Rightarrow (i)$**:** Let G act affinely on a compact convex subset C. Let μ be an invariant probability measure on G/Γ. Since Γ is amenable, the set C^{Γ} is closed, convex, non-empty. For $x_0 \in C^{\Gamma}$, the orbit map $G \to C : g \mapsto g.x_0$ is (right) Γ-invariant, so factors through G/Γ. So $x = \int_{G/\Gamma} y.x_0 \, d\mu(y) \in C$, and this x is a G-fixed point, by G-invariance of μ. □

2.4 Reiter's Property (P_1)

We denote by λ_G the left regular representation of G on the space of all functions $G \to \mathbb{C}$. We set $L^1(G)_{1,+} = \{f \in L^1(G) : f \geq 0, \|f\|_1 = 1\}$.

Theorem 2.2. *The following are equivalent:*

i) G is amenable;

ii) (Reiter's property (P_1), see [Rei52]) For every compact subset $Q \subset G$ and $\epsilon > 0$, there exists $f \in L^1(G)_{1,+}$ such that

$$\max_{x \in Q} \|\lambda_G(x)f - f\|_1 \leq \epsilon.$$

Proof: $(ii) \Rightarrow (i)$ Using the assumption, we find a net $(f_i)_{i \in I}$ in $L^1(G)_{1,+}$ such that $\lim_{i \in I} \|\lambda_G(x)f_i - f_i\|_1 = 0$ for every $x \in G$. Let m be any weak-* limit point of the f_i's in the set of means on G. Then m is a G-invariant mean on G.

$(i) \Rightarrow (ii)$ Recall that, for $f \in L^1(G)$, $\phi \in L^\infty(G)$:

$$(f \star \phi)(x) = \int_G f(y)\phi(y^{-1}x) \, dy = \int_G f(y)(\lambda_G(y)\phi)(x) \, dy$$

so that $f \star \phi \in L^\infty(G)$.

Let m be an invariant mean on G. Since m is continuous on $L^\infty(G)$, we have (Caution: see "Added on proof" at the end of the paper!)

(1) $$m(f \star \phi) = m(\phi)$$

if $f \in L^1(G)_{1,+}$.

Since $L^1(G)_{1,+}$ is weak-* dense in the space of all means on G, there exists a net $(f_i)_{i \in I}$ such that for every $\phi \in L^\infty(G)$:

$$\lim_{i \in I} \int_G f_i(y)\phi(y) \, dy = m(\phi).$$

For every $f \in L^1(G)_{1,+}$, we also have, because of (1):

$$\lim_{i \in I} \int_G f_i(y)(f \star \phi)(y)\,dy = m(\phi).$$

From this, we deduce $\lim_{i \in I}(f \star f_i - f_i) = 0$ in the weak topology of $L^1(G)$.

Consider now the space E of all functions $L^1(G)_{1,+} \to L^1(G)$, endowed with the pointwise norm topology. The set

$$\Sigma = \{(f \star g - g)_{f \in L^1(G)_{1,+}} : g \in L^1(G)_{1,+}\}$$

is convex in E, and its weak closure contains 0, by the previous observation. Since the weak closure of Σ coincides with its closure in the pointwise topology (a general fact from functional analysis, see Theorem 3.12 in [Rud73]), there exists a net $(g_j)_{j \in J}$ in $L^1(G)_{1,+}$ such that $\lim_{j \in J} \|f \star g_j - g_j\|_1 = 0$ for every $f \in L^1(G)_{1,+}$. Since $\|g_j\|_1 = 1$, this convergence is uniform on norm-compact subsets of $L^1(G)_{1,+}$. One such subset is $\{\lambda_G(x)f : x \in Q\}$. So fix $f_0 \in L^1(G)_{1,+}$, and find $j \in G$ large enough so that $\|\lambda(x)(f_0 \star g_j) - g_j\|_1 \le \frac{\epsilon}{2}$ for $x \in Q \cup \{1\}$. Set $f_{Q,\epsilon} = f_0 \star g_j$: then $f_{Q,\epsilon} \in L^1(G)_{1,+}$ and

$$\|\lambda_G(x)f_{Q,\epsilon} - f_{Q,\epsilon}\|_1$$

$$\le \|\lambda_G(x)(f_0 \star g_j) - g_j\|_1 + \|g_j - (f_0 \star g_j)\|_1 \le \epsilon$$

for every $x \in Q$. $\hfill\square$

2.5 Reiter's Property (P_2)

Definition 2.4. *A unitary representation π of G* **almost has invariant vectors,** *or* **weakly contains the trivial representation** *if, for every compact subset Q of G and every $\epsilon > 0$, there exists a non-zero $\xi \in \mathcal{H}_\pi$ such that*

$$\max_{g \in Q} \|\pi(g)\xi - \xi\| \le \epsilon\|\xi\|.$$

For a unitary representation π, we denote by $\infty\pi$ the (Hilbert) direct sum of countably many copies of π.

Theorem 2.3. *The following are equivalent:*

i) *G is amenable;*
ii) *(Reiter's property (P_2), see [Rei64]) The left regular representation λ_G almost has invariant vectors.*
iii) *The representation $\infty\lambda_G$ almost has invariant vectors.*

Proof: $(i) \Rightarrow (ii)$ Fix a compact subset $Q \subset G$ and $\epsilon > 0$. As G is amenable, by Reiter's property we find $f \in L^1(G)_{1,+}$ such that $\|\lambda_G(x)f - f\|_1 \le \epsilon$. Set $g = \sqrt{f}$. Then $g \in L^2(G)$ and $\|g\|_2 = 1$. Moreover, using $|a - b|^2 \le |a^2 - b^2|$ for $a, b \ge 0$, we get

$$\|\lambda_G(x)g - g\|_2^2 \leq \int_G |g(x^{-1}y)^2 - g(y)^2| \, dy$$

$$= \|\lambda_G(x)f - f\|_1 \leq \epsilon.$$

$(ii) \Rightarrow (iii)$ Obvious, since λ_G is a subrepresentation of $\infty\lambda_G$.

$(iii) \Rightarrow (i)$ We assume that $\infty\lambda_G$ almost has invariant vectors and prove in 3 steps that G satisfies Reiter's property (P_1), hence is amenable. So fix a compact subset $Q \subset G$, and $\epsilon > 0$; find a sequence $(f_n)_{n\geq 1}$ of functions, $f_n \in L^2(G)$, $\sum_{n=1}^{\infty} \|f_n\|_2^2 = 1$, such that $\sum_{n=1}^{\infty} \|\lambda_G(x)f_n - f_n\|_2^2 < \frac{\epsilon^2}{4}$ for $x \in Q$.

1) Replacing f_n with $|f_n|$, we may assume that $f_n \geq 0$.
2) Set $g_n = f_n^2$, so that $g_n \in L^1(G)$, $\sum_{n=1}^{\infty} \|g_n\|_1 = 1$, $g_n \geq 0$. For $x \in Q$, we have:

$$\sum_{n=1}^{\infty} \|\lambda_G(x)g_n - g_n\|_1 = \sum_{n=1}^{\infty} \int_G |f_n(x^{-1}y)^2 - f_n(y)^2| \, dy$$

$$= \sum_{n=1}^{\infty} \int_G |f_n(x^{-1}y) - f_n(y)|(f_n(x^{-1}y) + f_n(y)) \, dy$$

$$\leq \left(\sum_{n=1}^{\infty} \int_G |f_n(x^{-1}y) - f_n(y)|^2 \, dy \right)^{\frac{1}{2}} \times \left(\sum_{n=1}^{\infty} \int_G (f_n(x^{-1}y) + f_n(y))^2 \, dy \right)^{\frac{1}{2}}$$

$$\leq \left(\sum_{n=1}^{\infty} \|\lambda_G(x)f_n - f_n\|_2^2 \right)^{\frac{1}{2}} \times \left(2 \sum_{n=1}^{\infty} \int_G (f_n(x^{-1}y)^2 + f_n(y)^2) \, dy \right)^{\frac{1}{2}}$$

$$= 2 \left(\sum_{n=1}^{\infty} \|\lambda_G(x)f_n - f_n\|_2^2 \right)^{\frac{1}{2}} < \epsilon$$

where we have used consecutively the Cauchy-Schwarz inequality, $(a+b)^2 \leq 2(a^2 + b^2)$ for $a, b > 0$, and the fact that $\sum_{n=1}^{\infty} \|f_n\|_2^2 = 1$.

3) Set $F = \sum_{n=1}^{\infty} g_n$. Then $F \geq 0$ and $\|F\|_1 = \sum_{n=1}^{\infty} \|g_n\|_1 = 1$. Moreover, for $x \in Q$:

$$\|\lambda_G(x)F - F\|_1 \leq \sum_{n=1}^{\infty} \|\lambda_G(x)g_n - g_n\|_1 < \epsilon$$

by the previous step. This establishes property (P_1) for H. □

Finally we reach a result left unproved in section 2.2:

Corollary 2.1. *Closed subgroups of amenable groups are amenable.*

Proof: Let H be a closed subgroup of the amenable group G. Choose a measurable section s for $G \to H\backslash G$; so every $g \in G$ is written uniquely $g = hs(y)$, with $h \in H$, $y \in H\backslash G$. This gives an H-equivariant measurable identification $G \simeq H \times H\backslash G$, inducing a unitary map $L^2(G) \to L^2(H) \otimes L^2(H\backslash G)$ intertwining $\lambda_G|_H$ and $\lambda_H \otimes 1$. Choosing an orthonormal basis of $L^2(H\backslash G)$, we identify $\lambda_H \otimes 1$ with the direct sum of $[G : H]$ copies of λ_H, which we embed as a subrepresentation in $\infty \lambda_H$. This means that $\infty \lambda_H$ almost has invariant vectors, hence H is amenable by Theorem 2.3. $\square$

2.6 Amenability in Riemannian Geometry

In the Introduction, we mentioned that amenability became relevant in various fields of mathematics. In this section, independent of the rest of the Chapter, we substantiate this claim and indicate how amenability enters Riemannian geometry.

Let N be a complete Riemannian manifold. It carries a *Laplace operator*

$$\Delta_N = d^* d = -div \circ grad.$$

This operator is self-adjoint on $L^2(N)$, so it has a non-negative spectrum, and we denote by $\lambda_0(N)$ the bottom of its spectrum:

$$\lambda_0(N) = \inf\{\lambda \geq 0 : \lambda \in Spec_{L^2(N)}\Delta_N\}.$$

The following result was obtained by R. Brooks [Bro81].

Theorem 2.4. *Let M be a compact Riemannian manifold. Let $\tilde{M}$ be the universal cover of M, and $\pi_1(M)$ its fundamental group. The following are equivalent:*

i) $\pi_1(M)$ is amenable;
ii) $\lambda_0(\tilde{M}) = 0$. $\square$

Since $\pi_1(M)$ only depends of the topological structure of M, this shows in particular that the property $\lambda_0(\tilde{M}) = 0$ does *not* depend on the choice of a Riemannian structure on M.

Example 2.9. 1. $\lambda_0(\mathbf{R}^n) = 0$, which gives another proof of the amenability of $\mathbf{Z}^n$.
  2. If $\mathbb{H}^2$ denotes the Poincaré disk, with the metric of constant curvature -1, then $\lambda_0(\mathbb{H}^2) = \frac{1}{4}$, which gives another proof of the non-amenability of the surface group Γ_g, $g \geq 2$.

3 Measurable Ergodic Theory

3.1 Definitions and Examples

In this section, the context will be the following:

- G is a locally compact, σ-compact group;
- (X, μ) is a standard measure space;
- G is acting on (X, μ), i.e. we are given a measurable map $G \times X \to X :$ $(g, x) \mapsto gx$ which is an action such that, for every $g \in G$, the measure $g_* \mu$ is equivalent to μ (i.e. they have the same null sets); when this happens, we say that μ is **quasi-invariant**.

Definition 3.1. *The measure μ is **invariant** if $g_* \mu = \mu$ for every $g \in G$;*

Definition 3.2. *The action of G on (X, μ) is **ergodic** if every G-invariant measurable subset A is either null or co-null (i.e. $\mu(A) = 0$ or $\mu(X - A) = 0$).*

Example 3.1. (invariant measures on homogeneous spaces, see section 9 in [Wei65]) Let L be a closed subgroup of G; there always exists a quasi-invariant measure on G/L; there exists an invariant measure on G/L if and only if the restriction to L of the modular function of G, coincides with the modular function of L. Since the action of G on G/L is transitive, it is trivially ergodic.

Example 3.2. (irrational rotation) Take $X = S^1$, $\mu =$ normalized Lebesgue measure. Fix $\theta \in \mathbf{R} - \mathbb{Q}$. Let $G = \mathbb{Z}$ act on S^1 by powers of the irrational rotation T of angle $2\pi\theta$:

$$T(z) = e^{2\pi i\theta} z.$$

This action is measure-preserving, and ergodic.

To check ergodicity, it is convenient to appeal to Fourier series: if $A \subset S^1$ is T-invariant, let χ_A be its characteristic function, and

$$\chi_A(z) = \sum_{n=-\infty}^{+\infty} a_n z^n$$

be its Fourier expansion in $L^2(S^1)$. Then

$$T^*(\chi_A)(z) = \chi_A(T^{-1}z) = \sum_{n=-\infty}^{+\infty} e^{-2\pi i n\theta} a_n z^n.$$

By T-invariance, we must have $a_n = e^{-2\pi i n\theta} a_n$ for every $n \in \mathbb{Z}$, so $a_n = 0$ for $n \neq 0$, as θ is irrational; thus χ_A is constant, i.e. either $\chi_A = 0$ and $\mu(A) = 0$, or $\chi_A = 1$ and $\mu(A) = 1$.

Example 3.3. (linear action on tori) The linear action of $SL_n(\mathbb{Z})$ on $\mathbf{R}^n$ leaves $\mathbb{Z}^n$ invariant, so descends to an action on the n-torus $\mathbf{T}^n = \mathbf{R}^n/\mathbb{Z}^n$. Let μ be normalized Lebesgue measure on $\mathbf{T}^n$: since Lebesgue measure is $SL_n(\mathbf{R})$-invariant on $\mathbf{R}^n$, the measure μ is $SL_n(\mathbb{Z})$-invariant on $\mathbf{T}^n$. This action is ergodic.

To check ergodicity of this action, we use n-variable Fourier series: if $A \subset \mathbf{T}^n$ is $SL_n(\mathbb{Z})$-invariant, let χ_A be its characteristic function, and

$$\chi_A(z_1, \ldots, z_n) = \sum_{r \in \mathbb{Z}^n} a_r z^r$$

(where $z^r = z_1^{r_1} \ldots z_n^{r_n}$) be its Fourier expansion in $L^2(\mathbf{T}^n)$. For $g \in SL_n(\mathbb{Z})$, one has

$$(g\chi_A)(z) = \chi_A(g^{-1}z) = \sum_{r \in \mathbb{Z}^n} a_r z^{{}^t g^{-1} r} = \sum_{r \in \mathbb{Z}^n} a_{{}^t gr} z^r.$$

Since A is $SL_n(\mathbb{Z})$-invariant, we have $a_{{}^t gr} = a_r$ for every $r \in \mathbb{Z}^n$ and $g \in SL_n(\mathbb{Z})$; i.e. a_r is constant on $SL_n(\mathbb{Z})$-orbits on $\mathbb{Z}^n$. Notice that non-trivial orbits are infinite. Since $(a_r)_{r \in \mathbb{Z}^n} \in \ell^2(\mathbb{Z}^n)$, we must have $a_r = 0$ for $r \neq 0$. So χ_A is constant and we conclude as in Example 3.2.

Note that in example 3.3, $SL_n(\mathbb{Z})$ can be replaced by any subgroup with infinite non-trivial orbits on $\mathbb{Z}^n$. For $n = 2$, one can for example take the infinite cyclic subgroup generated by $\begin{pmatrix} 2 & 1 \\ 1 & 1 \end{pmatrix}$.

Definition 3.3. *A Borel space is* **countably separated** *if there exists a countable family of Borel subsets separating points (i.e. two distinct points can be put in two disjoint subsets of the countable family).*

For example, $\mathbf{R}^n$ is countably separated since we can take the collection of balls with rational centres and rational radii as a countable family separating points.

Proposition 3.1. *Let S be countably separated. If the G-action on (X, μ) is ergodic, any measurable G-invariant map $f : X \to S$, is almost everywhere constant.*

Proof: Let $(A_j)_{j \geq 1}$ be a sequence of Borel subsets separating points in S. Let $\mathcal{P}_n$ be the partition of S generated by $A_1, \ldots, A_n$. For every $B \in \mathcal{P}_n$, the set $f^{-1}(B)$ is G-invariant, so it is either null or co-null; moreover there exists a unique $B_n \in \mathcal{P}_n$ such that $f^{-1}(B_n)$ is co-null. The sequence $(B_n)_{n \geq 1}$ is decreasing, and the intersection $\bigcap_{n=1}^{\infty} B_n$ is reduced to one point s, as the A_j's separate points in S. So $f(x) = s$ almost everywhere. $\square$

The converse of Proposition 3.1 holds in the finite measure-preserving case. The following result, due to Koopman [Koo31], is known today under the name "Koopmanism".

Proposition 3.2. *Let G act on the probability space (X, μ), preserving μ. The following are equivalent:*

i) The action is ergodic;

ii) *Any G-invariant function in $L^2(X, \mu)$ is constant almost everywhere.*

iii) *Set $L_0^2(X, \mu) = \{f \in L^2(X, \mu) : \int_X f \, d\mu = 0\}$ (the orthogonal of constants in $L^2(X, \mu)$). The representation of G on $L_0^2(X, \mu)$ has no non-zero fixed vector.*

Proof: $(i) \Rightarrow (ii)$: Follows from Proposition 3.1.

$(ii) \Rightarrow (iii)$: Obvious.

$(iii) \Rightarrow (i)$: If a Borel subset A is G-invariant, set

$$\xi_A(x) = \begin{cases} 1 - \mu(A) & \text{if } x \in A \\ -\mu(A) & \text{if } x \notin A \end{cases}$$

ξ_A is a G-invariant function in $L_0^2(X, \mu)$, so $\xi_A = 0$; this gives the result. $\square$

Example 3.4. Proposition 3.2 false in infinite measure. Indeed, let $\mathbb{Z}$ act by translation on $\mathbf{R}$: the action is not ergodic (why?). However the only $\mathbb{Z}$-invariant function in $L^2(\mathbf{R})$ is the constant 0.

3.2 Moore's Ergodicity Theorem

We have seen that, if L is a closed subgroup of G, then the action of G on G/L is trivially ergodic. A more interesting situation arises by considering H, L, two closed subgroups of G. Question: when is the H-action on G/L ergodic? Moore's theorem gives the answer.

Theorem 3.1. *The following are equivalent:*

i) *The H-action on G/L is ergodic;*
ii) *The L-action on $H\backslash G$ is ergodic.*

Example 3.5. Let $\varGamma = SL_2(\mathbb{Z})$ act by fractional linear transformations on the real projective line $\mathbf{P}^1(\mathbf{R}) \simeq S^1$. Is the action ergodic? Write $\mathbf{P}^1(\mathbf{R}) = G/P$, where $G = SL_2(\mathbf{R})$ and $P = \left\{ \begin{pmatrix} * & * \\ 0 & * \end{pmatrix} \right\}$. By Theorem 3.1, ergodicity of $\varGamma$ on G/P (situation with no invariant measure) is equivalent to ergodicity of P on $\varGamma\backslash G$ (situation with a finite invariant measure).

Later we will see that these actions *are* ergodic.

Lemma 3.1. *Let H be a closed subgroup of G; let (X, μ) be a G-space. The following are equivalent:*

i) *H acts ergodically on X;*
ii) *G acts ergodically (via the diagonal action) on $X \times G/H$.*

Proof: $(i) \Rightarrow (ii)$ Contraposing, assume that G is not ergodic on $X \times G/H$. So find $A \subset X \times G/H$, neither null nor co-null, G-invariant. Let $p : X \times G/H \to G/H$ be the second projection, set $A_y = p^{-1}(y) \cap A$ for $y \in G/H$. Since p is G-equivariant, one has $A_{gy} = gA_y$. Since G acts transitively on G/H, this implies (by Fubini) that A_{eH} is neither null nor co-null. But A_{eH} is an H-invariant subset in X, so H is not ergodic on X.

$(ii) \Rightarrow (i)$ Contraposing, assume that H is not ergodic on X. So find $B \subset X$, neither null nor co-null, H-invariant. Choosing a measurable section s for $G \to G/H$, we may define

$$A = \{(x, y) \in X \times G/H : x \in s(y)B\}.$$

The set A is then G-invariant: indeed, for $g \in G$ and $(x, y) \in A$, we must check that $gx \in s(gy)B$. But $s(gy) = gs(y)h$ for some $h \in H$ so $gx \in gs(y)B = gs(y)hB = s(gy)B$. One sees easily that A is neither null or co-null.   □

Proof of Theorem 3.1: By lemma 3.1, the action of H on G/L is ergodic if and only if the action of G on $G/L \times H\backslash G$ is ergodic, if and only if the action of L on $H\backslash G$ is ergodic.   □

Definition 3.4. *Let Γ be a lattice in a semisimple Lie group G. Say that Γ is* **irreducible** *if, for any normal subgroup N of positive dimension in G, the image of Γ in G/N is dense.*

This definition is designed to eliminate examples of the form $\Gamma = \Gamma_1 \times \Gamma_2$ in $G = G_1 \times G_2$, with Γ_i a lattice in G_i $(i = 1, 2)$.

Example 3.6. Let σ be the non-trivial element of $Gal(\mathbb{Q}(\sqrt{2})/\mathbb{Q})$:

$$\sigma(r + s\sqrt{2}) = r - s\sqrt{2}.$$

Then $SL_n(\mathbb{Z}[\sqrt{2}])$ sits as a non-uniform irreducible lattice in $SL_n(\mathbf{R}) \times SL_n(\mathbf{R})$, via the embedding $g \mapsto (g, \sigma(g))$.

Here is Moore's ergodicity theorem [Moo66]:

Theorem 3.2. *Let G be a connected, semisimple Lie group with finite centre. Let Γ be an irreducible lattice in G, and H a closed subgroup in G. If H is not compact, then Γ acts ergodically on G/H.*

In the next section, we will deduce this result from the Howe-Moore vanishing theorem. Notice that Theorem 3.2 implies, in Example 3.5, that the action of Γ on G/P is ergodic.

3.3 The Howe-Moore Vanishing Theorem

Let π be a (strongly continuous) unitary representation of a locally compact group G, on a Hilbert space $\mathcal{H}$. Denote by $\mathcal{H}^{\pi(G)}$ the space of fixed vectors in $\mathcal{H}$.

Definition 3.5. π is a C_0-**representation** *if all coefficients of π vanish at infinity on G, i.e. $\lim_{g \to \infty} \langle \pi(g)\xi | \eta \rangle = 0$ for every $\xi, \eta \in \mathcal{H}$.*

Example 3.7. The left regular representation λ_G of G on $L^2(G)$ is C_0.

Indeed, if $\xi, \eta \in L^2(G)$ have compact support in G, then so does $g \mapsto \langle \lambda_G(g)\xi | \eta \rangle$. By density of $C_c(G)$ in $L^2(G)$, we conclude that every coefficient vanishes at infinity.

Example 3.8. C_0-representations have no finite-dimension subrepresentation.

The reason is: if σ is a finite-dimensional unitary representation, then the identity $1 = |\det \sigma(g)|$ prevents σ from being C_0. Observe that this implies in particular that a C_0-representation has no non-zero fixed vector.

Here is the Howe-Moore theorem ([HM79], Theorem 5.1).

Theorem 3.3. *Let $G = \prod_i G_i$ be a semisimple Lie group with finite centre and simple factors G_i's, and let π be a unitary representation of G. Assume that $\mathcal{H}^{\pi(G_i)} = 0$ for every i. Then π is C_0.*

From this we deduce Moore ergodicity.

Proof of Theorem 3.2: Let H be a closed non-compact subgroup of G. To prove that Γ is ergodic on G/H, by lemma 3.1 it is enough to prove that H is ergodic on G/Γ. So let π be the representation of G on $L^2_0(G/\Gamma)$. Take a function on G/Γ which is G_i-invariant, lift to a left-G_i, right-Γ invariant function on G, and project to a right-Γ invariant function on $G_i\backslash G$. Since Γ is irreducible, the image of Γ in $G_i\backslash G$ is dense, hence this function must be a.e. constant. This shows that $\pi(G_i)$ has no non-zero invariant function in $L^2_0(G/\Gamma)$. By Howe-Moore (Theorem 3.3), π is a C_0-representation. So $\pi|_H$ is C_0 as well, in particular (by Example 3.8) $\pi|_H$ has no non-zero invariant vector. By Proposition 3.2 (valid since we have a finite invariant measure on G/Γ), H is ergodic on G/Γ. □

We will give a complete proof of the Howe-Moore theorem in the case of $SL_2(\mathbf{R})$, and then indicate briefly how to pass from $SL_2(\mathbf{R})$ to a more general semisimple Lie group.

Let G be a locally compact group, and let $\alpha = (a_n)_{n \geq 1}$ be a sequence in G. Set

$$U_\alpha^+ = \{g \in G : 1 \text{ is an accumulation point of } (a_n^{-1} g a_n)_{n \geq 1}\}$$

and let N_α^+ be the subgroup generated by U_α^+.

The following result is known as Mautner's phenomenon.

Proposition 3.3. *Let π be a (strongly continuous) unitary representation of G. Let ξ, ξ_0 be vectors in $\mathcal{H}$ such that $\lim_{n \to \infty} \pi(a_n)\xi = \xi_0$ in the weak topology. Then $\pi(x)\xi_0 = \xi_0$ for every $x \in N_\alpha^+$.*

Proof: Fix $x \in U_\alpha^+$. Let $(a_{n_k})_{k \geq 1}$ be a subsequence of α such that $\lim_{n \to \infty} a_{n_k}^{-1} x a_{n_k} = 1$. For every $\eta \in \mathcal{H}$:

$$|\langle (\pi(x)\xi_0 - \xi_0)|\eta\rangle| = |\langle \pi(x)\xi_0|\eta\rangle - \langle \xi_0|\eta\rangle|$$

$$= \lim_{k \to \infty} |\langle \pi(x a_{n_k})\xi|\eta\rangle - \langle \pi(a_{n_k})\xi|\eta\rangle|$$

$$= \lim_{k \to \infty} |\langle \pi(a_{n_k}^{-1} x a_{n_k})\xi|\pi(a_{n_k}^{-1})\eta\rangle - \langle \xi|\pi(a_{n_k}^{-1})\eta\rangle|$$

$$\leq \lim_{k \to \infty} \|\pi(a_{n_k}^{-1} x a_{n_k})\xi - \xi\|.\|\eta\| = 0$$

by Cauchy-Schwarz. $\qquad \square$

Example 3.9. Let G be $SL_2(\mathbf{R})$, $a_n = \begin{pmatrix} e^{t_n} & 0 \\ 0 & e^{-t_n} \end{pmatrix}$, with $t_n \to +\infty$. It is easy to see that $N_\alpha^+ = N = \left\{ \begin{pmatrix} 1 & * \\ 0 & 1 \end{pmatrix} \right\}$.

Lemma 3.2. *Let π be a unitary representation of $G = SL_2(\mathbf{R})$. If a vector $\xi \in \mathcal{H}$ is N-invariant, then it is G-invariant.*

Proof: For a vector $\eta \in \mathcal{H}$ of norm 1, consider $\phi_\eta(g) = \langle \pi(g)\eta|\eta\rangle$, the associated coefficient function. For any closed subgroup $H \subset G$, the vector η is H-fixed if and only if $\phi|_H = 1$, if and only if ϕ is H-bi-invariant (by the equality case of the Cauchy-Schwarz inequality).

Here, ξ is N-invariant, so $\phi_\xi =: \phi$ is N-bi-invariant.

1st step: ξ is P-invariant. Indeed, by right N-invariance, ϕ descends to a function $\tilde{\phi}$ on $G/N \simeq \mathbf{R}^2 - \{0\}$, which is continuous and constant on orbits of N. In particular, $\tilde{\phi}$ is constant on lines parallel to the horizontal axis, and distinct from this axis. By continuity, $\tilde{\phi}$ is equal to 1 on the horizontal axis. Observing that this axis (minus $\{0\}$) is the P-orbit of $\begin{pmatrix} 1 \\ 0 \end{pmatrix}$, we get that $\phi|_P = 1$, i.e. ξ is P-fixed.

2nd step: Since ϕ is P-bi-invariant, ϕ descends to a function $\overline{\phi}$ on $G/P \simeq \mathbf{P}^1(\mathbf{R})$, which is continuous and constant on P-orbits. But there are exactly two P-orbits, namely $\{0\}$ and its complement. By continuity we have $\overline{\phi} \equiv 1$, so $\phi \equiv 1$ and ξ is G-fixed. $\qquad \square$

Proof of Theorem 3.3, case $G = SL_2(\mathbf{R})$: Set

$$A^+ = \left\{ a_t = \begin{pmatrix} e^t & 0 \\ 0 & e^{-t} \end{pmatrix} : t \geq 0 \right\}$$

and $K = SO(2)$. In view of the Cartan decomposition $G = KA^+K$, to show vanishing of coefficients it is enough to show that, for every $\xi, \eta \in \mathcal{H}$ one has $\lim_{t \to +\infty} \langle \pi(a_t)\xi | \eta \rangle = 0$. By compactness of closed balls in Hilbert spaces for the weak topology, we find an accumulation point ξ_0 of the $\pi(a_t)\xi$'s:

$$\lim_{n \to \infty} \pi(a_{t_n})\xi = \xi_0$$

in the weak topology. By Mautner's phenomenon (Proposition 3.3) and example 3.9, ξ_0 must be N-fixed. By lemma 3.2, the vector ξ_0 is also G-fixed. By assumption this implies $\xi_0 = 0$, so the only weak accumulation point of the $\pi(a_t)\xi$'s is 0. In other words $w - \lim_{t \to \infty} \pi(a_t)\xi = 0$, which amounts to the desired result. $\qquad\square$

Let us conclude by indicating how one can pass from $SL_2(\mathbf{R})$ to more general semisimple groups, say $SL_3(\mathbf{R})$. Here

$$A = \left\{ \begin{pmatrix} e^{t_1} & 0 & 0 \\ 0 & e^{t_2} & 0 \\ 0 & 0 & e^{t_3} \end{pmatrix} : t_1 + t_2 + t_3 = 0 \right\}.$$

Let π be a unitary representation of $SL_3(\mathbf{R})$, without non-zero fixed vector. Embed $SL_2(\mathbf{R})$ into $SL_3(\mathbf{R})$ in the three standard ways:

$$\begin{pmatrix} \star & \star & 0 \\ \star & \star & 0 \\ 0 & 0 & 1 \end{pmatrix}, \begin{pmatrix} \star & 0 & \star \\ 0 & 1 & 0 \\ \star & 0 & \star \end{pmatrix}, \begin{pmatrix} 1 & 0 & 0 \\ 0 & \star & \star \\ 0 & \star & \star \end{pmatrix}.$$

Claim: For each of these 3 copies of $SL_2(\mathbf{R})$, the restriction $\pi|_{SL_2(\mathbf{R})}$ has no non-zero fixed vector.

Taking this claim for granted, we see (by the Howe-Moore theorem in the case of $SL_2(\mathbf{R})$) that $\pi|_{SL_2(\mathbf{R})}$ is a C_0-representation. Since A is generated by its intersections with the three embeddings of $SL_2(\mathbf{R})$, we get for $a \in A$:

$$\lim_{a \to \infty} \langle \pi(a)\xi | \eta \rangle = 0$$

i.e. π is C_0.

Proof of the Claim: Assume that $\pi|_{SL_2(\mathbf{R})}$ has a non-zero fixed vector ξ, say for the first embedding of $SL_2(\mathbf{R})$. We are going to show that ξ is fixed under $SL_3(\mathbf{R})$.

We use the fact that $SL_3(\mathbf{R})$ is generated by elementary matrices $U_{ij}(t)$
$(t \in \mathbf{R}, i \neq j)$. Let us show that ξ is $U_{13}(\mathbf{R})$-invariant. Take $u = \begin{pmatrix} 1 & 0 & x \\ 0 & 1 & 0 \\ 0 & 0 & 1 \end{pmatrix}$,

and $\alpha = (a_n)_{n \geq 1}$, with $a_n = \begin{pmatrix} e^{t_n} & 0 & 0 \\ 0 & e^{-t_n} & 0 \\ 0 & 0 & 1 \end{pmatrix}$, for some sequence $t_n \to +\infty$.

Then $u \in N_\alpha^+$. Since $\pi(a_n)\xi = \xi$, we have $\pi(u)\xi = \xi$ by the Mautner phenomenon (Proposition 3.3). $\qquad\square$

4 Margulis' Super-rigidity Theorem

4.1 Statement

Recall that the **real rank** of a semisimple Lie group G, denoted by $\mathbf{R}-rk(G)$, is the dimension of a maximal split torus in G. For example:

$$\mathbf{R} - rk(SL_n(\mathbf{R})) = n - 1;$$

$$\mathbf{R} - rk(SO(p, q)) = \min\{p, q\}.$$

Theorem 4.1. *(see [Mar91], Theorem 5.6) Take:*

- G, *a connected, semisimple real algebraic group, with no compact factor, and* $\mathbf{R} - rk(G) \geq 2$;
- Γ *an irreducible lattice in* $G(\mathbf{R})$;
- k *a local field of characteristic 0 (i.e.* $k = \mathbf{R}$, $\mathbb{C}$ *or a finite extension of* $\mathbb{Q}_p$*) and* H *a simple, connected, algebraic* k-*group.*

Assume that $\pi : \Gamma \to H(k)$ *is a homomorphism with Zariski dense image. Then:*

- i) *If* $k = \mathbf{R}$ *and* $H(\mathbf{R})$ *is not compact, then* π *extends to a rational homomorphism* $G \to H$ *defined over* $\mathbf{R}$ *(hence induces* $G(\mathbf{R}) \to H(\mathbf{R})$*);*
- ii) *If* $k = \mathbb{C}$, *then either* $\overline{\pi(\Gamma)}$ *is compact, or* π *extends to a rational homomorphism* $G \to H$;
- iii) *If* k *is totally disconnected, then* $\overline{\pi(\Gamma)}$ *is compact.*

4.2 Mostow Rigidity

One of the most spectacular applications of Theorem 4.1 is Mostow's rigidity theorem [Mos73].

Theorem 4.2. *Let* G, G' *be connected semi-simple Lie groups with trivial centre, no compact factors, and suppose* $\Gamma \subset G$, $\Gamma' \subset G'$ *are lattices. Assume* Γ *irreducible in* G *and* $\mathbf{R}-rk(G) \geq 2$. *Let* $\pi : \Gamma \to \Gamma'$ *be an isomorphism. Then* π *extends to an isomorphism* $G \to G'$.

In other words, the lattice determines the ambient Lie group.

Proof, from Theorem 4.1: For each simple factor H_i' of G', find a structure of a simple real algebraic group such that $H_i' = H_i'(\mathbf{R})^\circ$. By the Borel density theorem [Bor60], Γ' is Zariski-dense in $G' = \mathcal{G}'(\mathbf{R})^\circ = \prod_i H_i'(\mathbf{R})^\circ$. Similarly, write $G = \mathcal{G}(\mathbf{R})^\circ$. By Theorem 4.1 (case $k = \mathbf{R}$), applied to each factor H_i', we may extend π to a rational homomorphism $\mathcal{G} \to \mathcal{G}'$. Since $\pi(\mathcal{G})$ is an algebraic subgroup of $\mathcal{G}'$, by Zariski-density we deduce $\pi(\mathcal{G}) = \mathcal{G}'$ and from that: $\dim_{\mathbf{R}} \pi(G) = \dim_{\mathbf{R}} G'$, so $\pi(G) = G'$ by connectedness. Set $N = Ker\,\pi$. Assume $N \neq \{1\}$. Since G has no center, then $\dim_{\mathbf{R}} N > 0$. Since Γ is an irreducible lattice, the image of Γ is dense in G/N, which implies that $\pi(\Gamma)$ is dense in G', contradicting discreteness of Γ'.   □

We may rephrase the Mostow rigidity theorem as follows.

Let M, M' be locally symmetric Riemannian manifolds, with finite volume, irreducible (in the sense that neither M nor M' is locally a Riemannian product), with rank ≥ 2. If $\pi_1(M) \simeq \pi_1(M')$, then M is isometric to M' (up to a rescaling of metrics).

4.3 Ideas to Prove Super-rigidity, $k = \mathbf{R}$

Lemma 4.1. *Suppose $P \subset G$ and $L \subset H$ are proper real algebraic subgroups, and there exists a rational Γ-equivariant map $\phi : G/P \to H/L$ defined over $\mathbf{R}$ (where Γ acts on H/L via π - explicitly: $\phi(\gamma.x) = \pi(\gamma).\phi(x)$). Then π extends to a rational homomorphism $G \to H$ defined over $\mathbf{R}$.*   □

Proof: Idea: look at the graph of π:

$$gr(\pi) = \{(\gamma, \pi(\gamma)) : \gamma \in \Gamma\} \subset G \times H,$$

and show that the Zariski closure $\overline{gr(\pi)}^Z$ is the graph of a homomorphism.

1st step: The projection of $\overline{gr(\pi)}^Z$ on the first factor G, is *onto*: this follows from the Borel density theorem.

2nd step: We have to show that, if $(g, h_1), (g, h_2) \in \overline{gr(\pi)}^Z$, then $h_1 = h_2$.
For this, let $R(G/P, H/L)$ be the set of all rational maps $G/P \to H/L$. Let $G \times H$ act on $R(G/P, H/L)$ by

$$((g, h).\psi)(x) = h.\psi(g^{-1}x)$$

($\psi \in R(G/P, H/L)$). By assumption, our ϕ is Γ-equivariant. In terms of the $G \times H$-action, this means that ϕ is $gr(\pi)$-invariant. Since ϕ is rational, this implies that ϕ is $\overline{gr(\pi)}^Z$-invariant. In particular, for every $x \in G/P$:

$$h_1.\phi(g^{-1}x) = h_2.\phi(g^{-1}x)$$

i.e. $h_1^{-1}h_2$ fixes $\phi(G/P)$ pointwise.

On the other hand, $\pi(\Gamma)$ stabilizes $\phi(G/P)$, so $\pi(\Gamma)$ also stabilizes $\overline{\phi(G/P)}^Z$. By Zariski density of $\pi(\Gamma)$, we deduce that H stabilizes $\overline{\phi(G/P)}^Z$. Since H acts transitively on H/L, this implies that $\overline{\phi(G/P)}^Z = H/L$ (i.e. $\phi(G/P)$ is Zariski-dense in H/L).

As a consequence, $h_1^{-1}h_2$ fixes H/L pointwise, so $h_1^{-1}h_2 \in \bigcap_{h \in H} hLh^{-1}$. The latter is a proper normal subgroup of H. Since H is assumed to be simple, this subgroup is $\{1\}$, i.e. $h_1^{-1}h_2 = 1$. This concludes the proof. $\qquad \square$

We will apply this when P is a minimal parabolic subgroup of G, defined over $\mathbf{R}$.

Example 4.1. $G = SL_n$, $G(\mathbf{R}) = SL_n(\mathbf{R})$, $P = \left\{ \begin{pmatrix} * & \cdots\cdots & * \\ 0 & \ddots & \vdots \\ \vdots & \ddots & \vdots \\ 0 \cdots & 0 & * \end{pmatrix} \right\}$.

Set $P_0 = P \cap G(\mathbf{R})^o$. Then $G(\mathbf{R})^o/P_0$ is Zariski dense in the flag variety G/P. So, if we have a rational map $\phi : G/P \to H/L$ defined on a Zariski dense subset of $G(\mathbf{R})^o/P_0$, which is Γ-equivariant as a map $G(\mathbf{R})^o/P_0 \to H/L$, then ϕ is Γ-equivariant.

It is therefore enough to find a proper real algebraic subgroup $L \subset H$ and a rational Γ-equivariant map $G(\mathbf{R})^o/P_0 \to H/L$, defined on a Zariski-dense subset of $G(\mathbf{R})^o/P_0$. This will be done in two steps.

1st step: There is a proper real algebraic subgroup $L \subset H$ and a measurable Γ-equivariant map $\phi : G(\mathbf{R})^o/P_0 \to H(\mathbf{R})/L(\mathbf{R})$.

2nd step: Any such measurable Γ-equivariant map agrees almost everywhere with a rational map.

We shall *not* elaborate on the second step, and refer instead to Chapter 5 in [Zim84]. Note however that *it is here that* $\mathbf{R} - rk(G) \geq 2$ *is used!* The proof of the 1st step appeals to the following result of Furstenberg (Theorem 15.1 in [Fur73]), which will be proved in the next subsection.

Recall that, for a compact space X, we denote by $M(X)$ the set of probability measures on X.

Proposition 4.1. *Let X be a compact, metrizable Γ-space. There exists a measurable Γ-equivariant map $\omega : G/P \to M(X)$, i.e. $\omega(\gamma x) = \gamma \omega(x)$ for all $\gamma \in \Gamma$ and almost all $x \in G/P$.*

Proof of the 1st step above: Let Q be a proper parabolic subgroup of H, defined over $\mathbf{R}$. Then $X = H(\mathbf{R})/Q(\mathbf{R})$ is a compact metrizable Γ-space.

By Proposition 4.1, we find $\omega : G/P \to M(X)$ measurable and Γ-equivariant. We then argue as follows:

- the orbit space $H(\mathbf{R})\backslash M(X)$ is countably separated (this is a result of Zimmer [Zim78]);
- Let $\overline{\omega} : G/P \to H(\mathbf{R})\backslash M(X)$ be the composition of ω with the quotient map $M(X) \to H(\mathbf{R})\backslash M(X)$. Then $\overline{\omega}(\gamma x) = \overline{\omega}(x)$ a.e. in $x \in G/P$. By Moore's ergodicity theorem 3.2, Γ acts ergodically on G/P. By Proposition 3.1, the map $\overline{\omega}$ is almost everywhere constant on G/P. This means that ω takes values essentially in a unique orbit $H(\mathbf{R})\mu_0 \in M(X)$.
- For every $\mu \in M(X)$, the stabilizer of μ in $H(\mathbf{R})$ is the set of real points of a proper, real algebraic subgroup of H. This is another result of Furstenberg [Fur63].

Set then $L = Stab_H(\mu_0)$. The map $H(\mathbf{R})\mu_0 \to H(\mathbf{R})/L(\mathbf{R})$ is $H(\mathbf{R})$-equivariant. Composing, we get a Γ-equivariant measurable map $\phi : G/P \to H(\mathbf{R})/L(\mathbf{R})$, defined on a Γ-invariant co-null set. $\qquad\square$

4.4 Proof of Furstenberg's Proposition 4.1 - Use of Amenability

We give Margulis' proof (see Theorem 4.5 in [Mar91]).

Denote by dg the Haar measure on G. Let $\Gamma \times G$ act on $G \times X$ by

$$(\gamma, g)(h, x) = (\gamma h g^{-1}, \gamma x).$$

The projection $p : G \times X \to G$ is $(\Gamma \times G)$-equivariant. Let Q be the set of non-negative Borel measures μ on $G \times X$ such that $p_*(\mu) = dg$ and $(\gamma, 1)_*\mu = \mu$ for every $\gamma \in \Gamma$. We make 3 observations.

- Q is non-empty. Indeed, fix D a Borel fundamental domain for Γ on G: for every $g \in G$, there exists a unique $\gamma_g \in \Gamma$ such that $g \in \gamma_g D$. Fix $x_0 \in X$ and define $\phi : G \to G \times X : g \mapsto (g, \gamma_g x_0)$. Then $(\gamma, 1)\phi(g) = (\gamma g, \gamma\gamma_g x_0) = \phi((\gamma, 1)g)$. So ϕ is measurable, Γ-equivariant, and $p \circ \phi = Id_G$. So $\phi_*(dg) \in Q$.
- Q is convex (clear) and compact in the weak-* topology. Indeed, if $(K_n)_{n \geq 1}$ is an increasing sequence of compact subsets of G such that $G = \bigcup_{n=1}^{\infty} K_n$. Since X is compact, so is $K_n \times X$, and therefore elements in Q are uniformly bounded on $K_n \times X$, namely $\mu(K_n \times X) \leq \int_{K_n} dg$ for $\mu \in Q$. So Q is bounded; since it is also weak-* closed, by Tychonov it is weak-* compact.
- Q is $(\Gamma \times G)$-invariant. This is because $(\gamma, h)_*(dg) = dg$, since G is unimodular.

Since P is amenable, there exists $\tau \in Q$ which is $(\{1\} \times P)$-invariant, hence also $(\Gamma \times P)$-invariant, by definition. As $p_*(\tau) = dg$, we may disintegrate τ over G:

$$\tau = \int_G (\delta_g \otimes \nu_g)\, dg$$

where $\nu_g \in M(X)$ and the field $g \mapsto \nu_g$ is measurable, and unique up to modification on a null set. Now

$$(\gamma, p)_*(\tau) = \int_G (\delta_{\gamma gp^{-1}} \otimes \gamma_* \nu_g)\, dg \qquad (h = \gamma gp^{-1})$$

$$= \int_G (\delta_h \otimes \gamma_* \nu_{\gamma^{-1}hp})\, dh.$$

By uniqueness: $\nu_g = \gamma_* \nu_{\gamma^{-1}gp}$ for almost every $g \in G$. In particular $\nu_{gp} = \nu_g$ for almost every $g \in G$ and every $p \in P$. So we may define a measurable map

$$\omega : G/P \to M(X) : gP \mapsto \nu_g$$

which is Γ-equivariant. $\qquad\qquad\qquad\qquad\qquad\qquad\qquad\qquad\qquad\qquad\square$

4.5 Margulis' Arithmeticity Theorem

Recall that two subgroups H_1, H_2 in the same group, are *commensurable* if their intersection $H_1 \cap H_2$ has finite index both in H_1 and H_2.

Definition 4.1. *Let G be a real, linear, semisimple Lie group with finite centre. A lattice Γ in G is **arithmetic** if there exists a semisimple algebraic $\mathbb{Q}$-group H and a surjective continuous homomorphism $\phi : H(\mathbf{R})^0 \to G$, with compact kernel, such that $\phi(H(\mathbb{Z}) \cap H(\mathbf{R})^0)$ is commensurable with Γ (here $H(\mathbf{R})^0$ is the connected component of identity in $H(\mathbf{R})$).*

Example 4.2. Let Φ be a quadratic form in $n + 1$ variables, with signature $(n, 1)$, and coefficients in a number field $k \subset \mathbf{R}$. We denote by SO_Φ the special orthogonal group of Φ: this is a simple algebraic group defined over k. Set $\Gamma = SO_\Phi(\mathcal{O})$, where $\mathcal{O}$ is the ring of integers of k.

a) $\Phi = x_1^2 + \ldots + x_n^2 - x_{n+1}^2$; here $k = \mathbb{Q}$ and $H = SO_\Phi$, so that $\Gamma = SO(n, 1)(\mathbb{Z})$ is a non-uniform arithmetic lattice in $SO_\Phi(\mathbf{R}) = SO(n, 1)$.

b) $\Phi = x_1^2 + \ldots + x_n^2 - \sqrt{2}x_{n+1}^2$; here $k = \mathbb{Q}(\sqrt{2})$ and $H = SO_\Phi \times SO_{\sigma(\Phi)}$, where σ is the non-trivial element of $Gal(k/\mathbb{Q})$. Then $\Gamma = SO_\Phi(\mathbb{Z}[\sqrt{2}])$ is a uniform arithmetic lattice in $SO_\Phi(\mathbf{R}) \simeq SO(n, 1)$.

c) $\Phi = x_1^2 + \ldots + x_n^2 - \delta x_{n+1}^2$ where $\delta > 0$ is a root of a cubic irreducible polynomial over $\mathbb{Q}$, having two positive roots δ, δ' and one negative root δ''. Here $k = \mathbb{Q}(\delta)$; let σ, τ be the embeddings of k into $\mathbf{R}$ defined by $\sigma(\delta) = \delta'$ and $\tau(\delta) = \delta''$. Then $H = SO_\Phi \times SO_{\sigma(\Phi)} \times SO_{\tau(\Phi)}$ and Γ is an irreducible, uniform, arithmetic lattice in $SO_\Phi(\mathbf{R}) \times SO_{\sigma(\Phi)}(\mathbf{R}) \simeq SO(n, 1) \times SO(n, 1)$.

Margulis'arithmeticity theorem is another spectacular application of super-rigidity (Theorem 4.1).

Theorem 4.3. *(see Chapter IX in [Mar91]) Let G be a connected semisimple Lie group with trivial centre, no compact factors, and $\mathbf{R} - rk(G) \geq 2$. Let $\Gamma \subset G$ be an irreducible lattice. Then Γ is arithmetic.*   □

Non-arithmetic lattices are known to exist in $SO(n, 1)$ for every $n \geq 2$ (Gromov-Piatetskii-Shapiro [GPS88]), and in $SU(n, 1)$ for $1 \leq n \leq 3$ (Deligne-Mostow [DM86]).

For other rank 1 groups, i.e. $Sp(n, 1)$ and the exceptional group $F_{4(-20)}$, super-rigidity and arithmeticity of lattices have been established by Corlette [Cor92] and Gromov-Schoen [GS92].

For a wealth of material on arithmetic groups, see [Bor69] and [WM].

Added on proof: It was pointed out to me by Gabriela Asli Nesin that equality (1) in the proof of Theorem 2.2 only holds for f left uniformly continuous and bounded, so that strictly speaking my proof is valid only for discrete groups.

To treat the general case, one may proceed as follows. Say that a mean m on $L^\infty(G)$ is *topologically left invariant* if $m(f \star \phi) = m(\phi)$ for every $\phi \in L^\infty(G)$, $f \in L^1(G)_{1,+}$. Observe then that, for every such f and such ϕ, the function $f \star \phi$ is left uniformly continuous and bounded. Fix $f_0 \in L^1(G)_{1,+}$. With m an invariant mean on $L^\infty(G)$, set $\tilde{m}(\phi) =: m(f_0 \star \phi)$. Then $\tilde{m}$ is a topological left invariant mean on $L^\infty(G)$ (for this, see the proof of Theorem G.3.1 in [BHV]). Replacing m by $\tilde{m}$, one may then repeat the end of the proof of Theorem 2.2.

References

[BHV]   B. Bekka, P. de la Harpe and A. Valette. *Kazhdan's Property (T).* Book to appear, Cambridge Univ. Press, 2008.

[BM00]   B. Bekka and M. Mayer. *Ergodic theory and topological dynamics of group actions on homogeneous spaces.* Cambridge Univ. Press, London Math. Soc. Lect. Note Ser. 269, 2000.

[Bor60]   A. Borel. Density properties for certain subgroups of semi-simple groups without compact components. *Ann. Math.*, 72:62–74, 1960.

[Bor69]   A. Borel. *Introduction aux groupes arithmétiques.* Hermann, Actu. sci. et industr. 1341, 1969.

[Bro81]   R. Brooks The fundamental group and the spectrum of the Laplacian. *Comment. Math. Helv.* 56:581–598, 1981.

[Cor92]   K. Corlette. Archimedean superrigidity and hyperbolic rigidity. *Ann. of Math.*, 135:165–182, 1992.

[Day50]   M. Day. Amenable groups. *Bull. Amer. Math. Soc.*, 56: 46, 1950.

[DM86]   P. Deligne and G.D. Mostow. Monodromy of hypergeometric functions and non-lattice integral monodromy. *Publ. Math. IHES*, 63:5–89, 1986.

[Fur63]   H. Furstenberg A Poisson formula for semisimple Lie groups *Annals of Math.* 77: 335-383, 1963.

[Fur73]   H. Furstenberg. Boundary theory and stochastic processes in homogeneous spaces. in: *Harmonic analysis on homogeneous spaces*, Symposia on Pure and Applied Math., Williamstown, Mass. 1972, Proceedings, 26: 193–229, 1973.

[GPS88]   M. Gromov and I. Piatetski-Shapiro. Nonarithmetic groups in Lobachevsky spaces. *Publ. Math. IHES*, 66:93–103, 1988.

[GS92]   M. Gromov and R. Schoen. Harmonic maps into singular spaces and *p*-adic superrigidity for lattices in groups of rank one. *Publ. Math. IHES*, 76:165–246, 1992.

[HM79]   R.E. Howe and C.C. Moore. Asymptotic properties of unitary representations. *Journal of Functional Analysis*, 32:72–96, 1979.

[Koo31]   B.O. Koopman. Hamiltonian systems and transformations in Hilbert spaces. *Proc. Nat. Acad. Sci. (USA)*, 17:315–318, 1931.

[Mar91]   G.A. Margulis. *Discrete subgroups of semisimple Lie groups*. Springer-Verlag, Ergeb. Math. Grenzgeb. 3 Folge, Bd. 17, 1991.

[Moo66]   C.C. Moore. Ergodicity of flows on homogeneous spaces. *Amer. J. Math.*, 88:154–178, 1966.

[Mos73]   G.D. Mostow. *Strong rigidity of locally symmetric spaces*. Annals of Math. studies 78, Princeton Univ. Press, 1973.

[vN29]   J. von Neumann. Zur allgemeinen Theorie des Masses. *Fund. Math.*, 13:73–116, 1929.

[Rei52]   H. Reiter. Investigations in harmonic analysis. *Trans. Amer. Math. Soc.*, 73:401–427, 1952.

[Rei64]   H. Reiter. Sur la propriété (P_1) et les fonctions de type positif. *C.R.Acad. Sci. Paris*, 258A:5134–5135, 1964.

[Rud73]   W. Rudin. *Functional analysis*. McGraw Hill, 1973.

[Wei65]   A. Weil. *L'intégration dans les groupes topologiques et ses applications*. Hermann, Paris, 1965.

[WM]   D. Witte-Morris. *Introduction to arithmetic groups*. Pre-book, february 2003.

[Zim78]   R.J. Zimmer. Induced and amenable actions of Lie groups. *Ann. Sci. Ec. Norm. Sup.* 11:407–428, 1978.

[Zim84]   R.J. Zimmer. *Ergodic theory and semisimple groups*. Birkhauser, 1984.

Unitary Representations and Complex Analysis

David A. Vogan, Jr [*]

Department of Mathematics, Massachusetts Institute of Technology, Cambridge,
Massachusetts 02139, USA
dav@math.mit.edu

1 Introduction

Much of what I will say depends on analogies between representation theory and linear algebra, so let me begin by recalling some ideas from linear algebra. One goal of linear algebra is to understand abstractly all possible linear transformations T of a vector space V. The simplest example of a linear transformation is multiplication by a scalar on a one-dimensional space. Spectral theory seeks to build more general transformations from this example. In the case of infinite-dimensional vector spaces, it is useful and interesting to introduce a topology on V, and to require that T be continuous. It often happens

[*] Supported in part by NSF grant DMS-9721441
    Mathematics Subject Classification: Primary 22E46

(as in the case when T is a differential operator acting on a space of functions) that there are many possible choices of V, and that choosing the right one for a particular problem can be subtle and important.

One goal of representation theory is to understand abstractly all the possible ways that a group G can act by linear transformations on a vector space V. Exactly what this means depends on the context. For topological groups (like Lie groups), one is typically interested in continuous actions on topological vector spaces. Using ideas from the spectral theory of linear operators, it is sometimes possible (at least in nice cases) to build such representations from irreducible representations, which play the role of scalar operators on one-dimensional spaces in linear algebra. Here is a definition.

Definition 1.1. *Suppose G is a topological group. A representation of G is a pair (π, V) with V a complete locally convex topological vector space, and π a homomorphism from G to the group of invertible linear transformations of V. We assume that the map*

$$G \times V \to V, \qquad (g, v) \mapsto \pi(g)v$$

is continuous.

An invariant subspace *for π is a closed subspace $W \subset V$ with the property that $\pi(g)W \subset W$ for all $g \in G$. The representation is said to be* irreducible *if there are exactly two invariant subspaces (namely V and 0).*

The flexibility in this definition—the fact that one does not require V to be a Hilbert space, or the operators $\pi(g)$ to be unitary—is a very powerful technical tool, even if one is ultimately interested only in unitary representations. Here is one reason. There are several important classes of groups (including reductive Lie groups) for which the classification of irreducible unitary representations is still an open problem. One way to approach the problem (originating in the work of Harish-Chandra, and made precise by Knapp and Zuckerman in [KZ77]) is to work with a larger class of "admissible" irreducible representations, for which a classification *is* available. The problem is then to identify the (unknown) unitary representations among the (known) admissible representations. Here is a formal statement.

Problem 1.2. Given an irreducible representation (π, V), is it possible to impose on V a Hilbert space structure making π a unitary representation? Roughly speaking, this question ought to have two parts.

(1.2)(A)   Does V carry a G-invariant Hermitian bilinear form $\langle , \rangle_\pi$?

Assuming that such a form exists, the second part is this.

(1.2)(B)   Is the form $\langle , \rangle_\pi$ positive definite?

The goal of these notes is to look at some difficulties that arise when one tries to make this program precise, and to consider a possible path around them. The difficulties have their origin exactly in the flexibility of Definition 1.1. Typically we want to realize a representation of G on a space of functions. If G acts on a set X, then G acts on functions on X, by

$$[\pi(g)f](x) = f(g^{-1} \cdot x).$$

The difficulty arises when we try to decide exactly which space of functions on X to consider. If G is a Lie group acting smoothly on a manifold X, then one can consider

$$C(X) = \text{continuous functions on } X,$$

$$C_c(X) = \text{continuous functions with compact support,}$$

$$C_c^\infty(X) = \text{compactly supported smooth functions.}$$

$$C^{-\infty}(X) = \text{distributions on } X.$$

If there is a reasonable measure on X, then one gets various Banach spaces like $L^p(X)$ (for $1 \leqslant p < \infty$), and Sobolev spaces. Often one can impose various other kinds of growth conditions at infinity. All of these constructions give topological vector spaces of functions on X, and many of these spaces carry continuous representations of G. These representations will not be "equivalent" in any simple sense (involving isomorphisms of topological vector spaces); but to have a chance of getting a reasonable classification theorem for representations, one needs to identify them.

When G is a reductive Lie group, Harish-Chandra found a notion of "infinitesimal equivalence" that addresses these issues perfectly. Inside every irreducible representation V is a natural dense subspace V_K, carrying an irreducible representation of the Lie algebra of G. (Actually one needs for this an additional mild assumption on V, called "admissibility.") Infinitesimal equivalence of V and W means algebraic equivalence of V_K and W_K as Lie algebra representations. (Some details appear in section 4.)

Definition 1.3. *Suppose G is a reductive Lie group. The* admissible dual *of G is the set $\widehat{G}$ of infinitesimal equivalence classes of irreducible admissible representations of G. The* unitary dual *of G is the set $\widehat{G}_u$ of unitary equivalence classes of irreducible unitary representations of G.*

Harish-Chandra proved that each infinitesimal equivalence class of admissible irreducible representations contains at most one unitary equivalence class of irreducible unitary representations. That is,

(1.4)
$$\widehat{G}_u \subset \widehat{G}.$$

This sounds like great news for the program described in Problem 1.2. Even better, he showed that the representation (π, V) is infinitesimally unitary if and only if the Lie algebra representation V_K admits a positive-definite invariant Hermitian form $\langle,\rangle_{\pi,K}$.

The difficulty is this. Existence of a continuous G-invariant Hermitian form $\langle,\rangle_\pi$ on V implies the existence of $\langle,\rangle_{\pi,K}$ on V_K; but the converse is not true. Since V_K is dense in V, there is at most one continuous extension of $\langle,\rangle_{\pi,K}$ to V, but the extension may not exist. In section 3, we will look at some examples, in order to understand why this is so. What the examples suggest, and what we will see in section 4, is that the Hermitian form can be defined only on appropriately "small" representations in each infinitesimal equivalence class. In the example of the various function spaces on X, compactly supported smooth functions are appropriately small, and will often carry an invariant Hermitian form. Distributions, on the other hand, are generally too large a space to admit an invariant Hermitian form.

Here is a precise statement. (We will write V^* for the space of continuous linear functionals on V, endowed with the strong topology (see section 8).)

Theorem 1.5. (Casselman, Wallach, and Schmid; see [Cas89], [Sch85], and section 4). *Suppose (π, V) is an admissible irreducible representation of a reductive Lie group G on a reflexive Banach space V. Define*

$$(\pi^\omega, V^\omega) = \text{analytic vectors in } V,$$

$$(\pi^\infty, V^\infty) = \text{smooth vectors in } V,$$

$$(\pi^{-\infty}, V^{-\infty}) = \text{distribution vectors in } V = \text{dual of } (V')^\infty, \text{ and}$$

$$(\pi^{-\omega}, V^{-\omega}) = \text{hyperfunction vectors in } V = \text{dual of } (V')^\omega.$$

Each of these four representations is a smooth representation of G in the infinitesimal equivalence class of π, and each depends only on that equivalence class. The inclusions

$$V^\omega \subset V^\infty \subset V \subset V^{-\infty} \subset V^{-\omega}$$

are continuous, with dense image.

Any invariant Hermitian form $\langle,\rangle_K$ on V_K extends uniquely to continuous G-invariant Hermitian forms $\langle,\rangle_\omega$ and $\langle,\rangle_\infty$ on V^ω and V^∞.

The assertions about Hermitian forms will be proven in Theorem 9.16.

The four representations appearing in Theorem 1.5 are called the *minimal globalization*, the *smooth globalization*, the *distribution globalization*, and the *maximal globalization* respectively. Unless π is finite-dimensional (so that all of the spaces in the theorem are the same) the Hermitian form will *not* extend continuously to the distribution or maximal globalizations $V^{-\infty}$ and $V^{-\omega}$.

We will be concerned here mostly with representations of G constructed using complex analysis, on spaces of holomorphic sections of vector bundles and generalizations. In order to use these constructions to get unitary representations, we need to do the analysis in such a way as to get the minimal or smooth globalizations; this will ensure that the Hermitian forms we seek will be defined on the representations. A theorem of Hon-Wai Wong (see [Won99] or Theorem 7.21 below) says that Dolbeault cohomology leads to the *maximal* globalizations in great generality. This means that there is no possibility of finding invariant Hermitian forms on these Dolbeault cohomology representations except in the finite-dimensional case.

We therefore need a way to modify the Dolbeault cohomology construction to produce minimal globalizations rather than maximal ones. Essentially we will follow ideas of Serre from [Ser55], arriving at realization of minimal globalization representations first obtained by Tim Bratten in [Bra97]. Because of the duality used to define the maximal globalization, the question amounts to this: how can one identify the topological dual space of a Dolbeault cohomology space on a (noncompact) complex manifold? The question is interesting in the simplest case. Suppose $X \subset \mathbb{C}$ is an open set, and $H(X)$ is the space of holomorphic functions on X. Make $H(X)$ into a topological vector space, using the topology of uniform convergence of all derivatives on compact sets. What is the dual space $H(X)'$?

This last question has a simple answer. Write $C_c^{-\infty}(X, \text{densities})$ for the space of compactly supported distributions on X. We can think of this as the space of compactly supported complex 2-forms (or $(1,1)$-forms) on X, with generalized function coefficients. (A brief review of these ideas will appear in section 8). More generally, write

$$A_c^{(p,q),-\infty}(X) = \text{compactly supported } (p,q)\text{-forms}$$

on X with generalized function coefficients.

The Dolbeault differential $\overline{\partial}$ maps (p,q)-forms to $(p, q+1)$ forms and preserves support; so

$$\overline{\partial} \colon A_c^{(1,0),-\infty}(X) \to A_c^{(1,1),-\infty}(X) = C_c^{-\infty}(X, \text{densities}).$$

Then (see [Ser55], Théorème 3)

(1.6) $$H(X)' \simeq A_c^{(1,1),-\infty}(X)/\overline{\overline{\partial}A_c^{1,0}(X)}.$$

Here the overline denotes closure. For X open in $\mathbb{C}$ the image of $\overline{\partial}$ is automatically closed, so the overline is not needed; but this formulation has an immediate extension to any complex manifold X (replacing 1 and 0 by the dimension n and $n-1$). Here is Serre's generalization.

Theorem 1.7. (Serre; see [Ser55], Théorème 2 or Theorem 8.13 below). *Suppose X is a complex manifold of dimension n, $\mathcal{V}$ is a holomorphic vector bundle on X, and Ω is the canonical line bundle (of $(n,0)$-forms on X). Define*

$$A^{0,p}(X, \mathcal{V}) = \text{smooth } \mathcal{V}\text{-valued } (0,p)\text{-forms on } X$$

$$A_c^{(0,p),-\infty}(X,\mathcal{V}) = \textit{compactly supported } \mathcal{V}\textit{-valued } (0,p)\textit{-forms}$$
$$\textit{with generalized function coefficients.}$$

Define the topological Dolbeault cohomology of X with values in $\mathcal{V}$ as

$$H_{top}^{0,p}(X,\mathcal{V}) = [\textit{kernel of } \overline{\partial} \textit{ on } A^{0,p}(X,\mathcal{V})]/\overline{\partial}A^{p-1,0}(X,\mathcal{V});$$

this is a quotient of the usual Dolbeault cohomology. It carries a natural locally convex topology. Similarly, define

$$H_{c,top}^{0,p}(X,\mathcal{V}) = [\textit{kernel of } \overline{\partial} \textit{ on } A_c^{(0,p),-\infty}(X,\mathcal{V})]/\overline{\overline{\partial}A_c^{(p-1,0),-\infty}(X,\mathcal{V})},$$

the topological Dolbeault cohomology with compact supports. Then there is a natural identification

$$H_{top}^{0,p}(X,\mathcal{L})^* \simeq H_{c,top}^{0,n-p}(X,\Omega \otimes \mathcal{L}^*).$$

Here $\mathcal{L}^*$ is the dual holomorphic vector bundle to $\mathcal{L}$.

When X is compact, then the subscript c adds nothing, and the $\overline{\partial}$ operators automatically have closed range. One gets in that case the most familiar version of Serre duality.

In Corollary 8.14 we will describe how to use this theorem to obtain Bratten's result, constructing minimal globalization representations on Dolbeault cohomology with compact supports.

Our original goal was to understand invariant bilinear forms on minimal globalization representations. Once the minimal globalizations have been identified geometrically, we can at least offer a language for discussing this problem using standard functional analysis. This is the subject of section 9.

There are around the world a number of people who understand analysis better than I do. As an algebraist, I cannot hope to estimate this number. Nevertheless I am very grateful to several of them (including Henryk Hecht, Sigurdur Helgason, David Jerison, and Les Saper) who helped me patiently with very elementary questions. I am especially grateful to Tim Bratten, for whose work these notes are intended to be an advertisement. For the errors that remain, I apologize to these friends and to the reader.

2 Compact Groups and the Borel-Weil Theorem

The goal of these notes is to describe a geometric framework for some basic questions in representation theory for noncompact reductive Lie groups. In order to explain what that might mean, I will recall in this section the simplest example: the Borel-Weil theorem describing irreducible representations of a compact group. Throughout this section, therefore, we fix a compact connected Lie group K, and a maximal torus $T \subset K$. (We will describe an example in a moment.) We fix also a K-invariant complex structure on the homogeneous space K/T. In terms of the structure theory of Lie algebras, this amounts to a choice of positive roots for the Cartan subalgebra $\mathfrak{t} = \mathrm{Lie}(T)_{\mathbb{C}}$ inside the complex reductive Lie algebra $\mathfrak{k} = \mathrm{Lie}(K)_{\mathbb{C}}$. For more complete expositions of the material in this section, we refer to [Kna86], section V.7, or [Hel94], section VI.4.3, or [Vog87], chapter 1.

Define

(2.1)(a) $\qquad\qquad \widehat{T} = \text{lattice of characters of } T;$

these are the irreducible representations of T. Each $\mu \in \widehat{T}$ may be regarded as a homomorphism of T into the unit circle, or as a representation $(\mu, \mathbb{C}_\mu)$ of T. Such a representation gives rise to a K-equivariant holomorphic line bundle

(2.1)(b) $\qquad\qquad \mathcal{L}_\mu \to K/T.$

Elements of $\widehat{T}$ are often called *weights*.

I do not want to recall the structure theory for K in detail, and most of what I say will make some sense without the details. With that warning not to pay attention, fix a simple root α of T in K, and construct a corresponding three-dimensional subgroup

(2.2)(a) $\qquad \phi_\alpha \colon SU(2) \to K, \qquad K_\alpha = \phi_\alpha(SU(2)), \qquad T_\alpha = K_\alpha \cap T.$

Then K_α/T_α is the Riemann sphere $\mathbb{CP}^1$, and we have a natural holomorphic embedding

(2.2)(b) $\qquad\qquad \mathbb{CP}^1 \simeq K_\alpha/T_\alpha \hookrightarrow K/T.$

The weight $\mu \in \widehat{T}$ is called *antidominant* if for every simple root α,

(2.2)(c) $\qquad \mathcal{L}_\mu|_{K_\alpha/T_\alpha}$ has non-zero holomorphic sections.

This is a condition on $\mathbb{CP}^1$, about which we know a great deal. The sheaf of germs of holomorphic sections of $\mathcal{L}_\mu|_{K_\alpha/T_\alpha}$ is $\mathcal{O}(-\langle \mu, \alpha^\vee \rangle)$; here $\alpha^\vee$ is the coroot for the simple root α, and $\langle \mu, \alpha^\vee \rangle$ is an integer. The sheaf $\mathcal{O}(n)$ on $\mathbb{CP}^1$ has non-zero sections if and only if $n \geqslant 0$. It follows that μ is antidominant if and only if for every simple root α,

(2.2)(d) $\qquad\qquad \langle \mu, \alpha^\vee \rangle \leqslant 0.$

Theorem 2.3. (Borel-Weil, Harish-Chandra; see [HC56], [Ser59]). *Suppose K is a compact connected Lie group with maximal torus T; use the notation of (2.1) and (2.2) above.*

(1) Every K-equivariant holomorphic line bundle on K/T is equivalent to $\mathcal{L}_\mu$, for a unique weight $\mu \in \widehat{T}$.

(2) *The line bundle $\mathcal{L}_\mu$ has non-zero holomorphic sections if and only if μ is antidominant.*

(3) *If μ is antidominant, then the space $\Gamma(\mathcal{L}_\mu)$ of holomorphic sections is an irreducible representation of K.*

(4) *This correspondence defines a bijection from antidominant characters of T onto $\widehat{K}$.*

As the references indicate, I believe that this theorem is due independently to Harish-Chandra and to Borel and Weil. Nevertheless I will follow standard practice and refer to it as the Borel-Weil theorem.

Before saying anything about a proof, we look at an example. Set

$$(2.4)(a) \qquad K = U(n) = n \times n \text{ complex unitary matrices}$$
$$= \{u = (u_1, \ldots, u_n) \mid u_i \in \mathbb{C}^n, \langle u_i, u_j \rangle = \delta_{i,j}\}.$$

Here we regard $\mathbb{C}^n$ as consisting of column vectors, so that the u_i are the columns of the matrix u; $\delta_{i,j}$ is the Kronecker delta. This identifies $U(n)$ with the set of orthonormal bases of $\mathbb{C}^n$. As a maximal torus, we choose

$$(2.4)(b) \qquad T = U(1)^n = \text{diagonal unitary matrices}$$
$$= \left\{ \begin{pmatrix} e^{i\phi_1} & & \\ & \ddots & \\ & & e^{i\phi_n} \end{pmatrix} \mid \phi_j \in \mathbb{R} \right\}.$$

As a basis for the lattice of characters of T, we can choose

$$(2.4)(c) \qquad \chi_j \begin{pmatrix} e^{i\phi_1} & & \\ & \ddots & \\ & & e^{i\phi_n} \end{pmatrix} = e^{i\phi_j},$$

the action of T on the jth coordinate of $\mathbb{C}^n$.

We want to understand the homogeneous space $K/T = U(n)/U(1)^n$. Recall that a *complete flag* in $\mathbb{C}^n$ is a collection of linear subspaces

$$(2.5)(a) \qquad F = (0 = F_0 \subset F_1 \subset \cdots \subset F_n = \mathbb{C}^n), \qquad \dim F_j = j.$$

The collection of all such complete flags is a complex projective algebraic variety

$$(2.5)(b) \qquad X = \text{complete flags in } \mathbb{C}^n,$$

of complex dimension $n(n-1)/2$. (When we need to be more precise, we may write $X^{GL(n)}$.) We claim that

$$(2.5)(c) \qquad U(n)/U(1)^n \simeq X.$$

The map from left to right is

$$(2.5)(d) \qquad (u_1, \ldots, u_n)U(1)^n \mapsto F = (F_j), \qquad F_j = \mathrm{span}(u_1, \ldots, u_j).$$

Right multiplication by a diagonal matrix replaces each column of u by a scalar multiple of itself; so the spans in this definition are unchanged, and the map is well-defined on cosets. For the map in the opposite direction, we choose an orthonormal basis u_1 of the one-dimensional space F_1; extend it by Gram-Schmidt to an orthonormal basis (u_1, u_2) of F_2; and so on. Each u_j is determined uniquely up to multiplication by a scalar $e^{i\phi_j}$, so the coset $(u_1, \ldots, u_n)U(1)^n$ is determined by F.

It is often useful to notice that the full general linear group $K_{\mathbb{C}} = GL(n, \mathbb{C})$ (the complexification of $U(n)$) acts holomorphically on X. For this action the isotropy group at the base point is the Borel subgroup $B_{\mathbb{C}}$ of upper triangular matrices:

$$X = K_{\mathbb{C}}/B_{\mathbb{C}}.$$

The fact that X is also homogeneous for the subgroup K corresponds to the group-theoretic facts

$$K_{\mathbb{C}} = KB_{\mathbb{C}}, \qquad K \cap B_{\mathbb{C}} = T.$$

Now the definition of X provides a number of natural line bundles on X. For $1 \leqslant j \leqslant n$, there is a line bundle $\mathcal{L}_j$ whose fiber at the flag F is the one-dimensional space F_j/F_{j-1}:

$$(2.6)(a) \qquad \mathcal{L}_j(F) = F_j/F_{j-1} \qquad (F \in X).$$

This is a $U(n)$-equivariant (in fact $K_{\mathbb{C}}$-equivariant) holomorphic line bundle. For any $\mu = (m_1, \ldots, m_n) \in \mathbb{Z}^n$, we get a line bundle

$$(2.6)(b) \qquad \mathcal{L}_\mu = \mathcal{L}_1^{m_1} \otimes \cdots \otimes \mathcal{L}_n^{m_n}$$

For example, if $p \leqslant q$, and F is any flag, then F_q/F_p is a vector space of dimension $q - p$. These vector spaces form a holomorphic vector bundle $\mathcal{V}_{q,p}$. Its top exterior power $\bigwedge^{q-p} \mathcal{V}_{q,p}$ is therefore a line bundle on X. Writing

$$(2.6)(c) \qquad \mu_{q,p} = (\underbrace{0, \ldots, 0}_{p \text{ terms}}, \underbrace{1, \ldots, 1}_{q-p \text{ terms}}, \underbrace{0, \ldots, 0}_{n-q \text{ terms}}),$$

we find

$$(2.6)(d) \qquad \mathcal{L}_{\mu_{q,p}} \simeq \bigwedge^{q-p} \mathcal{V}_{q,p}.$$

One reason for making all these explicit examples is that it shows how some of these bundles can have holomorphic sections. The easiest example is $\mathcal{L}_n$, whose fiber at F is the quotient space $\mathbb{C}^n/F_{n-1}$. Any element $v \in \mathbb{C}^n$ defines a section σ_v of $\mathbb{L}_n$, by the formula

$(2.6)(e)$ $\qquad\qquad \sigma_v(F) = v + F_{n-1} \in \mathbb{C}^n/F_{n-1} = \mathcal{L}_n(F).$

Notice that this works only for $\mathcal{L}_n$, and not for the other $\mathcal{L}_j$. In a similar way, taking $q = n$ in $(2.6)(c)$, we find that any element $\omega \in \bigwedge^{n-p} \mathbb{C}^n$ defines a section σ_ω of $\mathcal{L}_{\mu_{q,p}}$, by

$$\sigma_\omega(F) = \overline{\omega} \in \bigwedge^{n-p}(\mathbb{C}^n/F_p).$$

By multiplying such sections together, we can find non-zero holomorphic sections of any of the bundles $\mathcal{L}_\mu$, as long as

$(2.6)(e)$ $\qquad\qquad 0 \leqslant m_1 \leqslant \cdots \leqslant m_n.$

In case $m_1 = \cdots = m_n = m$, the sections we get are related to the function $\det^m$ on K (or $K_\mathbb{C}$). Since that function vanishes nowhere on the group, its inverse provides holomorphic sections of the bundle corresponding to $(-m, \cdots, -m)$. Multiplying by these, we finally have non-zero sections of $\mathcal{L}_\mu$ whenever

$(2.6)(f)$ $\qquad\qquad m_1 \leqslant \cdots \leqslant m_n.$

Here is what the Borel-Weil theorem says for $U(n)$.

Theorem 2.7. (Borel-Weil, Harish-Chandra). *Use the notation of (2.4)–(2.6).*

(1) *Every $U(n)$-equivariant holomorphic line bundle on the complete flag manifold X is equivalent to $\mathcal{L}_\mu = \mathcal{L}_1^{m_1} \otimes \cdots \otimes \mathcal{L}_n^{m_n}$, for a unique*

$$\mu = (m_1, \ldots, m_n) \in \mathbb{Z}^n.$$

(2) *The line bundle $\mathcal{L}_\mu$ has non-zero holomorphic sections if and only if μ is antidominant, meaning that*

$$m_1 \leqslant \cdots \leqslant m_n.$$

(3) *If μ is antidominant, then the space $\Gamma(\mathcal{L}_\mu)$ of holomorphic sections is an irreducible representation of $U(n)$.*

(4) *This correspondence defines a bijection from increasing sequences of integers onto $\widehat{U(n)}$.*

Here are some remarks about proofs for Theorems 2.3 and 2.7. Part (1) is very easy: making anything G-equivariant on a homogeneous space G/H is the same as making something H-equivariant. (Getting precise theorems of this form is simply a matter of appropriately specifying "anything" and "something," then following your nose.)

For part (2), "only if" is easy to prove using reduction to $\mathbb{CP}^1$: if μ fails to be antidominant, then there will not even be sections on some of those projective lines. The "if" part is more subtle. We proved it for $U(n)$ in $(2.6)(f)$, essentially by making use of a large supply of known representations of $U(n)$ (the exterior powers of the standard representation, the powers of the determinant character, and tensor products of these). For general K, one can do something similar: once one knows the existence of a representation of lowest weight μ, it is a simple matter to use matrix coefficients of that representation to construct holomorphic sections of $\mathcal{L}_\mu$. This is what Harish-Chandra did. I cannot tell from the account in [Ser59] exactly what argument Borel and Weil had in mind. In any case it is certainly possible to construct holomorphic sections of $\mathcal{L}_\mu$ (for antidominant μ) directly, using the Bruhat decomposition of K/T. It is easy to write a holomorphic section on the open cell (for any μ); then one can use the antidominance condition to prove that this section extends to all of K/T.

Part (3) and the injectivity in part (4) are both assertions about the space of intertwining operators

$$\mathrm{Hom}_K(\Gamma(\mathcal{L}_{\mu_1}), \Gamma(\mathcal{L}_{\mu_2})).$$

We will look at such spaces in more generality in section 9 (Corollary 9.13).

Finally, the surjectivity in part (4) follows from the existence of lowest weights for arbitrary irreducible representations. This existence is a fairly easy part of algebraic representation theory. I do not know of a purely complex analysis proof.

Before we abandon compact groups entirely, here are a few comments about how to generalize the linear algebra in (2.4)–(2.6). A *classical compact group* is (in the narrowest possible definition) one of the groups $U(n)$, $O(n)$ (of real orthogonal matrices), or $Sp(2n)$ (of complex unitary matrices also preserving a standard symplectic form on $\mathbb{C}^{2n}$). For each of these groups, there is a parallel description of K/T as a projective variety of certain complete flags in a complex vector space. One must impose on the flags certain additional conditions involving the bilinear form that defines the group. Here are some details.

Suppose first that $K = O(n)$, the group of linear transformations of $\mathbb{R}^n$ preserving the standard symmetric bilinear form

$$B(x, y) = \sum_{j=1}^{n} x_j y_j \qquad (x, y \in \mathbb{R}^n).$$

This form extends holomorphically to $\mathbb{C}^n$, where it defines the group

(2.8)(a) $$K_\mathbb{C} = O(n, \mathbb{C}).$$

If $W \subset \mathbb{C}^n$ is a p-dimensional subspace, then

$$W^\perp = \{y \in \mathbb{C}^n \mid B(x, y) = 0, \text{ all } x \in W\}$$

is a subspace of dimension $n - p$. We now define the *complete flag variety for* $O(n)$ to be

(2.8)(b)
$$X = X^{O(n)} = \{F = (F_j) \text{ complete flags} \mid F_p^\perp = F_{n-p}, \quad 0 \leqslant p \leqslant n\}.$$

Notice that this definition forces the subspaces F_q with $2q \leqslant n$ to satisfy $F_q^\perp \subset F_q$; that is, the bilinear form must vanish on these F_q. Such a subspace is called *isotropic*. Knowledge of the isotropic subspaces F_q (for $q \leqslant n/2$) determines the remaining subspaces, by the requirement $F_{n-q} = F_q^\perp$. We get an identification

(2.8)(c) $X^{O(n)} \simeq$ chains of isotropic subspaces $(F_q) = (F_0 \subset F_1 \subset \cdots)$,
with $\dim F_q = q$ for all $q \leqslant n/2$.

The orthogonal group is

(2.8)(d) $O(n) = \{v = (v_1, \ldots, v_n) \mid v_p \in \mathbb{R}^n, B(v_p, v_q) = \delta_{p,q}\}$,

the set of orthonormal bases of $\mathbb{R}^n$. As a maximal torus T in K, we can take $SO(2)^{[n/2]}$, embedded in an obvious way. We claim that

(2.8)(e) $O(n)/SO(2)^{[n/2]} \simeq X^{O(n)}$.

The map from left to right is
(2.8)(f)
$(v_1, \ldots, v_n)SO(2)^n \mapsto F = (F_p), \qquad F_p = \operatorname{span}(v_1 + iv_2, \ldots, v_{2p-1} + iv_{2p})$.

We leave to the reader the verification that this is a well-defined bijection, and an extension of the ideas in (2.6) to this setting. (Notice only that $O(n)$ is not connected, that correspondingly X has two connected components, and that the irreducibility assertion in Theorem 2.3(3) can fail.) The space X is also homogeneous for the (disconnected) reductive algebraic group $K_{\mathbb{C}} = O(n, \mathbb{C})$, the isotropy group being a Borel subgroup of the identity component.

Finally, consider the standard symplectic form on $\mathbb{C}^{2n}$,

$$\omega(x, y) = \sum_{p=1}^{n} x_p y_{n+p} - x_{n+p} y_p.$$

The group of linear transformations preserving this form is

(2.9)(a) $K_{\mathbb{C}} = Sp(2n, \mathbb{C})$;

the corresponding compact group may be taken to be

$$K = K_{\mathbb{C}} \cap U(2n).$$

(Often it is easier to think of K as a group of $n \times n$ matrices with entries in the quaternions. This point of view complicates slightly the picture of $K_{\mathbb{C}}$, and so I will not adopt it.) Just as for the symmetric form B, we can define

$$W^\perp = \{y \in \mathbb{C}^{2n} \mid \omega(x, y) = 0, \text{ all } x \in W\};$$

if W has dimension p, then $W^\perp$ has dimension $2n - p$. The *complete flag variety for* $Sp(n)$ is

(2.9)(b)
$$X = X^{Sp(2n)} = \{F = (F_j) \text{ complete flags} \mid F_p^\perp = F_{2n-p}, \quad 0 \leqslant p \leqslant 2n\}.$$

Again the definition forces F_q to be isotropic for $q \leqslant n$, and we can identify

(2.9)(c) $X \simeq$ chains of isotropic subspaces $(F_q) = (F_0 \subset F_1 \subset \cdots)$,
with $\dim F_q = q$ for all $q \leqslant n$.

The complex symplectic group $K_\mathbb{C} = Sp(2n, \mathbb{C})$ acts holomorphically on the projective variety X.

The complex symplectic group is

(2.9)(d)
$$Sp(2n, \mathbb{C}) = \{v = (v_1, \ldots, v_{2n}) \mid v_p \in \mathbb{C}^{2n}, \omega(v_p, v_q) = \delta_{p, q-n} \quad (p \leqslant q)\},$$

This identifies $K_\mathbb{C}$ with the collection of standard symplectic bases for $\mathbb{C}^{2n}$. The compact symplectic group is identified with standard symplectic bases that are also orthonormal for the standard Hermitian form $\langle, \rangle$ on $\mathbb{C}^{2n}$:

(2.9)(e) $Sp(2n) = \{(v_1, \ldots, v_{2n}) \mid v_p \in \mathbb{C}^{2n}, \ \omega(v_p, v_q) = \delta_{p, q-n},$
$$\langle v_p, v_q \rangle = \delta_{p, q} \quad (p \leqslant q)\},$$

As a maximal torus in K, we choose the diagonal subgroup

(2.9)(f) $$T = \left\{ \begin{pmatrix} e^{i\phi_1} & & & & & \\ & \ddots & & & & \\ & & e^{i\phi_n} & & & \\ & & & e^{-i\phi_1} & & \\ & & & & \ddots & \\ & & & & & e^{-i\phi_n} \end{pmatrix} \right\} \simeq U(1)^n.$$

We claim that

(2.9)(g) $$Sp(2n)/U(1)^n \simeq X^{Sp(2n)}.$$

The map from left to right is

(2.9)(h)
$$(v_1, \ldots, v_{2n})U(1)^n \mapsto F = (F_p), \qquad F_p = \text{span}(v_1, \ldots, v_p) \quad (0 \leqslant p \leqslant n).$$

Again we leave to the reader the verification that this is a well-defined bijection, and the task of describing the equivariant line bundles on X.

3 Examples for $SL(2, \mathbb{R})$

In this section we will present some examples of representations of $SL(2, \mathbb{R})$, in order to develop some feeling about what infinitesimal equivalence, minimal globalizations, and so on look like in examples. More details can be found in [Kna86], pages 35–41.

In fact it is a little simpler for these examples to consider not $SL(2, \mathbb{R})$ but the isomorphic group

$$(3.1)(a) \qquad G = SU(1,1) = \left\{ \begin{pmatrix} \alpha & \beta \\ \bar{\beta} & \bar{\alpha} \end{pmatrix} \mid \alpha, \beta \in \mathbb{C}, |\alpha|^2 - |\beta|^2 = 1 \right\}.$$

This is the group of linear transformations of $\mathbb{C}^2$ preserving the standard Hermitian form of signature $(1,1)$, and having determinant 1. We will be particularly interested in a maximal compact subgroup:

$$(3.1)(b) \qquad K = \left\{ \begin{pmatrix} e^{i\theta} & 0 \\ 0 & e^{-i\theta} \end{pmatrix} \mid \theta \in \mathbb{R} \right\}.$$

The group G acts on the open unit disc by linear fractional transformations:

$$(3.2)(a) \qquad \begin{pmatrix} \alpha & \beta \\ \bar{\beta} & \bar{\alpha} \end{pmatrix} \cdot z = \frac{\bar{\alpha} z + \bar{\beta}}{\beta z + \alpha}.$$

It is not difficult to check that this action is transitive:

$$(3.2)(b) \qquad D = \{ z \mid |z| < 1 \} = G \cdot 0 \simeq G/K.$$

The last identification comes from the fact that K is the isotropy group for the action at the point 0.

The action of G on D preserves complex structures. Setting

$$(3.2)(c) \qquad V^{-\omega} = \text{holomorphic functions on } D,$$

we therefore get a representation π of G on $V^{-\omega}$ by

$$(3.2)(d) \qquad [\pi(g)f](z) = f(g^{-1} \cdot z) = f\left(\frac{\alpha z - \bar{\beta}}{-\beta z + \bar{\alpha}} \cdot \right)$$

The representation $(\pi, V^{-\omega})$ is not irreducible, because the one-dimensional closed subspace of constant functions is invariant. Nevertheless (as we will see in section 4) the Casselman-Wallach-Schmid theory of distinguished globalizations still applies. As the notation suggests, $V^{-\omega}$ is a maximal globalization for the corresponding Harish-Chandra module

$$(3.2)(e) \qquad V^K = \text{polynomials in } z$$

of K-finite vectors.

In order to describe other (smaller) globalizations of V^K, we can control the growth of functions near the boundary circle of D. The most drastic possibility is to require the functions to extend holomorphically across the boundary of D:

(3.3)(a) V^ω = holomorphic functions on $\overline{D}$

This space can also be described as the intersection (over positive numbers ϵ) of holomorphic functions on discs of radius $1 + \epsilon$. Restriction to the unit circle identifies V^ω with real analytic functions on the circle whose negative Fourier coefficients all vanish. There is a natural topology on V^ω, making it a representation of G by the action π of (3.2)(d). As the notation indicates, this representation is Schmid's minimal globalization of V^K.

A slightly larger space is

(3.3)(b) V^∞ = holomorphic functions on D with smooth boundary values.

More or less by definition, V^∞ can be identified with smooth functions on the circle whose negative Fourier coefficients vanish. The identification topologizes V^∞, and it turns out that the resulting representation of G is the Casselman-Wallach smooth globalization of V^K. Larger still is

(3.3)(c) $V^{(2)}$ = holomorphic functions on D with L^2 boundary values.

This is a Hilbert space, the square-integrable functions on the circle whose negative Fourier coefficients vanish. The representation of G on this Hilbert space is continuous but not unitary (because these linear fractional transformations of the circle do not preserve the measure). Of course there are many other function spaces on the circle that can be used in a similar way; I will mention only

(3.3)(d)

$V^{-\infty}$ = holomorphic functions on D with distribution boundary values.

This is the Casselman-Wallach distribution globalization of V^K.

We therefore have

$$V^\omega \subset V^\infty \subset V^{(2)} \subset V^{-\infty} \subset V^{-\omega}.$$

These inclusions of representations are continuous with dense image. Holomorphic functions on the disc all have Taylor expansions

$$f(z) = \sum_{n=0}^{\infty} a_n z^n.$$

We can describe each space by conditions on the coefficients a_n; these descriptions implicitly specify the topologies very nicely.

$$V^{-\omega} \leftrightarrow \{(a_n) \mid \sum_{n=0}^{\infty} |a_n|(1-\epsilon)^n < \infty, \quad 0 < \epsilon \leqslant 1\}.$$

$$V^{-\infty} \leftrightarrow \{(a_n) \mid \text{for some } N > 0,\ |a_n| < C_N(1+n)^N\}.$$

$$V^{(2)} \leftrightarrow \{(a_n) \mid \sum_{n=0}^{\infty} |a_n|^2 < \infty\}.$$

$$V^{\infty} \leftrightarrow \{(a_n) \mid \text{for every } N > 0,\ |a_n| < C_N(1+n)^{-N}\}.$$

$$V^{\omega} \leftrightarrow \{(a_n) \mid \sum_{n=0}^{\infty} |a_n|(1+\epsilon)^n < \infty, \text{ some } \epsilon > 0\}.$$

4 Harish-Chandra Modules and Globalization

In this section we will recall very briefly some general facts about representations of real reductive groups. The first problem is to specify what groups we are talking about. A Lie algebra (over any field of characteristic zero) is called *semisimple* if it is a direct sum of non-abelian simple Lie algebras. It is natural to define a real Lie group to be *semisimple* if it is connected, and its Lie algebra is semisimple. Such a definition still allows some technically annoying examples (like the universal cover of $SL(2,\mathbb{R})$, which has no non-trivial compact subgroups). Accordingly there is a long tradition of working with connected semisimple groups having finite center. There are several difficulties with that. As we will see, there are many results relating the representation theory of G to representation theory of subgroups of G; and the relevant subgroups are rarely themselves connected and semisimple. Another difficulty comes from the demands of applications. One of the most important applications of representation theory for Lie groups is to automorphic forms. In that setting the most fundamental example is $GL(n,\mathbb{R})$, a group which is neither connected nor semisimple.

Most of these objections can be addressed by working with algebraic groups, and considering always the group of real points of a connected reductive algebraic group defined over $\mathbb{R}$. (The group $GL(n)$ is a connected algebraic group, even though its group of real points is disconnected as a Lie group.) The difficulty with this is that it still omits some extremely important examples. Some of the most interesting representation theory lives on the nonlinear double cover $Mp(2n,\mathbb{R})$ of the algebraic group $Sp(2n,\mathbb{R})$ (consisting of linear transformations of $\mathbb{R}^{2n}$ preserving a certain symplectic form). The "oscillator representation" of this group is fundamental to mathematical physics, to the theory of automorphic forms, and to classical harmonic analysis. (Such an assertion needs to be substantiated, and I won't do that; but here at least are some interesting references: [Wei64], [How88], [How89].)

So we want to include at least finite covering groups of real points of connected reductive algebraic groups. At some point making a definition along these lines becomes quite cumbersome. I will therefore follow the path taken by Knapp in [Kna86], and take as the definition of reductive a property that usually appears as a basic structure theorem. The definition is elementary and short, and it leads quickly to some fundamental facts about the groups. One can object that it does not extend easily to groups over other local fields, but for the purposes of these notes that will not be a problem.

The idea is that the most basic example of a reductive group is the group $GL(n, \mathbb{R})$ of invertible $n \times n$ real matrices. We will recall a simple structural fact about $GL(n)$ (the polar decomposition of Proposition 4.2 below). Then we will define a reductive group to be (more or less) any subgroup of some $GL(n)$ that inherits the polar decomposition.

If $g \in G = GL(n, \mathbb{R})$, define

$$(4.1)(a) \qquad\qquad \theta g = {}^t g^{-1}$$

the inverse of the transpose of g. The map θ is an automorphism of order 2, called the *Cartan involution* of $GL(n)$. Write $O(n) = GL(n)^\theta$ for the subgroup of fixed points of θ. This is the group of $n \times n$ real orthogonal matrices, the *orthogonal group*. It is compact.

Write $\mathfrak{gl}(n, \mathbb{R}) = \mathrm{Lie}(GL(n, \mathbb{R}))$ for the Lie algebra of $GL(n)$ (the space of all $n \times n$ real matrices). The automorphism θ of G differentiates to an involutive automorphism of the Lie algebra, defined by

$$(4.1)(b) \qquad\qquad (d\theta)(X) = -{}^t X.$$

Notice that if X happens to be invertible, then $(d\theta)(X)$ and $\theta(X)$ are both defined, and they are *not* equal. Despite this potential for confusion, we will follow tradition and abuse notation by writing simply θ for the differential of θ. The -1-eigenspace of θ on the Lie algebra is

$$(4.1)(c) \qquad\qquad \mathfrak{p}_0 = n \times n \text{ symmetric matrices.}$$

Proposition 4.2. (Polar or Cartan decomposition for $GL(n, \mathbb{R})$). *Suppose $G = GL(n, \mathbb{R})$, $K = O(n)$, and $\mathfrak{p}_0$ is the space of $n \times n$ symmetric matrices. Then the map*

$$O(n) \times \mathfrak{p}_0 \to GL(n) \qquad (k, X) \mapsto k \exp(X)$$

is an analytic diffeomorphism of $O(n) \times \mathfrak{p}_0$ onto $GL(n)$.

Definition 4.3. A linear reductive group *is a subgroup $G \subset GL(n, \mathbb{R})$ such that*

(1) G is closed (and therefore G is a Lie group).
(2) G has finitely many connected components.
(3) G is preserved by the Cartan involution θ of $GL(n, \mathbb{R})$ (cf. (4.1)(a)).

Of course the last requirement means simply that the transpose of each element of G belongs again to G. The restriction of θ to G (which we still write as θ) is called the Cartan involution *of G. Define*

$$K = G \cap O(n) = G^\theta,$$

a compact subgroup of G. Write

$$\mathfrak{g}_0 = \mathrm{Lie}(G) \subset \mathfrak{gl}(n, \mathbb{R}).$$

Finally, define

$$\mathfrak{s}_0 = \textit{symmetric matrices in } \mathfrak{g}_0,$$

the −1-eigenspace of θ.

Proposition 4.4. (Cartan decomposition for linear real reductive groups). *Suppose $G \subset GL(n, \mathbb{R})$ is a linear reductive group, $K = G \cap O(n)$, and $\mathfrak{s}_0$ is the space of symmetric matrices in the Lie algebra of G. Then the map*

$$K \times \mathfrak{s}_0 \to G, \qquad (k, X) \mapsto k \exp(X)$$

is an analytic diffeomorphism of $K \times \mathfrak{s}_0$ onto G.

One immediate consequence of this proposition is that K is a maximal compact subgroup of G; that is, that any subgroup of G properly containing K must be noncompact.

Here is a result connecting this definition with a more traditional one.

Proposition 4.5. *Suppose $\mathbf{H}$ is a reductive algebraic group defined over $\mathbb{R}$, and $\pi\colon \mathbf{H} \to \mathbf{GL}(\mathbf{V})$ is a faithful representation defined over $\mathbb{R}$. Then we can choose a basis of $V = \mathbf{V}(\mathbb{R})$ in such a way that the corresponding embedding*

$$\pi\colon \mathbf{H}(\mathbb{R}) \to GL(n, \mathbb{R})$$

has image a linear reductive group in the sense of Definition 4.3.

Conversely, suppose G is a linear reductive group in the sense of Definition 4.3. Then we can choose $\mathbf{H}$ and π as above in such a way that

$$\pi(\mathbf{H}(\mathbb{R}))_0 = G_0;$$

that is, these two groups have the same identity component.

Here at last is the main definition.

Definition 4.6. *A* real reductive group *is a Lie group $\widetilde{G}$ endowed with a surjective homomorphism*

$$\pi\colon \widetilde{G} \to G \subset GL(n, \mathbb{R})$$

onto a linear reductive group, such that $\ker \pi$ is finite. Use the differential of π to identify

$$\widetilde{\mathfrak{g}}_0 = \mathrm{Lie}(\widetilde{G}) \simeq \mathrm{Lie}(G) = \mathfrak{g}_0 \subset \mathfrak{gl}(n, \mathbb{R}).$$

This identification makes θ into an automorphism $\widetilde{\theta}$ of $\widetilde{\mathfrak{g}}_0$. Define

$$\widetilde{K} = \pi^{-1}(K) \subset \widetilde{G},$$

a compact subgroup (since π has finite kernel). Finally, define

$$\widetilde{\mathfrak{s}}_0 = -1\text{-}eigenspace \ of \ \widetilde{\theta} \ on \ \widetilde{\mathfrak{g}}_0.$$

In the next statement we will use tildes to distinguish elements of $\widetilde{G}$ and the exponential map of $\widetilde{G}$ for clarity; this is also helpful in writing down the (very easy) proof based on Proposition 4.5. But thereafter we will drop all the tildes.

Proposition 4.7. (Cartan decomposition for real reductive groups). *Suppose $\widetilde{G}$ is a real reductive group as in Definition 4.6. Then the map*

$$\widetilde{K} \times \mathfrak{s}_0 \to \widetilde{G}, \qquad (\widetilde{k}, \widetilde{X}) \mapsto \widetilde{k} \exp^{\sim}(\widetilde{X})$$

is an analytic diffeomorphism of $\widetilde{K} \times \widetilde{\mathfrak{s}}_0$ onto $\widetilde{G}$. Define a diffeomorphism $\widetilde{\theta}$ of $\widetilde{G}$ by

$$\widetilde{\theta}(\widetilde{k} \exp^{\sim}(\widetilde{X})) = \widetilde{k} \exp^{\sim}(-\widetilde{X}).$$

Then $\widetilde{\theta}$ is an automorphism of order two, with fixed point group $\widetilde{K}$.

We turn now to the problem of exploiting this structure for understanding representations of a reductive group G. Recall from Definition 1.1 the notion of representation of G.

Definition 4.8. *Suppose (π, V) is a representation of G. A vector $v \in V$ is said to be* smooth *(respectively* analytic*) if the map*

$$G \to V, \qquad g \mapsto \pi(g) \cdot v$$

is smooth (respectively analytic). Write V^∞ (respectively V^ω) for the space of smooth (respectively analytic) vectors in V. When the group G is not clear from context, we may write for example $V^{\infty, G}$.

Each of V^∞ and V^ω is a G-stable subspace of V; we write π^∞ and π^ω for the corresponding actions of G. Each of these representations differentiates to a representation $d\pi$ of the Lie algebra $\mathfrak{g}_0$, and hence also of the enveloping algebra $U(\mathfrak{g})$. (We will always write

$$(4.9) \qquad\qquad \mathfrak{g}_0 = \mathrm{Lie}(G), \qquad \mathfrak{g} = \mathfrak{g}_0 \otimes_\mathbb{R} \mathbb{C},$$

and use analogous notation for other Lie groups.) Each of V^∞ and V^ω has a natural complete locally convex topology, making the group representations continuous. In the case of V^∞, this topology can be given by seminorms

$$v \mapsto \rho(d\pi(u)v)$$

with ρ one of the seminorms defining the topology of V, and $u \in U(\mathfrak{g})$. Since the enveloping algebra has countable dimension, it follows at once that V^∞ is Fréchet (topologized by countably many seminorms) whenever V is Fréchet. The condition that the function $\pi(g)v$ be real analytic may be expressed in terms of the existence of bounds on derivatives of the function: that if $X_1, X_2, \ldots, X_m$ is a basis of $\mathfrak{g}_0$, and $g_0 \in G$, then there should exist $\epsilon > 0$ and a neighborhood U of g_0 so that for any seminorm ρ on V,

$$\rho(d\pi(X^I)\pi(g)v) \leqslant C_\epsilon I! \, \epsilon^{-|I|}$$

for all multiindices $I = (i_1, \ldots, i_m)$ and all $g \in U$. Here we use standard multiindex notation, so that

$$X^I = X_1^{i_1} \cdots X_m^{i_m}, \qquad |I| = \sum_{j=1}^{m} i_j,$$

and so on. This description suggests how to define the topology on V^ω as an inductive limit (over open coverings of G, with positive numbers $\epsilon(U)$ attached to each set in the cover).

It is a standard theorem (due to Gårding, and true for any Lie group) that V^∞ is dense in V. I am not certain in what generality the density of V^ω in V is known; we will recall (in Theorem 4.13) Harish-Chandra's proof of this density in enough cases for our purposes.

One of Harish-Chandra's fundamental ideas was the use of relatively easy facts in the representation theory of compact groups to help in the study of representations of G. Here are some basic definitions.

Definition 4.10. *Suppose (π, V) is a representation of a compact Lie group K. A vector $v \in V$ is said to be K-finite if it belongs to a finite-dimensional K-invariant subspace. Write V^K for the space of all K-finite vectors in V.*

Suppose (μ, E_μ) is an irreducible representation of K. (Then E_μ is necessarily finite-dimensional, and carries a K-invariant Hilbert space structure.) The μ-isotypic subspace $V(\mu)$ is the span of all copies of E_μ inside V.

Proposition 4.11. *Suppose (π, V) is a representation of a compact Lie group K. Then*

$$V^K \subset V^{\omega, K} \subset V^{\infty, K} \subset V;$$

V^K *is dense in V. There is an algebraic direct sum decomposition*

$$V^K = \sum_{\mu \in \widehat{K}} V(\mu).$$

Each subspace $V(\mu)$ is closed in V, and so inherits a locally convex topology. There is a unique continuous operator

$$P(\mu)\colon V \to V(\mu)$$

commuting with K and acting as the identity on $V(\mu)$. For any $v \in V$ and $\mu \in \widehat{K}$, we can therefore define

$$v_\mu = P(\mu)v \in V(\mu).$$

If $v \in V^{\infty,K}$, then

$$v = \sum_{\mu \in \widehat{K}} v_\mu,$$

an absolutely convergent series.
Finally, define

$$V^{-K} = \prod_{\mu \in \widehat{K}} V(\mu),$$

the algebraic direct product. The operators $P(\mu)$ define an embedding

$$V \hookrightarrow V^{-K}, \qquad v \mapsto \prod v_\mu.$$

There are natural complete locally convex topologies on V^K and V^{-K} making all the inclusions here continuous, but we will have no need of this.

Returning to the world of reductive groups, here is Harish-Chandra's basic definition. The definition will refer to

$$\mathfrak{Z}^G(\mathfrak{g}) = \mathrm{Ad}(G)\text{-invariant elements of } U(\mathfrak{g}).$$

If G is connected, this is just the center of the enveloping algebra. Schur's lemma suggests that $\mathfrak{Z}^G(\mathfrak{g})$ ought to act by scalars on an irreducible representation of G, but there is no general way to make this suggestion into a theorem. (Soergel gave an example of an irreducible Banach representation of $SL(2,\mathbb{R})$ in which $\mathfrak{Z}^G(\mathfrak{g})$ does not act by scalars.) Nevertheless the suggestion is correct for most representations arising in applications, so Harish-Chandra made it into a definition.

Definition 4.12. *Suppose G is real reductive with maximal compact subgroup K, and (π, V) is a representation of G. We say that π is admissible if π has finite length, and either of the following equivalent conditions is satisfied:*

(1) for each $\mu \in \widehat{K}$, the isotypic space $V(\mu)$ is finite-dimensional.
(2) each $v \in V^\infty$ is contained in a finite-dimensional subspace preserved by $d\pi(\mathfrak{Z}^G(\mathfrak{g}))$.

The assumption of "finite length" means that V has a finite chain of closed invariant subspaces in which successive quotients are irreducible. Harish-Chandra actually called the first condition "admissible" and the second "quasisimple," and he proved their equivalence. It is the term admissible that has become standard now, perhaps because it carries over almost unchanged to the setting of p-adic reductive groups. Harish-Chandra also proved that a unitary representation of finite length is automatically admissible.

Theorem 4.13. (Harish-Chandra). *Suppose (π, V) is an admissible representation of a real reductive group G with maximal compact subgroup K.*

(1) $V^K \subset V^\omega \subset V^\infty$.
(2) The subspace V^K is preserved by the representations of $\mathfrak{g}$ and K.
(3) There is a bijection between the set

$$\{\, closed\ G\text{-}stable\ subspaces\ W \subset V \,\}$$

and the set

$$\{\, arbitrary\ (\mathfrak{g}, K)\text{-}stable\ subspaces\ W^K \subset V^K \,\}.$$

Here W corresponds to its subspace W^K of K-finite vectors, and W^K corresponds to its closure W in V.

The structure carried by V^K is fundamental, and has a name of its own.

Definition 4.14. *Suppose G is a real reductive group with complexified Lie algebra $\mathfrak{g}$ and maximal compact subgroup K. A $(\mathfrak{g}, K)$-module is a vector space X endowed with actions of $\mathfrak{g}$ and of K, subject to the following conditions.*

(1) Each vector in X belongs to a finite-dimensional K-stable subspace, on which the action of K is continuous.
(2) The differential of the action of K (which exists by the first condition) is equal to the restriction of the action of $\mathfrak{g}$.
(3) The action map
$$\mathfrak{g} \times X \to X, \qquad (Z, x) \mapsto Z \cdot x$$
is equivariant for the actions of K. (Here K acts on $\mathfrak{g}$ by Ad.)

Harish-Chandra's Theorem 4.13 implies that if (π, V) is an admissible irreducible representation of G, then V^K is an irreducible $(\mathfrak{g}, K)$-module. We call V^K the Harish-Chandra module of π. We say that two such representations (π, V) and (ρ, W) are infinitesimally equivalent *if $V^K \simeq W^K$ as $(\mathfrak{g}, K)$-modules.*

Theorem 4.15. (Harish-Chandra). *Every irreducible $(\mathfrak{g}, K)$-module arises as the Harish-Chandra module of an irreducible admissible representation of G on a Hilbert space.*

(Actually Harish-Chandra proved this theorem only for linear reductive groups G. The general case was completed by Lepowsky.) In light of this theorem and the preceding definitions, we define

$$\widehat{G}_{adm} = \text{infinitesimal equivalence classes of}$$
$$\text{irreducible admissible representations}$$
$$= \text{equivalence classes of irreducible } (\mathfrak{g}, K)\text{-modules.}$$

A continuous group representation with Harish-Chandra module X is called a *globalization of X*.

We fix now a finite length $(\mathfrak{g}, K)$-module

$$(4.16)(a) \qquad\qquad X = \sum_{\mu \in \widehat{K}} X(\mu).$$

In addition, we fix a Hilbert space globalization

$$(4.16)(b) \qquad\qquad (\pi^{Hilb}, X^{Hilb})$$

(For irreducible X, such a globalization is provided by Theorem 4.15. For X of finite length, the existence of X^{Hilb} is due to Casselman.) For the purposes of the theorems and definitions that follow, any reflexive Banach space globalization will serve equally well; with minor modifications, one can work with a reflexive Fréchet representation of moderate growth (see [Cas89], Introduction). The Hilbert (or Banach) space structure restricts to a norm

$$(4.16)(c) \qquad\qquad \| \ \|_\mu \colon X(\mu) \to \mathbb{R}$$

We will need also the dual Harish-Chandra module

$$X^{dual} = K\text{-finite vectors in the algebraic dual of } X.$$

The contragredient representation of G on the dual space $(X^{Hilb})'$, defined by

$$(4.16)(d) \qquad\qquad (\pi^{Hilb})'(g) = {}^t(\pi^{Hilb}(g^{-1}))$$

has Harish-Chandra module X^{dual}. (We will discuss the transpose of a linear map and duality in general in more detail in section 8.) In particular, $(X^{Hilb})'(\mu) = X(\mu)'$ inherits the norm

$$(4.16)(e) \qquad\qquad \| \ \|'_\mu \colon X(\mu) \to \mathbb{R}$$

from $(X^{Hilb})'$. This is unfortunately *not* precisely the dual norm to $\| \ \|_\mu$, but the difference can be controlled[1]: there is a constant $C \geqslant 1$ so that

---

[1] The difference is that the dual norm to $\| \ \|_\mu$ involves the size of a linear functional only on elements of $X(\mu)$, whereas $\| \ \|'_\mu$ involves the size of a linear functional on the whole space. The second inequality in $(4.16)(f)$ is obvious. The constant in the first inequality is an estimate for the norm of the projection operator $P(\mu)$ from Proposition 4.11. The estimate comes from the standard formula for $P(\mu)$ as an integral over K of $\pi(k)$ against the character of μ.

(4.16)(f) $$(C \cdot \dim \mu)^{-1} \| \; \|_{\mu}' \leqslant (\| \; \|_{\mu})' \leqslant \| \; \|_{\mu}'.$$

A Hilbert space globalization is technically valuable in the subject, but it has a very serious weakness: it is not canonically defined, even up to a bounded operator. More concretely, suppose that X happens to be unitary, so that there *is* a canonical Hilbert space globalization $X^{Hilb\sim}$ (coming from the unitary structure). The nature of the problem is that the infinitesimal equivalence of X^{Hilb} and $X^{Hilb\sim}$ need not be implemented by a bounded operator from X^{Hilb} to $X^{Hilb\sim}$. A consequence is that the invariant Hermitian form on X giving rise to the unitary structure need not be defined on all of X^{Hilb}: we cannot hope to look for unitary structures by looking for Hermitian forms on random Hilbert space globalizations. Here is the technical heart of the work of Wallach, Casselman, and Schmid addressing this problem.

Theorem 4.17. (Casselman-Wallach; see [Cas89]). *In the setting of (4.16), the norm $\| \; \|_{\mu}$ is well-defined up to a polynomial in $|\mu|$ (which means the length of the highest weight of the representation μ of K): if $\| \; \|_{\mu}^{\sim}$ is the collection of norms arising from any other Hilbert (or reflexive Banach) globalization of X, then there are a positive integer M and a constant C_M so that*

$$\| \; \|_{\mu} \leqslant C_M (1 + |\mu|)^M \| \; \|_{\mu}^{\sim}.$$

The proof of this result given by Wallach and Casselman is quite complicated and indirect; indeed it is not entirely easy even to *extract* the result from their papers. It would certainly be interesting to find a more direct approach: beginning with the Harish-Chandra module X, to construct the various norms $\| \; \|_{\mu}$ (defined up to inequalities like those in Theorem 4.17); and then to prove directly that the topological vector spaces constructed in Theorems 4.18, 4.20 carry smooth representations of G. The first step in this process (defining the norms) is perhaps not very difficult. The second seems harder.

Theorem 4.18. (Casselman-Wallach; see [Cas89]). *In the setting of (4.16), the space of smooth vectors of X^{Hilb} is*

$$X^{Hilb,\infty} = \left\{ \sum_{\mu \in \widehat{K}} x_{\mu} \mid x_{\mu} \in X(\mu), \; \|x_{\mu}\|_{\mu} \text{ rapidly decreasing in } |\mu| \right\}.$$

Here "rapidly decreasing" means that for every positive integer M there is a constant C_N so that

$$\|x_{\mu}\|_{\mu} \leqslant C_N (1 + |\mu|)^{-N}.$$

The minimum possible choice of C_N defines a seminorm; with these seminorms, $X^{Hilb,\infty}$ is a nuclear Fréchet space.

Regarded as a collection of sequences of elements chosen from $X(\mu)$, the space of smooth vectors and its topology are independent of the choice of the globalization X^{Hilb}.

Definition 4.19. *Suppose that X is any Harish-Chandra module of finite length. The* Casselman-Wallach smooth *globalization of X is the space of smooth vectors in any Hilbert space globalization of X. We use Theorem 4.18 to identify it as a space of sequences of elements of X, and denote it X^∞.*

There is a parallel description of analytic vectors.

Theorem 4.20. (Schmid; see [Sch85]). *In the setting of (4.16), the space of analytic vectors of X^{Hilb} is*

$$X^{Hilb,\omega} = \left\{ \sum_{\mu \in \widehat{K}} x_\mu \mid x_\mu \in X(\mu), \|x_\mu\|_\mu \text{ exponentially decreasing in } |\mu| \right\}.$$

Here "exponentially decreasing" means that there are an $\epsilon > 0$ and a constant C_ϵ so that

$$\|x_\mu\|_\mu \leqslant C_\epsilon (1 + \epsilon)^{-|\mu|}.$$

The minimum choice of C_ϵ defines a Banach space structure on a subspace, and $X^{Hilb,\omega}$ has the inductive limit topology, making it the dual of a nuclear Fréchet space.

Regarded as a collection of sequences of elements chosen from $X(\mu)$, the space of analytic vectors and its topology are independent of the choice of the globalization X^{Hilb}.

Definition 4.21. *Suppose that X is any Harish-Chandra module of finite length. Schmid's* minimal *or* analytic *globalization of X is the space of analytic vectors in any Hilbert space globalization of X. We use Theorem 4.18 to identify it as a space of sequences of elements of X, and denote it X^ω.*

Finally, we will need the duals of these two constructions.

Definition 4.22. *Suppose that X is any Harish-Chandra module of finite length. The* Casselman-Wallach distribution *globalization of X is the continuous dual of the space of smooth vectors in any Hilbert space globalization of X^{dual}. We can use Theorem 4.18 to identify it as a space of sequences of elements of X, and denote it $X^{-\infty}$. Explicitly,*

$$X^{-\infty} = \left\{ \sum_{\mu \in \widehat{K}} x_\mu \mid x_\mu \in X(\mu), \|x_\mu\|_\mu \text{ slowly increasing in } |\mu| \right\}.$$

Here "slowly increasing" means that there is a positive integer N and a constant C_N so that

$$\|x_\mu\|_\mu \leqslant C_N (1 + |\mu|)^N.$$

This exhibits $X^{-\infty}$ as an inductive limit of Banach spaces, and the dual of the nuclear Fréchet space $X^{dual,\infty}$.

Definition 4.23. *Suppose that X is any Harish-Chandra module of finite length. Schmid's maximal or hyperfunction globalization of X is the continuous dual of the space of analytic vectors in any Hilbert space globalization of X^{dual}. We can use Theorem 4.19 to identify it as a space of sequences of elements of X, and denote it $X^{-\omega}$. Explicitly,*

$$X^{-\omega} = \left\{ \sum_{\mu \in \widehat{K}} x_\mu \mid x_\mu \in X(\mu), \right.$$
$$\left. \|x_\mu\|_\mu \ \text{less than exponentially increasing in } |\mu| \right\}.$$

Here "less than exponentially increasing" means that for every $\epsilon > 0$ there is a constant C_ϵ so that

$$\|x_\mu\|_\mu \leqslant C_\epsilon (1 + \epsilon)^{|\mu|}.$$

This exhibits $X^{-\omega}$ as a nuclear Fréchet space.

When we wish to emphasize the K-finite nature of X, we can write the space as $X^K = \sum_{\mu \in \widehat{K}} X(\mu)$. It is also convenient to write

$$X^{-K} = \prod_{\mu \in \widehat{K}} X(\mu) = (X^{dual})^*.$$

Our various globalizations now appear as sequence spaces, with gradually weakening conditions on the sequences:

(4.24) $$X^K \subset X^\omega \subset X^\infty \subset X^{-\infty} \subset X^{-\omega} \subset X^{-K}.$$

The conditions on the sequences are: almost all zero, exponentially decreasing, rapidly decreasing, slowly increasing, less than exponentially increasing, and no condition. All of the conditions are expressed in terms of the norms chosen in (4.16), and the naturality of the definitions depends on Theorem 4.17. We could also insert our Hilbert space X^{Hilb} in the middle of the list (between X^∞ and $X^{-\infty}$), corresponding to sequences in ℓ^2. I have not done this because that sequence space *does* depend on the choice of X^{Hilb}.

Even the spaces X^K and X^{-K} carry natural complete locally convex topologies (still given by the sequence structure); the representations of $\mathfrak{g}$ and K are continuous for these topologies. (The group G will not act on either of them unless X is finite-dimensional.)

5 Real Parabolic Induction and the Globalization Functors

In order to get some feeling for the various globalization functors defined in section 4, we are going to compute them in the setting of parabolically induced representations. Logically this cannot be separated from the definition of the functors: the proof by Wallach and Casselman of Theorem 4.15 proceeds by embedding arbitrary representations in parabolically induced representations, and computing there. But we will ignore these subtleties, taking the results of section 4 as established.

Throughout this section, G will be a real reductive group with Cartan involution θ and maximal compact subgroup K, as in Definition 4.6. We want to construct representations using parabolic subgroups of G, so the first problem is to say what a parabolic subgroup is. In part because of the possible disconnectedness of G, there are several possible definitions. We want to take advantage of the fact that the complexified Lie algebra $\mathfrak{g}$ (cf. (4.9)) is a complex reductive Lie algebra, for which lots of structure theory is available.

Definition 5.1. *A* real parabolic subgroup *of the real reductive group* G *is a Lie subgroup* $P \subset G$ *with the property that* $\mathfrak{p} = \mathrm{Lie}(P)_{\mathbb{C}}$ *is a parabolic subalgebra of* $\mathfrak{g}$. *Write*

$$\mathfrak{u} = \text{nil radical of } \mathfrak{p};$$

because this is an ideal preserved by all automorphisms of $\mathfrak{p}$ *as a real Lie algebra, it is the complexification of an ideal* $\mathfrak{u}_0$ *of* $\mathfrak{p}_0$. *Let* U *be the connected Lie subgroup of* P *with Lie algebra* $\mathfrak{u}_0$; *it is a nilpotent Lie group, normal in* P.

This is the most liberal possible definition of real parabolic subgroup. The most restrictive would require in addition that P be the normalizer of $\mathfrak{p}$ (under the adjoint action) in G.

The quotient Lie algebra $\mathfrak{p}/\mathfrak{u}$ is reductive, and is always represented by a subalgebra (a *Levi factor*) of $\mathfrak{p}$. But the Levi factor is not unique, and picking a good one is often a slightly delicate matter. In the present setting this problem is solved for us. Because θ is an automorphism of G, θP is another parabolic subgroup. Define

$$(5.2) \qquad\qquad L = P \cap \theta P,$$

a θ-stable Lie subgroup of G.

Proposition 5.3. *Suppose* P *is a parabolic subgroup of the real reductive group* G, *and* $L = P \cap \theta P$.

(1) The subgroups P, L, *and* U *are all closed in* G.
(2) Multiplication defines a diffeomorphism

$$L \times U \to P, \qquad (l, u) \mapsto lu.$$

In particular, $L \simeq P/U$.
(3) L *is a reductive subgroup of* G, *with Cartan involution* $\theta|_L$.
(4) The exponential map is a diffeomorphism from $\mathfrak{u}_0$ *onto* U.

(5) Every element of G is a product (not uniquely) of an element of K and an element of P: $G = KP$. Furthermore

$$P \cap K = L \cap K$$

is a maximal compact subgroup of L and of P. Consequently there are diffeomorphisms of homogeneous spaces

$$G/P \simeq K/L \cap K, \qquad G/K \simeq P/L \cap K.$$

The first of these is K-equivariant, and the second P-equivariant.
(6) The map

$$K \times (\mathfrak{l}_0 \cap \mathfrak{s}_0) \times U \to G, \qquad (k, X, u) \mapsto k \cdot \exp(X) \cdot u$$

is an analytic diffeomorphism.

The last assertion interpolates between the Iwasawa decomposition (the case when P is a minimal parabolic subgroup) and the Cartan decomposition (the case when P is all of G, or more generally when P is open in G). We saw in section 2 (after 2.5) an example of the diffeomorphism in (5), with $G = GL(n, \mathbb{C})$, P the Borel subgroup of upper triangular matrices, and $K = U(n)$. In this case L is the group of diagonal matrices in $GL(n, \mathbb{C})$, and $L \cap K = U(1)^n$.

How does one find parabolic subgroups? The easiest examples are "block upper-triangular" subgroups of $GL(n, \mathbb{R})$. I will assume that if you've gotten this far, those subgroups are more or less familiar, and look only at more complicated reductive groups.

It's better to ask instead how to find the homogeneous spaces G/P, in part because construction of representations by induction really takes place on the whole homogeneous space and not just on the isotropy group P. A good answer is that one begins with the corresponding homogeneous spaces related to the complex Lie algebra $\mathfrak{g}$, and looks for appropriate orbits of G on those spaces. We will give some more details about this approach in section 6; but here is one example.

Suppose p and q are non-negative integers, and $n = p + q$. The standard Hermitian form of signature (p, q) on $\mathbb{C}^n$ is

$$\langle v, w \rangle_{p,q} = \sum_{j=1}^{p} v_j \overline{w}_j - \sum_{k=1}^{q} v_{p+k} \overline{w}_{p+k}.$$

The group $G = U(p, q)$ of complex linear transformations preserving this form is a real reductive group, with maximal compact subgroup

(5.4)(a) $K = U(p) \times U(q).$

The group G does not have obvious "block upper-triangular" subgroups; but here is a way to make a parabolic. Fix a non-negative integer $r \leqslant p, q$, and define

(5.4)(b) $\qquad f_j = e_j + ie_{p+j}, \quad g_j = e_j - ie_{p+j} \qquad (1 \leqslant j \leqslant r)$

The subspace

$$I_r = \text{span}(f_1, \dots, f_r)$$

is an r-dimensional isotropic plane (for the form $\langle v, w \rangle_{p,q}$), and so is its complex conjugate

$$\overline{I}_r = \text{span}(g_1, \dots, g_r).$$

Define

(5.4)(c) $\qquad P_r = $ stabilizer of I_r in $U(p,q)$;

this will turn out to be a parabolic subgroup of $U(p,q)$. One checks easily that

$$\theta P_r = \text{stabilizer of } \overline{I}_r \text{ in } U(p,q).$$

Writing $\mathbb{C}^{p,q}$ for $\mathbb{C}^n$ endowed with the Hermitian form $\langle v, w \rangle_{p,q}$, we find a natural vector space decomposition

$$\mathbb{C}^{p,q} = I_r \oplus \overline{I}_r \oplus \mathbb{C}^{p-r,q-r}.$$

This decomposition provides an embedding

(5.4)(d) $\qquad GL(r, \mathbb{C}) \times U(p-r, q-r) \hookrightarrow U(p,q):$

a matrix g in $GL(r, \mathbb{C})$ acts as usual on the basis $\{f_j\}$ of I_r, by ${}^t\overline{g}^{-1}$ on the basis $\{g_j\}$ of $\overline{I}_r$, and trivially on $\mathbb{C}^{p-r,q-r}$. Now it is easy to check that

(5.4)(e) $\qquad L_r = P_r \cap \theta P_r = GL(r, \mathbb{C}) \times U(p-r, q-r).$

(I will leave to the reader the problem of describing the group U_r explicitly. As a hint, my calculations indicate

$$\dim U_r = r(2[(p-r) + (q-r)] + r).$$

If my calculations are incorrect, please disregard this hint.)

The example shows that

(5.4)(f) $\qquad G/P_r \simeq r$-dimensional isotropic subspaces of $\mathbb{C}^{p,q}$.

As r varies from 1 to $\min(p,q)$, we get in this way all the maximal proper parabolic subgroups of $U(p,q)$. Smaller parabolic subgroups can be constructed directly in similar ways, or by using the following general structural fact.

Proposition 5.5. *Suppose G is a real reductive group and $P = LU$ is a parabolic subgroup. Suppose $Q_L = M_L N_L$ is a parabolic subgroup of L. Then*

$$Q = Q_L U = M_L(N_L U)$$

is a parabolic subgroup of G, with Levi factor M_L and unipotent radical $N = N_L U$. This construction defines a bijection

parabolic subgroups of $L \leftrightarrow$ parabolic subgroups of G containing P.

If you believe that the P_r are all the maximal parabolics in $U(p, q)$, and if you know about parabolic subgroups of $GL(r, \mathbb{C})$, then you see that the conjugacy classes of parabolic subgroups of $U(p, q)$ are parametrized by sequences (possibly empty) $\mathbf{r} = (r_1, \ldots, r_s)$ of positive integers, with the property that

$$r = \sum r_j \leqslant \min(p, q).$$

The Levi subgroup $L_{\mathbf{r}}$ of $P_{\mathbf{r}}$ is

$$L_r = GL(r_1, \mathbb{C}) \times \cdots \times GL(r_s, \mathbb{C}) \times U(p - r, q - r).$$

Parallel analyses can be made for all the classical groups (although the possibilities for disconnectedness can become quite complicated).

We turn now to representation theory.

Definition 5.6. *Suppose* $P = LU$ *is a parabolic subgroup of the reductive group* G. *A representation* (τ, Y) *of* P *is called* admissible *if its restriction to* L *is admissible (Definition 4.12). In this case the* Harish-Chandra module of Y *is the* $(\mathfrak{p}, L \cap K)$-module $Y^{L \cap K}$ *of* $L \cap K$-finite vectors in Y.

The notation here stretches a bit beyond what was defined in section 4, but I hope that is not a serious problem. The easiest way to get admissible representations of P is from admissible representations of L, using the isomorphism $L \simeq P/U$. That is, we extend a representation of L to P by making U act trivially. Any irreducible admissible representation of P is of this form. On any admissible representation (τ, Y) of P, the group U must act unipotently, in the following strong sense: there is a finite chain

$$0 = Y_0 \subset Y_1 \subset \cdots \subset Y_m = Y$$

of closed P-invariant subspaces, with the property that U acts trivially on each subquotient Y_i/Y_{i-1}. One immediate consequence is that the action of U is analytic on all of Y, so that (for example) the P-analytic vectors for Y are the same as the L-analytic vectors.

The main reason for allowing representations of P on which U acts nontrivially is for the Casselman-Wallach proof of Theorem 4.17. They show that (for P minimal) any admissible representation of G can be embedded in a representation induced from an admissible representation of P. This statement is not true if one restricts to representations trivial on U.

So how do we pass from a representation of P to a representation of G? Whenever G is a topological group, H a closed subgroup, and (τ, Y) a representation of H, the *induced representation* of G is defined on a space like

(5.7)(a) $X = \{f : G \to Y \mid f(xh) = \tau(h)^{-1} f(x) \quad (x \in G, h \in H)\}.$

The group G acts on such functions by left translation:

(5.7)(b) $(\pi(g)f)(x) = f(g^{-1}x).$

To make the definition precise, one has to decide exactly *which* functions to use, and then to topologize X so as to make the representation continuous. Depending on exactly what structures are available on G, H, and Y, there are many possibilities: continuous functions, smooth functions, analytic functions, measurable functions, integrable functions, distributions, and many more.

To be more precise in our setting, let us fix

(5.8)(a) $$Y^{L \cap K} = \text{admissible } (\mathfrak{p}, L \cap K)\text{-module};$$

by "admissible" we mean that $Y^{L \cap K}$ should have finite length as an $(\mathfrak{l}, L \cap K)$-module. The theory of globalizations in section 4 extends without difficulty to cover admissible $(\mathfrak{p}, L \cap K)$-modules. This means first of all that $Y^{L \cap K}$ is the Harish-Chandra module of an admissible Hilbert space representation

(5.8)(b) $$(\tau^{Hilb}, Y^{Hilb}).$$

Using this Hilbert space representation, we can construct the subrepresentations of smooth and analytic vectors, and dually the distribution and hyperfunction vectors (duals of the smooth and analytic vectors in a Hilbert globalization of $Y^{L \cap K, dual}$). In the end, just as in (4.24), we have

(5.8)(c) $$Y^{L \cap K} \subset Y^\omega \subset Y^\infty \subset Y^{Hilb} \subset Y^{-\infty} \subset Y^{-\omega} \subset Y^{-L \cap K}.$$

These are complete locally convex topological vector spaces; the inclusions are continuous with dense image. All but the first and last carry irreducible representations of P, which we denote τ^ω, etc. When $Y^{L \cap K}$ is finite-dimensional (as is automatic for P minimal), all of these spaces are the same.

We now want to use these representations of P and the general idea of (5.7) to construct representations of G. That is, we want to begin with one of the representations Y of (5.8)(a), and define an appropriate space of functions

(5.9)(a) $$X = \{f \colon G \to Y \mid f(xp) = \tau(p)^{-1} f(x) \quad (x \in G, p \in P)\}.$$

What we will use constantly is Proposition 5.3(5). This provides an identification

(5.9)(b) $$X \simeq \{f \colon K \to Y \mid f(kl) = \tau(l)^{-1} f(k) \quad (k \in K, l \in L \cap K\}.$$

The description of X in (5.9)(a) is called the "induced picture"; we may write X_{ind} to emphasize that. The description in (5.9)(b) is the "compact picture," and may be written X_{cpt}. The great advantage of the first picture is that the action of G (by left translation) is apparent. The great advantage of the second is that many questions of analysis come down to the compact group K. Eventually we will need to understand at least the action of the Lie algebra $\mathfrak{g}$ in the compact picture; a formula appears in (5.14)(d). For the moment, notice that an element of X_{ind} is continuous (respectively measurable) if and only if the corresponding element of X_{cpt} is continuous (respectively measurable). If the representation τ is smooth (respectively analytic), then the same is true of smooth (respectively analytic) functions.

The classical setting for induction is unitary representations. In the setting of (5.7), suppose Y is a Hilbert space, with Hilbert space norm $\|\cdot\|_Y$ preserved by H. We will choose the space X in (5.7)(a) to consist of certain measurable functions from G to Y. If f is such a function, then

$$(5.10)(a) \qquad\qquad g \mapsto \|f(g)\|_Y$$

is a non-negative real-valued measurable function on G. Because of the transformation law on f imposed in (5.7)(a), this function is actually right-invariant under H:

$$(5.10)(b) \qquad \|f(gh)\|_Y = \|f(g)\|_Y \qquad (g \in G, h \in H)$$

We want to find a Hilbert space structure on some of these functions f. A natural way to do that is to require $\|f(g)\|_Y$ to be square-integrable in some sense. Because of the H-invariance in (5.9)(b), what is natural is to integrate over the homogeneous space G/H. That is, we define a Hilbert space norm on these functions by

$$(5.10)(c) \qquad\qquad \|f\|_X^2 = \int_{G/H} \|f(g)\|_Y^2 \, d\overline{g}.$$

Here $d\overline{g}$ is some measure on G/H; the integrand is actually a function on G/H by (5.10)(b). The Hilbert space for the G representation is then

$$(5.10)(d) \qquad X = \{f \text{ as in (5.7)(a) measurable, } \|f\|_X < \infty\}.$$

The group G will preserve this Hilbert space structure (that is, the representation will be unitary) if $d\overline{g}$ is a G-invariant measure.

Let us see how to use this idea and our Hilbert space representation Y^{Hilb} of P to construct a Hilbert space representation of G. There are two difficulties. First, the representation Y^{Hilb} need not be unitary for P, so (5.10)(b) need not hold: the function $\|f(g)\|_{Y^{Hilb}}$ need not descend to G/P. Second, the homogeneous space G/P carries no nice G-invariant measure (unless P is open in G); so we cannot hope to get a unitary representation of G even if τ^{Hilb} is unitary.

Mackey found a very general way to address the second problem, essentially by tensoring the representation τ by a certain one-dimensional character of P defining the bundle of "half-densities" on G/P. This is the source of a strange exponential term (for example the "ρ" in section VII.1 of [Kna86]) in many formulas for induced representations. There is a long-winded explanation in Chapter 3 of [Vog87]. Because we will not be using parabolic induction to construct unitary representations, we will ignore this problem (and omit the "ρ" from the definition of parabolic induction). If the action of G changes the measure $d\bar{g}$ in a reasonable way, we can still hope that G will act by bounded operators on the Hilbert space of (5.10)(d).

The first problem is more serious, since it seems to prevent us even from writing down an integral defining a Hilbert space. The function we want to integrate is defined on all of G, but it is dangerous to integrate over G: if the representation of P *were* unitary, the function would be constant on the cosets of P, so the integral (at least with respect to Haar measure on G) would not converge. This suggests using instead of Haar measure some measure on G that decays at infinity in some sense. (One might at first be tempted to use the delta function, assigning the identity element of G the measure 1 and every other element the measure zero. This certainly takes care of convergence problems, but this measure behaves so badly under translation by G that G fails to act continuously on the corresponding Hilbert space).

A reasonable resolution is hiding in Proposition 5.3(5).

Proposition 5.11. *Suppose P is a parabolic subgroup of the real reductive group G, and (τ^{Hilb}, Y^{Hilb}) is an admissible representation of P on a Hilbert space. Define*

$$X^{Hilb}_{cpt} = \{f: K \to Y^{Hilb} \text{ measurable } |$$

$$f(kl) = \tau(l)^{-1}f(k), \int_K \|f(k)\|^2_{Y^{Hilb}} dk < \infty\}.$$

Here dk is the Haar measure on K of total mass 1; the norm on X^{Hilb}_{cpt} is the square root of the integral in the definition. Define X^{Hilb}_{ind} to be the corresponding space of functions on G, using the identification in (5.9). Then X^{Hilb}_{ind} is preserved by left translation by G. The corresponding representation π^{Hilb} of G is continuous and admissible; its restriction to K is unitary.

It may seem strange that we have obtained a unitary representation of K even though we did not assume that τ^{Hilb} was unitary on $L \cap K$. This is possible because we have integrated over K rather than over $K/L \cap K$. If we apply this proposition with $P = G$ (so that τ^{Hilb} is a representation of G), then $X^{Hilb} = Y^{Hilb}$ as a topological vector space, but the Hilbert space structures $\| \cdot \|^2_{Y^{Hilb}}$ and $\| \cdot \|^2_{X^{Hilb}}$ are different: the latter is obtained by averaging the former over K.

We now have a Hilbert space globalization of a Harish-Chandra module for G, so the machinery of section 4 can be applied. To begin, it is helpful to write down the Harish-Chandra module for G explicitly. This is

(5.12)(a) $X^K = \{f: G \to Y^{Hilb} \mid f(xp) = \tau(p)^{-1}f(x), \text{ and } f \text{ left } K\text{-finite}\}.$

In order to understand this as a vector space, it is most convenient to use the "compact picture" of (5.9)(b):

(5.12)(b)
$$X^K_{cpt} = \{f: K \to Y^{Hilb} \mid f(kl) = \tau(l)^{-1}f(k), \text{ and } f \text{ left } K\text{-finite}\}.$$

Now a function f in X^K_{cpt} can transform on the left according to a representation μ of K only if it transforms on the right according to representations of $L \cap K$ appearing in the restriction of the dual of μ. It follows that the functions in X^K_{cpt} must take values in $Y^{L \cap K}$. (This is not true of the corresponding functions in the induced picture (5.12)(a).) Therefore

(5.12)(c)
$$X^K_{cpt} = \{f \colon K \to Y^{L \cap K} \mid f(kl) = \tau(l)^{-1} f(k), \text{ and } f \text{ left } K\text{-finite}\}.$$

As in (5.9), the drawback of this description of X^K is that the action of the Lie algebra $\mathfrak{g}$ is not as clear as in (5.12)(a).

We turn next to the determination of X^∞, the space of smooth vectors in X^{Hilb}. Recall that "smooth" refers to the differentiability of the action of G, not directly to smoothness as functions on G. What is more or less obvious (from standard theorems saying that functions on compact manifolds with lots of L^2 derivatives are actually smooth) is this:

(5.13)(a) $f \in X^{Hilb}_{cpt}$ is smooth for the representation of K if and only if

it is smooth as a function on K with values in Y^{Hilb}.

Smoothness of a function on K may be tested by differentiating by Lie algebra elements either on the left or on the right. Because of the transformation property imposed under $L \cap K$ on the right, it therefore follows that the K-smooth vectors in X^{Hilb} must take values in Y^∞:

(5.13)(b)
$$X^{Hilb, K\text{-smooth}} = \{f \colon K \to Y^\infty \mid f(kl) = \tau(l)^{-1} f(k), \text{ and } f \text{ smooth on } K\}.$$

(Implicitly there is a Fréchet topology here, with seminorms like

$$\sup_{k \in K} \nu(\lambda(u) \cdot f);$$

here $u \in U(\mathfrak{k})$ is acting by differentiation on the left (this is λ), and ν is a seminorm defining the topology of Y^∞.) We will show that $X^{Hilb, K\text{-smooth}}$ is precisely the set of smooth vectors of X^{Hilb}. In order to do that, we must show that the left translation action of G on this space (as a subspace of X^{Hilb}) is smooth. This means that we need to describe explicitly the action of the Lie algebra $\mathfrak{g}$ in the compact picture.

So suppose $Z \in \mathfrak{g}$. The action of Z is by differentiation on the left:

(5.14)(a) $d\pi(Z)f = \lambda(Z)f$ $(f \in X^{Hilb})$.

Now differentiation on the left by an element Z of the Lie algebra (which we have written $\lambda(Z)$) is related to differentiation on the right (written $\rho(Z)$) by the adjoint action:

(5.14)(b) $[\lambda(Z)f](g) = [\rho(-\operatorname{Ad}(g^{-1})Z)f](g).$

We are interested in the restriction of f to K. By Proposition 5.3(6), any Lie algebra element $W \in \mathfrak{g}$ has a unique decomposition

(5.14)(c) $\qquad W = W_{\mathfrak{k}} + W_{\mathfrak{p}}, \qquad (W_{\mathfrak{k}} \in \mathfrak{k}, W_{\mathfrak{p}} \in \mathfrak{l} \cap \mathfrak{s} + \mathfrak{u} \subset \mathfrak{p}.$

We apply this decomposition to the element $-\operatorname{Ad}(k^{-1})Z$ in (5.14)(b), and use the transformation property of f on the right under τ. The conclusion is

(5.14)(d) $[d\pi(Z)f](k) = [\rho((-\operatorname{Ad}(k^{-1})Z))_{\mathfrak{k}}f](k) + [d\tau(\operatorname{Ad}(k^{-1})Z))_{\mathfrak{p}}(f(k))].$

This is a kind of first order differential operator on functions on K with values in Y: the first term is a first derivative, and the second (zeroth order) term is just a linear operator on the values of f. We can if we like move the derivative back to the left:

(5.14)(e)
$[d\pi(Z)f](k) = [\lambda(\operatorname{Ad}(k)(((-\operatorname{Ad}(k^{-1})Z))_{\mathfrak{k}})f](k) + [d\tau(Ad(k^{-1})Z))_{\mathfrak{p}}(f(k))].$

The space of K-smooth vectors in X^{Hilb} was defined by seminorms involving the left action of $U(\mathfrak{k})$, which is analogous to constant coefficient differential operators. We have seen in (5.14)(e) that the action of $\mathfrak{g}$ is given by something like variable coefficient differential operators on K. Because the coefficient functions are smooth and bounded on K, this proves that the action of G on the K-smooth vectors of X^{Hilb} is in fact differentiable. That is,

(5.15)(a) $\qquad X^{\infty} = \{f: K \to Y^{\infty} \mid f(kl) = \tau(l)^{-1}f(k), \; f \text{ smooth on } K\}.$

A parallel argument identifies the analytic vectors

(5.15)(b) $\qquad X^{\omega} = \{f: K \to Y^{\omega} \mid f(kl) = \tau(l)^{-1}f(k), \; f \text{ analytic on } K\}.$

Finally, there are the distribution and hyperfunction globalizations to consider. Each of these requires a few more soft analysis remarks. For example, if V is reflexive topological vector space with dual space V^*, then the space of "generalized functions" on a manifold M with values in V is by definition

$$C^{-\infty}(M, V) = [C_c^{\infty}(M, V^* \otimes (\text{densities on } M))]^*,$$

the topological dual of the space of compactly supported smooth "test densities" on M with values in V^*. (Topologies on the dual space are discussed in section 8; we will be interested most of all in the strong dual topology.) We can then define
(5.15)(c)
$X^{-\infty} = \{f: K \to Y^{-\infty} \mid f(kl) = \tau(l)^{-1}f(k), \; f \text{ generalized function on } K\}.$

This is the Casselman-Wallach distribution globalization of X. Similarly, we can make sense of
(5.15)(d)
$X^{-\omega} = \{f: K \to Y^{-\omega} \mid f(kl) = \tau(l)^{-1}f(k), \; f \text{ hyperfunction on } K\},$

Schmid's maximal globalization of X. We have in the end a concrete version of (4.24):

(5.16)(a) $X^K \subset X^\omega \subset X^\infty \subset X^{Hilb} \subset X^{-\infty} \subset X^{-\omega} \subset X^{-K}$.

This time each space may be regarded as "functions" on K with values in $Y^{-L\cap K}$, with weakening conditions on the functions: first K-finite, then analytic, then smooth, then L^2, then distribution-valued, and so on. (Beginning with $X^{-\infty}$, these are not literally "functions" on K.) It is natural and convenient to write
(5.16)(b)
$$X^K = (\mathrm{Ind}_P^G)^K(Y), \qquad X^\omega = (\mathrm{Ind}_P^G)^\omega(Y), \qquad X^\infty = (\mathrm{Ind}_P^G)^\infty(Y),$$

and so on.

6 Examples of Complex Homogeneous Spaces

In this section we will begin to examine the complex homogeneous spaces for reductive groups that we will use to construct representations. We are going to make extensive use of the structure theory for complex reductive Lie algebras, and for that purpose it is convenient to have at our disposal a complex reductive group. (This means a complex Lie group that is also a reductive group in the sense of Definition 4.6.)

Definition 6.1. *A* complexification *of G is a complex reductive group $G_\mathbb{C}$, endowed with a Lie group homomorphism*

$$j\colon G \to G_\mathbb{C},$$

subject to the following conditions.

(1) The map j has finite kernel.
(2) The corresponding Lie algebra map

$$dj\colon \mathfrak{g}_0 \to \mathrm{Lie}(G_\mathbb{C})$$

identifies $\mathfrak{g}_0$ as a real form of $\mathrm{Lie}(G_\mathbb{C})$. More explicitly, this means that

$$\mathrm{Lie}(G_\mathbb{C}) = dj(\mathfrak{g}_0) \oplus i\,dj(\mathfrak{g}_0),$$

with i the complex multiplication on the complex Lie algebra $\mathrm{Lie}(G_\mathbb{C})$. Using this, we identify $\mathrm{Lie}(G_\mathbb{C})$ with the complexified Lie algebra $\mathfrak{g}$ henceforth.
(3) The Cartan involutions of G and $G_\mathbb{C}$ are compatible via the map j.

It is possible to construct a complexification that actually contains the linear reductive group im(π) in Definition 4.6, so that j may be taken to be the composition of an inclusion with the finite covering π. The complexification of G is not unique, but the ambiguity will cause us no problems. If G is the group of real points of a reductive algebraic group, we can of course take for $G_\mathbb{C}$ the group of complex points; this is perhaps the most important case.

We need notation for the maximal compact subgroup of $G_\mathbb{C}$. It is fairly common to refer to this group as U (perhaps in honor of the case of $U(n) \subset GL(n, \mathbb{C})$). Since we will also be discussing parabolic subgroups and their unipotent radicals, the letter U will not be convenient. So we will write

$$C_G = \text{maximal compact subgroup of } G_\mathbb{C}.$$

Hypothesis (3) in Definition 6.1 guarantees that

$$K = C_G \cap G.$$

The complex homogeneous spaces we want will be coverings of (certain) open orbits of G on (certain) complex homogeneous spaces for $G_\mathbb{C}$. Here first are the homogeneous spaces for $G_\mathbb{C}$ that we want.

Definition 6.2. *In the setting of (6.1), a partial flag variety for $G_\mathbb{C}$ is a homogeneous space*

$$X = G_\mathbb{C}/Q_\mathbb{C},$$

with $Q_\mathbb{C}$ a parabolic subgroup of $G_\mathbb{C}$ (Definition 5.1). (Recall that this means

$$\mathfrak{q} = \text{Lie}(Q_\mathbb{C}) \subset \text{Lie}(G_\mathbb{C}) = \mathfrak{g}$$

is a parabolic subalgebra). It will sometimes be helpful to write

$$Q_\mathbb{C}^{max} = \{g \in G_\mathbb{C} \mid \text{Ad}(g)\mathfrak{q} = \mathfrak{q}\},$$
$$Q_\mathbb{C}^{min} = \text{connected subgroup with Lie algebra } \mathfrak{q}$$
$$= \text{identity component of } Q_\mathbb{C}^{max}.$$

It follows from standard structure theory for complex groups that

$$Q_\mathbb{C}^{max} \cap \text{ identity component of } G_\mathbb{C} = Q_\mathbb{C}^{min}.$$

Each element of the partial flag variety $X^{min} = G_\mathbb{C}/Q_\mathbb{C}^{max}$ may be identified with a parabolic subalgebra of $\mathfrak{g}$, by

$$gQ_\mathbb{C}^{max} \mapsto \text{Ad}(g)(\mathfrak{q}).$$

Each element of $X^{max} = G_\mathbb{C}/Q_\mathbb{C}^{min}$ may be identified with a pair consisting of a parabolic subalgebra of $\mathfrak{g}$ and a connected component of $G_\mathbb{C}$.

Write $L_\mathbb{C}$ for the Levi factor of $Q_\mathbb{C}$ defined in Definition 5.1, and

$$C_L = C_G \cap Q_{\mathbb{C}}.$$

Then Proposition 5.3(5) says that

$$X = C_G/C_L,$$

a compact homogeneous space for C_G.

Theorem 6.3. (Wolf [Wol69]). *Suppose $G_{\mathbb{C}}$ is a complexification of the real reductive group G, and*

$$X = G_{\mathbb{C}}/Q_{\mathbb{C}} = C_G/C_L$$

is a partial flag variety for $G_{\mathbb{C}}$ (Definition 6.2). Then X is a compact complex manifold. The group G acts on X with finitely many orbits; so the finitely many open orbits of G on X are complex homogeneous spaces for G.

Up to covering, the spaces on which we wish to construct representations of G are certain of these open orbits. It remains to say which ones. For that, it is helpful to think about what an arbitrary G orbit on X can look like. We may as well look only at the orbit of the base point $eQ_{\mathbb{C}}$. This G-orbit is

$$(6.4)(a) \qquad G \cdot (eQ_{\mathbb{C}}) \simeq G/H, \qquad (H = G \cap Q_{\mathbb{C}}).$$

Let us compute the Lie algebra of the isotropy group. Write bar for the complex conjugation defining the real form $\mathfrak{g}_0 = \mathrm{Lie}(G)$ of $\mathfrak{g}$:

$$(6.4)(b) \qquad \overline{A + iB} = A - iB \qquad (A, B \in \mathfrak{g}_0).$$

Then bar is an involutive automorphism of $\mathfrak{g}$, with fixed points $\mathfrak{g}_0$. It follows that $\overline{\mathfrak{q}}$ is another parabolic subalgebra of $\mathfrak{g}$, and that the complexified Lie algebra $\mathfrak{h}$ of H is

$$(6.4)(c) \qquad \mathfrak{h} = \mathfrak{q} \cap \overline{\mathfrak{q}}.$$

So understanding $\mathfrak{h}$ means understanding the intersection of the two parabolic subalgebras $\mathfrak{q}$ and $\overline{\mathfrak{q}}$. The key to analyzing this in general is the fact that the intersection of any two parabolic subalgebras must contain a Cartan subalgebra; this is essentially equivalent to the Bruhat decomposition. (In our case it is even true that the intersection of $\mathfrak{q}$ and $\overline{\mathfrak{q}}$ must contain the complexification of a Cartan subalgebra of $\mathfrak{g}_0$; but we will not use this.) Once one has chosen a Cartan in both parabolics, the analysis of the intersection comes down to combinatorics of sets of roots. There are many interesting possibilities, but we will be looking only at two extreme cases. One extreme is $\mathfrak{q} = \overline{\mathfrak{q}}$. In this case H is a real parabolic subgroup of G. This is the case we looked at in section 5. The following definition describes the opposite extreme. (The terminology "nice" is entirely artificial, and not to be taken seriously.)

Definition 6.5. *Suppose G is a real reductive group with complexified Lie algebra $\mathfrak{g}$. A parabolic subalgebra $\mathfrak{q} \subset \mathfrak{g}$ is called* nice *if $\mathfrak{q} \cap \bar{\mathfrak{q}}$ is a Levi subalgebra $\mathfrak{l}$ of $\mathfrak{q}$. In this case the group*

$$L^{max} = \{g \in G \mid \mathrm{Ad}(g)\mathfrak{q}) = \mathfrak{q}\} = j^{-1}(Q_{\mathbb{C}}^{max})$$

is a real reductive subgroup of G. The G orbit

$$X_0^{min} = G \cdot (eQ_{\mathbb{C}}) \simeq G/L^{max} \subset G_{\mathbb{C}}/Q_{\mathbb{C}}^{max} = X^{min}$$

is open, and therefore inherits a G-invariant complex structure.
Define

$$L^{min} = \text{identity component of } L^{max}.$$

A real Levi factor for $\mathfrak{q}$ is by definition any subgroup L such that $L^{min} \subset L \subset L^{max}$. A measurable complex partial flag variety for G is by definition any homogeneous space $X = G/L$, endowed with the complex structure pulled back by the covering map

$$X_0 = G/L \rightarrow G/L^{max} = X_0^{min} \subset X^{min}.$$

(An explanation of the term "measurable" may be found in [Wol69].)
We say that $\mathfrak{q}$ is very nice *if it is nice, and in addition $\mathfrak{q}$ is preserved by the complexified Cartan involution θ. In this case every real Levi factor L is also preserved by θ, so that $L \cap K$ is a maximal compact subgroup of L.*

Obviously the condition of being "nice" is constant on $\mathrm{Ad}(G)$-orbits of parabolic subalgebras. It turns out that every nice parabolic subalgebra is conjugate by $\mathrm{Ad}(G)$ to a very nice one; so we may confine our attention to those.

If G is a compact group, then every parabolic subalgebra $\mathfrak{q}$ of $\mathfrak{g}$ is very nice, and measurable complex partial flag varieties for G are exactly the same thing as partial flag varieties for the (canonical) complexification of G. We will begin to look at some noncompact examples in a moment.

Proposition 6.6. *Suppose G is a real reductive group, and $\mathfrak{q}$ is a very nice parabolic subalgebra of $\mathfrak{g}$ (Definition 6.5). Let L be a real Levi factor for $\mathfrak{q}$, so that*

$$X_0 = G/L$$

is a measurable complex partial flag variety for G. Then $L \cap K$ is a real Levi factor for the (automatically nice) parabolic subalgebra $\mathfrak{q} \cap \mathfrak{k}$ of $\mathfrak{k}$, so

$$Z = K/L \cap K$$

is a (compact partial) flag variety for K and for $K_{\mathbb{C}}$. The inclusion

$$Z = K/L \cap K \hookrightarrow G/L = X_0$$

is holomorphic, and meets every connected component of X_0 exactly once.

We turn now to some examples for classical groups. Recall from section 2 that a classical complex group $G_{\mathbb{C}}$ is a group of linear transformations of $\mathbb{C}^n$, perhaps preserving some standard symmetric or skew-symmetric bilinear form. A real form G is the subgroup defined by some kind of reality condition on the matrices. Just as we saw in section 2 for complete flag varieties, a partial flag variety for $G_{\mathbb{C}}$ will be a space of partial flags in $\mathbb{C}^n$, subject to some conditions involving the bilinear form defining the group. We need to analyze the orbits of G on such flags, which is usually a matter of linear algebra.

Here is an example. Suppose $G = GL(n, \mathbb{R})$ and $G_{\mathbb{C}} = GL(n, \mathbb{C})$. A partial flag variety for $G_{\mathbb{C}}$ is determined by a collection $\mathbf{m}$ of integers

$$(6.7)(a) \qquad 0 = m_0 < m_1 < \cdots < m_r = n.$$

The variety $X_{\mathbf{m}}$ is the collection of all possible partial flags

$$(6.7)(b) \quad X_{\mathbf{m}} = \{F = (F_j) \mid 0 = F_0 \subset F_1 \subset \cdots \subset F_r = \mathbb{C}^n, \quad \dim F_j = m_j\}.$$

Here each F_j is a linear subspace of $\mathbb{C}^n$. There is a standard flag F^{std}, with

$$(6.7)(c) \qquad F_j^{std} = \mathbb{C}^{m_j} \subset \mathbb{C}^n,$$

embedded in the first m_j coordinates. The group $G_{\mathbb{C}}$ acts transitively on $X_{\mathbf{m}}$; the isotropy group at the base point F^{std} consists of block upper triangular matrices

$$(6.7)(d) \qquad P_{\mathbf{r}}^{std} = \left\{ \begin{pmatrix} A_1 & * & * \\ 0 & \ddots & * \\ 0 & 0 & A_r \end{pmatrix} \mid A_j \in GL((m_j - m_{j-1}), \mathbb{C}) \right\}.$$

So how does one understand the orbits of $GL(n, \mathbb{R})$ on this space? If F is a flag of type $\mathbf{m}$, then so is $\overline{F}$. (If W is a subspace of $\mathbb{C}^n$, then $\overline{W}$ consists of all the complex conjugates of vectors in W.) The collection of dimensions

$$(6.7)(e) \qquad \dim_{\mathbb{C}}(F_j \cap \overline{F}_k)$$

is obviously constant on $GL(n, \mathbb{R})$ orbits in $X_{\mathbf{m}}$; and it is not very hard to show that these invariants specify the $GL(n, \mathbb{R})$ orbits completely. It is also not so difficult to describe exactly what sets of dimensions are possible. Roughly speaking, the open orbits of $GL(n, \mathbb{R})$ should be "generic," and so should be characterized by having all the dimensions in (6.7)(e) as small as possible. It is an excellent exercise for the reader to work this out in detail for complete flags; the conclusion in that case is that the $GL(n, \mathbb{R})$ orbits correspond to elements of order 2 in the symmetric group S_n. (For complete flags in general split real groups G, there is a surjective map from G orbits to elements of order 2 in the Weyl group, but the map can have non-trivial fibers.)

We will analyze instead the much simpler case

(6.8)(a) $$\mathbf{m} = (0, m, n).$$

In this case

(6.8)(b) $$X_m = \{F \subset \mathbb{C}^n \mid \dim F = m\}$$

is the Grassmann variety of m-planes in $\mathbb{C}^n$. The unique invariant of a $GL(n, \mathbb{R})$ orbit on X_m is the integer

(6.8)(c) $$r = \dim(F \cap \overline{F}) \leqslant m.$$

The integer r is a measure of the "reality" of F: F is the complexification of a real subspace if and only if $r = m$. The $GL(n, \mathbb{R})$ orbit corresponding to $r = m$ is the real Grassmann variety of m-planes in $\mathbb{R}^n$, and the isotropy groups are examples of the real parabolic subgroups studied in section 5.

We are interested now in the opposite case, when r is as small as possible. How small is that? The constraint comes from the fact that $\dim(F + \overline{F}) \leqslant n$. The sum has dimension $2m - r$, so we find $2m - r \leqslant n$, or equivalently $r \geqslant 2m - n$. The conclusion is that possible values of r are

(6.8)(d) $$\min(0, 2m - n) \leqslant r \leqslant m.$$

The unique open orbit of $GL(n, \mathbb{R})$ on X_m corresponds to the smallest possible value of r; it is

(6.8)(e) $\quad X_{m,0} = \{F \subset \mathbb{C}^n \mid \dim F = m, \quad \dim(F \cap \overline{F}) = \min(0, 2m - n)\},$

a complex homogeneous space for $GL(n, \mathbb{R})$.

For definiteness, let us now concentrate on the case

(6.8)(f) $$2m \geqslant n.$$

In this case we are looking at subspaces $F \subset \mathbb{C}^n$ such that

(6.8)(g) $$0 \subset \underbrace{F \cap \overline{F}}_{\text{dimension } 2m - n} \subset \underbrace{F}_{\text{dimension } m} \subset F + \overline{F} = \mathbb{C}^n.$$

Let us now look at the corresponding parabolic subalgebra $\mathfrak{q}$, the stabilizer of F. We can choose a basis of $\mathbb{C}^n$ so that

(6.8)(h)
$$F \cap \overline{F} = \text{span of middle } 2m - n \text{ basis vectors}$$
$$F = \text{span of first } m \text{ basis vectors}$$
$$\overline{F} = \text{span of last } m \text{ basis vectors}$$

In these coordinates, we compute

$$\mathfrak{q} = \left\{ \begin{pmatrix} * & * & * \\ * & * & * \\ 0 & 0 & * \end{pmatrix} \right\}, \qquad \overline{\mathfrak{q}} = \left\{ \begin{pmatrix} * & 0 & 0 \\ * & * & * \\ * & * & * \end{pmatrix} \right\}.$$

Here the blocks correspond to the first $n - m$, middle $2m - n$, and last $n - m$ coordinates. The intersection of these two parabolic subalgebras is

$$\mathfrak{q} \cap \overline{\mathfrak{q}} = \left\{ \begin{pmatrix} * & 0 & 0 \\ * & * & * \\ 0 & 0 & * \end{pmatrix} \right\}.$$

The nil radical of this Lie algebra is

$$\left\{ \begin{pmatrix} 0 & 0 & 0 \\ * & 0 & * \\ 0 & 0 & 0 \end{pmatrix} \right\},$$

which has dimension $2(n - m)(2m - n)$. It follows that $\mathfrak{q} \cap \overline{\mathfrak{q}}$ is *not* reductive when $n/2 < m < n$: in these cases, the complex homogeneous space $X_{m,0}$ is not "measurable" in the sense of Definition 6.5. (The same conclusion applies to the cases $0 < m < n/2$.)

We now look more closely at the case $n = 2m$. Recall that a *complex structure* on a real vector space V is a linear map J such that $J^2 = -I$.

Proposition 6.9. *Suppose $n = 2m$ is a positive even integer. Define*

$$X = X_{m,0} = \{F \subset \mathbb{C}^{2m} \mid \dim F = m, \quad F \cap \overline{F} = 0\}.$$

Then X is a measurable complex partial flag variety for $GL(2m, \mathbb{R})$. Its points may be identified with complex structures on $\mathbb{R}^{2m}$; the identification sends a complex structure J to the $+i$-eigenspace of J acting on $(\mathbb{R}^{2m})_{\mathbb{C}} = \mathbb{C}^{2m}$.

The isotropy group at a subspace F corresponding to the complex structure J_F consists of all linear automorphisms of $\mathbb{R}^{2m}$ commuting with the complex structure J_F; that is, of complex-linear automorphisms of the corresponding m-dimensional complex vector space. In particular, if we choose as a base point of $X_{m,0}$ the standard complex structure, then the isotropy group is

$$L = GL(m, \mathbb{C}) \subset GL(2m, \mathbb{R}).$$

This base point is "very nice" in the sense of Definition 6.5. The corresponding $O(n)$ orbit is

$$Z = \underbrace{O(2m)/U(m)}_{\dim_{\mathbb{C}} = (m^2 - m)/2} \subset \underbrace{GL(2m, \mathbb{R})/GL(m, \mathbb{C}) = X}_{\dim_{\mathbb{C}} = m^2}.$$

This compact subvariety consists of all orthogonal complex structures on $\mathbb{R}^{2m}$ (those for which multiplication by i preserves length).

We conclude this section with an easier example: the case of $U(p, q)$. We begin with non-negative integers p and q, and write $n = p + q$. There is a standard Hermitian form

$$(6.10)(a) \qquad \langle v, w \rangle_{p,q} = \sum_{i=1}^{p} v_i \overline{w}_i - \sum_{j=1}^{q} v_{p+j} \overline{w}_{p+j}$$

of signature (p, q) on $\mathbb{C}^n$. The indefinite unitary group of signature (p, q) is
(6.10)(b)
$$U(p, q) = \{g \in GL(n, \mathbb{C}) \mid \langle g \cdot v, g \cdot w \rangle_{p,q} = \langle v, w \rangle_{p,q} \quad (v, w \in \mathbb{C}^n)\}$$

Just as in the case of $U(n)$, it is easy to check that every $n \times n$ complex matrix Z can be written uniquely as $Z = A + iB$, with A and B in Lie($U(p, q)$). It follows that $GL(n, \mathbb{C})$ is a complexification of G. Let us fix a partial flag variety $X_{\mathbf{m}}$ as in (6.7), and try to understand the orbits of $U(p, q)$ on $X_{\mathbf{m}}$. Consider a flag $F = (F_j)$ in $X_{\mathbf{m}}$. The orthogonal complement $F_k^{\perp}$ (with respect to the form $\langle \cdot \cdot \rangle_{p,q}$) is a subspace of dimension $n - m_k$; we therefore get a partial flag consisting of the subspaces $F_j \cap F_k^{\perp}$ inside F_j. The dimensions of these subspaces are invariants of the $U(p, q)$ orbit of F. We are interested in open orbits, where the dimensions are as small as possible. The minimum possible dimensions are

$$(6.10)(c) \qquad \dim F_j \cap F_k^{\perp} = \begin{cases} m_j - m_k, & k \leqslant j \\ 0, & j \leqslant k. \end{cases}$$

Looking in particular at the case $k = j$, we see that on an open orbit, $F_j \cap F_j^{\perp} = 0$. This means that the restriction of $\langle \cdot \cdot \rangle_{p,q}$ to F_j will be a non-degenerate Hermitian form, which will therefore have some signature $(p(F_j), q(F_j)) = (p_j, q_j)$. These non-negative integers must satisfy the conditions
(6.10)(d)
$$p_j + q_j = m_j, \qquad 0 = p_0 \leqslant p_1 \leqslant \cdots \leqslant p_r = p, \qquad 0 = q_0 \leqslant q_1 \leqslant \cdots \leqslant q_r = q.$$

These sequences $(\mathbf{p}, \mathbf{q})$ are invariants of the $U(p, q)$ orbit of F. Conversely, if F' is any other flag giving rise to the same sequence of signatures, then it is easy to find an element of $U(p, q)$ carrying F to F'. The following proposition summarizes this discussion, and some easy calculations.

Proposition 6.11. *Suppose $X_{\mathbf{m}}$ is a partial flag variety for $GL(n, \mathbb{C})$ as in (6.7). The open orbits of $U(p, q)$ on $X_{\mathbf{m}}$ are in one-to-one correspondence with pairs of sequences $(\mathbf{p}, \mathbf{q})$ as in (6.10)(d). Write $X_{\mathbf{p},\mathbf{q}}$ for the corresponding orbit. Each of these orbits is measurable (Definition 6.5). The corresponding real Levi factor (Definition 6.5) is isomorphic to*

$$\prod_{j=1}^{r} U(p_j - p_{j-1}, q_j - q_{j-1}).$$

The orbit of $K = U(p) \times U(q)$ through a very nice point is isomorphic to

$$\left[U(p) / \prod_{j=1}^{r} U(p_j - p_{j-1}) \right] \times \left[U(q) / \prod_{j=1}^{r} U(q_j - q_{j-1}) \right] \simeq X_{\mathbf{p}} \times X_{\mathbf{q}}.$$

This is a compact complex subvariety of $X_{\mathbf{p},\mathbf{q}}$.

7 Dolbeault Cohomology and Maximal Globalizations

The central idea in these notes is this: we want to construct representations of a real reductive group G by starting with a measurable complex flag variety $X = G/L$ (Definition 6.5) and using G-equivariant holomorphic vector bundles on X. For G compact connected, the Borel-Weil theorem (Theorem 2.3) says that all irreducible representations of G arise in this way, as spaces of holomorphic sections of holomorphic line bundles. In order to get some feeling for what to expect about noncompact groups, we look first at the example of $U(1,1)$. In the language of Proposition 6.11, let us take $r = 2$ and consider the complete flag variety for $GL(2,\mathbb{C})$, corresponding to

$$(7.1)(a) \qquad\qquad \mathbf{m} = (0,1,2).$$

Explicitly, $X_{\mathbf{m}}$ is just the projective space $\mathbb{CP}^1$ of lines in $\mathbb{C}^2$. We identify

$$(7.1)(b) \qquad (\mathbb{C} \cup \infty) \simeq X_{\mathbf{m}}, \quad z \mapsto \text{ line through } \begin{pmatrix} 1 \\ z \end{pmatrix}$$

We consider the open $U(1,1)$ orbit $X_{\mathbf{p},\mathbf{q}}$ with

$$\mathbf{p} = (0,1,1), \qquad \mathbf{q} = (0,0,1).$$

Explicitly, these are the lines in $\mathbb{C}^2$ on which the Hermitian form $\langle \cdot, \cdot \rangle_{1,1}$ is strictly positive. Because

$$\left\langle \begin{pmatrix} 1 \\ z \end{pmatrix}, \begin{pmatrix} 1 \\ z \end{pmatrix} \right\rangle_{1,1} = 1 - |z|^2,$$

it follows that the identification of (7.1)(b) gives

$$(7.1)(c) \qquad\qquad X_{\mathbf{p},\mathbf{q}} \simeq \{z \in \mathbb{C} \mid |z| < 1\},$$

the unit disc. The action of $U(1,1)$ on the disc is by linear fractional transformations as in (3.2); the reason is

$$\begin{pmatrix} \alpha & \beta \\ \bar\beta & \bar\alpha \end{pmatrix} \begin{pmatrix} 1 \\ z \end{pmatrix} = \begin{pmatrix} \beta z + \alpha \\ \bar\alpha z + \bar\beta \end{pmatrix} = c \cdot \begin{pmatrix} 1 \\ (\bar\alpha z + \bar\beta)/(\beta z + \alpha) \end{pmatrix}.$$

The standard base point is the origin $z = 0$, where the isotropy group is $U(1) \times U(1)$. It follows that equivariant holomorphic line bundles on $X_{\mathbf{p},\mathbf{q}}$ are in one-to-one correspondence with characters

$$(7.1)(d) \qquad\qquad \mu = (m_1, m_2) \in (U(1) \times U(1))^\widehat{} \simeq \mathbb{Z}^2.$$

Write $\mathcal{L}_\mu$ for the holomorphic line bundle corresponding to μ. Because μ extends to a holomorphic character of the group of complex upper triangular matrices, $\mathcal{L}_\mu$ extends to a $GL(2,\mathbb{C})$-equivariant holomorphic line bundle on the Riemann sphere $X_{\mathbf{m}}$.

The most straightforward analogy with the Borel-Weil theorem suggests defining

(7.1)(e) $H_\mu =$ holomorphic sections of $\mathcal{L}_\mu$.

If we endow this space with the topology of uniform convergence on compact sets, then it is a complete topological vector space, and the action π_μ of $G = U(1,1)$ by left translation is continuous.

Proposition 7.2. *The representation π_μ of $U(1,1)$ is always infinite-dimensional. It is irreducible unless μ is antidominant; that is, unless $m_1 \leqslant m_2$. If μ is antidominant, there is exactly one proper closed G-invariant subspace: the $(m_2 - m_1 + 1)$-dimensional space of sections extending holomorphically to the entire Riemann sphere $X_\mathbf{m}$.*

Here are some hints about proofs. Any holomorphic line bundle on the disc is holomorphically (although not equivariantly!) trivial, so the space of sections may be identified with holomorphic functions on the disc; this is certainly infinite-dimensional. For μ anti-dominant, Theorem 2.3 provides a finite-dimensional subspace of sections extending to the Riemann sphere. The dimension calculation is a standard fact about $U(2)$, and the invariance of this subspace is clear.

For the remaining assertions, examining Taylor series expansions shows that every $U(1) \times U(1)$ weight of H_μ is of the form $(m_1 + k, m_2 - k)$, with k a non-negative integer; and that each of these weights has multiplicity one. Now one can apply facts about Verma modules for $\mathfrak{gl}(2)$ to finish.

Proposition 7.2 is a bit discouraging with respect to the possibility of extending Theorem 2.3 to noncompact groups. The case of $U(2,1)$ is even worse. Let us look at

(7.3)(a) $\mathbf{m} = (0,1,2,3)$, $\mathbf{p} = (0,1,1,2)$, $\mathbf{q} = (0,0,1,1)$.

Then $X_{\mathbf{p},\mathbf{q}}$ consists of complete flags F with the property that the Hermitian form is positive on F_1 and of signature $(1,1)$ on F_2. The isotropy group at the standard base point is $U(1)^3$, and its characters are given by triples

(7.3)(b) $\mu = (m_1, m_2, m_3) \in \mathbb{Z}^3$

Write $\mathcal{L}_\mu$ for the corresponding equivariant holomorphic line bundle on $X_{\mathbf{p},\mathbf{q}}$ (which automatically extends to be $GL(3, \mathbb{C})$-equivariant on $X_\mathbf{m}$) and (π_μ, H_μ) for the representation of $U(2,1)$ on its space of holomorphic sections.

Proposition 7.4. *In the setting of (7.3), the representation π_μ of $U(2,1)$ is zero unless μ is antidominant; that is, unless $m_1 \leqslant m_2 \leqslant m_3$. In that case it is finite-dimensional, and all holomorphic sections extend to the full flag variety $X_\mathbf{m}$.*

Again one can use Taylor series to relate the Harish-Chandra module of H_μ to a highest weight module. What one needs to know is that if V is an irreducible highest weight module for $\mathfrak{gl}(3)$, and the non-simple root $\mathfrak{gl}(2)$ subalgebra acts in a locally finite way, then V is finite-dimensional. Again we omit the details.

The behavior in Proposition 7.4 is typical. Holomorphic sections of vector bundles on measurable complex partial flag varieties rarely produce anything except finite-dimensional representations of G. One way to understand this is that the varieties fail to be Stein, so we should not expect to understand them looking only at holomorphic sections: we must also consider "higher cohomology." We begin with a brief review of Dolbeault cohomology.

Suppose X is a complex manifold, with complexified tangent bundle $T_\mathbb{C}X$. The complex structure on X provides a decomposition

$$(7.5)(a) \qquad\qquad T_\mathbb{C}X = T^{1,0} \oplus T^{0,1}$$

into holomorphic and antiholomorphic tangent vectors. These may be understood as the $+i$ and $-i$ eigenspaces of the complex structure map J (defining "multiplication by i" in the real tangent space.) The two subspaces are interchanged by complex conjugation. The space $T^{0,1}$ consists of the tangent vectors annihilating holomorphic functions: the Cauchy-Riemann equations are in $T^{0,1}$.

There is a terminological dangerous bend here. One might think that a smooth section of $T^{1,0}$ should be called a "holomorphic vector field," but in fact this terminology should be reserved only for *holomorphic* sections (once those are defined). We will call a smooth section a *vector field of type* $(1,0)$. On $\mathbb{C}$, the vector field

$$x\frac{\partial}{\partial z} = \frac{x}{2}\left(\frac{\partial}{\partial x} - i\frac{\partial}{\partial y}\right)$$

is of type $(1,0)$, but is not holomorphic. If we replace the coefficient function x by 1 (or by any holomorphic function), we get a holomorphic vector field.

Write

$$(7.5)(b) \qquad A^m = \text{complex-valued differential forms of degree } m \text{ on } X,$$

$$= \sum_{p+q=m} A^{p,q}.$$

Here $A^{p,q}$ consists of differential forms that vanish on sets of p' type $(1,0)$ vector fields and q' type $(0,1)$ vector fields unless $p' = p$ and $q' = q$. The de Rham differential

$$d\colon A^m \to A^{m+1}$$

satisfies

$$(7.5)(c) \qquad\qquad d(A^{p,q}) \subset A^{p+1,q} \oplus A^{p,q+1}.$$

This follows by inspection of the formula

$$d\omega(Y_0, \ldots, Y_m) = \sum_{i=0}^{m} (-1)^i Y_i \cdot \omega(Y_0, \ldots, \widehat{Y_i}, \ldots, Y_m)$$

$$+ \sum_{i<j} (-1)^{i+j} \omega([Y_i, Y_j], Y_0, \ldots, \widehat{Y_i}, \ldots, \widehat{Y_j}, \ldots, Y_m),$$

and the fact the Lie bracket of two type $(1,0)$ (respectively type $(0,1)$) vector fields is type $(1,0)$ (respectively type $(0,1)$). Now the decomposition in (7.5)(c) allows us to write $d = \partial + \bar{\partial}$, with

(7.5)(d) $\qquad\qquad \partial \colon A^{p,q} \to A^{p+1,q}, \qquad \bar{\partial} \colon A^{p,q} \to A^{p,q+1}$

The fact that $d^2 = 0$ implies that

(7.5)(e) $\qquad\qquad \partial^2 = \bar{\partial}^2 = 0, \qquad \partial\bar{\partial} + \bar{\partial}\partial = 0.$

If we try to write explicit formulas for ∂ and $\bar{\partial}$, the only difficulty arises from terms involving $[Y, \overline{Z}]$, with Y a vector field of type $(1,0)$ and $\overline{Z}$ of type $(0,1)$. The bracket is again a vector field, so it decomposes as

$$[Y, \overline{Z}] = [Y, \overline{Z}]_{1,0} + [Y, \overline{Z}]_{0,1}.$$

The first summand will appear in a formula for ∂, and the second in a formula for $\bar{\partial}$. One way to avoid this unpleasantness is to notice that if Y is actually a holomorphic vector field, then the first summand $[Y, \overline{Z}]_{1,0}$ is automatically zero; one can take this as a definition of a holomorphic vector field on X. If $\overline{Z}$ is antiholomorphic, then the second summand vanishes.

Here are the formulas that emerge.

Proposition 7.6. *Suppose X is a complex manifold, $\omega \in A^{p,q}$ is a complex-valued differential form of type (p,q) (cf. (7.5)), $(Y_0, \ldots Y_p)$ are holomorphic vector fields, and $(\overline{Z}_0, \ldots, \overline{Z}_q)$ are antiholomorphic vector fields. Then*

$$\partial\omega(Y_0, \ldots, Y_p, \overline{\mathbf{Z}}) = \sum_{i=0}^{p} (-1)^i Y_i \cdot \omega(Y_0, \ldots, \widehat{Y_i}, \ldots, Y_p, \overline{\mathbf{Z}})$$

$$+ \sum_{i<j} (-1)^{i+j} \omega([Y_i, Y_j], Y_0, \ldots, \widehat{Y_i}, \ldots, \widehat{Y_j}, \ldots, Y_p, \overline{\mathbf{Z}}).$$

In this formula,

$$\overline{\mathbf{Z}} = \overline{Z}_1, \ldots, \overline{Z}_q.$$

Similarly,

$$(-1)^p \bar{\partial}\omega(\mathbf{Y}, \overline{Z}_1, \ldots, \overline{Z}_q) = \sum_{i=0}^{q} (-1)^i \overline{Z}_i \cdot \omega(\mathbf{Y}, \overline{Z}_1, \ldots, \widehat{\overline{Z}_i}, \ldots, \overline{Z}_q)$$

$$+ \sum_{i<j}(-1)^{i+j}\omega(\mathbf{Y},[\overline{Z}_i,\overline{Z}_j],\overline{Z}_0,\ldots,\widehat{\overline{Z}_i},\ldots,\widehat{\overline{Z}_j},\ldots,\overline{Z}_q).$$

Here

$$\mathbf{Y} = Y_1,\ldots,Y_p.$$

Definition 7.7. *Suppose X is a complex manifold. The (p,q)-Dolbeault cohomology of X is by definition*

$$H^{p,q}(X) = (kernel\ of\ \overline{\partial}\ on\ A^{p,q})/(image\ of\ \overline{\partial}\ from\ A^{p,q-1}).$$

This makes sense because of (7.5)(e). The space $A^{p,0}$ consists of smooth sections of the bundle Ω^p of holomorphic p-forms on X; and it is easy to check that

$$H^{p,0} = kernel\ of\ \overline{\partial}\ on\ A^{p,0}$$
$$= holomorphic\ p\text{-}forms\ on\ X.$$

In particular,

$$H^{0,0} = holomorphic\ functions\ on\ X.$$

Suppose now that $\mathcal{V}$ is a holomorphic vector bundle on X. One cannot apply the de Rham differential to forms with values in a bundle, because there is no canonical way to differentiate sections of a bundle by a vector field. However, we *can* apply type $(0,1)$ vector fields canonically to smooth sections of a holomorphic vector bundle. Here is how this looks locally. Suppose $\overline{Z}$ is a type $(0,1)$ vector field (near $x \in X$), and v is a smooth section of $\mathcal{V}$ (defined near x). Choose a basis $(v_1,\ldots,v_d)$ of holomorphic sections of $\mathcal{V}$ (still near x) and write

(7.8)(a) $$v = \sum g_i v_i,$$

with g_i smooth on X (near x). Finally, define

(7.8)(b) $$\overline{Z} \cdot v = \sum (\overline{Z} \cdot g_i)v_i.$$

Why is this well-defined? If we choose a different basis $(v'_1,\ldots,v'_d)$, then it differs from the first by an invertible matrix B_{ij} of holomorphic functions on X (near x):

$$v_i = \sum_j B_{ij}v'_j.$$

If we expand v in the new basis, the coefficient functions g'_i are

$$g'_j = \sum_i g_i B_{ij}.$$

Applying the vector field $\overline{Z}$ and using the Leibnitz rule gives

$$\overline{Z} \cdot g'_j = \sum_i [(\overline{Z} \cdot g_i) B_{ij} + g_i (\overline{Z} \cdot B_{ij})].$$

The second terms all vanish, because $\overline{Z}$ is a type $(0,1)$ vector field and B_{ij} is a holomorphic function. What remains says (after multiplying by v'_j and summing over j) that

$$\sum_j (\overline{Z} \cdot g'_j) v'_j = \sum_i (\overline{Z} \cdot g_i) v_i;$$

that is, that our definition of $\overline{Z} \cdot v$ is well-defined.

What follows from (7.8) is that the Dolbeault $\overline{\partial}$ operator can be defined on (p,q) forms with values in a holomorphic vector bundle on X. Here is an explicit account. Write

(7.9)(a) $A^{p,q}(\mathcal{V}) = $ smooth (p,q) forms on X with values in $\mathcal{V}$.

An element of this space attaches to p type $(1,0)$ vector fields and q type $(0,1)$ vector fields a smooth section of $\mathcal{V}$. The Dolbeault operator

(7.9)(b) $\overline{\partial}: A^{p,q}(\mathcal{V}) \to A^{p,q+1}(\mathcal{V})$

is defined by the formula in Proposition 7.6, with the terms of the form

(7.9)(c) $\overline{Z}_i \cdot$ (smooth section of $\mathcal{V}$)

defined by (7.8). If we need to be more explicit, we may write this operator as $\overline{\partial}^{p,q}(\mathcal{V})$. Just as in (7.5)(e), we have

$$\overline{\partial}^2 = 0.$$

Definition 7.10. *Suppose X is a complex manifold, and $\mathcal{V}$ is a holomorphic vector bundle on X. The (p,q)-Dolbeault cohomology of X with coefficients in $\mathcal{V}$ is by definition*

$$H^{p,q}(X, \mathcal{V}) = (\text{kernel of } \overline{\partial} \text{ on } A^{p,q}(\mathcal{V}))/(\text{image of } \overline{\partial} \text{ from } A^{p,q-1}(\mathcal{V})).$$

This makes sense because of (7.5)(e). The space $A^{0,0}(\mathcal{V})$ consists of smooth sections of $\mathcal{V}$, and

$$H^{0,0}(X, \mathcal{V}) = \text{kernel of } \overline{\partial} \text{ on } A^{0,0}(\mathcal{V})$$
$$= \text{holomorphic p-forms on } X.$$

In particular,

$$H^{0,0} = \text{holomorphic sections of } \mathcal{V}.$$

As a first application, we can understand the dependence of Dolbeault cohomology on p. Recall that Ω^p is the bundle of holomorphic p-forms on X. It is easy to see that

$$A^{p,q}(\mathcal{V}) \simeq A^{0,q}(\Omega^p \otimes \mathcal{V}),$$

and that this isomorphism respects the $\bar\partial$ operators (up to a factor of $(-1)^p$). It follows that

$$H^{p,q}(X, \mathcal{V}) \simeq H^{0,q}(X, \Omega^p \otimes \mathcal{V}).$$

Here is the central fact about Dolbeault cohomology.

Theorem 7.11. (Dolbeault, Serre [Ser55]). *Suppose $\mathcal{V}$ is a holomorphic vector bundle on a complex manifold X. Write $\mathcal{O}_\mathcal{V}$ for the sheaf of germs of holomorphic sections of $\mathcal{V}$. Then there is a canonical isomorphism*

$$H^{0,q}(X, \mathcal{V}) \simeq H^q(X, \mathcal{O}_\mathcal{V}).$$

On the right is the Čech cohomology of X with coefficients in the sheaf $\mathcal{O}_\mathcal{V}$.

It may be helpful to see how Dolbeault cohomology looks on a homogeneous space. For this we can allow G to be any Lie group and L any closed subgroup. Write

$$(7.12)(a) \qquad X = G/L, \qquad \mathfrak{g} = \mathrm{Lie}(G)_{\mathbb{C}} \supset \mathrm{Lie}(L)_{\mathbb{C}} = \mathfrak{l}.$$

A G-invariant complex structure on G/L corresponds to a complex Lie subalgebra $\mathfrak{q} \subset \mathfrak{g}$ satisfying

$$(7.12)(b) \qquad \mathrm{Ad}(L)\mathfrak{q} = \mathfrak{q}, \qquad \mathfrak{q} + \bar{\mathfrak{q}} = \mathfrak{g}, \qquad \mathfrak{q} \cap \bar{\mathfrak{q}} = \mathfrak{l}.$$

In terms of the decomposition in (7.5)(a), $\mathfrak{q}$ corresponds to the antiholomorphic tangent vectors:

$$(7.12)(c) \qquad T^{0,1}_{eL}(G/L) = \mathfrak{q}/\mathfrak{l}, \qquad T^{1,0}_{eL}(G/L) = \bar{\mathfrak{q}}/\mathfrak{l}.$$

(All of this is described for example in [TW71] or in [Vog87], Proposition 1.19.) A complex-valued smooth vector field on G/L may be identified with a smooth function

$$(7.12)(d) \qquad Y : G \to \mathfrak{g}/\mathfrak{l}, \qquad Y(gl) = \mathrm{Ad}(l)^{-1}Y(g) \quad (l \in L, g \in G)$$

(cf. (5.7)(a)). In this identification, vector fields of type $(0,1)$ are those taking values in $\mathfrak{q}/\mathfrak{l}$. Smooth functions on G/L correspond to smooth functions

$$f : G \to \mathbb{C}, \qquad f(gl) = f(g).$$

The vector field Y acts on f by

$$(7.12)(e) \qquad (Y \cdot f)(g) = [\rho(Y(g)) \cdot f](g).$$

That is, we differentiate f on the right by the Lie algebra element $Y(g)$. (Of course $Y(g)$ is only a coset of $\mathfrak{l}$, but that is harmless since f is invariant on the right by L. The condition on Y in (7.12)(d) forces the new function $Y \cdot f$ also to be right invariant by L.)

From the identification in (7.12)(d), it is not hard to deduce an identification of the smooth m-forms on G/L:

$$(7.13)(a) \qquad A^m(G/L) \simeq \mathrm{Hom}_L\left(\bigwedge\nolimits^m(\mathfrak{g}/\mathfrak{l}), C^\infty(G)\right).$$

Here L acts on the exterior algebra by Ad, and on the smooth functions by right translation. The decomposition

$$\mathfrak{g}/\mathfrak{l} = \overline{\mathfrak{q}}/\mathfrak{l} \oplus \mathfrak{q}/\mathfrak{l}$$

(which follows from (7.12)(b)) gives

$$\bigwedge\nolimits^m(\mathfrak{g}/\mathfrak{l}) = \sum_{p+q=m} \bigwedge\nolimits^p(\overline{\mathfrak{q}}/\mathfrak{l}) \otimes \bigwedge\nolimits^q(\mathfrak{q}/\mathfrak{l}),$$

and a corresponding decomposition of the m-forms. The pieces are exactly the (p, q) forms of (7.5)(b):

$$(7.13)(b) \qquad A^{p,q}(G/L) \simeq \mathrm{Hom}_L\left(\bigwedge\nolimits^p(\overline{\mathfrak{q}}/\mathfrak{l}) \otimes \bigwedge\nolimits^q(\mathfrak{q}/\mathfrak{l}), C^\infty(G)\right).$$

Writing formulas for the operators ∂ and $\overline{\partial}$ in this setting is slightly unpleasant, because the description of vector fields in (7.12)(d) does not obviously hand us any holomorphic or antiholomorphic vector fields. We will sweep this problem under the rug for the moment, by not writing formulas yet.

A smooth equivariant vector bundle $\mathcal{V}$ on G/L is the same thing as a smooth representation (τ, V) of L; the correspondence is

$$(7.13)(c) \qquad \mathcal{V} \mapsto (\text{fiber of } \mathcal{V} \text{ at } eL), \qquad V \mapsto G \times_L V.$$

The space of smooth sections of $\mathcal{V}$ may be identified with smooth functions

$$(7.13)(d) \qquad f: G \to V, \qquad f(gl) = \tau(l)^{-1} f(g) \quad (l \in L, g \in G).$$

This description makes sense for infinite-dimensional vector bundles. What does it mean for $\mathcal{V}$ to be a holomorphic vector bundle? Certainly this ought to amount to imposing some additional structure on the representation (τ, V) of L. Here is the appropriate definition, taken from [TW71].

Definition 7.14. *Suppose L is a Lie group with complexified Lie algebra $\mathfrak{l}$. Assume that $\mathfrak{q}$ is a complex Lie algebra containing $\mathfrak{l}$, and the adjoint action of L extends to*

$$\mathrm{Ad}: L \to \mathrm{Aut}(\mathfrak{q}),$$

with differential the Lie bracket of $\mathfrak{l}$ on $\mathfrak{q}$. A $(\mathfrak{q}, L)$-representation is a complete locally convex vector space V, endowed with a smooth representation τ of L and a continuous Lie algebra action (written just with a dot). These are required to satisfy

(1) The q action extends the differential of τ: if $Y \in \mathfrak{l}$ and $v \in V$, then

$$d\tau(Y)v = Y \cdot v.$$

(2) For $l \in L$, $Z \in \mathfrak{q}$, and $v \in V$, we have

$$\tau(l)(Z \cdot v) = (\mathrm{Ad}(l)Z) \cdot (\tau(l)v).$$

For $l \in L_0$, condition (2) is a consequence of condition (1).

Condition (2) can also be formulated as requiring that the action map

$$\mathfrak{q} \times V \to V, \qquad (Z, v) \to Z \cdot v$$

is L-equivariant. This entire definition is formally very close to that of a $(\mathfrak{g}, K)$-module in Definition 4.14, except that we have no finiteness assumption on the L representation.

Proposition 7.15. *Suppose G/L is a homogeneous space for Lie groups, and that $\mathfrak{q}$ defines an invariant complex structure (cf. (7.12)). Then passage to the fiber at eL defines a bijective correspondence from G-equivariant holomorphic vector bundles $\mathcal{V}$ on G/L, to $(\mathfrak{q}, L)$-representations (τ, V) (Definition 7.14). Suppose $\overline{U}$ is an open subset of G/L, and U its inverse image in G. Then holomorphic sections of $\mathcal{V}$ on $\overline{U}$ correspond to smooth functions*

$$f : U \to V$$

satisfying the transformation law

$$f(gl) = \tau(l)^{-1}f(g) \qquad (l \in L, g \in U)$$

and the differential equations

$$(\rho(Z)f)(g) = Z \cdot (f(g)) \qquad (Z \in \mathfrak{q}, g \in U).$$

Here ρ is the right regular representation of the Lie algebra on smooth functions.

To be more honest and precise: this result is certainly true for finite-dimensional bundles (where it is proved in [TW71]). I have not thought carefully about the appropriate abstract definition of infinite-dimensional holomorphic vector bundles; but that definition needs to be arranged so that Proposition 7.15 is true.

The transformation law in Proposition 7.15 is just what describes a smooth section of $\mathcal{V}$ (cf. (7.13)(d)). For $Z \in \mathfrak{l}$, the differential equation is a consequence of the transformation law. The differential equations for other elements of $\mathfrak{q}$ are the Cauchy-Riemann equations.

Lie algebra cohomology was invented for the purpose of studying de Rham cohomology of homogeneous spaces. It is therefore not entirely surprising that Dolbeault cohomology (which we described in (7.5) as built from de Rham cohomology) is also related to Lie algebra cohomology. To state the result, we need one more definition.

Definition 7.16. *Suppose V is a $(\mathfrak{q}, L)$-representation (Definition 7.14). The complex defining Lie algebra cohomology is*

$$C^m(\mathfrak{q}; V) = \mathrm{Hom}\left(\bigwedge{}^m \mathfrak{q}, V\right).$$

The differential is

$$d\omega(Z_0, \ldots, Z_m) = \sum_{i=0}^{m} (-1)^i Z_i \cdot \omega(Z_0, \ldots, \widehat{Z}_i, \ldots, Z_m)$$

$$+ \sum_{i<j} (-1)^{i+j} \omega(Z_0, \ldots, \widehat{Z}_i, \ldots, \widehat{Z}_j, \ldots, Z_m)$$

The Lie algebra cohomology of $\mathfrak{q}$ with coefficients in V is by definition

$$H^m(\mathfrak{q}; V) = (\textit{kernel of } d \textit{ on } C^m(\mathfrak{q}; V))/(\textit{image of } d \textit{ from } C^{m-1}(\mathfrak{q}; V)).$$

We now consider the subspace

$$C^m(\mathfrak{q}, L; V) = \mathrm{Hom}_L\left(\bigwedge{}^m \mathfrak{q}/\mathfrak{l}, V\right).$$

We are imposing two conditions: that ω vanish on the ideal generated by $\mathfrak{l}$ in the exterior algebra, and that the linear map ω respect the action of L (by Ad on the domain and τ on the range). The differential d respects the second condition; and in the presence of the second condition, it respects the first as well. We can therefore define the relative Lie algebra cohomology of $\mathfrak{q}$ with coefficients in V *as*

$$H^m(\mathfrak{q}, L; V) = (\textit{kernel of } d \textit{ on } C^m(\mathfrak{q}, L; V))/(\textit{image of } d \textit{ from } C^{m-1}(\mathfrak{q}, L; V)).$$

This cohomology is most often considered in the case when L is compact. One reason is that when L is not compact, taking L invariants (as in Hom_L in the definition of the relative complex) is not an exact functor, and should really only be considered along with its derived functors. This difficulty will come back to haunt us in section 9, but for now we ignore it.

Here now is Kostant's description of Dolbeault cohomology for equivariant bundles.

Proposition 7.17. (Kostant [Kos61], (6.3.5); see also [Won99], section 2). *Suppose G/L is a homogeneous space for Lie groups, and that $\mathfrak{q}$ defines an invariant complex structure (cf. (7.12)). Suppose (τ, V) is a $(\mathfrak{q}, L)$-representation (Definition 7.14), and $\mathcal{V}$ the corresponding G-equivariant holomorphic vector bundle on G/L (Proposition 7.15). We regard $C^\infty(G, V)$ as a $(\mathfrak{q}, L)$-representation by the "tensor product" of the right regular action on functions with the action on V. Explicitly, the representation τ_r of L is*

$$[\tau_r(l)f](g) = \tau(l)f(gl^{-1}).$$

The action of $Z \in \mathfrak{q}$ is

$$[Z \cdot f](g) = [\rho(Z)f](g) + Z \cdot (f(g)).$$

The first term is the right regular action of $\mathfrak{g}$ on functions, and the second is the action of $\mathfrak{q}$ on V.

Then the space of smooth $(0, q)$ forms on G/L with values in V is

$$A^{0,q}(G/L, V) \simeq \mathrm{Hom}_L\left(\bigwedge^q(\mathfrak{q}/\mathfrak{l}), C^\infty(G, V)\right) = C^q(\mathfrak{q}, L; C^\infty(G, V)),$$

(Definition 7.16). This identifies the Dolbeault differential $\bar\partial$ with the relative Lie algebra cohomology differential d, and so

$$H^{0,q}(G/L, V) \simeq H^q(\mathfrak{q}, L; C^\infty(G, V)).$$

To talk about (p, q) Dolbeault cohomology, we can use the fact mentioned before Theorem 7.11. This involves the bundle Ω^p of holomorphic p-forms on X; so we need to understand Ω^p in the case of $X = G/L$. This is an equivariant vector bundle, so it corresponds to a certain representation of L: the pth exterior power of the holomorphic cotangent space $(T_{eL}^{1,0})^*$. According to (7.12)(c), this is

$$\bigwedge^p(\bar{\mathfrak{q}}/\mathfrak{l}).$$

To specify the holomorphic structure, we need a representation of $\mathfrak{q}$ on this space, extending the adjoint action of $\mathfrak{l}$. This we can get from the natural isomorphism

$$\bar{\mathfrak{q}}/\mathfrak{l} \simeq \mathfrak{g}/\mathfrak{q},$$

which is a consequence of (7.12)(b). That is, in the correspondence of Proposition 7.15,

(7.18)(a) $$\Omega^p \leftrightarrow \bigwedge^p(\mathfrak{g}/\mathfrak{q})^*,$$

with the obvious structure of $(\mathfrak{q}, L)$-representation on the right. A consequence of this fact, Proposition 7.17, and the fact before Theorem 7.11 is

(7.18)(b) $$H^{p,q}(G/L, V) \simeq H^q(\mathfrak{q}, L; C^\infty(G, \bigwedge^p(\mathfrak{g}/\mathfrak{q})^* \otimes V)).$$

With Dolbeault cohomology in our tool box, we can now make the idea at the beginning of this section a little more precise. Beginning with a measurable complex flag variety $X = G/L$ and a G-equivariant holomorphic vector bundle V over X, we want to consider representations of G on Dolbeault cohomology spaces $H^{p,q}(X, V)$. First of all, notice that G acts by translation on the forms $A^{p,q}(V)$ (cf. (7.9)), and that this action respects $\bar\partial$. It follows that we get a linear action of G on the Dolbeault cohomology. To have a representation, of course we need a topological vector space structure. The space of V-valued differential forms (for any smooth vector bundle on any smooth manifold) naturally has such a structure; in our case, the forms are described in Proposition 7.17 as a closed subspace of the (complete locally convex) space $C^\infty(G, V)$ tensored with a finite-dimensional space. (This shows in particular that if V is Fréchet, then so is $A^{p,q}(\mathcal{V})$.) With respect to this topology, any differential operator is continuous; so in particular the $\bar\partial$ operator is continuous, and its kernel is a closed subspace of $A^{p,q}(\mathcal{V})$. It is also clear that the action of G on $A^{p,q}(\mathcal{V})$ is continuous.

We can impose on $H^{p,q}(X, \mathcal{V})$ the quotient topology coming from the kernel of $\bar\partial$: a subset of the cohomology is open (or closed) if and only if its preimage in the kernel of $\bar\partial$ is open (or closed). The action of G is clearly continuous for this quotient topology. The difficulty is that the closure of the point 0 in the quotient topology is equal to

$$(\text{closure of the image of } \bar\partial)/(\text{image of } \bar\partial).$$

In particular, the topology is Hausdorff only if the image of $\bar\partial$ is closed. This difficulty is essentially the only difficulty: if W is a complete locally convex Hausdorff space and U is a closed subspace, then the quotient topology on W/U is complete and locally convex Hausdorff. (In these notes "Hausdorff" is part of the definition of "locally convex"; I have mentioned it explicitly here only for emphasis.)

Here is a summary of this discussion.

Proposition 7.19. *Suppose $X = G/L$ is a measurable complex flag variety for the real reductive group G (Definition 6.5), and that $\mathcal{V}$ is the holomorphic vector bundle on X attached to a $(\mathfrak{q}, L)$-representation (τ, V). Endow the Dolbeault cohomology $H^{p,q}(X, \mathcal{V})$ with the quotient topology as above, and define*

$$H^{p,q}_{top}(X, \mathcal{V}) = \text{maximal Hausdorff quotient of } H^{p,q}(X, \mathcal{V})$$

$$= \text{kernel of } \bar\partial/\text{closure of image of } \bar\partial.$$

Then $H^{p,q}_{top}(X, \mathcal{V})$ carries a smooth representation of G (by translation of forms).

Serious geometers find the notion of $H^{p,q}_{top}$ in this result to be anathema. Many of the long exact sequences (that make life worth living in sheaf theory) are lost on this quotient. Nevertheless, representation theory seems to demand this quotient. We will make use of it once more in section 8, to formulate Serre's duality theorem for Dolbeault cohomology.

In the end our examples will offer no conclusive evidence about the value of the notion of $H^{p,q}_{top}(X, \mathcal{V})$. We will recall next a theorem of Hon Wai Wong which says that in all of the cases we will consider, the operator $\bar\partial$ has closed range.

The first definition is analogous to Definition 5.6.

Definition 7.20. *Suppose G is a real reductive group, $\mathfrak{q}$ is a very nice parabolic subalgebra of the complexified Lie algebra $\mathfrak{g}$, and L is a Levi factor of for $\mathfrak{q}$ (Definition 6.5). A $(\mathfrak{q}, L)$ representation (τ, V) (Definition 7.14) is said to be* admissible *if the representation τ of L is admissible (Definition 4.12). In this case the Harish-Chandra module of V is the $(\mathfrak{q}, L \cap K)$-module $V^{L \cap K}$ of $L \cap K$-finite vectors in V.*

Because $\mathfrak{q} = \mathfrak{l} \oplus \mathfrak{u}$, with $\mathfrak{u}$ an L-stable ideal in $\mathfrak{q}$, every admissible representation (τ, V) of L extends canonically to an admissible $(\mathfrak{q}, L)$ representation, by making $\mathfrak{u}$ act by zero. If (τ, V) is irreducible for L, then this is the only possible extension. But if the representation of L is reducible, then other extensions exist, and even arise in practice.

Theorem 7.21. *(Wong [Won99], Theorem 2.4). In the setting of Definition 7.20, assume that the admissible representation V is the maximal globalization of the underlying $(\mathfrak{q}, L \cap K)$ module. Let $\mathcal{V}$ be the G-equivariant holomorphic vector bundle on $X = G/L$ attached to V (Proposition 7.15). Then the $\bar{\partial}$ operator for Dolbeault cohomology has closed range, so that each of the spaces $H^{p,q}(X, \mathcal{V})$ carries a smooth representation of G. Each of these representations is admissible, and is the maximal globalization of its underlying Harish-Chandra module.*

Wong goes on to explain how these Harish-Chandra modules are constructed from $V^{L \cap K}$, by a process called "cohomological parabolic induction." We will say only a little about this.

This theorem should be compared to Proposition 5.11, to which it bears some formal resemblance. In detail it is unfortunately much weaker. With real parabolic induction, using any globalization on L led to a globalization of the same Harish-Chandra module on G. In the present setting that statement may be true, but Wong's methods seem not to prove it.

Another difference is that in section 5 we were able to get many different globalizations just by varying the kinds of functions we used. The situation here is quite different. It is perfectly possible to consider (for example) the Dolbeault complex with generalized function coefficients instead of smooth functions. But the resulting Dolbeault cohomology turns out to be exactly the same. (This is certainly true if the vector bundle $\mathcal{V}$ is finite-dimensional, and it should be possible to prove a version for infinite-dimensional bundles as well.)

One goal of this section was to find a reasonable extension of the Borel-Weil Theorem to noncompact reductive groups. Theorem 2.3 suggested that one might look at bundles that are "antidominant" in some sense; but Propositions 7.2 and 7.4 suggested that antidominant is not such a good choice for noncompact G. I will dispense with further illuminating examples, and instead pass directly to the definition we want.

Suppose therefore that

(7.22)(a) $$\mathfrak{q} = \mathfrak{l} + \mathfrak{u}$$

is a very nice parabolic subalgebra (Definition 6.5), and L is a real Levi factor for $\mathfrak{q}$. Write

$$(7.22)(b) \qquad X = G/L, \qquad \dim_{\mathbb{C}} X = \dim_{\mathbb{C}}(\mathfrak{u}) = n$$

$$(7.22)(c) \qquad Z = K/L \cap K, \qquad \dim_{\mathbb{C}} Z = \dim_{\mathbb{C}}(\mathfrak{u} \cap \mathfrak{k}) = s$$

Because of (7.18)(b), we are going to need

$$(7.22)(d) \qquad 2\rho(\mathfrak{u}) = \text{representation of } L \text{ on } \bigwedge^{n}(\mathfrak{g}/\mathfrak{q})^{*}$$

This is a one-dimensional character of L, so its differential (which we also write as $2\rho(\mathfrak{u})$) is a linear functional on the Lie algebra $\mathfrak{l}$:

$$(7.22)(e) \qquad 2\rho(\mathfrak{u}): \mathfrak{l} \to \mathbb{C}, \qquad 2\rho(\mathfrak{u})(Y) = \text{trace of } \operatorname{ad}(Y) \text{ on } (\mathfrak{g}/\mathfrak{q})^{*}$$

This linear functional is of course divisible by two, so we can define $\rho(\mathfrak{u}) \in \mathfrak{l}^{*}$, even though the group character $2\rho(\mathfrak{u})$ may have no square root. Any G-invariant symmetric nondegenerate bilinear form on $\mathfrak{g}$ provides an L-equivariant identification

$$(7.22)(f) \qquad (\mathfrak{g}/\mathfrak{q})^{*} \simeq \mathfrak{u},$$

which allows for some simplifications in the formulas for $2\rho(\mathfrak{u})$.

Let $(\tau^{L \cap K}, V^{L \cap K})$ be an irreducible Harish-Chandra module for L, and $(\tau^{\omega}, V^{\omega})$ its maximal globalization. Regard V^{ω} as $(\mathfrak{q}, L)$-representation by making $\mathfrak{u}$ act by zero. Let

$$(7.22)(g) \qquad \mathcal{V}^{\omega} = G \times_L V$$

be the associated holomorphic vector bundle on X (Proposition 7.15).

We want to write a condition on τ, more or less analogous to "antidominant" in Theorem 2.3, that will force Dolbeault cohomology with coefficients in $\mathcal{V}^{\omega}$ to be well-behaved. For this purpose, a little bit of structure theory in the enveloping algebra is needed. Put

$$(7.23)(a) \qquad \mathfrak{z}(\mathfrak{g}) = \text{center of } U(\mathfrak{g})$$

The group G acts by algebra automorphisms on $\mathfrak{z}(\mathfrak{g})$; the G_0 action is trivial, so G/G_0 is a finite group of automorphisms. We need the fixed point algebra

$$(7.23)(b) \qquad \mathfrak{z}^{G}(\mathfrak{g}) = \{u \in U(\mathfrak{g}) \mid \operatorname{Ad}(g)u = u, \text{ all } g \in G\} \subset \mathfrak{z}(\mathfrak{g}).$$

(The first algebra appeared already in the definition of admissible representations in Definition 4.12.) These algebras are described by the Harish-Chandra isomorphism. For that, fix a Cartan subalgebra

(7.23)(c) $\hspace{5cm}$ $\mathfrak{h} \subset \mathfrak{l}$

The Weyl group of $\mathfrak{h}$ in $\mathfrak{g}$ (generated by root reflections) is written

(7.23)(d) $\hspace{4cm}$ $W = W(\mathfrak{g}, \mathfrak{h}).$

The Harish-Chandra isomorphism is

(7.23)(e) $\hspace{4cm}$ $\xi^{\mathfrak{g}} : \mathfrak{Z}(\mathfrak{g}) \xrightarrow{\sim} S(\mathfrak{h})^{W(\mathfrak{g}, \mathfrak{h})}.$

The disconnectedness of G provides a slightly larger group

(7.23)(e) $\hspace{3.5cm}$ $W^G(\mathfrak{g}, \mathfrak{h}) \subset \mathrm{Aut}(\mathfrak{g}),$

still acting as automorphisms of the root system. (We omit the definition in general. In the special case that every component of G has an element normalizing $\mathfrak{h}$, then W^G is generated by $W(\mathfrak{g}, \mathfrak{h})$ and the automorphisms coming from $\mathrm{Ad}(G)$.) The group W^G contains $W(\mathfrak{g}, \mathfrak{h})$ as a normal subgroup, and there is a natural surjective homomorphism

(7.23)(f) $\hspace{3.5cm}$ $G/G_0 \twoheadrightarrow W^G(\mathfrak{g}, \mathfrak{h})/W(\mathfrak{g}, \mathfrak{h}).$

Now G/G_0 acts on $\mathfrak{Z}(\mathfrak{g})$, and W^G/W acts on $S(\mathfrak{h})^W$. These two actions are compatible via the Harish-Chandra isomorphism of (7.23)(e) and (7.23)(f). In particular, we get an isomorphism

(7.23)(g) $\hspace{3.5cm}$ $\xi^G : \mathfrak{Z}^G(\mathfrak{g}) \xrightarrow{\sim} S(\mathfrak{h})^{W^G(\mathfrak{g}, \mathfrak{h})}.$

It follows from (7.23) that there is a bijection

(7.24)(a) $\quad$ (algebra homomorphisms $\mathfrak{Z}^G(\mathfrak{g}) \to \mathbb{C}$) $\quad \longleftrightarrow \quad$ $\mathfrak{h}^*/W^G(\mathfrak{g}, \mathfrak{h}).$

The connection with representation theory is this. On any irreducible admissible representation (π, U) of G, the algebra $\mathfrak{Z}^G(\mathfrak{g})$ must act by scalars. Consequently there is an element

(7.24)(b) $\hspace{4cm}$ $\lambda = \lambda(\pi) \in \mathfrak{h}^*$

(defined up to the action of W^G) with the property that

(7.24)(c) $\hspace{3cm}$ $\pi(z) = \xi^G(z)(\lambda) \qquad (z \in \mathfrak{Z}^G(\mathfrak{g}).$

We call $\lambda(\pi)$ the *infinitesimal character* of π.

$\quad$ The notion of dominance that we need for the representation τ will be defined in terms of the infinitesimal character of τ. To put the result in context, here is a basic fact about how the Dolbeault cohomology construction of Theorem 7.21 affects infinitesimal characters.

Proposition 7.25. *In the setting of Theorem 7.21, assume that the $(\mathfrak{q}, L)$-representation (τ, V) has infinitesimal character $\lambda_L(\tau) \in \mathfrak{h}^*$. Write $n = \dim_{\mathbb{C}} X$, and $\rho(\mathfrak{u}) \in \mathfrak{h}^*$ for the restriction of the linear functional in (7.22)(e). Then each G-representation $H^{0,q}(X, \mathcal{V})$ has infinitesimal character $\lambda_L - \rho(\mathfrak{u})$, and each G representation $H^{n,q}(X, \mathcal{V})$ has infinitesimal character $\lambda_L + \rho(\mathfrak{u})$.*

The second assertion (about (n, q)-cohomology) is an immediate consequence of the first and (7.18)(b): tensoring a representation of L with $\bigwedge^n(\mathfrak{g}/\mathfrak{q})^*$ adds $2\rho(\mathfrak{u})$ to its infinitesimal character. The first assertion is a version of the Casselman-Osborne theorem relating the action of $\mathfrak{Z}^G(\mathfrak{g})$ to cohomology. (One can use the description (7.18)(b) of Dolbeault cohomology. The action of $\mathfrak{Z}^G(\mathfrak{g})$ in that picture is by differentiation on the left on functions on G. Because we are considering central elements, this is equal to differentiation on the right, which is where the $(\mathfrak{q}, L)$-cohomology is computed. We omit the elementary details.)

Of course the weight λ_L in Proposition 7.25 is defined only up to $W^L(\mathfrak{l}, \mathfrak{h})$.

Definition 7.26. *Suppose $\mathfrak{q} = \mathfrak{l} + \mathfrak{u}$ is a Levi decomposition of a parabolic subalgebra in the complex reductive Lie algebra $\mathfrak{g}$, and $\mathfrak{h} \subset \mathfrak{l}$ is a Cartan subalgebra. A weight $\lambda \in \mathfrak{h}^*$ is called* weakly dominant *with respect to $\mathfrak{u}$ if for every coroot $\alpha^\vee$ corresponding to a root of $\mathfrak{h}$ in $\mathfrak{u}$, $\langle \alpha^\vee, \lambda \rangle$ is not a strictly negative real number. That is,*

$$\langle \alpha^\vee, \lambda \rangle \geqslant 0 \qquad or \qquad \langle \alpha^\vee, \lambda \rangle \text{ is not real.}$$

We say that λ is strictly dominant *if (still for every such coroot)*

$$\langle \alpha^\vee, \lambda \rangle > 0 \qquad or \qquad \langle \alpha^\vee, \lambda \rangle \text{ is not real.}$$

The set of coroots $\alpha^\vee$ for roots of $\mathfrak{h}$ in $\mathfrak{u}$ is permuted by $W^L(\mathfrak{l}, \mathfrak{h})$, so these condtions depend only on the $W^L(\mathfrak{l}, \mathfrak{h})$-orbit of λ.

The terminology here is far from standard. One common variant is to require only that $\langle \alpha^\vee, \lambda \rangle$ never be a negative integer. That kind of hypothesis is not sufficient for the assertions about unitarity in Theorem 7.27.

Theorem 7.27. *In the setting of Proposition 7.25, assume also that $\lambda_L + \rho(\mathfrak{u})$ is weakly dominant for $\mathfrak{u}$ (Definition 7.26). Recall that $Z = K/L \cap K \subset X$ is a compact complex subvariety, and set $s = \dim_{\mathbb{C}}(Z)$.*

(1) $H^{n,q}(X, \mathcal{V}) = 0$ unless $q = s$.
(2) If $L = L_{max}$ (Definition 6.5) and V is an irreducible representation of L, then $H^{n,s}(X, \mathcal{V})$ is irreducible or zero.
(3) If the Harish-Chandra module of V admits an invariant Hermitian form, then the Harish-Chandra module of $H^{n,s}(X, \mathcal{V})$ admits an invariant Hermitian form.
(4) If the Harish-Chandra module of V is unitary, then the Harish-Chandra module of $H^{n,s}(X, \mathcal{V})$ is unitary.

Suppose now that $\lambda_L + \rho(\mathfrak{u})$ is strictly dominant for $\mathfrak{u}$.

(5) *If $L = L_{max}$, then the representation V of L is irreducible if and only if $H^{n,s}(X, \mathcal{V})$ is irreducible or zero.*

(6) *The Harish-Chandra module of V admits an invariant Hermitian form if and only if the Harish-Chandra module of $H^{n,s}(X, \mathcal{V})$ admits an invariant Hermitian form.*

(7) *The Harish-Chandra module of V is unitary if and only if the Harish-Chandra module of $H^{n,s}(X, \mathcal{V})$ is unitary.*

This summarizes some of the main results of [Vog84], translated into the language of Dolbeault cohomology using [Won99].

Theorem 7.27 is in many respects a valuable analogue of the Borel-Weil theorem for noncompact groups. One annoying feature is that the statement does not contain the Borel-Weil theorem as a special case. If G is compact, then Theorem 7.27 concerns top degree cohomology and dominant V, whereas Theorem 2.3 concerns degree zero cohomology and antidominant V. In order to round out the motivation appropriately, here is an alternate version of Theorem 2.3 addressing this incompatibility.

Theorem 7.28. (Borel-Weil, Harish-Chandra; see [HC56], [Ser59]). *Suppose K is a compact connected Lie group with maximal torus T; use the notation of (2.1) and (2.2) above, and put $n = \dim_{\mathbb{C}}(K/T)$.*

(1) *The infinitesimal character of the representation $(\mu, \mathbb{C}_\mu)$ of T is given by the differential of $d\mu \in \mathfrak{t}^*$ of μ.*

(2) *The weight $d\mu + \rho$ is strictly dominant (Definition 7.26) if and only if μ is dominant in the sense of (2.2).*

(3) *The top degree Dolbeault cohomology $H^{n,n}(K/T, \mathcal{L}_\mu)$ is non-zero if and only if μ is dominant. In that case, the Dolbeault cohomology space is an irreducible representation of K.*

(4) *This correspondence defines a bijection from dominant characters of T onto $\widehat{K}$.*

Here we say that μ is dominant if and only if the inverse character $-\mu$ is antidominant (cf. (2.2)); that is, if and only if

$$\langle \mu, \alpha^\vee \rangle \geqslant 0$$

for every simple root α of T in K.

8 Compact Supports and Minimal Globalizations

Theorem 7.27 provides a large family of group representations with unitary Harish-Chandra modules. It is entirely natural to look for something like a pre-Hilbert space structure on these group representations, that might be completed to a unitary group representation. Theorem 7.21 guarantees that each representation provided by Theorem 7.27 is the maximal globalization of its Harish-Chandra module. As explained in the introduction, we will see in Theorem 9.16 that maximal globalizations never admit G-invariant pre-Hilbert space structures (unless they are finite-dimensional). We need something analogous to Theorem 7.21 that produces instead *minimal* globalizations. Because of Definition 4.23, this means that we need to identify the dual of the topological vector space $H^{p,q}(X, \mathcal{V})$. Let us first examine this in the setting of Definition 7.10, with $\mathcal{V}$ a holomorphic vector bundle on the complex manifold X. The definition involves the topological vector spaces $A^{p,q}(\mathcal{V})$ of smooth (p, q)-forms on X with values in $\mathcal{V}$ (cf. (7.9)), and the Dolbeault operators

$$(8.1)(a) \qquad A^{p,q-1}(\mathcal{V}) \xrightarrow{\bar{\partial}} A^{p,q}(\mathcal{V}) \xrightarrow{\bar{\partial}} A^{p,q+1}(\mathcal{V}).$$

The first point is to identify the topological duals of these three spaces. The space $C_c^{-\infty}(\mathcal{W})$ of compactly supported distribution sections of a vector bundle $\mathcal{W}$ is by definition the topological dual of the space $C^{\infty}(\mathcal{W} \otimes \mathcal{D})$, with $\mathcal{D}$ the bundle of densities on the manifold. Because our manifold X is complex, it is orientable; so the bundle of densities is just the bundle of top degree differential forms on X. Top degree forms are (n, n)-forms (cf. (7.5)(b)), and it follows easily that

$$(8.1)(b) \qquad A^{p,q}(\mathcal{V})^* \simeq A_c^{(n-p,n-q),-\infty}(\mathcal{V}^*).$$

Any continuous linear map $T\colon E \to F$ between topological vector spaces has a transpose ${}^tT\colon F^* \to E^*$. The Dolbeault operators in (8.1)(a) therefore give rise to transposes
(8.1)(c)

$$A_c^{(n-p,n-q-1),-\infty}(\mathcal{V}^*) \xrightarrow{{}^t\bar{\partial}} A_c^{(n-p,n-q),-\infty}(\mathcal{V}^*) \xrightarrow{{}^t\bar{\partial}} A_c^{(n-p,n-q+1),-\infty}(\mathcal{V}^*).$$

Calculating in coordinates shows that (up to a sign depending on p and q, which according to [Ser55], page 19 is $(-1)^{p+q+1}$) this transpose map "is" just the $\bar{\partial}$ operator for the Dolbeault complex for $\mathcal{V}^*$, applied to compactly supported distribution sections. (The way this calculation is done is to regard compactly supported smooth forms $A_c^{(n-p,n-q+1),\infty}(\mathcal{V}^*)$ as linear functionals on $A^{p,q}(\mathcal{V})$, by pairing the $\mathcal{V}$ and $\mathcal{V}^*$ and integrating the resulting (compactly supported smooth) (n, n)-form over X. Comparing the effects of $\bar{\partial}$ and ${}^t\bar{\partial}$ on $A_c^{(n-p,n-q+1),\infty}(\mathcal{V}^*)$ now amounts to integrating by parts.)

Using this new complex, we can formulate an analogue of Definition 7.10.

Definition 8.2. *Suppose X is a complex manifold of complex dimension n, and $\mathcal{V}$ is a holomorphic vector bundle on X. The* compactly supported (p, q)-Dolbeault cohomology *of X with coefficients in $\mathcal{V}$ is by definition*

$$H_c^{p,q}(X, \mathcal{V}) =$$

$$(kernel\ of\ \bar{\partial}\ on\ A_c^{(p,q),-\infty}(\mathcal{V}))/(image\ of\ \bar{\partial}\ from\ A_c^{(p,q-1),-\infty}(\mathcal{V})).$$

At least if $\mathcal{V}$ is finite-dimensional, this is the Čech cohomology with compact supports of X with coefficients in the sheaf $\mathcal{O}_{\Omega_p \otimes \mathcal{V}}$ of holomorphic p-forms with values in $\mathcal{V}$. Just as in Proposition 7.19, there is a natural quotient topology on this cohomology, and we can define

$$H_{c,top}^{p,q}(X, \mathcal{V}) = maximal\ Hausdorff\ quotient\ of\ H_c^{p,q}(X, \mathcal{V})$$

$$= kernel\ of\ \bar{\partial}/closure\ of\ image\ of\ \bar{\partial}.$$

In order to discuss transposes and duality, we need to recall a little about topologies on the dual E^* of a complete locally convex space E. Details may be found for example in [Tre67], Chapter 19. For any subset $B \subset E$, and any $\epsilon > 0$, we can define

$$(8.3)(a) \qquad W_\epsilon(B) = \{\lambda \in E^* \mid \sup_{e \in B} |\lambda(e)| \leqslant \epsilon\} \subset E^*.$$

This is a subset of E^* containing 0. The topologies we want on E^* are defined by requiring certain of these subsets to be open. The *weak topology* on E^* is defined to have neighborhood basis at the origin consisting of the sets $W_\epsilon(B)$ with $B \subset E$ finite. (Another way to say this is that the weak topology is the coarsest one making all the evaluation maps $\lambda \mapsto \lambda(e)$ continuous.) We write E_{wk}^* for E^* endowed with the weak topology. (Treves writes E_σ^*, and Bourbaki writes E_s^*; more precisely, each uses a prime instead of a star for the continuous dual.) The *topology of compact convergence* on E^* is defined to have neighborhood basis at the origin consisting of the sets $W_\epsilon(B)$ with $B \subset E$ compact. We write E_{cpt}^* for this topological space; Treves writes E_c^*. The *strong topology* is defined to have neighborhood basis at the origin consisting of the sets $W_\epsilon(B)$ with $B \subset E$ bounded. (Recall that B is *bounded* if for every neighborhood U of 0 in E, there is a scalar $r \in \mathbb{R}$ so that $B \subset rU$.) We write E_{str}^* for this topological space; Treves and Bourbaki write E_b^*. Because a finite set is automatically compact, and a compact set is automatically bounded, it is clear that the topologies

$$(8.3)(b) \qquad\qquad weak,\ compact\ convergence,\ strong$$

are listed in increasing strength; that is, each has more open sets than the preceding ones. For any of these three topologies on dual spaces, the transpose of a continuous linear map is continuous ([Tre67], Corollary to Proposition 19.5). If E is a Banach space, then the usual Banach space structure on E^* defines the strong topology.

For most of the questions we will consider, statements about the strong topology on E^* are the strongest and most interesting. Here is an example. We can consider the double dual space $(E^*)^*$; what this is depends on the chosen topology on E^*. Strengthening the topology on E^* allows more continuous linear functionals, so

(8.4)(a) $$(E^*_{wk})^* \subset (E^*_{cpt})^* \subset (E^*_{str})^*.$$

Each of these spaces clearly includes E (the evaluation maps at an element of E being continuous on E^* in all of our topologies). In fact

(8.4)(b) $$(E^*_{wk})^* = E$$

([Tre67], Proposition 35.1); this equality is a statement about sets, not topologies. Asking for similar statements for the other two topologies on E^* asks for more; the most that one can ask is

Definition 8.5 . (see [Tre67], Definition 36.2). *The (complete locally-convex) topological vector space E is called* reflexive *if the natural inclusion*

$$E \hookrightarrow (E^*_{str})^*$$

is an isomorphism of topological vector spaces.

For us, reflexivity will arise in the following way.

Definition 8.6. (see [Tre67], Definition 34.2). *The (complete locally convex) topological vector space E is called a* Montel *space if every closed and bounded subset $B \subset E$ is compact.*

Proposition 8.7. ([Tre67], Corollary to Proposition 36.9, and Corollary 3 to Proposition 50.2). *A Montel space is reflexive. A complete nuclear space is Montel. In particular, the analytic, smooth, distribution, and hyperfunction globalizations of any finite-length Harish-Chandra module (cf. section 4) are all reflexive.*

Topological vector spaces that we define as dual spaces, like distribution spaces and the maximal globalization, will usually be endowed with the strong topology.

We are interested in the dual space of Dolbeault cohomology, which is a quotient of subspaces of a simple space of forms. We therefore need to know how to compute dual spaces of subspaces and quotients of topological vector spaces.

Proposition 8.8. *Suppose E is a complete locally convex topological vector space, and $M \subset E$ is a closed subspace. Endow M with the subspace topology, and E/M with the quotient topology (whose open sets are the images of the open sets in E.) Write $i\colon M \to E$ for the inclusion, $q\colon E \to E/M$ for the quotient map, and $M^\perp \subset E^*$ for the subspace of linear functionals vanishing on M.*

(1) *Every continuous linear functional λ_M on M (endowed with the subspace topology) extends to a continuous linear functional λ on E. That is, the transpose map*

$$^t i \colon E^* \to M^*$$

is surjective, with kernel equal to $M^\perp$.

(2) *Suppose that E is reflexive. Then the vector space isomorphism*

$$^t i \colon E^*/M^\perp \xrightarrow{\sim} M^*$$

is a homeomorphism from the quotient of the strong topology on E^* to the strong topology on M^*.

(3) *If E/M is endowed with the quotient topology, then the continuous linear functionals are precisely those on E that vanish on M. That is, the transpose map*

$$^t q \colon (E/M)^* \to E^*$$

is injective, with image equal to $M^\perp$.

(4) *Suppose that E and $M^\perp$ are reflexive. Then the vector space isomorphism*

$$^t q \colon (E/M)^* \xrightarrow{\sim} M^\perp$$

is a homeomorphism from the strong topology on $(E/M)^*$ onto the subspace topology on $M^\perp$ induced by the strong topology on E^*.

The first assertion is the Hahn-Banach Theorem (see for example [Tre67], Chapter 18). The second may be found in [Bou87], Corollary to Theorem 1 in section IV.2.2. The third is more or less obvious. For the fourth, applying the second assertion to $M^\perp \subset E^*$ gives a homeomorphism

$$(M^\perp)^*_{str} \simeq E/M.$$

Now take duals of both sides, and use the reflexivity of $M^\perp$.

Finally, we need a few general remarks about transpose maps (to be applied to $\overline{\partial}$). So suppose that

(8.9)(a) $$T \colon E \to F$$

is a continuous linear map of complete locally convex topological vector spaces, and

(8.9)(b) $$^t T \colon F^* \to E^*$$

is its transpose. The kernel of T is a closed subspace of E, so the quotient $E/\ker T$ is a complete locally convex space in the quotient topology. The image of T is a subspace of F, but not necessarily closed; its subspace topology is locally convex, and the completion of $\operatorname{im} T$ may be identified with its closure in F. We have a continuous bijection

(8.9)(c) $$E/\ker T \to \operatorname{im} T,$$

but this need not be a homeomorphism. We now have almost obvious identifications

(8.9)(d) $$\ker {}^tT = \text{linear functionals on } F \text{ vanishing on } \operatorname{im} T$$
$$= (F/\overline{\operatorname{im} T})^* \subset F^*$$

(8.9)(e) $$\operatorname{im} {}^tT = \text{linear functionals on } E \text{ vanishing on } \ker T,$$
$$\text{and extending continuously from } \operatorname{im} T \text{ to } F$$
$$= (\operatorname{im} T)^* = (\overline{\operatorname{im} T})^* \subset E^*$$

(8.9)(f) $$\overline{\operatorname{im} {}^tT} = \text{linear functionals on } E \text{ vanishing on } \ker T$$
$$= (E/\ker T)^* \subset E^*$$

The question of when these vector space isomorphisms respect topologies is addressed by Proposition 8.8.

Lemma 8.10. *In the setting of (8.9)(a), assume that the map (8.9)(c) is a homeomorphism. Then ${}^tT$ has closed range.*

This is immediate from the descriptions in (8.9)(e) and (8.9)(f), together with the Hahn-Banach theorem. A famous theorem of Banach gives a sufficient condition for (8.9)(c) to be a homeomorphism:

Theorem 8.11. ([Tre67], Theorem 17.1). *In the setting of (8.9)(a), assume that E and F are Fréchet spaces. Then (8.9)(c) is a homeomorphism if and only if T has closed range.*

We can now say something about duals of cohomology spaces.

Proposition 8.12. *Suppose that*

$$E \xrightarrow{T} F \xrightarrow{S} G$$

is a complex of continuous linear maps of complete locally convex topological vector spaces, so that $S \circ T = 0$. Define

$$H = \ker S/\operatorname{im} T,$$

endowed with the quotient topology. This may be non-Hausdorff, and we define the maximal Hausdorff quotient

$$H_{top} = \ker S/\overline{\operatorname{im} T},$$

a complete locally convex topological vector space. Define the transpose complex

$$G^* \overset{{}^tS}{\to} F^* \overset{{}^tT}{\to} E^*$$

with cohomology

$$^tH = \ker{}^tT / \operatorname{im}{}^tS$$

and maximal Hausdorff quotient

$$^tH_{top} = \ker{}^tT / \overline{\operatorname{im}{}^tS}.$$

(1) *There is a continuous linear bijection*

$$F^* / \overline{\operatorname{im}{}^tS} \to (\ker S)^*.$$

This is a homeomorphism if F is reflexive.

(2) *The map in (1) restricts to a continuous linear bijection*

$$^tH_{top} = \ker{}^tT / \overline{\operatorname{im}{}^tS} \to (\operatorname{im}T)^{\perp} \subset (\ker S)^*.$$

This is a homeomorphism if F is reflexive.

(3) *There is a continuous linear bijection*

$$H^* = (H_{top})^* \to (\operatorname{im}T)^{\perp} \subset (\ker S)^*.$$

This is a homeomorphism if $\ker S$ and $(\operatorname{im}T)^{\perp}$ are both reflexive.

(4) *There is a linear bijection*

$$H^* = (H_{top})^* \to {}^tH_{top}.$$

This is a homeomorphism if F, its subspace $\ker S$, and $(\operatorname{im}T)^{\perp} \subset (\ker S)^*$ are all reflexive.

(5) *Assume that E, F, and G are nuclear Fréchet spaces, and that*

$$H \simeq H_{top}, \qquad {}^tH \simeq {}^tH_{top}$$

are Hausdorff. The linear isomorphism

$$H^* = (H_{top})^* \simeq {}^tH_{top}$$

of (4) is a homeomorphism.

(6) *Assume that E, F, and G are nuclear Fréchet spaces, and that T and S have closed range. Then ^tT and ^tS also have closed range, so the cohomology spaces*

$$H \simeq H_{top}, \qquad {}^tH \simeq {}^tH_{top}$$

are Hausdorff. The linear isomorphism

$$H^* \simeq {}^tH$$

of (4) is a homeomorphism.

Proof. Parts (1)–(3) are more or less immediate from Proposition 8.8, in light of (8.9). Part (4) simply combines (2) and (3). For (5), we need to know the reflexivity of the three spaces mentioned in (4). A subspace of a nuclear space is nuclear, and therefore reflexive (Proposition 8.7). This shows that F and $\ker S$ are reflexive. The dual of a nuclear Fréchet space is nuclear ([Tre67], Proposition 50.6), so $(\ker S)^*$ is nuclear; so its closed subspace $(\operatorname{im} T)^\perp$ is nuclear, and therefore reflexive. For (6), the assertion about closed range follows from Banach's Theorem 8.11, and the fact that the cohomology is Hausdorff follows at once.

From these generalities in hand, we get immediately a description of the topological dual of Dolbeault cohomology.

Theorem 8.13. (Serre [Ser55], Théorème 2). *Suppose X is a complex manifold of dimension n, and $\mathcal{V}$ is a smooth holomorphic nuclear Fréchet vector bundle on X. Write $\mathcal{V}^*$ for the topological dual bundle. Write $H^{p,q}(X, \mathcal{V})$ for the (p, q) Dolbeault cohomology of X with coefficients in $\mathcal{V}$, endowed with the (possibly non-Hausdorff) topological vector space structure defined before Proposition 7.19, and $H^{p,q}_{top}(X, \mathcal{V})$ for its maximal Hausdorff quotient. Similarly define $H^{p,q}_c(X, \mathcal{V}^*)$, the Dolbeault cohomology with compact support (and generalized function coefficients), and its maximal Hausdorff quotient $H^{p,q}_{c,top}(X, \mathcal{V}^*)$ as in Definition 8.2. Then there is a natural topological isomorphism*

$$H^{p,q}_{top}(X, \mathcal{V})^* \simeq H^{n-p,n-q}_{c,top}(X, \mathcal{V}^*).$$

If the Dolbeault cohomology operators for $\mathcal{V}$ have closed range, then the same is true for the Dolbeault operators on compactly supported $\mathcal{V}^*$-valued forms with generalized function coefficients, and

$$H^{p,q}(X, \mathcal{V})^* \simeq H^{n-p,n-q}_c(X, \mathcal{V}^*).$$

The main point is that the space of smooth sections of a smooth nuclear Fréchet bundle is a nuclear Fréchet space; it is easy to imitate [Ser55], section 8, to define a countable collection of seminorms giving the topology. With this fact in hand, Theorem 8.13 is a special case of Proposition 8.12, (4)–(6) (together with (8.1) and Definition 8.2).

Corollary 8.14. (cf. Bratten [Bra97], Theorem on page 285). *In the setting of Definition 7.20, suppose X is the complex manifold G/L, and assume that the admissible representation V is the minimal globalization of the underlying $(\mathfrak{q}, L \cap K)$-module. Let $A^{p,q}_c(X, \mathcal{V})$ be the Dolbeault complex for $\mathcal{V}$ with generalized function coefficients of compact support (cf. (8.1)(c)). Then the $\bar{\partial}$ operator has closed range, so that each of the corresponding cohomology spaces $H^{p,q}_c(X, \mathcal{V})$ carries a smooth representation of G (on the dual of a nuclear Fréchet space). Each of these representations of G is admissible, and is the minimal globalization of its underlying Harish-Chandra module.*

This is immediate from Wong's Theorem 7.21, Serre's Theorem 8.13, and the duality relationship between minimal and maximal globalizations (Definitions 4.21 and 4.23). The theorem proved by Bratten is slightly different: he defines a "sheaf of germs of holomorphic sections" $\mathcal{A}(X, \mathcal{V})$, and proves a parallel result for the sheaf cohomology with compact support on X with coefficients in $\mathcal{A}(X, \mathcal{V})$. When V is finite-dimensional, the two results are exactly the same, since it is easy to check that Dolbeault cohomology (with compactly supported generalized function coefficients) computes sheaf cohomology in that case.

For infinite-dimensional V, comparing Corollary 8.14 with Bratten's results in [Bra97] is more difficult. In these notes I have avoided many subtleties by speaking only about the Dolbeault complex, and not about sheaf cohomology. Part of the point of page 317 of Bratten's paper is that I have in the past (for example in Conjecture 6.11 of [Vog87]) glossed over the difficulty of connecting sheaf and Dolbeault cohomology for infinite-dimensional bundles.

In the same way, we can translate Theorem 7.27 into this setting. For context, we should remark that Proposition 7.25 (computing infinitesimal characters of Dolbeault cohomology representations) applies equally to Dolbeault cohomology with compact support. The weight $\lambda_L - \rho(\mathfrak{u})$ appearing in the next corollary is therefore the infinitesimal character of the representation $H_c^{0,r}(X, \mathcal{V})$.

Corollary 8.15. *In the setting of Definition 7.20, recall that $Z = K/L \cap K$ is an s-dimensional compact complex submanifold of the n-dimensional complex manifold $X = G/L$. Write $r = n - s$ for the codimension of Z in X. Assume that V is an admissible $(\mathfrak{q}, L)$-module of infinitesimal character $\lambda_L \in \mathfrak{h}^*$ (cf. (7.24)), and that V is the minimal globalization of the underlying $(\mathfrak{q}, L \cap K)$-module. Assume that $\lambda_L - \rho(\mathfrak{u})$ is weakly antidominant for $\mathfrak{u}$; that is, that $-\lambda_L + \rho(\mathfrak{u})$ is weakly dominant. Then*

(1) $H_c^{0,q}(X, \mathcal{V}) = 0$ unless $q = r$.

(2) If $L = L_{max}$ (Definition 6.5) and V is an irreducible representation of L, then $H_c^{0,r}(X, \mathcal{V})$ is irreducible or zero.

(3) If the Harish-Chandra module of V admits an invariant Hermitian form, then the Harish-Chandra module of $H_c^{0,r}(X, \mathcal{V})$ admits an invariant Hermitian form.

(4) If the Harish-Chandra module of V is unitary, then the Harish-Chandra module of $H_c^{0,r}(X, \mathcal{V})$ is unitary.

Suppose now that $\lambda_L - \rho(\mathfrak{u})$ is strictly antidominant for $\mathfrak{u}$.

(5) If $L = L_{max}$, then the representation V of L is irreducible if and only if $H_c^{0,r}(X, \mathcal{V})$ is irreducible or zero.

(6) The Harish-Chandra module of V admits an invariant Hermitian form if and only if the Harish-Chandra module of $H_c^{0,r}(X, \mathcal{V})$ admits an invariant Hermitian form.

(7) The Harish-Chandra module of V is unitary if and only if the Harish-Chandra module of $H_c^{0,r}(X, \mathcal{V})$ is unitary.

These statements follow immediately from Theorem 7.27 and Theorem 8.13.

We will see in section 9 that the Hermitian forms of Corollary 8.15(6) automatically extend continuously to $H_c^{0,r}(X, \mathcal{V})$.

To conclude this section, notice that in the setting of the Borel-Weil Theorem (Theorem 2.3), we have $X = Z = K/T$, so $r = 0$; Theorem 2.3 is therefore "compatible" with Corollary 8.15.

9 Invariant Bilinear Forms and Maps between Representations

In Theorem 7.21 and Corollary 8.14, we have identified many representations with spaces related to smooth functions and distributions on manifolds. In this section, we will use these realizations to describe Hermitian forms on the representations. This is a three-step process. First, we will see (in Definition 9.6) how to

(9.1)(a) understand a Hermitian form on one representation as a special kind of linear map between two representations.

Describing Hermitian forms therefore becomes a special case of describing linear maps. For the representations we are considering, this amount to describing linear maps between function spaces. The second step (Theorem 9.8) is to

(9.1)(b) understand spaces of linear maps between function spaces as topological tensor products of function spaces.

The third step (which we will deal with more or less case by case) is to

(9.1)(c) understand tensor products of function spaces as function spaces on a product.

The second and third steps are closely connected to the Schwartz kernel theorem for distributions, and rely on the theory of nuclear spaces that Grothendieck developed to explain and generalize Schwartz's theorem.

Before embarking on the technical details, we record the elementary ideas that we will be trying to generalize. So suppose for a moment that A and B are finite sets, say with n elements and m elements respectively. Define

$$(9.2)(a) \qquad V_A = \{\text{complex-valued functions on } A\} \simeq \mathbb{C}^n,$$

$$(9.2)(b) \qquad V_A^* = \{\text{complex-valued measures on } A\} \simeq \mathbb{C}^n,$$

and similarly for B. The space V_A^* is naturally identified with the dual space of V_A (as the notation indicates), by

$$\lambda(f) = \int_A f \, d\lambda = \sum_{a \in A} f(a)\lambda(a);$$

in the second formula, the measure λ has been identified with the linear combination of delta functions (unit masses at points of A) $\sum \lambda(a)\delta_a$. For motivating the ideas above, we are meant to be thinking of V_A as the space of smooth functions on the manifold A, and of V_A^* as distributions on A. In this setting, a version of (9.1)(b) is

(9.2)(c) $\qquad\qquad \operatorname{Hom}_{\mathbb{C}}(V_A^*, V_B) \simeq V_A \otimes V_B.$

The natural map from right to left is

$$f \otimes g \to T_{f \otimes g}, \qquad T_{f \otimes g}(\lambda) = \lambda(f)g.$$

A version of (9.1)(c) is

(9.2)(d) $\qquad\qquad V_A \otimes V_B \simeq V_{A \times B}.$

The composite map

(9.2)(e) $\qquad\qquad V_{A \times B} \xrightarrow{\sim} \operatorname{Hom}_{\mathbb{C}}(V_A^*, V_B)$

is

$$h \to K_h, \qquad [K_h(\lambda)](b) = \int_A h(x, b) \, d\lambda(x).$$

The operator K_h is a kernel operator, and (9.2)(e) is an example of the Schwartz kernel theorem.

One lesson that can be extracted even from this very simple example is that some of the easiest linear maps to understand are those going *from* spaces of distributions *to* spaces of functions. By rearranging the example slightly, we could also have found a nice description of the linear maps *from* a space of functions *to* a space of distributions.

As a second kind of warming up, here are two versions of the Schwartz kernel theorem that we will be imitating in the steps (9.1)(b) and (9.1)(c) above. In order to state these theorems, we will follow Schwartz and write $\mathcal{D}'(M)$ for the space of distributions on the smooth manifold M with arbitrary support; that is, the continuous dual of $C_c^\infty(M)$. (Elsewhere we have written this as $C^{-\infty}(M, \mathcal{D})$, with $\mathcal{D}$ the bundle of smooth densities on M.)

Theorem 9.3. (Schwartz kernel theorem; see [Tre67], Theorem 51.7). *Suppose X and Y are smooth manifolds. Then the space $L(C_c^\infty(Y), \mathcal{D}'(X))$ (of continuous linear maps from compactly supported smooth functions on Y to distributions on X) may be identified with $\mathcal{D}'(X \times Y)$. The identification sends a distribution h on $X \times Y$ to the kernel operator*

$$K_h \colon C_0^\infty(Y) \to \mathcal{D}'(X), \qquad [K_h(\phi)](\psi) = h(\psi \otimes \phi).$$

Here on the left we are describing the distribution $K_h(\phi)$ by evaluating it on a test function $\psi \in C_c^\infty(X)$. On the right, we regard $\psi \otimes \phi$ as a test function on $X \times Y$ (to which the distribution h may be applied) by

$$(\psi \otimes \phi)(x, y) = \psi(x)\phi(y).$$

Formally, the kernel operator in the theorem may be written

$$K_h(\phi)(x) = \int_Y h(x, y)\phi(y).$$

This equation makes sense as written if $h = H(x,y)dx\,dy$, with dx and dy smooth measures on X and Y, and H a continuous function on $X \times Y$. In this case

$$K_h(\phi) = f(x)\,dx, \qquad f(x) = \int_Y H(x,y)\phi(y)\,dy.$$

Again following Schwartz, write $\mathcal{E}'(M)$ for the space of distributions with compact support (what we have written elsewhere as $C_c^{-\infty}(M, \mathcal{D})$.) For us a useful variant of the kernel theorem will be

Theorem 9.4. ([Tre67], page 533). *Suppose X and Y are smooth manifolds. Then the space $L(\mathcal{E}'(Y), C_c^\infty(X))$ (of continuous linear maps from compactly supported distributions on Y to smooth functions on X) may be identified with $C^\infty(X \times Y)$. The identification sends $h \in C^\infty(X \times Y)$ to the kernel operator*

$$K_h \colon \mathcal{E}'(Y) \to C^\infty(X), \qquad [K_h(\lambda)](x) = \lambda(h(x, \cdot)).$$

We begin now with the machinery of linear maps and invariant Hermitian forms.

Definition 9.5. *Suppose E and F are complete locally convex topological vector spaces. Write $L(E, F)$ for the vector space of continuous linear maps from E to F. There are a number of important topologies on $L(E, F)$, but (by virtue of omitting proofs) we will manage with only one: the strong topology of uniform convergence on bounded subsets of E (cf. [Tre67], page 337). (The definition is a straightforward generalization of the case $F = \mathbb{C}$ described in (8.3) above.) Write $L_{str}(E, F)$ for the topological vector space of linear maps with this topology. This is a locally convex space, and it is complete if E is bornological; this holds in particular if E is Fréchet or the dual of a nuclear Fréchet space.*

Definition 9.6. *Suppose E is a complete locally convex topological vector space. The Hermitian dual E^h of E consists of the continuous conjugate-linear functionals on E:*

$$E^h = \{\lambda \colon E \to \mathbb{C}, \lambda(av + bw) = \bar{a}\lambda(v) + \bar{b}\lambda(w) \quad (a, b \in \mathbb{C}, \; v, w \in E)\}.$$

These are the complex conjugates of the continuous linear functionals on E, so there is a conjugate-linear identification $E^* \simeq E^h$. We use this identification to topologize E^h (cf. (8.3)); most often we will be interested in the strong topology E^h_{str}. In particular, we use the strong topology to define the double Hermitian dual, and find a natural continuous linear embedding

$$E \hookrightarrow (E^h)^h,$$

which is a topological isomorphism exactly when E is reflexive.

Any continuous linear map $T: E \to F$ has a Hermitian transpose

$$T^h: F^h \to E^h, \qquad T^h(\lambda)(e) = \lambda(Te).$$

The map $T \to T^h$ is conjugate-linear. In case $S \in L(E, F^h)$, we will also write

$$S^h \in L(F, E^h)$$

for the restriction of the Hermitian transpose to $F \subset (F^h)^h$.

A Hermitian pairing *between E and F is a separately continuous map*

$$\langle , \rangle: E \times F \to \mathbb{C}$$

that is linear in the first variable and conjugate linear in the second. It is immediate that such pairings are naturally in bijection with $L(E, F^h)$. The correspondence is

$$\langle , \rangle_T \quad \leftrightarrow \quad T: E \to F^h, \qquad T(e)(f) = \langle e, f \rangle_T.$$

In case $E = F$, we say that the pairing is a Hermitian form *on E if in addition*

$$\langle e, f \rangle = \overline{\langle f, e \rangle}.$$

In terms of the corresponding linear map $T \in L(E, E^h)$, the condition is $T = T^h$. (Here we restrict T^h to $E \subset (E^h)^h$.) The Hermitian form is said to be positive definite *if*

$$\langle e, e \rangle > 0, \text{ all non-zero } e \in E.$$

Of course one can speak about bilinear pairings between E and F, which correspond to $L(E, F^*)$.

For tensor products we will make only a few general remarks, referring for details to [Tre67].

Definition 9.7. *Suppose E and F are complete locally convex topological vector spaces. A "topological tensor product" of E and F is defined by imposing on the algebraic tensor product $E \otimes F$ a locally convex topology, and completing with respect to that topology. We will be concerned only with the* projective *tensor product. If p is a seminorm on E and q a seminorm on F, then we can define a seminorm $p \otimes q$ on $E \otimes F$ by*

$$p \otimes q(x) = \inf_{x = \sum e_i \otimes f_i} \sum_i p(e_i) q(f_i) \qquad (x \in E \otimes F).$$

The projective topology on $E \otimes F$ is that defined by the family of seminorms $p \otimes q$, where p and q vary over seminorms defining the topologies of E and F. The projective tensor product of E and F is the completion in this topology; it is written

$$E \widehat{\otimes}_\pi F.$$

A characteristic property of this topology is that for any complete locally convex topological vector space G, $L(E \widehat{\otimes}_\pi F, G)$ may be identified with G-valued jointly continuous bilinear forms on $E \times F$.

Here is Grothendieck's general solution to the problem posed as (9.1)(b) above.

Theorem 9.8. ([Tre67], Proposition 50.5). *Suppose E and F are complete locally convex topological vector spaces. Assume that*

(1) E is barreled ([Tre67], page 346).
(2) E^* is nuclear and complete.

(Both of these conditions are automatic if E is nuclear Fréchet or the dual of a nuclear Fréchet space.) Then the natural isomorphism

$$E^* \otimes F \simeq \text{finite rank continuous linear maps from } E \text{ to } F$$

extends to a topological isomorphism

$$E^* \otimes_\pi F \simeq L_{str}(E, F).$$

To translate this into representation-theoretic language, we need a lemma.

Lemma 9.9. *Suppose X_1 and X_2 are Harish-Chandra modules of finite length for reductive groups G_1 and G_2.*

(1) $X_1 \otimes X_2$ is a Harish-Chandra module of finite length for $G_1 \times G_2$.
(2) The minimal globalization of $X_1 \otimes X_2$ is the projective tensor product of the minimal globalizations of X_1 and X_2:

$$X_1^\omega \otimes_\pi X_2^\omega \simeq (X_1 \otimes X_2)^\omega.$$

(3) The smooth globalization of $X_1 \otimes X_2$ is the projective tensor product of the smooth globalizations of X_1 and X_2:

$$X_1^\infty \otimes_\pi X_2^\infty \simeq (X_1 \otimes X_2)^\infty.$$

(4) The distribution globalization of $X_1 \otimes X_2$ is the projective tensor product of the distribution globalizations of X_1 and X_2:

$$X_1^{-\infty} \otimes_\pi X_2^{-\infty} \simeq (X_1 \otimes X_2)^{-\infty}.$$

(5) The maximal globalization of $X_1 \otimes X_2$ is the projective tensor product of the maximal globalizations of X_1 and X_2:

$$X_1^{-\omega} \otimes_\pi X_2^{-\omega} \simeq (X_1 \otimes X_2)^{-\omega}.$$

Proof. The assertion in (1) is elementary. For the rest, fix Hilbert space globalizations X_i^{Hilb}, with orthonormal bases $\{e_i^m\}$ of K_i-finite vectors. Then $X_1^{Hilb} \otimes_\pi X_2^{Hilb}$ is a Hilbert space globalization of $X_1 \otimes X_2$, with orthonormal basis $\{e_1^m \otimes e_2^n\}$. (The projective tensor product of two Hilbert spaces is topologically the same as the Hilbert space tensor product.) Now all of the canonical globalizations in sight are sequence spaces. For example,

$$X_1^\infty = \left\{ \sum a_m e_1^m \mid a_m \in \mathbb{C}, |a_m \cdot m^k| \leqslant C_k, \text{ all } k \geqslant 0 \right\}$$

(cf. Theorem 4.18). The assertion in (3) amounts to the statement that the projective tensor product of the space of rapidly decreasing sequences on $\mathbb{N}$ with itself is the space of rapidly decreasing sequences on $\mathbb{N} \times \mathbb{N}$. This is an easy exercise (using the seminorms implicit in the definition of rapidly decreasing; compare [Tre67], Theorem 51.5). The remaining cases can be treated in exactly the same way.

Here is an abstract representation-theoretic version of the Schwartz kernel theorem.

Corollary 9.10. *Suppose X_1 and X_2 are Harish-Chandra modules of finite length for G. Write X_1^{dual} for the K-finite dual Harish-Chandra module, X_1^∞ for its smooth globalization, and so on as in section 4.*

(1) There is a natural identification

$$\operatorname{Hom}_{K \times K\text{-finite}}(X_1, X_2) \simeq X_1^{dual} \otimes X_2.$$

This is a Harish-Chandra module of finite length for $G \times G$.

(2) There is a natural identification (as representations of $G \times G$)

$$L_{str}(X_1^\omega, X_2^{-\omega}) \simeq (X_1^{dual})^{-\omega} \otimes_\pi X_2^{-\omega} \simeq (X_1^{dual} \otimes X_2)^{-\omega}.$$

That is, the space of continuous linear maps from the minimal globalization of X_1 to the maximal globalization of X_2 may be identified with the maximal globalization of a Harish-Chandra module for $G \times G$.

(3) There is a natural identification (as representations of $G \times G$)

$$L_{str}(X_1^\infty, X_2^{-\infty}) \simeq (X_1^{dual})^{-\infty} \otimes_\pi X_2^{-\infty} \simeq (X_1^{dual} \otimes X_2)^{-\infty}.$$

That is, the space of continuous linear maps from the smooth globalization of X_1 to the distribution globalization of X_2 may be identified with the distribution globalization of a Harish-Chandra module for $G \times G$.

Proof.   The assertions in (1) are elementary. The first isomorphism in (2) is Theorem 9.8 (bearing in mind the fact from Definition 4.23 that $(X^\omega)^* = (X^{dual})^{-\omega})$. The second is Lemma 9.9(2). Part (3) is identical.

We want to express this corollary more geometrically in the presence of geometric realizations of the representations X_i. As a warmup, we consider the situation of Proposition 5.11.

Corollary 9.11. *Suppose P_1 and P_2 are parabolic subgroups of the reductive groups G_1 and G_2 (Definition 5.1), and that E_i is an admissible Harish-Chandra module for P_i (Definition 5.6). Write*

$$X_i = (\text{Ind}_{P_i}^{G_i})^{K_i}(E_i)$$

for the induced Harish-Chandra module for G_i as in (5.12), and describe their various canonical globalizations as in (5.15).

(1) The space $L_{str}(X_1^\omega, X_2^{-\omega})$ of maps from the minimal globalization to the maximal one may be identified with

$$(\text{Ind}_{P_1 \times P_2}^{G_1 \times G_2})^{-\omega}(L_{str}(E_1^\omega, E_2^{-\omega})),$$

the space of hyperfunction sections of the bundle on $(G_1 \times G_2)/(P_1 \times P_2)$ induced by the corresponding space of linear maps between representations of P_i.

(2) The space $L_{str}(X_1^\infty, X_2^{-\infty})$ of maps from the smooth globalization to the distribution one may be identified with

$$(\text{Ind}_{P_1 \times P_2}^{G_1 \times G_2})^{-\infty}(L_{str}(E_1^\infty, E_2^{-\infty})),$$

the space of distribution sections of the bundle on $(G_1 \times G_2)/(P_1 \times P_2)$ induced by the corresponding space of linear maps between representations of P_i.

The second observation was first made by Bruhat, who used it to begin the analysis of reducibility of induced representations. Here is the idea.

Corollary 9.12. (Bruhat [Bru56], Théorème 6;1). *In the setting of Corollary 9.11, assume that $G_1 = G_2 = G$. The the space of G-intertwining operators from X_1^∞ to $X_2^{-\infty}$ may be identified with the space of G_Δ-invariant generalized function sections of the bundle on $(G \times G)/(P_1 \times P_2)$ induced by $L_{str}(E_1^\infty, E_2^{-\infty})$. This space can in turn be identified with the space of continuous linear maps*

$$\text{Hom}_{P_1}(E_1^\infty, X_2^{-\infty}),$$

or with

$$\text{Hom}_{P_2}(X_1^\infty, E_2^{-\infty}).$$

The first displayed formula (which is a version of Frobenius reciprocity) identifies intertwining operators with distributions on G/P_2 having a certain transformation property under P_1 on acting on the left. Bruhat proceeds to analyze such distributions using the (finite) decomposition of G/P_1 into P_2 orbits; equivalently, using the (finite) decomposition of $(G \times G)/(P_1 \times P_2)$ into G_Δ orbits.

Here is the corresponding result for Dolbeault cohomology.

Corollary 9.13. *Suppose* $\mathfrak{q}_1$ *and* $\mathfrak{q}_2$ *are very nice parabolic subalgebras for the reductive groups* G_1 *and* G_2, *with Levi factors* L_1 *and* L_2 *(Definition 6.5), and that* E_i *is an admissible* $(\mathfrak{q}_i, L_i \cap K_i)$-*module (Definition 7.20). Set*

$$n_i = \dim_{\mathbb{C}} Y_i = G_i/L_i.$$

Write E_1^ω *for the minimal globalization of* E_1, *and* $E_2^{-\omega}$ *for the maximal globalization of* E_2. *These define holomorphic vector bundles*

$$\mathcal{E}_1^\omega \to Y_1 = G_1/L_1, \qquad \mathcal{E}_2^{-\omega} \to Y_2 = G_2/L_2.$$

Define

$$X_1^{p,\omega} = H_c^{0,n_1-p}(Y_1, \mathcal{E}_1^\omega)$$

(Definition 8.2), the compactly supported Dolbeault cohomology of Y_1 *with coefficients in* $\mathcal{E}_1^\omega$. *This is an admissible representation of* G_1, *the minimal globalization of the underlying Harish-Chandra module* X_1^{p,K_1} *(Corollary 8.14). Similarly, define*

$$X_2^{q,-\omega} = H^{n_2,q}(Y_2, \mathcal{E}_2^{-\omega})$$

(Definition 8.2), the Dolbeault cohomology of Y_2 *with coefficients in* $\mathcal{E}_2^{-\omega}$. *This is an admissible representation of* G_2, *the maximal globalization of the underlying Harish-Chandra module* X_2^{q,K_2} *(Theorem 7.21).*

(1) The space of continuous linear maps

$$E_{12}^{-\omega} = L_{str}(E_1^\omega, E_2^{-\omega})$$

is an admissible $(\mathfrak{q}_1 \times \mathfrak{q}_2, L_1 \times L_2)$-*representation (Definition 7.20), and is the maximal globalization of its underlying Harish-Chandra module. Write*

$$\mathcal{E}_{12}^{-\omega} \to Y_1 \times Y_2$$

for the corresponding holomorphic bundle.

(2) There is a natural identification

$$L_{str}(X_1^{\cdot,\omega}, X_2^{\cdot,-\omega}) \simeq H^{n_1+n_2,\cdot}(Y_1 \times Y_2, \mathcal{E}_{12}^{-\omega}).$$

Here in each case the dot $\cdot$ *indicates that one should sum over the possible indices in question. More precisely,*

$$\sum_{p+q=m} L_{str}(X_1^{p,\omega}, X_2^{q,-\omega}) \simeq H^{n_1+n_2,m}(Y_1 \times Y_2, \mathcal{E}_{12}^{-\omega}).$$

Proof. The cohomology on $Y_1 \times Y_2$ is computed using a complex of forms

$$\mathcal{A}^{n_1+n_2,m}(Y_1 \times Y_2, \mathcal{E}_{12}^{-\omega}).$$

The fibers of $\mathcal{E}_{12}^{-\omega}$ are tensor products

$$(E_1^{dual})^{-\omega} \otimes E_2^{-\omega}$$

(Corollary 9.10(2)). Using this fact, the group-equivariant description of forms in Proposition 7.17, and standard ideas about tensor products of function spaces (cf. [Tre67], Theorem 51.6), one can prove that

$$\mathcal{A}^{n_1+n_2,m}(Y_1 \times Y_2, \mathcal{E}_{12}^{-\omega}) \simeq \sum_{p+q=m} \mathcal{A}^{n_1,p}(Y_1, \mathcal{E}_1^{dual,-\omega}) \otimes_\pi \mathcal{A}^{n_2,q}(Y_2, \mathcal{E}_2^{-\omega})$$

That is, the complex for Dolbeault cohomology on $Y_1 \times Y_2$ is the projective tensor product of the complexes for Y_1 and Y_2. Now one needs a Künneth formula for tensor products of nice complexes. (Recall that we know from Wong's Theorem 7.21 that the $\bar{\partial}$ operators are topological homomorphisms, and all the spaces here are nuclear Fréchet.) We leave this step as an exercise for the reader.

Corollary 9.14. *In the setting of Corollary 9.13, assume that $G_1 = G_2 = G$. Then the space of G-intertwining operators*

$$\sum_{p+q=m} \mathrm{Hom}_G(X_1^{p,\omega}, X_2^{q,-\omega})$$

may be identified with the space of G_Δ-invariant Dolbeault cohomology classes in $H^{n_1+n_2,m}(Y_1 \times Y_2, \mathcal{E}_{12}^{-\omega})$.

Because the coefficient bundle $\mathcal{E}_{12}^{-\omega}$ is a tensor product, the Dolbeault cohomology on $Y_1 \times Y_2$ has a natural quadruple grading; that is, each term of the bidegree has a bidegree, reflecting the degrees on Y_1 and Y_2. We could therefore write

$$(9.15) \qquad \mathrm{Hom}_G(X_1^{p,\omega}, X_2^{q,-\omega}) \simeq H^{(n_1,n_2),(p,q)}(Y_1 \times Y_2, \mathcal{E}_{12}^{-\omega}).$$

Just as in the setting of Corollary 9.12, the group G_Δ acts on $Y_1 \times Y_2$ with finitely many orbits. Everything about the analysis of this setting is slightly more complicated than in Corollary 9.12; even the Frobenius reciprocity isomorphisms described there are replaced by spectral sequences. Nevertheless one should be able to find some reasonable and interesting statements. We leave this task to the reader (with some suggestions in section 10).

It is now a simple matter to apply this result to the description of invariant Hermitian forms on Dolbeault cohomology representations. We begin with some general facts about Hermitian forms on representations.

Theorem 9.16. *Suppose X^K is a Harish-Chandra module of finite length for G. Write $X^{dual,K}$ for the K-finite dual Harish-Chandra module, and $X^{herm,K}$ for the K-finite Hermitian dual; this is the same real vector space as $X^{dual,K}$, with the conjugate complex structure. Write X^ω and $(X^{herm})^\omega$ for the minimal globalizations, and so on as in section 4.*

(1) *The algebraic Hermitian dual of X is isomorphic to $(X^{herm})^{-K}$. Accordingly there is a natural identification*

$$\mathrm{Hom}(X, (X^{herm})^{-K}) \simeq \text{(Hermitian pairings on } X).$$

The conjugate linear automorphism of order two given by Hermitian transpose corresponds to interchanging variables and taking complex conjugate on Hermitian pairings. Hermitian forms on X correspond to the fixed points of this automorphism. We have

$$\mathrm{Hom}_{\mathfrak{g},K}(X, (X^{herm})^{-K}) \simeq \text{(invariant Hermitian pairings on } X).$$

Any linear map on the left must take values in $(X^{herm})^K$.

(2) *There is a natural identification of the continuous Hermitian dual*

$$(X^\omega)^h = (X^{herm})^{-\omega}.$$

Accordingly there is a natural identification

$$L_{str}(X^\omega, (X^{herm})^{-\omega}) \simeq \text{(continuous Hermitian pairings on } X^\omega.)$$

This restricts to

$$\mathrm{Hom}_{G,cont}(X^\omega, (X^{herm})^{-\omega}) \simeq \text{(invariant Hermitian pairings on } X^\omega).$$

Any linear map on the left must take values in $(X^{herm})^\omega$.

(3) *There is a natural identification of the continuous Hermitian dual*

$$(X^\infty)^h = (X^{herm})^{-\infty}.$$

Accordingly there is a natural identification

$$L_{str}(X^\infty, (X^{herm})^{-\infty}) \simeq \text{(continuous Hermitian pairings on } X^\infty).$$

This restricts to

$$\mathrm{Hom}_{G,cont}(X^\infty, (X^{herm})^{-\infty}) \simeq \text{(invariant Hermitian pairings on } X^\infty).$$

Any linear map on the left must take values in $(X^{herm})^\infty$.

(4) *Restriction of linear transformations defines isomorphisms of the (finite-dimensional) spaces*

$$\mathrm{Hom}_{G,cont}(X^\infty, (X^{herm})^{-\infty}) \simeq \mathrm{Hom}_{G,cont}(X^\omega, (X^{herm})^{-\omega})$$
$$\simeq \mathrm{Hom}_{\mathfrak{g},K}(X, (X^{herm})^{-K}).$$

(5) *Any* $(\mathfrak{g}, K)$-*invariant Hermitian form on the admissible Harish-Chandra module* X^K *extends continuously to the minimal and smooth globalizations* X^ω *and* X^∞.

(6) *Assume that* X^K *is irreducible. Then* X^K *admits a non-zero invariant Hermitian form if and only if* X^K *is equivalent to the (irreducible) Harish-Chandra module* $(X^{herm})^K$. *Such a form has a unique continuous extension to* X^ω *and to* X^∞; *it has no continous extension to* $X^{-\infty}$ *or to* $X^{-\omega}$ *unless* X^K *is finite-dimensional.*

In all cases "continuous" Hermitian pairing means "separately continuous." It turns out that the separately continuous forms here are automatically continuous; see for example [Tre67], Theorem 41.1.)

Proof.   All the assertions in (1) are easy. The first assertion in (2) is essentially Definition 4.23 (with some complex conjugations inserted). The second then follows from the remarks in Definition 9.7. The third isomorphism is an obvious consequence. The final assertion in (2) is a special case of the "functoriality of minimal globalization" established in [Sch85]. Part (3) is proved in exactly the same way, using the Casselman-Wallach results. For part (4), we can change $-\infty$ to ∞, $-\omega$ to ω, and $-K$ to K by parts (1), (2), and (3). Then these isomorphisms are again "functoriality of globalization." Part (5) restates (4) using the facts in (2) and (3). For part (6), the irreducibility of $(X^{herm})^K$ is elementary, so the assertion about forms on X^K amounts to (1) and Schur's lemma. The existence of extensions to X^ω and to X^∞ is (5).

For the non-existence of extensions to (say) $X^{-\infty}$, one can prove exactly as in (3) that invariant Hermitian pairings on $X^{-\infty}$ correspond to continuous G-equivariant linear maps

$$\mathrm{Hom}_{G,cont}(X^{-\infty}, (X^{herm})^\infty).$$

A G-map of admissible group representations must restrict to a $(\mathfrak{g}, K)$-map of the underlying Harish-Chandra modules; and in our setting that map (if it is non-zero) has to be an isomorphism. From section 4, it follows that a non-zero map T must restrict to an isomorphism

$$T^\infty : X^\infty \to (X^{herm})^\infty.$$

The sequence space descriptions of the globalizations in section 4 show that such an isomorphism cannot extend continuously to $X^{-\infty}$: a sequence (x_μ) would necessarily (by continuity) map to the sequence $(T^\infty(x_\mu))$. This sequence is rapidly decreasing if and only if (x_μ) is rapidly decreasing (by the Casselman-Wallach uniqueness theorem for smooth globalization). If X^K is infinite-dimensional, then $X^{-\infty}$ must include slowly increasing sequences (x_μ) that are not rapidly decreasing, so $T(x_\mu)$ cannot be defined. The argument for $X^{-\omega}$ is identical.

We turn finally to Hermitian forms on Dolbeault cohomology representations. So suppose $\mathfrak{q} = \mathfrak{l} + \mathfrak{u}$ is a very nice parabolic subalgebra for the reductive group G, with Levi factor L. Put

(9.17)(a) $$Y = G/L, \qquad n = \dim_{\mathbb{C}} Y.$$

Suppose E is an admissible $(\mathfrak{q}, L \cap K)$-module (Definition 7.20), with minimal globalization E^ω. Define

(9.17)(b) $$E^{herm} = L \cap K\text{-finite Hermitian dual of } E,$$

(cf. Definition 9.6); this is naturally an admissible $(\bar{\mathfrak{q}}, L \cap K)$-module. By Theorem 9.16(2), the maximal globalization of E^{herm} is precisely the (continuous) Hermitian dual of E^ω:

(9.17)(c) $$(E^{herm})^{-\omega} \simeq (E^\omega)^h.$$

Define

(9.17)(d) $$F = \mathrm{Hom}_{L \cap K \times L \cap K\text{-finite}}(E, E^{herm}) \simeq E^{dual} \otimes E^h,$$

a space of Hermitian forms on E (cf. Definition 9.6 and Corollary 9.10(1)). This is an admissible $(\mathfrak{q} \times \bar{\mathfrak{q}}, L \cap K \times L \cap K)$-module. Its maximal globalization is

(9.17)(e) $$F^{-\omega} = L_{str}(E^\omega, (E^\omega)^h) \simeq (E^{dual})^{-\omega} \otimes_\pi (E^h)^{-\omega},$$

the space of (separately continuous) Hermitian pairings on the minimal globalization E^ω (cf. Theorem 9.16(2) and Corollary 9.10(2)). The space $F^{-\omega}$ carries a conjugate-linear involution that we will write as bar. On Hermitian pairings τ, it is defined by

(9.17)(f) $$\bar{\tau}(e, f) = \overline{\tau(f, e)}.$$

On linear maps, it is Hermitian transpose (Definition 9.6). In the tensor product (the last isomorphism of (9.17)(e)) it simply interchanges the factors. (This makes sense because E^h is the same real vector space as E^{dual}, with the opposite complex structure. With respect to the $\mathfrak{q} \times \bar{\mathfrak{q}}$ action, we have

(9.17)(g) $$\overline{(X, Y) \cdot \tau} = (\bar{Y}, \bar{X}) \cdot \bar{\tau} \qquad (X \in \mathfrak{q}, Y \in \bar{\mathfrak{q}}).$$

There is a similar formula for the $L \times L$ representation.

Corollary 9.18. *In the setting of (9.17), define G representations*

$$X^{p,\omega} = H_c^{0,n-p}(Y, \mathcal{E}^\omega),$$

which are minimal globalizations of the underlying Harish-Chandra modules X^p. Write Y^{op} for G/L with the opposite complex structure (defined by $\bar{\mathfrak{q}}$ instead of $\mathfrak{q}$).

(1) The Hermitian dual of $X^{p,\omega}$ is

$$(X^{p,h})^{-\omega} \simeq H^{n,p}(Y^{op}, (\mathcal{E}^h)^{-\omega}).$$

(2) The space of separately continuous Hermitian pairings on $X^{p,\omega}$ is

$$H^{(n,n),(p,p)}(Y \times Y^{op}, \mathcal{F}^{-\omega});$$

here the coefficient bundle is induced by the representation $F^{-\omega}$ of (9.17)(e).
(3) The space of G-invariant Hermitian forms on $X^{p,\omega}$ may be identified with the space of G_Δ-invariant real Dolbeault cohomology classes in

$$H^{(n,n),(p,p)}(Y \times Y^{op}, \mathcal{F}^{-\omega}).$$

Almost all of this is a formal consequence of Corollary 9.14, Theorem 9.16, and the definitions. One point that requires comment is the reference to "real" cohomology classes in (3). Suppose X is any complex manifold, and X^{op} the opposite complex manifold: this is the same as X as a smooth manifold, and the holomorphic vector fields on one are the antiholomorphic vector fields on the other. Complex conjugation carries (p, q)-forms on X to (p, q) forms on X^{op}, and respects $\bar{\partial}$. Therefore complex conjugation defines a conjugate-linear isomorphism of order two

$$H^{p,q}(X) \simeq H^{p,q}(X^{op}).$$

Beginning with this idea, and the automorphism bar of $F^{-\omega}$, one finds a conjugate linear isomorphism of order two

$$H^{(a,b),(c,d)}(Y \times Y^{op}, \mathcal{F}^{-\omega}) \to H^{(b,a),(d,c)}(Y \times Y^{op}, \mathcal{F}^{-\omega}).$$

(The terms like (a, b) in the bidegree are transposed when we use the isomorphism

$$(Y \times Y^{op})^{op} \simeq Y \times Y^{op},$$

which interchanges the factors.) A "real" cohomology class is one fixed by this isomorphism.

What do the identifications of Hermitian pairings in Corollary 9.18 look like? An element

(9.19)(a) $$v \in X^{p,\omega} = H_c^{0,n-p}(Y, \mathcal{E}^\omega)$$

is represented by a compactly supported $(0, n-p)$-form $\tilde{v}$ on Y, with values in the bundle $\mathcal{E}^\omega$. (I will write $\tilde{v}$ as if it were a smooth function, even though it actually has generalized function coefficients.) We can identify $\tilde{v}$ as a function

(9.19)(b) $$G \to \operatorname{Hom}\left(\bigwedge\nolimits^{n-p} \bar{\mathfrak{u}}, E^\omega\right),$$

satisfying a transformation law on the right under L (cf. Proposition 7.17). If $\tilde{w}$ is another such representative, then $\tilde{v} \otimes \overline{\tilde{w}}$ is a function

(9.19)(c) $$G \times G \to \operatorname{Hom}\left(\bigwedge\nolimits^{n-p} \bar{\mathfrak{u}} \otimes \bigwedge\nolimits^{n-p} \mathfrak{u}, E^\omega \otimes \overline{E^\omega}\right).$$

Suppose now that τ is a cohomology class as in Corollary 9.18(2). A representative $\tilde{\tau}$ may be identified with a smooth map

$$(9.19)(d) \qquad G \times G \to \mathrm{Hom}\left(\bigwedge^{p} \bar{\mathfrak{u}} \otimes \bigwedge^{p} \mathfrak{u}, F^{-\omega}\right),$$

with $F^{-\omega}$ the space of continuous Hermitian pairings on E^{ω}. Consequently the formal product $\tilde{\tau} \wedge (\tilde{v} \otimes \tilde{w})$ is a $(2n, 2n)$ form on $Y \times Y^{op}$ taking values in

$$(9.19)(e) \qquad F^{-\omega} \otimes E^{\omega} \otimes \overline{E^{\omega}}$$

At each point of G, we can apply the form value to the two vector values:

$$(9.19)(f) \qquad F^{-\omega} \otimes E^{\omega} \otimes \overline{E^{\omega}} \to \mathbb{C}, \qquad \phi \otimes e \otimes \overline{f} \mapsto \phi(e, f).$$

This defines a complex-valued $(2n, 2n)$-form that we might sensibly denote $\tilde{\tau}(\tilde{v}, \tilde{w})$. This form is compactly supported because v and w are. It has generalized function coefficients, meaning that it is defined as an element of the dual space of smooth functions on $Y \times Y^{op}$. We may therefore integrate it (that is, pair it with the function 1) and define

$$(9.19)(g) \qquad \langle v, w \rangle_{\tau} = \int_{Y \times Y^{op}} \tilde{\tau}(\tilde{v}, \tilde{w})$$

This is the identification in Corollary 9.18(2).

Here is a construction of unitary representations.

Corollary 9.20. *In the setting of Definition 7.20, recall that $Z = K/L \cap K$ is an s-dimensional compact complex submanifold of the n-dimensional complex manifold $X = G/L$. Write $r = n - s$ for the codimension of Z in X. Assume that $V^{(2)}$ is an irreducible unitary representation of L of infinitesimal character $\lambda_L \in \mathfrak{h}^*$ (cf. (7.24)). Write V^{ω} for the subspace of analytic vectors in $V^{(2)}$. Regard V^{ω} as a $(\mathfrak{q}, L)$-module by making $\mathfrak{u}$ act by zero, and let $\mathcal{V}^{\omega}$ be the corresponding holomorphic bundle on X. Assume that $\lambda_L - \rho(\mathfrak{u})$ is weakly antidominant for $\mathfrak{u}$; that is, that $-\lambda_L + \rho(\mathfrak{u})$ is weakly dominant. Then*

(1) $H_c^{0,q}(X, \mathcal{V}^{\omega}) = 0$ unless $q = r$.
(2) If $L = L_{max}$ (Definition 6.5), then $H_c^{0,r}(X, \mathcal{V})$ is irreducible or zero.
(3) The representation $H_c^{0,r}(X, \mathcal{V}^{\omega})$ admits a natural continuous positive definite invariant Hermitian form. Completing $H_c^{0,r}(X, \mathcal{V}^{\omega})$ with respect to this form defines a unitary representation of G.

Suppose in addition that $\lambda_L - \rho(\mathfrak{u})$ is strictly antidominant for $\mathfrak{u}$. Then $H_c^{0,r}(X, \mathcal{V}^{\omega})$ is not zero.

This result is immediate from Corollary 9.18 and Corollary 8.15.

10 Open Questions

My original goal in these notes was to write down (explicitly and geometrically) the pre-unitary structures provided by Corollary 9.20. According to Corollary 9.18, this amounts to

Question 10.1. In the setting of Corollary 9.20, write F^ω for the space of continuous Hermitian pairings on the analytic vectors V^ω for the unitary representation $V^{(2)}$ of L. Regard F^ω as a smooth $L \times L$ representation, and write $\mathcal{F}^\omega$ for the corresponding holomorphic bundle on $Y \times Y^{op}$. Corollaries 9.18 and 9.20 provide a distinguished G_Δ-invariant Dolbeault cohomology class in

$$H^{(n,n),(s,s)}(Y \times Y^{op}, \mathcal{F}^\omega),$$

with s the complex dimension of $K/L \cap K$. This class is non-zero if $\lambda_L - \rho(\mathfrak{u})$ is strictly anti-dominant for $\mathfrak{u}$. The problem is to give a simple geometric description of this class; perhaps to write down a representative $(2n, 2s)$ form. The space of forms F^ω contains a distinguished line corresponding to the invariant form on V^ω. The difficulty is that this form is only L_Δ-invariant, so the line does not define a one-dimensional subbundle of $\mathcal{F}^\omega$ (except along the diagonal in $Y \times Y^{op}$).

In the case of Verma modules, construction of the Shapovalov form depends entirely on understanding the universal mapping property of Verma modules. In our setting, Question 10.1 should be related to questions of Frobenius reciprocity for Dolbeault cohomology representations, and these are in any case of interest in their own right.

Question 10.2. Suppose E is an admissible Harish-Chandra module for L, with maximal globalization $E^{-\omega}$ and minimal globalization E^ω. Regard these representations of L as $(\mathfrak{q}, L)$-modules, by making $\mathfrak{u}$ act by zero. If X is any smooth admissible representation of G, we would like to calculate

$$\mathrm{Hom}_G(X, H^{n,p}(Y, \mathcal{E}^\omega)).$$

This should be related to (in fact equal to if L is compact)

$$\mathrm{Hom}_L(H^{n-p}(\mathfrak{u}, X), E^{-\omega}).$$

We will offer a more precise statement in Conjecture 10.3. What appears in the second formula is the cohomology of the Lie algebra $\mathfrak{u}$ with coefficients in X. This is at least formally a representation of L; it is not clear how to define a nice topology.

Similarly, we would like to calculate

$$\mathrm{Hom}_G(H^{0,q}_c(Y, \mathcal{E}^\omega), X).$$

This should be related to (equal to if L is compact)

$$\mathrm{Hom}_L(E^\omega, H^q(\mathfrak{u}, X)).$$

Conjecture 10.3. (cf. [KV95], Theorem 5.120). Suppose X^K is an admissible Harish-Chandra module for G, with canonical globalizations X^g (for $g = \omega$, $g = \infty$, and so on). Suppose $\mathfrak{q} = \mathfrak{l} + \mathfrak{u}$ is a very nice parabolic subalgebra of $\mathfrak{g}$ with Levi factor L (Definition 6.5). It is known that the Lie algebra cohomology $H^p(\mathfrak{u}, X^K)$ is an admissible Harish-Chandra module for L. Here are the conjectures.

(1) The $\mathfrak{u}$-cohomology complexes

$$\mathrm{Hom}\left(\bigwedge^p \mathfrak{u}, X^g\right)$$

have the closed range property, so that the cohomology spaces inherit nice locally convex topologies. That L acts continuously on these cohomology spaces is easy.

(2) The representations of L on these cohomology spaces are canonical globalizations:

$$H^p(\mathfrak{u}, X^g) \simeq [H^p(\mathfrak{u}, X^K)]^g$$

for $g = \omega$, $g = \infty$, and so on.

(3) In the setting of Question 10.2, there are two first quadrant spectral sequences with E_2 terms

$$\mathrm{Ext}^r_L(H^{n-t}(\mathfrak{u}, X^\omega), E^{-\omega})$$

and

$$\mathrm{Ext}^a_G(X^\omega, H^{n,b}(Y, \mathcal{E}^{-\omega}))$$

with a common abutment.

(4) There are two first quadrant spectral sequences with E_2 terms

$$\mathrm{Ext}^r_L(E^\omega, H^t(\mathfrak{u}, X^{-\omega}))$$

and

$$\mathrm{Ext}^a_G(H^{0,b}_c(Y, \mathcal{E}^\omega), X^{-\omega}).$$

Statement (1) makes sense with X^g replaced by any smooth globalization of X^K; one should ask only that the cohomology be some smooth globalization of the right Harish-Chandra module. I have not thought carefully about this, but I know no reason for it to fail. The specific version in (1) here ought to be fairly easy to prove, however.

Similarly, statements (3) and (4) should probably be true with X^ω and $X^{-\omega}$ replaced by any smooth globalization of X^K. The specific versions here are those most closely related to the construction of forms in Corollary 9.20.

In the case of the minimal globalization (the case $g = \infty$), generalizations of statements (1) and (2) of these conjectures have been established by Tim Bratten in [Bra98]; he considers arbitrary parabolic subalgebras endowed with real θ-stable Levi subalgebras. The generalization of (1) for the maximal globalization follows easily by a duality argument.

Bratten has pointed out that the generalization of (2) cannot extend to the case of real parabolic subalgebras and distribution or maximal globalizations. Here is one reason. Suppose that $P = LU$ is a real parabolic subgroup of G, and that X^K is an irreducible Harish-Chandra module with maximal globalization $X^{-\omega}$. Harish-Chandra's subquotient theorem guarantees that $H^0(\mathfrak{u}, X^{-\omega}) \neq 0$. (Simply embed the dual representation, which is a minimal globalization, in a space of analytic sections of a bundle on G/P. Then evaluation of sections at the identity coset eP defines a U-invariant continuous linear functional on the dual representation; that it, a U-invariant vector in $X^{-\omega}$. The same argument applies to the distribution globalization.) But $H^0(\mathfrak{u}, X^K)$ is equal to zero in almost all cases, contradicting the analogue of (2).

The spaces Ext_L and Ext_G are in the category of continuous representations of G. In order to interpret Conjecture 10.3, it would be helpful to have

Conjecture 10.4. Suppose X^K and Y^K are admissible Harish-Chandra modules for G, with canonical globalizations X^g and Y^g, for $g = \omega$, $g = \infty$, and so on. Then the standard complex

$$\mathrm{Hom}_K \left(\bigwedge{}^{\!p} (\mathfrak{g}/\mathfrak{k}, L(X^\omega, Y^{-\omega})) \right)$$

has the closed range property. Its cohomology is isomorphic to $\mathrm{Ext}_{\mathfrak{g}, K}(X^K, Y^K)$. The same result holds with ω replaced by ∞.

Results at least very close to this may be found in [BW80], Chapter 9; I have not checked whether this statement follows from their results.

References

[Bou87] N. Bourbaki. *Topological vector spaces. Chapters 1–5.* Springer-Verlag, Berlin-Heidelberg-New York, 1987.

[Bra97] T. Bratten. Realizing representations on generalized flag manifolds. *Compositio math.*, 106:283–319, 1997.

[Bra98] T. Bratten. A comparison theorem for Lie algebra homology groups. *Pacific J. Math.*, 182:23–36, 1998.

[Bru56] F. Bruhat. Sur les représentations induites des groupes de Lie. *Bull. Soc. Math. France*, 84:97–205, 1956.

[BW80] A. Borel and N. Wallach. *Continuous Cohomology, Discrete Subgroups, and Representations of Reductive Groups.* Princeton University Press, Princeton, New Jersey, 1980.

[Cas89] W. Casselman. Canonical extensions of Harish-Chandra modules to representations of G. *Can. J. Math.*, 41:385–438, 1989.

[HC56] Harish-Chandra. Representations of semisimple Lie groups. V. *Amer. J. Math.*, 78:1–41, 1956.

[Hel94] S. Helgason. *Geometric Analysis on Symmetric Spaces*, volume 39 of *Mathematical Surveys and Monographs*. American Mathematical Society, Providence, Rhode Island, 1994.

[How88] R. Howe. The oscillator semigroup. In *The Mathematical Heritage of Hermann Weyl*, volume 48 of *Proceedings of Symposia in Pure Mathematics*, pages 61–132. American Mathematical Society, Providence, Rhode Island, 1988.

[How89] R. Howe. Transcending classical invariant theory. *J. Amer. Math. Soc.*, 2:535–552, 1989.

[Kna86] A. Knapp. *Representation Theory of Semisimple Groups: An Overview Based on Examples*. Princeton University Press, Princeton, New Jersey, 1986.

[Kos61] B. Kostant. Lie algebra cohomology and the generalized Borel-Weil theorem. *Ann. of Math.*, 74:329–387, 1961.

[KV95] A. Knapp and D. Vogan. *Cohomological Induction and Unitary Representations*. Princeton University Press, Princeton, New Jersey, 1995.

[KZ77] A. Knapp and G. Zuckerman. Classification theorems for representations of semisimple Lie groups. In *Non-commutative Harmonic Analysis*, volume 587 of *Lecture Notes in Mathematics*, pages 138–159. Springer-Verlag, Berlin-Heidelberg-New York, 1977.

[Sch85] W. Schmid. Boundary value problems for group invariant differential equations. In *The mathematical heritage of Élie Cartan*, pages 311–321. Astérisque, 1985.

[Ser55] J. P. Serre. Un théorème de dualité. *Comment. Math. Helv.*, 29:9–26, 1955.

[Ser59] J. P. Serre. Représentations linéaires et espaces homogènes kählériens des groupes de Lie compacts. In *Séminaire Bourbaki, 6ième année: 1953/54*. Secrétariat mathématique, Paris, 1959.

[Tre67] F. Treves. *Topological vector spaces, distributions, and kernels*. Academic Press, New York, 1967.

[TW71] J. Tirao and J. Wolf. Homogeneous holomorphic vector bundles. *Indiana Univ. Math. J.*, 20:15–31, 1970/1971.

[Vog84] D. Vogan. Unitarizability of certain series of representations. *Ann. of Math.*, 120:141–187, 1984.

[Vog87] D. Vogan. *Unitary Representations of Reductive Lie Groups*. Annals of Mathematics Studies. Princeton University Press, Princeton, New Jersey, 1987.

[Wei64] A. Weil. Sur certaines groupes d'opérateurs unitaires. *Acta Math.*, 111:143–211, 1964.

[Wol69] J. Wolf. The action of a real semisimple group on a complex flag manifold. i. orbit structure and holomorphic arc components. *Bull. Amer. Math. Soc.*, 75:1121–1237, 1969.

[Won99] H. Wong. Cohomological induction in various categories and the maximal globalization conjecture. *Duke Math. J.*, 96:1–27, 1999.

Quantum Computing and Entanglement
for Mathematicians

Nolan R. Wallach [*]

University of California, San Diego, USA
nwallach@ucsd.edu

These notes are an expanded form of lectures presented at the C.I.M.E. summer school in representation theory in Venice, June 2004. The sections of this article roughly follow the five lectures given. The first three lectures (sections) are meant to give an introduction to an audience of mathematicians (or mathematics graduate students) to quantum computing. No attempt is given to describe an implementation of a quantum computer (it is still not absolutely clear that any exist). There are also some simplifying assumptions that have been made in these lectures. The short introduction to quantum mechanics in the first section involves an interpretation of measurement that is still being debated which involves the "collapse of the wave function" after a measurement. This interpretation is not absolutely necessary but it simplifies the discussion of quantum error correction. The next two sections give an introduction to quantum algorithms and error correction through examples including fairly complete explanations of Grover's (unordered search) and

---

[*] Research partially supported by ARO and NSF grants.

Shor's (period search and factorization) algorithms and the quantum perfect (five qubit) code. The last two sections present applications of representation and Lie theory to the subject. We have emphasized the applications to entanglement since this is the most mathematical part of recent research in the field and this is also the main area to which the author has made contributions. The material in subsections 5.1 and 5.3 appears in this article for the first time.

1 The Basics

In his seminal paper [F1] Richard Feynman introduced the idea of a computer based on quantum mechanics. Of course, all modern digital computers involve transistors that are by their very nature quantum mechanical. However, the quantum mechanics only plays a role in the theory that explains why the transistor switches. The actual switch in the computer is treated as if it were mechanical. In other words as if it were governed by classical mechanics. Feynman had something else in mind. The basic operations of a quantum computer would involve the allowable transformations of quantum mechanics, that is, unitary operators and measurements. The analogue of bit strings for a quantum computer are superpositions of bit strings (we will make this precise later) and the analogue of a computational step (for example the operation NOT on one bit) is a unitary operator on the Hilbert space of bit strings (say of a fixed length). The reason that Feynman thought that there was a need for such a "computer" is that quantum mechanical phenomena are extremely difficult (if not impossible) to model on a digital computer. The reason why the field of quantum computing has blossomed into one of the most active parts of the sciences is the work of Peter Shor [S1] that showed that on a (hypothetical) quantum computer there are polynomial time algorithms for factorization and discrete logarithms. Since most of the security of the internet is based on the assumption that these two problems are computationally hard (that is, the only known algorithms are superpolynomial in complexity) this work has attracted an immense amount of attention and trepidation. In these lectures we will discuss a model for computation based on this idea and discuss its power, ways in which it differs from standard computation and its limitations. Before we can get started we need to give a crash course in quantum mechanics.

1.1 Basic Quantum Mechanics

The *states* of a quantum mechanical system are the unit vectors of a Hilbert space, V, over $\mathbb{C}$ ignoring phase. In other words the states are the elements of the projective space of all lines through the origin in V. If $v, w \in V$ then we write $\langle v|w \rangle$ for the inner product of v with w. We will follow the physics convention so the form is conjugate linear in v and linear in w. Following Dirac a vector gives rise to a *"bra"*, $\langle v|$ and a *"ket"* $|v\rangle$ the latter is exactly the same as v the former is the linear functional that takes the value $\langle v|w\rangle$ on w. Thus if v is a state then $\langle v|v\rangle = 1$. In these lectures most Hilbert spaces will be finite dimensional. For the moment we will assume that $\dim V < \infty$. An observable is a self adjoint operator, A, on V. Thus A has a spectral decomposition

$$V = \bigoplus_{\lambda \in \mathbb{R}} V_\lambda$$

with $A_{|V_\lambda} = \lambda I$. We can write this as follows. The spaces V_λ are orthogonal relative to the Hilbert space structure. Thus we can define the orthogonal projection $P_\lambda : V \to V_\lambda$. Then we have $A = \sum \lambda P_\lambda$. If v is a state then we set $v_\lambda = P_\lambda v$. A measurement of the state v with respect to an observable A yields a number λ that is an eigenvalue of A with the probability $\|v_\lambda\|^2$. This leads to the following problem. If we do another measurement almost instantaneously we should get a value close to λ. Thus one would expect the probability to be very close to 1 for the state to be in V_λ. In the standard formulation of quantum mechanics this is "explained" by the collapse of the wave function. That is, a measurement by an apparatus corresponding to the observable A has two effects. The first is an eigenvalue, λ of A (the measurement) with probability $\|v_\lambda\|^2$ and the second is that the state has collapsed to

$$\frac{v_\lambda}{\|v_\lambda\|}.$$

This is one of the least intuitive aspects of quantum mechanics. It has been the subject of much philosophical discussion. We will not enter into this debate and will merely take this as an axiom for our system.

If we have a quantum mechanical system then in addition to the Hilbert space V we have a self adjoint operator H the *Hamiltonian* of the system. The evolution of a state in this system is governed by *Schroedinger's equation*

$$\frac{d\phi}{dt} = iH\phi.$$

Thus if we have the initial condition $\phi(0) = v$ then

$$\phi(t) = e^{itH}v.$$

Thus the basic dynamics is the operation of unitary operators. If U is a unitary operator on V then $|Uv\rangle = U\,|v\rangle$ and $\langle Uv| = \langle v|\,U^{-1}$. This is the only consistent way to have $\langle Uv|Uv\rangle = \langle v|v\rangle$ for a unitary operator.

Of course, these finite dimensional Hilbert spaces do not exist in isolation. The state of the entire universe, u, is a state in a Hilbert space, U, governed by the Schroedinger equation with Hamiltonian H_U. We we will simplify the situation and think of the finite dimensional space V as a tensor factor of U that is

$$U = V \otimes E$$

with E standing for the *environment*. This is not a tremendous assumption since in practice the part of the universe that will have a real effect on V is given by this tensor product. Now, the Hamiltonian H_U will not preserve the tensor product structure. Thus, even though we are attempting to do only operations on states in V the environment will cause the states to change in ways that are beyond the control of the experiment that we might be attempting to do on states in V. Thus if we prepare a state on which we will do a quantum mechanical operation, that is, by applying a unitary transformation or doing a measurement we can only assume that the state will not "morph" into a quite different state for a very short time. This uncontrolled change of the state is called *decoherence* caused by the environment.

The fact that our small Hilbert space V is not completely isolated from the rest of the universe is the reason why it is more natural to use density matrices as the basic states. A *density matrix (operator)* is a self adjoint operator T on V that is positive semi-definite and has trace 1. In this context a state $v \in V$ would then be called a *pure state* and a density matrix a *mixed state*. If v is a pure state then its density matrix is $|v\rangle \langle v|$. We note that this operator is just the projection onto the line corresponding to the pure state v. Thus we can identify the pure states with the mixed states that have rank 1. If T is a mixed state then T transforms under a unitary operator by $T \mapsto kTk^{-1}$ if k is unitary. If we have a pure state u in U then it naturally gives rise to a mixed state on V which is called the *reduced density matrix* and is defined as follows. Let $\{e_i\}$ be an orthonormal basis of E. Then $u = \sum v_i \otimes e_i$ with $v_i \in V$. The reduced density matrix is $\sum |v_i\rangle \langle v_i|$. More generally, if T is a mixed state on U then it gives rise to a mixed state $\mathrm{Tr}_2(T)$ on V by the formula

$$\langle w|\mathrm{Tr}_2(T)|v\rangle = \sum_i \langle w \otimes e_i|T|v \otimes e_i\rangle \,.$$

This mixed state is the reduced density matrix. One checks easily that since unitary operators don't necessarily preserve the tensor product structure that a unitary transformation of the state, T, will not necessarily entail a unitary transformation of the reduced density matrix. We will mainly deal with pure states in these lectures. However, we should realize that this is a simplification of what nature allows us to see.

1.2 Bits

Although it is not mandatory we will look upon digital computing as the manipulation of bit strings. That is, we will only consider fixed sequences of 0's and 1's. One bit is either 0 or it is 1. Two bits can have one of four values $00, 01, 10, 11$. These four strings can be looked upon as the expansion in base 2 of the integers $0, 1, 2, 3$, they can be looked upon as representatives of the integers mod 4, or they can be considered to be the standard basis of the vector space $\mathbb{Z}_2 \times \mathbb{Z}_2$. In general, an n-bit computer can manipulate bit strings of length n. We will call n the *word length* of our computer. Most personal computers now have word length 32 (soon 64). We will not be getting into the subtleties of computer science in these lectures. Also, we will not worry about the physical characteristics of the machines that are needed to do bit manipulations. A computer also can hold a certain number of words in its *memory*. There are various forms of memory (fast, somewhat fast, less fast, slow) but we will ignore the differences. We will look upon a computer program as a sequence of steps (usually encoded by bit strings of length equal to the word length) which implement a certain set of rules that we will call the *algorithm*. The first step inputs a bit string into memory. Each succeeding step operates on a sequence of words in memory that came from the operation of the preceding step and produces another sequence of words, which may or may not replace some of the words from the previous step and may or may not put words into new memory locations. If properly designed the program will have rules that terminate it and under each of the rules an output of bit strings. That is the actual computation. There are, of course, other ways the program might terminate, for example it runs out of memory, it is terminated by the operating system for attempting to access protected memory locations, or even that it is terminated by the user out of impatience. In these cases there is no (intended) output except possibly an error message.

This is the von Neumann model of computation. The key is that the computer does one step of a program at a time. Most computers can actually do several steps at one time. But this is because the computers are actually several von Neumann computers working simultaneously. For example, a computer might have an adder and a multiplier that can work independently. Or it might have several central processors that communicate with each other and attempt to do program steps simultaneously. These modifications will only lead to a parallelism that is determined by the number of processors and can only lead to a constant speed-up of a computation. For example, assume we have 10 von Neumann computers searching through a sequence of N elements with the task of finding one with a specified property. For example you have $N - 1$ red chips and 1 white one. The program might be set to divide the sequence into 10 subsequences each of size $\frac{N}{10}$ and then each processor is assigned the job of searching through one part. In the worst case each processor will have to evaluate $\frac{N}{10}$ elements. So we see a speed up of a factor of 10 over using one processor in this simple problem (slightly less since the worst case with one processor is $N - 1$).

We will come back to a few more aspects of digital computing as we develop a model for quantum computation.

1.3 Qubits

The simplest description of the basic objects to be manipulated by a quantum computer of word length n are complex superpositions of bit strings of length

n. Since a bit string is a sequence of numbers and the coefficients of the superpositions can also be some of these numbers we will use the ket notation for the bit strings as pure states. These superpositions will be called *qubits*. Thus one *qubit* is an element of the two dimensional vector space over $\mathbb{C}$ with states

$$a \left|0\right\rangle + b \left|1\right\rangle$$

and $|a|^2 + |b|^2 = 1$. We will be dealing with qubits quantum mechanically so we ignore phase (multiples by complex numbers of norm 1). Thus our space of qubits is one dimensional projective space over $\mathbb{C}$. We will think of this qubit as being in state $\left|0\right\rangle$ with probability $|a|^2$ and in state $\left|1\right\rangle$ with probability $|b|^2$. Although this is a vast simplification we will take the simplest one step operation on a qubit to be a unitary operator (projective unitary operator to be precise).

Contrasting this with bits we see that on the set of bits $\{0,1\}$ there are exactly 2 basic reversible operations: the identity map and NOT that interchanges 0 and 1. In the case of qubits we have a 3 dimensional continuum of basic operations that can be done. There is only one caveat. After doing these operations which are difficult to impossible classically we must do a measurement to retrieve a bit. This measurement will yield 0 or 1 with some probability. Thus in a very real sense going to qubits and allowing unitary transformations has not helped at all.

An element of 2 qubit space will be of the form

$$u = a \left|00\right\rangle + b \left|01\right\rangle + c \left|10\right\rangle + d \left|11\right\rangle$$

with $|a|^2 + |b|^2 + |c|^2 + |d|^2 = 1$. We interpret this as u is in state $\left|00\right\rangle$ with probability $|a|^2$, in state $\left|01\right\rangle$ with probability $|b|^2$, etc. Similarly for n qubits. The steps in a quantum computation will be unitary transformations. However, each unitary transformation given in a step will have to be broken up into basic transformations that we can construct with a known and hopefully small cost (time and storage).

A quantum program starts with an n qubit state, u_0, the input, and then does a sequence of unitary transformations T_j on the state so the steps are $u_1 = T_1 u_0, \ldots, u_m = T_m u_{m-1}$, and a rule for termination and at termination a measurement. The output is the measurement of the state to which the measured state has collapsed.

References

[F1] R. P. Feynman, *Simulating physics with computers*, Int. J. Theor. Phys., **21** (1982).

[S1] P. W. Shor, *Algorithms for quantum computation, discrete logarithms and factorizing*, Proceedings 35th annual symposium on foundations of computer science, IEEE Press, 1994.

2 Quantum Algorithms

In the last lecture we gave simple models for a classical and a quantum computation. In this lecture we will give a very simple example of a quantum algorithm that implements something that is impossible to do on a Von Neumann computer. We will next give a more sophisticated example of a quantum algorithm (Grover's algorithm [G1]) that does an unstructured search of N objects of the type described in the last lecture in $\sqrt{N}$ steps. At the end of the lecture we will introduce the quantum (fast) Fourier transform and explain why on a (hypothetical) quantum computer it is exponentially faster than the Fast Fourier transform

2.1 Quantum Parallelism

Suppose that we are studying a function, f, on bit strings that takes the values 1 and -1 and assume that it takes only one step on a classical computer to calculate its value given a bit string. For example the function that takes value 1 if the last bit is 0, -1 if it is 1. We will think of bit strings of length n as binary expansions of the numbers $0, 1, \ldots, 2^n - 1$. Thus our n qubit space, V, has the orthonormal basis $|0\rangle, |1\rangle, \ldots, |N-1\rangle$ with $N = 2^n - 1$. We can replace f by the unitary operator defined by $T|j\rangle = f(j)|j\rangle$. T operates on a state $v \in V$,

$$v = \sum_{j=0}^{N-1} a_j |j\rangle, \quad \sum_{j=0}^{N-1} |a_j|^2 = 1$$

by

$$Tv = \sum_{j=0}^{N-1} f(j)a_j |j\rangle.$$

Quantum mechanically this means that we have calculated $f(j)$ with the probability $|a_j|^2$. In other words the calculation of T on this superposition seems to have calculated all of the values of $f(j)$ simultaneously if all of the $|a_j| > 0$ in one quantum step. In a sense we have, but the rub is that if we do a measurement then all we have after a measurement is $f(j)|j\rangle$ with probability $|a_j|^2$ and since we ignore phase the value the object we are calculating is lost. Perhaps it would be better to decide that we will operate quantum mechanically and then read the coordinates classically? I assert that we will still not be able to make direct use of this parallelism. The reason is that we are only interested in very big N. In this situation the set of states, $\sum_{j=0}^{N-1} a_j |j\rangle$, with $|a_j|^2$ all about the same size have a complement in the sphere of extremely small volume. This implies that most of the states will have probabilities, $|a_j|^2 \sim \frac{1}{N}$. If n is, say, 1000 then all the coordinates will be too small to measure classically. We can see this as follows. We consider the unit sphere in real N dimensional space. Let ω_N be the $O(N)$ invariant volume element on S^{N-1} that is normalized so that

$$\int_{S^{N-1}} \omega_N = 1.$$

We write a state in the form $v = \cos\theta u + \sin\theta\,|N-1\rangle$ with $-\frac{\pi}{2} \leq \theta \leq \frac{\pi}{2}$. With u an element of the unit sphere in $N-1$ dimensional space. Then we have

$$\omega_N = c_N \cos\theta^{n-2}\omega_{N-1} \wedge d\theta$$

and $c_N \sim C\sqrt{N}$ with C independent of N. The set of all v in the sphere with last coordinate a_{N-1} that satisfies $|a_{N-1}|^2 \geq r^2 > \frac{1}{N} + \varepsilon$ with $\varepsilon > 0$ has volume at most

$$C\sqrt{N}(1-r^2)^{\frac{N}{2}-1} = C\sqrt{N}(1-\varepsilon)^{\frac{N}{2}-1}(1 - \frac{\frac{1}{1-\varepsilon}}{N})^{\frac{N}{2}-1}$$

which is extremely small for N large.

The upshot is that a quantum algorithm must contain a method of increasing the size of the coefficient of the desired output so that when a measurement is made will have the output with high probability.

2.2 The Tensor Product Structure of n-qubit Space

Recall that the standard (sometimes called the *computational*) basis of the space of 2 qubits is $|00\rangle, |01\rangle, |10\rangle, |11\rangle$. A physicist would also write $|0\rangle\,|1\rangle = |01\rangle$. We mathematicians would rather think that the multiplication is a tensor product. That is $|0\rangle, |1\rangle$ form the standard basis of $\mathbb{C}^2$. Then

$$|0\rangle \otimes |0\rangle, |0\rangle \otimes |1\rangle, |1\rangle \otimes |0\rangle, |1\rangle \otimes |1\rangle$$

form an orthonormal basis of $\mathbb{C}^2 \otimes \mathbb{C}^2$ with the tensor product Hilbert space structure. In other words we identify $|ab\rangle$ with $|a\rangle \otimes |b\rangle$. In this form the original bit strings are fully decomposable that is are tensor products of n elements in $\mathbb{C}^2$. We will call an n-qubit state a product state if it is of the form

$$v_1 \otimes v_2 \otimes \cdots \otimes v_n$$

with $\|v_i\|^2 = 1$ for $i = 1, \ldots, n$. One very important product state is the uniform state ($N = 2^n$):

$$v = \frac{1}{\sqrt{N}} \sum_{j=0}^{N-1} |j\rangle.$$

To see that it is indeed a product state we set $u = \frac{1}{\sqrt{2}}(|0\rangle + |1\rangle)$. Then $v = u \otimes \cdots \otimes u$ (n-fold tensor product). This formula also shows that the uniform state can be constructed in $n = \log_2(N)$ steps. This can be seen by making an apparatus that implements the one qubit unitary transformation (called the *Hadamard transformation*)

$$H = \frac{1}{\sqrt{2}} \begin{bmatrix} 1 & 1 \\ -1 & 1 \end{bmatrix}.$$

It has the property that $H|0\rangle = \frac{|0\rangle + |1\rangle}{\sqrt{2}}, H|1\rangle = \frac{|0\rangle - |1\rangle}{\sqrt{2}}$. We write $H(k)$ for $I \otimes \cdots \otimes H \otimes \cdots I$ with all factors one qubit operations and all factors but one the identity and in the k-th factor the Hadamard transformation. Thus on a quantum computer that can implement a one qubit Hadamard transformation in constant time can construct the uniform state in logarithmic time. We will actually over simplify the model and assume that all one qubit operations can be implemented is one step on a quantum computer, Then

$$u = H(1)H(2) \cdots H(n)|0\rangle.$$

With this in mind we can give our first quantum algorithm. Set up an apparatus that corresponds to an observable, A, with simple spectrum. Here is the algorithm:

Make a uniform state v.

Measure A.

v collapses to $|j\rangle$ with j between 0 and $N-1$ with probability $\frac{1}{N}$.

In other words we are generating truly random numbers. The complexity of this algorithm is n. On a digital computer the best one can do is generate pseudo random numbers. The classical algorithms involve multiplication and division. Thus they are slightly more complex. However they do not generate random numbers and no deterministic algorithm can (since the numbers will satisfy the property that they are given by the algorithm).

2.3 Grover's Algorithm

We return to unstructured search. We assume that we have a function, f, on n-bit strings that takes the value -1 on exactly one string and 1 on all of the others. We assume that given a bit string the calculation of the value is one step (in computer science f might be called an *oracle*). Here is Grover's algorithm:

Form the uniform state $u = \frac{1}{\sqrt{N}} \sum |j\rangle$. Let T be the unitary transformation defined by $T|j\rangle = f(j)|j\rangle$. Let S be the orthogonal reflection about u. That is

$$S_u(v) = v - 2\langle u|v\rangle u.$$

Then S_u is a unitary operator that in theory can be implemented quantum mechanically with logarithmic complexity (indeed Grover gave a formula for S_u involving the order of n Hadamard transformations). If the number of bits is 2 (we are searching a list of 4 elements) then we observe

$$STu = -\left|j\right\rangle$$

with $f(j) = -1$. Thus one quantum operation and one measurement yields the answer. Whereas classically in the worst case we would have had to calculate f three times and then printed the answer.

The general algorithm is just an iteration of this step. $u_0 = u$ and $u_{m+1} = STu_m$. A calculation using trigonometry shows that after $[4\pi\sqrt{N}]$ steps the coefficient of $\left|j\right\rangle$ with $f(j) = -1$ has absolute value squared .99. (Here $[x]$ is the maximum of the set of integers less than or equal to x). Thus with almost certainty a measurement at this step in the iteration will yield the answer.

2.4 The Quantum Fourier Transform

Interpreted as a map of $L^2(\mathbb{Z}/N\mathbb{Z})$ to itself the fast Fourier transform can be interpreted as a unitary operator on this Hilbert space. In general, if G is a finite abelian group of order $|G|$ then we define the Hilbert space $L^2(G)$ to be the space of a complex valued functions on G with inner product

$$\langle f|g\rangle = \sum_{x \in G} \overline{f(x)}g(x).$$

Let $\widehat{G}$ denote the set of unitary characters of G. Then it is standard that the set $\{\frac{1}{\sqrt{|G|}}\chi | \chi \in \widehat{G}\}$ is an orthonormal basis of $L^2(G)$. If $G = \mathbb{Z}/N\mathbb{Z} = \mathbb{Z}_N$ and if we set $\chi_m(n) = e^{\frac{2\pi i n m}{N}}$ for $m = 0, \dots, N-1$ then we can define

$$\mathcal{F}(f)(m) = \left\langle \frac{1}{\sqrt{N}}\chi_m | f \right\rangle = \frac{1}{\sqrt{N}} \sum_{n=0}^{N-1} f(n)\chi_m(n)^{-1}$$

and so

$$f(n) = \sum_{m=0}^{N-1} \left\langle \frac{1}{\sqrt{N}}\chi_m | f \right\rangle \frac{1}{\sqrt{N}}\chi_m(n) = \frac{1}{\sqrt{N}} \sum_{m=0}^{N-1} \mathcal{F}(f)(m)\chi_m(n).$$

As in the case of the fast Fourier transform we will take $N = 2^n$. The standard orthonormal basis of $L^2(\mathbb{Z}_N)$ is the set of delta functions $\{\delta_m | m = 0, \dots, N-1\}$ with $\delta_m(x) = 1$ if $x = m$ and 0 otherwise. We will identify these delta functions with the computational basis, that is $\left|m\right\rangle = \delta_m$. We therefore have

$$\mathcal{F}\left|m\right\rangle = \frac{1}{\sqrt{N}} \sum_{j=0}^{N-1} \chi_m(j)^{-1}\left|j\right\rangle.$$

The linear extension to n qubit space is the quantum Fourier transform. The discussion above makes it obvious that this is a unitary operator. What is less obvious is that we can devise a quantum algorithm to implement this operator as (essentially) a tensor product of one qubit operators (which we are assuming are easily implemented on our hypothetical quantum computer). We will conclude this section with the factorization (due to Shor [S2]) that suggests a fast quantum algorithm

If $0 \leq j \leq N - 1$ then we write $j = \sum_{i=0}^{n-1} j_i 2^i$ with $j_i \in \{0, 1\}$ so that with our convention $|j\rangle = |j_{n-1}j_{n-2}\cdots j_0\rangle$. If $0 \leq m \leq N - 1$ then

$$\frac{m}{N} = \sum_{i=1}^{n} m_{n-i} 2^{-i}$$

and since

$$2^{-k}j = \sum_{l=0}^{k-1} j_l 2^{-k+l} + u_{kj}$$

with $u_{kj} \in \mathbb{Z}$. We have

$$e^{2\pi i \frac{m}{N} j} = e^{2\pi i \sum_{k=1}^{n} m_{n-k} \sum_{l=0}^{k-1} j_l 2^{-k+l}}.$$

This leads to the following factorization

$$\mathcal{F}|j\rangle = u_n(j) \otimes u_{n-1}(j) \otimes \cdots \otimes u_1(j)$$

with

$$u_k(j) = \frac{|0\rangle + e^{2\pi i \sum_{l=0}^{k-1} j_l 2^{-k+l}} |1\rangle}{\sqrt{2}}.$$

References

[G] L. K. Grover, *Quantum mechanics helps in searching for a needle in a haystack*, Phys. Rev. Let., **79** 1997.

[S2] P. K. Shor, *Polynomial time algorithms for prime factorization and discrete logarithms on a quantum computer*, SIAM J. Comp., **26** 1484–1509,1997.

3 Factorization and Error Correction

In this section we will study the complexity of the quantum Fourier transform and indicate its relationship with Shor's factorization algorithm. We will also discuss the role of error correction in quantum computing and describe a quantum error correcting code.

3.1 The Complexity of the Quantum Fourier Transform

Recall that our simplified model takes a one qubit unitary operator to be one computational step this is a simplification but the one qubit operators that will come into the rest of the discussion of the quantum Fourier transform are provably of constant complexity. We will also be using some two qubit operations which are also each of constant complexity. In addition we assume an implementation of the total flip, τ

$$v_1 \otimes v_2 \otimes \cdots \otimes v_n \mapsto v_n \otimes v_{n-1} \otimes \cdots \otimes v_1$$

One can show that the complexity of this operation is a multiple of n. We will show how to implement the transformation

$$|j\rangle \mapsto u_n(j) \otimes u_{n-1}(j) \otimes \cdots \otimes u_1(j) = \mathcal{F}|j\rangle$$

with

$$u_k(j) = \frac{|0\rangle + e^{2\pi i \sum\limits_{l=0}^{k-1} j_l 2^{-k+l}} |1\rangle}{\sqrt{2}}.$$

To describe the steps in the implementation we need the notion of a *controlled* one qubit operation. Let $U \in U(2)$ we define a unitary operator, C_U, on $\mathbb{C}^2 \otimes \mathbb{C}^2$ as follows

$$C_U |j_1 j_2\rangle = (U |j_1\rangle) \otimes |j_2\rangle$$

if $j_2 = 1$ and

$$C_U |j_1 j_2\rangle = |j_1 j_2\rangle$$

if $j_2 = 0$. We call j_2 the *control bit*. If we are operating on n qubits and applying a controlled U operation with the control in the k-th factor and the operation in the l-th factor then we will write $C_U^{l,k}$ (the reader should be warned that this is not standard notation). Thus

$$C_U^{23} |0110\rangle = |0\rangle \otimes U |1\rangle \otimes |1\rangle \otimes |0\rangle$$

and

$$C_U^{23} |0100\rangle = |0100\rangle.$$

If U is easily implemented then controlled U is also easily implemented. We define

$$U_k = \begin{bmatrix} 1 & 0 \\ 0 & e^{\frac{2\pi i}{2^k}} \end{bmatrix}$$

and recall that the Hadamard operator acting on the k-th qubit was denoted $H(k)$ in section 2. We will now describe an operator the implements the quantum Fourier transform. It will be a product $\tau \circ A_n A_{n-1} \cdots A_1 \circ \tau$ with

$$A_1 = C_{U_n}^{1,n} C_{U_{n-1}}^{1,n-1} \cdots C_{U_2}^{1,2} H(1), \ldots$$

$$A_k = C^{k,n}_{U_k} C^{k,n-1}_{U_{k-1}} \cdots C^{k,k+1}_{U_2} H(k), \ldots, A_n = H(n).$$

We note that in this expression the operator A_k changes the k-th qubit but doesn't depend on the value of the j-th qubit for $j < k$. We leave it to the reader to expand the product and see that it works. The operator A_k is a product of k operators that we can assume are implemented in constant time. Thus the complexity of the transform is a constant times $\frac{n(n+1)}{2}$. This is exponentially faster than the classical fast Fourier transform which has complexity Nn.

Shor introduced this transform in order to give a generalization of an algorithm of Deutch (cf. [NC]). Shor's algorithm finds the period of a function with an unknown period with complexity a power of the number of bits involved. A very nice exposition of the period finding algorithm can be found in [NC]. We will give a different approach here.

We begin with a periodic integer valued function, f of unknown period L which we know is less than 2^n. We will work in $2n$ qubit space and consider the values of the function to be between 0 and $N-1$ with $N = 2^{2n}$ (this is not necessary but is not a real restriction since we will know the range of values of the function and for useful application we should be able to take n very large). We can thus think of f as defining a unitary map from $\left(\bigotimes^{2n} \mathbb{C}^2\right) \otimes \left(\bigotimes^{2n} \mathbb{C}^2\right)$ to itself by

$$F\left(|x\rangle \otimes |y\rangle\right) = |x\rangle \otimes |f(x) + y\rangle.$$

Here the addition is modulo 2^{2n}. We will think of each of the tensor factors of $2n$ qubits as a register. The first step in the algorithm is to construct the uniform state

$$\frac{1}{N} \sum_{1 \le x,y \le N-1} |x\rangle \otimes |y\rangle$$

and then do a measurement in the second register (factor) getting

$$\frac{1}{\sqrt{N}} \sum_{1 \le x \le N-1} |x\rangle \otimes |y_o\rangle.$$

We now apply F and get

$$\frac{1}{\sqrt{N}} \sum_{1 \le x \le N-1} |x\rangle \otimes |f(x) + y_o\rangle.$$

Set $g(x) = f(x) + y_o$. g is also a periodic function of period L projected into $\mathbb{Z}_{2^{2n}}$. Thus it doesn't matter what y_o occurred. We now measure the second register and get

$$\frac{1}{\sqrt{|\{x | g(x) = g(x_o)\}|}} \sum_{g(x)=g(x_o)} |x\rangle \otimes |g(x_o)\rangle.$$

We write $N = LM + r$ with $0 \leq r < L$ then $|\{x|g(x) = g(x_o)\}| = M$ or $M + 1$. Since we are assuming that N is large compared to L in formulas we will use the approximation $|\{x|g(x) = g(x_o)\}| = M$. We will also ignore the congruence modulo 2^{2n}. Thus after the measurement in the second register we have

$$\left(\frac{1}{\sqrt{M}} \sum_s |x_o + sL\rangle\right) \otimes |g(x_o)\rangle$$

where we may take as an approximation the sum over all $0 \leq s \leq M - 1$. We now apply the quantum Fourier transform in the first register getting

$$\frac{1}{\sqrt{NM}} \sum_{s=0}^{M-1} \sum_{x=0}^{N-1} \exp(\frac{2\pi i x}{N}(x_o + sL)) |x\rangle \otimes |g(x_o)\rangle .$$

If we now do a measurement in both registers we will obtain $|x\rangle \otimes |g(x_o)\rangle$ with probability

$$\frac{1}{MN}|\sum_{s=0}^{M-1} \exp(\frac{2\pi i x}{N}(x_o + sL))|^2.$$

We now observe that if q is a strictly positive integer and c is a real number then we have

$$\sum_{j=0}^{q-1} \exp(2\pi i j c)$$

is equal to

$$\frac{1 - \exp(2\pi i q c)}{1 - \exp(2\pi i c)}$$

if c is not an integer and it is equal to q if $c \in \mathbb{Z}$. Thus observing that $MN = M^2\left(\frac{N}{M}\right)$ and we are approximating $\frac{N}{M}$ by L we find that if $c = xL/N$ is not an integer then the probability of having collapsed to $|x\rangle \otimes |g(x_o)\rangle$ is approximately

$$\frac{1}{M^2 L}\left|\frac{1 - \exp(2\pi i M c)}{1 - \exp(2\pi i c)}\right|^2.$$

And the probability of obtaining x with xL/N an integer is $\frac{1}{\sqrt{L}}$. We therefore see that after a measurement it is most probable that the state will have collapsed to $|x\rangle \otimes |g(x_o)\rangle$ with xL/N very close to being an integer. That is, if the integral part of $\frac{N}{x}$ is an integer times L. One then checks if the outcome is a period by substitution. If it is a period (i.e. a multiple of L) we must make sure that we have found the minimal period. After on the order of $\log(L)$ applications of this method one would have determined L with probability close to 1

We note that this algorithm is probabilistic as are all known quantum algorithms. This method of Shor is a special case of a larger class of algorithms known as "hidden subgroup" problems. Here one starts with a group G, an unknown subgroup H and a function f on G such that f is constant on the cosets of H. The problem is to construct an efficient algorithm to find H. This has been done for G finite and commutative, for H a normal subgroup and for some two step solvable groups.

3.2 Reduction of Factorization to Period Search

We will now describe the method Shor uses to reduce the problem of factorization to period search for which he had devised a fast quantum algorithm. Consider an integer N for which we want to find a nontrivial factor. We may assume that it is odd and composite. Chose a number $1 < y < N - 1$ randomly. If the greatest common divisor (gcd) of N and y is not one then we are done. We can therefore assume that $\gcd(y, N) = 1$. Hence y is invertible as an element of $\mathbb{Z}_N$ (under multiplication). Consider $f(m) = y^m \bmod N$. Then since the group of invertible elements of the ring $\mathbb{Z}_N$ is a finite group the function will have a minimum period. We can thus use Shor's algorithm to find the period, T. If T is even we assert that $y^{\frac{T}{2}} + 1$ and N have a common factor larger than 1. We can thus use the Euclidean algorithm (which is easy classically) to find a factor of N. Before we demonstrate that this works consider $N = 30$ and $y = 11$. Then $f(0) = 1, f(1) = 11, 11^2 = 121 = 1 \bmod 30$, so $f(2) = 1 = f(0)$. Thus $T = 2$. Now $11^1 + 1 = 12$. The greatest common divisor of 12 and 30 is 6.

We will now prove the assertion about the greatest common divisor. We first note that

$$(y^{\frac{T}{2}} + 1)^2 = y^T + 2y^{\frac{T}{2}} + 1.$$

But $y^T = 1 + m \cdot N$ by the definition of T. Thus $(y^{\frac{T}{2}} + 1)^2 \equiv 2(y^{\frac{T}{2}} + 1) \bmod N$. Hence

$$(y^{\frac{T}{2}} + 1)^2 - 2(y^{\frac{T}{2}} + 1)$$

is evenly divisible by N. We therefore see that

$$\left((y^{\frac{T}{2}} + 1) - 2 \right) \left(y^{\frac{T}{2}} + 1 \right) = \left(y^{\frac{T}{2}} - 1 \right) \left(y^{\frac{T}{2}} + 1 \right)$$

is evenly divisible by N. Hence, if $y^{\frac{T}{2}} + 1$ and N have no common factor then $y^{\frac{T}{2}} - 1$ is evenly divisible by N. This would imply that $\frac{T}{2}$ (which is smaller than T) satisfies

$$f(x + \frac{T}{2}) = f(x).$$

This contradicts the choice of T as the minimal period. This is still not enough to get a non-trivial divisor of N. We must still show that y can be chosen so that N doesn't divide $y^{\frac{T}{2}} + 1$ and that we can choose y so that T is even. Neither can be done with certainty. What can be proved is that if N is not a pure prime power then the probability of choosing $1 < y < N - 1$ such that $\gcd(y, N) = 1$, f has even period and N doesn't divide $y^{\frac{T}{2}} + 1$ is at least

$\frac{3}{4}$. The proof of this would take us too far afield, a good reference is [NC]. We note that classically the test whether a number, N, is a pure power of a number $a > 2$ and if so to calculate the number is polynomial in the number of bits of N. The upshot is that a quantum computer will factor a number with very large probability (if the algorithm is done say 10 times then the probability of success would be 0.999999 in polynomial time).

3.3 Error Correction

So far we have ignored several of the difficulties that we had indicated in section 1 having to do with two problems that are caused by the environment. The first is that we can only really look at mixed states since we cannot compute the actual action of the environment and the second is the decoherence caused by the dynamics of the total system. We will assume that our quantum computations are divided into steps that take so little time that our initial pure states remain close enough to being pure states that we can ignore the first difficulty. For the second we will look at the decoherence over this small period as a small error. For most of the systems that are proposed the most likely error is a one qubit error. Thus as in classical error correction we will show how to set up a quantum error correcting code that corrects a one qubit error. The standard procedure is to encode a qubit as an element of a two dimensional subspace of a higher qubit space.

That is we take V to be the space of n qubits and we take u_0 and u_1 orthonormal in V and assign

$$a \left|0\right\rangle + b \left|1\right\rangle \mapsto a u_0 + b u_1.$$

The right hand side will be called the *encoded* qubit. The question is what is the most likely error if we transmit the encoded qubit? The generally accepted answer is that it would be a transformation of the form

$$E = I \otimes \cdots \otimes A \otimes \cdots \otimes I$$

with all factors the identity except for an A in the k-th factor and this A is a fairly arbitrary linear map on 1-qubit space that is close to the identity. The problem is to fix the error which means change $E\left(a u_0 + b u_1\right)$ to $a u_0 + b u_1$ without knowing which qubit has an error, what the error is and not collapsing the wave function of the unknown qubit. Classically one can transmit one bit in terms of 3 bits. $0 \mapsto 000, 1 \mapsto 111$. The most likely error is a NOT in one bit. To fix such an error one reads the sum of the entries of this possibly erroneous output and if it is at most 1 then change it to 000 if it is at least 2 change it to 111. This will correct exactly one NOT in any position. Quantum mechanically we must correct a continuum of possible errors. This seems to be impossible and if it were impossible then quantum computation looked impossible also since decoherence would set in before we could do any useful computation. As usual, Shor [S3] found a method. We will describe a later development that yielded a quantum analog of a perfect code (such as the three bit classical error correction scheme described above).

We will describe a special class of error correcting codes that are known as *orthogonal codes* (or non-degenerate codes). In fact, Shor's original example was not an orthogonal code, but we feel that these codes are easier for mathematicians to understand. We will need some additional notation.

If $X, Y \in M_2(\mathbb{C})$ then we define $\langle X | Y \rangle = \frac{1}{2}\text{tr}(X^*Y)$ ($X^* = X^\dagger$ to a physicist, is the Hermitian adjoint of X). Given $j = 1, \ldots, n$ we define $F_j : M_2(\mathbb{C}) \to \text{End}(\bigotimes^n \mathbb{C}^2)$ by

$$F_j(A) = I \otimes \cdots \otimes A \otimes \cdots \otimes I$$

where all of the factors on the right hand side are I except for the k-th term which is A. We say that an isometry, $T : \mathbb{C}^2 \to \bigotimes^n \mathbb{C}^2$ defines an *orthogonal code space* if it has the following properties:

1. The maps $T_j : M_2(\mathbb{C}) \otimes \mathbb{C}^2 \to \bigotimes^n \mathbb{C}^2$ given by $T_j(X \otimes v) = F_j(X)T(v)$ are isometries (onto their images) for $i = 1, \ldots, n$.

2. If $V = \{X \in M_2(\mathbb{C}) | tr(X) = 0\}$. Then the sum

$$Z = T(\mathbb{C}^2) \oplus \bigoplus_{1 \le j \le n} T_j(V \otimes \mathbb{C}^2)$$

is an orthogonal direct sum.

We will now show how to correct a one qubit error if we have an orthogonal code. Let X_1, X_2, X_3 be an orthonormal basis of V consisting of invertible elements. For example we could choose the Pauli matrices

$$\begin{bmatrix} 1 & 0 \\ 0 & -1 \end{bmatrix}, \begin{bmatrix} 0 & 1 \\ -1 & 0 \end{bmatrix}, \begin{bmatrix} 0 & i \\ i & 0 \end{bmatrix}.$$

We write $\bigotimes^n \mathbb{C}^2 = Z \oplus Z'$ with Z' the orthogonal complement to Z. Let A be an observable that acts by distinct scalars as indicated $\lambda_0 I$ on $T(\mathbb{C}^2), \lambda_{ij} I$ on $T_j(X_i \otimes \mathbb{C}^2), 1 \le j \le n, 1 \le i \le 3$ and μI on Z'. If we start with $T(v)$ and it has incurred an error and we have w rather than $T(v)$ then we do a measurement of A on w. If the measurement is μ then with high probability the error wasn't a one qubit error. Otherwise we assume a one qubit error then there is j such that $w = T_j(X \otimes v)$. $X = aI + bX_1 + cX_2 + dX_2$. Thus with probability 1 the eigenvalue will be one of $\lambda_0, \lambda_{1j}, \lambda_{2j}, \lambda_{3j}$. If it is λ_0 then w will have collapsed to v. If it is λ_{ij} then if w collapses to z then $F_j(X_i^{-1})z = T(v)$ we have thus corrected the error.

Obviously, to use this idea we must have a way of finding T. We note first of all that $\dim Z \le 2^n$ and $\dim Z = 6n + 2$. If $6n + 2 \le 2^n$ then $n \ge 5$ and if $n = 5$ then $2^5 = 6 \cdot 5 + 2$. Thus the smallest n that we could use would be $n = 5$. We will now give conditions on a map T that are equivalent to having an orthogonal code. If $w \in \bigotimes^n \mathbb{C}^2$ then $w = \sum w_j |j\rangle$. Let $0 \le p < q < n$ be two bit positions. Then we form a $4 \times 2^{n-2}$ matrix as follows. The $i = i_0 + i_1 2$, $j = j_0 + j_1 2 + \ldots + j_{n-3} 2^{n-3}$ entry is w_k where

$$k = j_0 + \cdots + j_{p-1}2^{p-1} + i_0 2^p + j_p 2^{p+1} + \cdots +$$
$$j_{q-2}2^{q-1} + i_1 2^q + j_{q-1}2^{q+1} + \cdots + j_{n-3}2^{n-1}.$$

If $p = 0$, $q = 1$ this is just $k = i_0 + i_1 2 + 2^2(j_0 + j_1 2 + \cdots + j_{n-3}2^{n-3})$. Let $W(p, q, w)$ denote this matrix. We have

Theorem 1. $T : \mathbb{C}^2 \to \bigotimes^n \mathbb{C}^2$ *defines an orthogonal code if and only if*

$$W(p, q, T\,|i\rangle)W(p, q, T\,|j\rangle)^* = \frac{1}{4}\delta_{i,j}I.$$

For all p, q and $i, j \in \{0, 1\}$.

This can be written as a system of quadratic equations. If we put them into Mathematica for $n = 5$ the first solution is given as follows: We define

$$\langle j_1 j_2 j_3 j_4 j_5 \rangle = |j_1 j_2 j_3 j_4 j_5\rangle + |j_5 j_1 j_2 j_3 j_4\rangle + |j_4 j_5 j_1 j_2 j_3\rangle$$
$$+ |j_3 j_4 j_5 j_1 j_2\rangle + |j_2 j_3 j_4 j_5 j_1\rangle.$$

Then set

$$T\,|0\rangle = \frac{1}{4}\left(|0000\rangle + \langle 11000 \rangle - \langle 10100 \rangle - \langle 11110 \rangle\right)$$

and

$$T\,|1\rangle = \frac{1}{4}\left(|11111\rangle + \langle 00111 \rangle - \langle 01011 \rangle - \langle 00001 \rangle\right).$$

Because of the symmetry it is easy to check that the condition of the theorem is satisfied. This code was originally found by other methods (c.f. [KL]).

References

[NC] Michael Nielson and Isaac Chang, *Quantum Computation and Quantum Information*, Cambridge University Press, Cambridge, 2000.

[S3] P. Shor, *A scheme for reducing decoherence in quantum computer memory*, Phys. Rev. A, **52** 1995.

[KL] E. Kroll and R. Laflamme, *A theory of quantum error correcting codes*, Phys. Rev. A, **50** 900–911, 1997.

4 Entanglement

As we have seen the only non-trivial reversible one bit operation is NOT which interchanges 0 and 1. We have made the simplifying assumption that all one qubit unitary operators are easily implementable on a quantum computer. The operation NOT gives rise to the unitary operator in one qubit with matrix

$$\begin{bmatrix} 0 & 1 \\ 1 & 0 \end{bmatrix}$$

relative to the computational basis $|0\rangle$, $|1\rangle$. We will say that a transformation of n bits that is given by applying either NOT or the identity to each bit is a *classical local transformation*. A *quantum local transformation* in n qubits is a unitary operator of the form

$$A_1 \otimes A_2 \otimes \cdots \otimes A_n$$

where $A_i \in U(2)$. There is a major distinction between the classical and the quantum cases. The classical local transformations act transitively on the set of all n bit bit strings. Whereas the quantum local transformations act transitively only in the case when $n = 1$. For example there is no local transformation that takes the state

$$\frac{|00\rangle + |11\rangle}{\sqrt{2}}$$

to $|00\rangle$ (see the next section for a proof). We will call a state that is not a product state (not in the orbit of $|00\rangle$ under local transformations) an *entangled* state. The two code words of the five bit error correcting code are entangled. Furthermore, entanglement explains some of the apparent paradoxes that appeared in the early thought experiments of quantum mechanics. It is also basic to quantum teleportation (a subject that we will not be covering in these sections). In this section we will study the orbit structure of the local transformations on the pure states and in particular functions that help to separate these orbits: the measures of entanglement. We will emphasize methods that allow one to determine if two states are related by a local transformation and to determine the extent of the entanglement of a state.

4.1 Measures of Entanglement

We will first look at the example of an entangled state: $\frac{|00\rangle+|11\rangle}{\sqrt{2}}$. One way that one can see that it is entangled is by observing that if we act on $\mathbb{C}^2 \otimes \mathbb{C}^2$ by $G = SL(2,\mathbb{C}) \times SL(2,\mathbb{C})$ by the tensor product action, then G leaves invariant a symmetric form, the tensor product of the symplectic forms on each of the $\mathbb{C}^2$ factors that are $SL(2,\mathbb{C})$ invariant. This form, (,), is given by

$$(|00\rangle, |11\rangle) = (|11\rangle, |00\rangle = 1,$$
$$(|01\rangle, |10\rangle) = (|10\rangle, |01\rangle) = -1$$

and all the other products are 0. We note that $(\frac{|00\rangle+|11\rangle}{\sqrt{2}}, \frac{|00\rangle+|11\rangle}{\sqrt{2}}) = 1$ and $(|00\rangle, |00\rangle) = 0$ so there can't be a local transformation taking one to another since the function $\phi(u) = |(u, u)|$ is invariant under local transformations. It is an example of a *measure of entanglement*. Indeed, one can prove that a state, u, in 2 qubits is entangled if and only if $\phi(u) > 0$. Another property enjoyed by this function is that for all pure 2 qubit states $\phi(u) \leq 1$ and $\phi(u) = 1$ if and only if u is in the orbit of $\frac{|00\rangle + |11\rangle}{\sqrt{2}}$ under local transformations. To prove the upper bound we consider $u = a\,|00\rangle + b\,|01\rangle + c\,|10\rangle + d\,|11\rangle$. Then $\phi(u) = 2(ad - bc)$. Since $2|a||d| \le |a|^2 + |d|^2$ we have $|\phi(u)| \le |a|^2 + |b|^2 + |c|^2 + |d|^2 = 1$. Although it is not hard to prove the assertion about the orbit directly we will use a result of Kempf and Ness [KN] which is useful in other contexts. For those of you who are unfamiliar with semisimple Lie groups take G to be the product of n copies of $SL(2, \mathbb{C})$ and K to be n copies of $SU(2)$.

Theorem 2. *Let G be a semisimple Lie group over $\mathbb{C}$ and let K be a maximal compact subgroup of G. Let (π, V) be a finite dimensional holomorphic representation of G with the K-invariant Hilbert space structure $\langle\ |\ \rangle$. Let $v \in V$ and $m = \inf\{\langle \pi(g)v | \pi(g)v \rangle\,| g \in G\}$. Then if $u \in \pi(G)v$ and $\langle u | u \rangle = m$ then $\pi(K)u = \{w \in \pi(G)v | \langle w | w \rangle = m\}$. Furthermore the infimum is actually attained if and only if the orbit $\pi(G)v$ is closed.*

In words this says that the elements of minimal norm in a G orbit form a single K-orbit.

We will now give an idea of the proof. We note that $Lie(G) = Lie(K) + iLie(K)$. We therefore have

$$G = K \exp(iLie(K)).$$

If $X \in iLie(K)$ then $d\pi(X)^* = d\pi(X)$. Thus

$$\frac{d^2}{dt^2}\,\langle \pi(\exp tX)v | \pi(\exp tX)v \rangle$$
$$= 4\,\langle d\pi(X)\pi(\exp tX)v | d\pi(X)\pi(\exp tX)v \rangle \ge 0$$

With equality if and only if $d\pi(X)v = 0$. Everything follows from this.

We will now show how the Kempf-Ness result applies to our situation for 2 qubits. We first note that relative to $G = SL(2, \mathbb{C}) \times SL(2, \mathbb{C})$ the space $V = \mathbb{C}^2 \otimes \mathbb{C}^2$ has the following orbit structure. For each $\lambda \in \mathbb{C} - \{0\}$ the set $M_\lambda = \{w \in V | (v, v) = \lambda\}$ is a single orbit. The other orbits are $\pi(G)\,|00\rangle$ and $\{0\}$. The union of the latter two is M_0. We set $u_0 = \frac{|00\rangle + |11\rangle}{\sqrt{2}}$. We note that we have $M_\lambda = z\pi(G)u_0$ with $z^2 = \lambda$. We therefore see that the elements in the unit sphere that maximize ϕ are contained in the set of elements the form $w = e^{i\theta}\frac{\pi(g)u_0}{\|\pi(g)u_0\|}$, $g \in G$. For such a w we have $\phi(w) = \frac{1}{\|\pi(g)u_0\|^2}$. Thus maximizing ϕ on the unit sphere means (up to phase) minimizing the norm on $\pi(G)u_0$. The Kempf-Ness theorem implies that this subset of $\pi(G)u_0$ is $\pi(K)u_0$. This completes the proof of the assertion.

We note that the group of local transformations on $\bigotimes^n \mathbb{C}^2$ is the image of $S^1 \times SU(2)^n$ with S^1 the circle group acting by scalar multiplication and $SU(2)^n$ acting by the tensor product action (i.e. by local transformations). We will therefore concentrate on invariants for $SU(2)^n$. We also note that if we consider $\phi(u)^2$ rather than $\phi(u)$ then it is a polynomial function on $\mathbb{C}^2 \otimes \mathbb{C}^2$ as a real vector space. We will only consider measures of entanglement that are polynomials invariant under $K = SU(2)^n$ on $\bigotimes^n \mathbb{C}^2$ as a real vector space. We will use the term measure of entanglement for such a polynomial. We will denote the algebra of such polynomials by $\mathcal{P}_{\mathbb{R}}(\bigotimes^n \mathbb{C}^2)^K$. These are exactly what we need to separate the K-orbits.

Theorem 3. *If $u, v \in \bigotimes^n \mathbb{C}^2$ then $u \in \pi(K)v$ if and only if $f(u) = f(v)$ for all $f \in \mathcal{P}_{\mathbb{R}}(\bigotimes^n \mathbb{C}^2)^K$.*

We also note that if we look at the action of the circle group by multiplication on $\bigotimes^n \mathbb{C}^2$ we can define a $\mathbb{Z}$-grading on $\mathcal{P}_{\mathbb{R}}(\bigotimes^n \mathbb{C}^2)^K$ by $f \in \mathcal{P}_{\mathbb{R}}^j(\bigotimes^n \mathbb{C}^2)^K$ if $f \in \mathcal{P}_{\mathbb{R}}(\bigotimes^n \mathbb{C}^2)^K$ and $f(zu) = z^j f(u)$ for all $z \in S^1$ and $u \in \bigotimes^n \mathbb{C}^2$.

Theorem 4. *If $u, v \in \bigotimes^n \mathbb{C}^2$ then $u \in S^1\pi(K)v$ if and only if $f(u) = f(v)$ for all $f \in \mathcal{P}_{\mathbb{R}}^0(\bigotimes^n \mathbb{C}^2)^K$.*

Both of these theorems are consequences of the following result.

Theorem 5. *Let U be a compact Lie group. Let (ρ, W) be a finite dimensional representation of U on a real Hilbert space. Let $\mathcal{P}(W)^U$ be the algebra of all complex valued polynomials on W that are invariant under U. If $u, v \in W$ then $u \in \rho(U)v$ if and only if $f(u) = f(v)$ for all $f \in \mathcal{P}(W)^U$.*

Proof. The necessity is obvious. Since $v \mapsto \|v\|^2$ is in $\mathcal{P}(W)^U$ we will prove that if $\|v\| = \|u\| = r > 0$ and $f(u) = f(v)$ for all $f \in \mathcal{P}(W)^U$ then $u \in \rho(U)v$. The Stone-Weierstrauss theorem implies that the restriction of $\mathcal{P}(W)$ to the sphere of radius r, S_r, is uniformly dense in the space of continuous functions on S_r. Suppose that $\rho(U)v \cap \rho(U)u$ is empty then Uryson's Lemma implies there is a continuous function φ on S_r such that $\varphi|_{\rho(U)v} \equiv 1$ and $\varphi|_{\rho(U)u} \equiv 0$. The uniform density implies that there exists an $f \in \mathcal{P}(W)$ such that $|f(x) - \varphi(x)| < \frac{1}{4}$ for all $x \in S_r$. Let du denote normalized invariant measure on U. We define $\overline{f}(x) = \int_U f(\rho(z)x)dz$. Then $\overline{f} \in \mathcal{P}(W)^U$. We have

$$|\overline{f}(v) - 1| = \left| \int_U f(\rho(z)v)dz - 1 \right| \leq \int_U |f(\rho(z)v)dz - \varphi(\rho(z)v)|dz \leq \frac{1}{4}$$

hence $|\overline{f}(v)| > \frac{3}{4}$ similarly $|\overline{f}(u)| < \frac{1}{4}$. This proves the theorem.

4.2 Three Qubits

These results make it reasonable to assert that the orbit of $\frac{|00\rangle + |11\rangle}{\sqrt{2}}$ under local transformations consists of the most entangled two qubit states. In the case of 3 qubits there is a similar result. First the ring of invariant (complex polynomials) on $\mathbb{C}^2 \otimes \mathbb{C}^2 \otimes \mathbb{C}^2$ under the tensor product action of

$$G = SL(2, \mathbb{C}) \times SL(2, \mathbb{C}) \times SL(2, \mathbb{C})$$

is generated by one element, f, of degree 4 (here we will be stating several results without proof in this case the details can be found in [GrW]). We can define it as follows: if $v \in \mathbb{C}^2 \otimes \mathbb{C}^2 \otimes \mathbb{C}^2$ then we can write it as

$$v = |0\rangle \otimes v_0 + |1\rangle \otimes v_1$$

with $v_0, v_1 \in \mathbb{C}^2 \otimes \mathbb{C}^2$. If we use the symmetric form defined above we have

$$f(v) = \det \begin{bmatrix} (v_0, v_0) & (v_0, v_1) \\ (v_1, v_0) & (v_1, v_1) \end{bmatrix}.$$

As in the case of two qubits most of the orbits under G are described by the values of f. Here we set $M_\lambda = \{v \in \mathbb{C}^2 \otimes \mathbb{C}^2 \otimes \mathbb{C}^2 | f(v) = \lambda\}$. Then if $\lambda \neq 0$ we have M_λ consists of a single orbit. If $\lambda = 0$ then there are 6 orbits in M_0. We note that $f(\frac{|000\rangle + |111\rangle}{\sqrt{2}}) = \frac{1}{4}$. Thus $M_\lambda = zG\left(\frac{|000\rangle + |111\rangle}{\sqrt{2}}\right)$ with $z^4 = 4\lambda$. We will now describe the orbits in M_0. First there is the open orbit in this quartic given as the orbit of

$$w_0 = \frac{|001\rangle + |010\rangle + |100\rangle}{\sqrt{3}}.$$

If we remove this orbit from M_0 then there are three open orbits in what remains. They are the orbits of $\frac{|000\rangle + |011\rangle}{\sqrt{2}}$, $\frac{|000\rangle + |101\rangle}{\sqrt{2}}$ and $\frac{|000\rangle + |110\rangle}{\sqrt{2}}$. If in addition these are removed then what we have left is the union of 0 and the product states (which form a single orbit).

One can show by an argument similar to that in two qubits that if u is a state then $|f(u)| \leq \frac{1}{4}$ and if $u_o = \frac{|000\rangle + |111\rangle}{\sqrt{2}}$ then $f(u_o) = \frac{1}{4}$. Since the set where f is non-zero is exactly the set of all elements $\mathbb{C}^\times G\left(\frac{|000\rangle + |111\rangle}{\sqrt{2}}\right)$ we see that if u is a state with $|f(u)| \neq 0$ then $u = \frac{gu_o}{\|gu_o\|}$ with $g \in G$. Thus $f(u) = f\left(\frac{gu_o}{\|gu_o\|}\right) = \frac{1}{\|gu_o\|^4} f(gu_o) = \frac{1}{4\|gu_o\|^4}$. Thus the set of states with $|f(u)| = \frac{1}{4}$ are exactly the elements that minimize the value of $\|gu_o\|^4$ for $g \in G$. Thus Theorem 2 implies:

Proposition 6. If $K = S^1 SU(2) \times SU(2) \times SU(2)$ then

$$\{v \in \mathbb{C}^2 \otimes \mathbb{C}^2 \otimes \mathbb{C}^2 \mid |f(v)| = \frac{1}{4}, \|v\| = 1\} = K\left(\frac{|000\rangle + |111\rangle}{\sqrt{2}}\right).$$

Thus one value of one invariant is enough to determine if a state can be gotten from $\frac{|000\rangle + |111\rangle}{\sqrt{2}}$ by local transformations. For example

$$v = \frac{|111\rangle + |001\rangle + |010\rangle + |100\rangle}{2}$$

has the property that $f(v) = \frac{1}{4}$. So it can be obtained by a local transformation from $\frac{|000\rangle + |111\rangle}{\sqrt{2}}$.

So far we have been analyzing only one polynomial measure of entanglement. There is the natural problem of determining a generating set for these measures. To do this it is useful to reduce the problem to a problem involving complex algebraic groups and complex polynomials. The basic idea is that if G is a simply connected semi-simple Lie group over $\mathbb{C}$ then G is a linear algebraic group. If K is a maximal compact subgroup of G and if (ρ, V) is a finite dimensional unitary representation of K then ρ extends to a regular representation of G on V. The real polynomials on V are the complex polynomials in both the bra and the ket vectors. The ket vectors give a copy of V as a complex vector space whereas the ket vectors give a copy of the complex dual representation of V. This implies that the algebra $\mathcal{P}_R(V)^K$ is naturally isomorphic with $\mathcal{P}(V \oplus V^*)^G$. In the case when we are dealing with qubits the representation of $G = SL(2)^n$ on $\bigotimes^n \mathbb{C}^2$ is self dual. We are thus looking at the problem of determining the invariants of G acting on two copies of $\bigotimes^n \mathbb{C}^2$ by the diagonal action. We analyze this problem for two and three qubits.

4.3 Measures of Entanglement for Two and Three Qubits

We first look at 2 qubits and continue the discussion begun in the previous subsection. As we have observed $G = SL(2, \mathbb{C}) \times SL(2, \mathbb{C})$ leaves invariant a symmetric bilinear form on $\mathbb{C}^2 \otimes \mathbb{C}^2$. A dimension count shows that the image of G on $\mathbb{C}^2 \otimes \mathbb{C}^2$ is the full orthogonal group for this form. Thus the action on $\mathbb{C}^2 \otimes \mathbb{C}^2$ can be interpreted as the action of $SO(4, \mathbb{C})$ on $\mathbb{C}^4$. We are thus looking at the invariants of $SO(4, \mathbb{C})$ on two copies of $\mathbb{C}^4$. Classical invariant theory implies that the algebra of invariants is generated by the three polynomials $\alpha(v \oplus w) = (v, v)$, $\beta(v \oplus w) = (v, w)$ and $\gamma(v \oplus w) = (w, w)$, This implies

Lemma 7. *The algebra of measures of entanglement in 2 qubits is the set of polynomials in (v, v), $\langle v|v\rangle$ and $\overline{(v, v)}$.*

Thus in this case we were using the only "interesting" measure, since we are only considering states which are assumed to satisfy $\langle v|v\rangle = 1$.

The situation is different for three qubits. We will describe a set of generators in this case that was determined in [MW1] our method is a modification which is an outgrowth of joint work with H. Kraft. As above we look upon $\mathbb{C}^2 \otimes \mathbb{C}^2 \otimes \mathbb{C}^2$ as $\mathbb{C}^2 \otimes \mathbb{C}^4$ and $G = SL(2, \mathbb{C}) \times SL(2, \mathbb{C}) \times SL(2, \mathbb{C})$ acting as $SL(2, \mathbb{C}) \times SO(4, \mathbb{C})$. For the moment we will ignore the $SL(2, \mathbb{C})$ factor and look at $I \otimes SO(4, \mathbb{C})$ acting on two copies of $\mathbb{C}^2 \otimes \mathbb{C}^4$. If we consider only the action of $SO(4, \mathbb{C})$ then we are looking at its action on 4 copies of $\mathbb{C}^4$. We look at this as $SO(4, \mathbb{C})$ acting on $X \in M_4(\mathbb{C})$ under right multiplication by the transpose of the matrix. Then the invariants for $SO(4, \mathbb{C})$ are generated by the matrix entries of XX^T (the upper T stands for transpose) and $\det(X)$ (for these results and others stated without proof in this subsection please see

[GW]). We now look at the action of the remaining $SL(2,\mathbb{C})$. The $SL(2,\mathbb{C})$ is acting on the left on the matrix via multiplication by the block diagonal matrix

$$h = \begin{bmatrix} g & 0 \\ 0 & g \end{bmatrix}.$$

Thus the $SL(2,\mathbb{C})$ factor is acting on the generators of the $SO(4,\mathbb{C})$ invariants trivially on $\gamma = \det X$ (an invariant under the full G of degree 4) and via $hXX^T h^T$ with h as above, We write XX^T in block form

$$\begin{bmatrix} A & B \\ B^T & C \end{bmatrix}$$

then the $SL(2,\mathbb{C})$ is acting on the components via $A \mapsto gAg^T, B \mapsto gBg^T, C \mapsto gCg^T$. We note that A and C are symmetric and completely general and B is an arbitrary 2×2 matrix which we can write as

$$a \begin{bmatrix} 0 & 1 \\ -1 & 0 \end{bmatrix} + Z$$

with Z a general two by two symmetric matrix. The coefficient a defines an invariant for G of degree 2 on the qubits which we will call α. The rest of the action is by three copies of the action of $SL(2,\mathbb{C})$ on the symmetric 2×2 matrices. Using the trace form we see that this is just the action of $SO(3,\mathbb{C})$ on three copies of $\mathbb{C}^3$. Again we look upon this as the action of $SO(3,\mathbb{C})$ on $Y = M_3(\mathbb{C})$ via left multiplication. The invariants in this case are generated by $\beta = \det Y$ (an invariant of degree 6 on the qubits) and the matrix coefficients of $Y^T Y$ which yield 6 invariants of degree 4. The upshot is the invariants are generated by an invariant of degree 2 (α), an invariant of degree 4 (γ), an invariant of degree 6 (β) and 6 invariants of degree 4 (the matrix coefficients of $Y^T Y$), $\mu_1, \ldots, \mu_6$. We note that the invariants γ and β have the property that their squares are invariant under $O(3) \times O(4)$. Thus γ^2 and β^2 are in the algebra generated by α and $\mu_1, \ldots, \mu_6$. We can also see from the invariant theory of $SO(3)$ that the functions $\alpha, \mu_1, \ldots, \mu_6$ are algebraically independent. We therefore see that the full ring of invariants is $\mathbb{C}[\alpha, \mu_1, \ldots, \mu_6] \oplus \mathbb{C}[\alpha, \mu_1, \ldots, \mu_6]\beta \oplus \mathbb{C}[\alpha, \mu_1, \ldots, \mu_6]\gamma \oplus \mathbb{C}[\alpha, \mu_1, \ldots, \mu_6]\beta\gamma$.

References

[GrW] Benedict H. Gross and Nolan R. Wallach, *On quaternionic discrete series and their continuations*, J. Reine Angew. Math. **481** (1996), 73–123.

[GW] Roe Goodman and Nolan R. Wallach, *Representations and invariants of the classical groups*, Cambridge University Press, Cambridge, 1998.

[MW1] David Meyer and Nolan Wallach, *Invariants for multiple qubits: the case of 3 qubits*. Mathematics of quantum computation, 77–97, Comput. Math. Ser., Chapman & Hall/CRC, Boca Raton, FL, 2002.

[KN] George Kempf and Linda Ness, *The length of vectors in representation spaces.* Algebraic geometry (Proc. Summer Meeting, Univ. Copenhagen, Copenhagen, 1978), pp. 233–243, Lecture Notes in Math., 732, Springer, Berlin, 1979.

5 Four and More Qubits

In the cases of 2 and 3 qubits it is fairly clear what the maximally entangled states should be or at least there are just a few candidates for that honor. We will see that there is an immense variety of states that are highly entangled in the case of 4 qubits. This and the calculation of Hilbert series for measures of entanglement (see subsection 5.2) indicate that the search for all measures of entanglement or the complete description of the orbit structure for arbitrary numbers of qubits will be so hard and complicated as to become useless. However the case of 4 qubits gives some indications of how to find more invariants. Also, methods similar to the Kempf-Ness theorem can be used to prove uniqueness theorems (for example the theorem of Rains [R] that implies that the 5 bit error correcting code we discussed earlier is unique up to local transformations).

As it turns out the orbit structure under

$$G = SL(2,\mathbb{C}) \times SL(2,\mathbb{C}) \times SL(2,\mathbb{C}) \times SL(2,\mathbb{C})$$

on $\mathbb{C}^2 \otimes \mathbb{C}^2 \otimes \mathbb{C}^2 \otimes \mathbb{C}^2$ can be determined using the results of Kostant and Rallis [KR]. Since it fits in their theory in case of the symmetric pair $(SO(4,4), SO(4) \times SO(4))$. We will now describe the outgrowth of this theory purely in terms of qubits.

5.1 Four Qubits

We are therefore analyzing the action of $G = SL(2) \times SL(2) \times SL(2) \times SL(2)$ on the space $V = \mathbb{C}^2 \otimes \mathbb{C}^2 \otimes \mathbb{C}^2 \otimes \mathbb{C}^2$ via the tensor product action

$$(g_1, g_2, g_3, g_4)(v_1 \otimes v_2 \otimes v_3 \otimes v_4) = g_1 v_1 \otimes g_2 v_2 \otimes g_3 v_3 \otimes g_4 v_4$$

We first note that if $H = SL(2) \times SL(2)$ and if $W = \mathbb{C}^2 \otimes \mathbb{C}^2$ and if we have H act on W by the tensor product action then there is a H-invariant non-degenerate symmetric bilinear form, $(\ldots, \ldots)$, on W given as follows

$$(v \otimes w, x \otimes y) = \omega(v, x)\omega(w, y).$$

Here $\omega((x_1, y_1), (x_2, y_2)) = x_1 y_2 - x_2 y_1$. This form allows us to define a linear map, T, of V onto $End(W)$ in the following way

$$T(v_1 \otimes v_2 \otimes v_3 \otimes v_4)(w_1 \otimes w_2) = \omega(v_3, w_1)\omega(v_4, w_2)v_1 \otimes v_2.$$

We look upon G as $H \times H$. Thus if $g = (h_1, h_2)$ then

$$T(gv)(w) = h_1 T(v)(h_2^{-1} w).$$

If $A \in End(W)$ then we define $A^{\#}$ by $(Aw_1, w_2) = (w_1, A^{\#} w_2)$. We note that if $h \in H$ then $h^{\#} = h^{-1}$. This implies that

$$
\begin{aligned}
T(gv)T(gv)^{\#} &= h_1 T(v) h_2^{-1} (h_1 T(v) h_2^{-1})^{\#} \\
&= h_1 T(v) h_2^{-1} h_2 T(v)^{\#} h_1^{-1} = h_1 T(v) T(v)^{\#} h_1^{-1}.
\end{aligned}
$$

We therefore have invariants $f_{2j}(v) = tr((T(v)T(v)^{\#})^j)$, $j = 1, 2, \ldots$ and $g_4(v) = \det(T(v))$.

Theorem 8. *The ring of invariants under the action of G on V is generated by the algebraically independent elements f_2, f_4, g_4, f_6.*

The following discussion gives a sketch of a proof.

We will use qubit notation for elements of V. Thus V has a basis consisting of elements $|i_0 i_1 i_2 i_3\rangle$ with $i_j = 0, 1$. We set

$$v_1 = \frac{1}{2}(|0000\rangle + |1111\rangle + |0011\rangle + |1100\rangle),$$

$$v_2 = \frac{1}{2}(|0000\rangle + |1111\rangle - |0011\rangle - |1100\rangle),$$

$$v_3 = \frac{1}{2}(|1010\rangle + |0101\rangle + |0110\rangle + |1001\rangle),$$

$$v_4 = \frac{1}{2}(|1010\rangle + |0101\rangle - |0110\rangle - |1001\rangle).$$

These states can be described in terms of the Bell states for 2 qubits. Let $u_{\pm} = \frac{|00\rangle \pm |11\rangle}{\sqrt{2}}$ and $v_{\pm} = \frac{|01\rangle \pm |10\rangle}{\sqrt{2}}$ then

$$v_1 = u_+ \otimes u_+, \quad v_2 = u_- \otimes u_-, \quad v_3 = v_+ \otimes v_+, \quad v_4 = v_- \otimes v_-.$$

We note that if $v = x_1 v_1 + x_2 v_2 + x_3 v_3 + x_4 v_4$ then

$$
T(v) = \begin{bmatrix}
\frac{x_1 - x_2}{2} & 0 & 0 & \frac{x_1 + x_2}{2} \\
0 & \frac{x_4 - x_3}{2} & -\frac{x_3 + x_4}{2} & 0 \\
0 & -\frac{x_3 + x_4}{2} & \frac{x_4 - x_3}{2} & 0 \\
\frac{x_1 + x_2}{2} & 0 & 0 & \frac{x_1 + x_2}{2}
\end{bmatrix}.
$$

Hence

$$f_{2j}(v) = \sum x_i^{2j}$$

and

$$g_4(v) = x_1 x_2 x_3 x_4.$$

We note that this implies that the functions f_2, f_4, g_4, f_6 are algebraically independent. Set $\mathfrak{a} = \{v = x_1 v_1 + x_2 v_2 + x_3 v_3 + x_4 v_4 | x_j \in \mathbb{C}\}$ and $\mathfrak{a}' = \{v = x_1 v_1 + x_2 v_2 + x_3 v_3 + x_4 v_4 | x_i \neq \pm x_j \text{ for } i \neq j\}$. One can check that the map $G \times \mathfrak{a}' \to V$ given by $g, v \longmapsto gv$ is regular. Furthermore, if $x \in \mathfrak{a}'$ then the set of $g \in G$ such that $gx = x$ is finite. Since $\dim G = 12$ and $\dim \mathfrak{a} = 4$ we see that if f is a G invariant polynomial then f is completely determined by its restriction to $\mathfrak{a}$ (since $G\mathfrak{g}'$ has interior). We also note that if $N = \{g \in G | g\mathfrak{a} = \mathfrak{a}\}$ then the group $W = N_{|\mathfrak{a}}$ is the subgroup of the group generated by the linear maps given by the permutations of v_1, v_2, v_3, v_4 and those that involve an even number of sign changes. For example,

$$\left(\begin{bmatrix} 0 & i \\ i & 0 \end{bmatrix}, \begin{bmatrix} 1 & 0 \\ 0 & 1 \end{bmatrix}, \begin{bmatrix} 0 & i \\ i & 0 \end{bmatrix}, \begin{bmatrix} 1 & 0 \\ 0 & 1 \end{bmatrix} \right)$$

corresponds to $v_1 \to v_3, v_2 \to v_4, v_3 \to v_1, v_4 \to v_2$,

$$\left(\frac{1}{\sqrt{2}} \begin{bmatrix} 1 & 1 \\ -1 & 1 \end{bmatrix}, \frac{1}{\sqrt{2}} \begin{bmatrix} 1 & 1 \\ -1 & 1 \end{bmatrix}, \frac{1}{\sqrt{2}} \begin{bmatrix} 1 & 1 \\ -1 & 1 \end{bmatrix}, \frac{1}{\sqrt{2}} \begin{bmatrix} 1 & 1 \\ -1 & 1 \end{bmatrix} \right)$$

corresponds to $v_1 \to v_1, v_2 \to -v_3, v_3 \to -v_2, v_4 \to v_4$. Thus W is the subgroup of the group of signed permutations with an even number of sign changes. One can check directly that every invariant under W is a polynomial in $(f_2)_{|\mathfrak{a}}$, $(f_4)_{|\mathfrak{a}}$, $(g_4)_{|\mathfrak{a}}$, $(f_6)_{|\mathfrak{a}}$ This completes the sketch of the proof of the theorem.

Remark 9. *This result is an explicit form of the Chevalley restriction theorem for the group* $SO(4,4)$.

We will now relate the space $\mathfrak{a}$ to the orbit structure. For this we need another construct. If $v, w \in \mathbb{C}^2$ then we write vw for the product of v, w in $S^2(\mathbb{C}^2)$. We set

$$[u_1 \otimes u_2 \otimes u_3 \otimes u_4, w_1 \otimes w_2 \otimes w_3 \otimes w_4]_i =$$

$$\left(\prod_{j \neq i} \omega(u_j, w_j) \right) u_i w_i, i = 1, 2, 3, 4.$$

We say that $v, w \in V$ commute if $[v, w]_i = 0$ for $i = 1, 2, 3, 4$. We note that $[v_i, v_j]_k = 0$ for $i, j, k = 1, 2, 3, 4$. We also observe that if $v, w \in V$ and $g = (g_1, \ldots, g_4) \in G$ then $[gv, gw]_i = g_i[v, w]_i$ with the latter given by the action of $SL(2)$ on $S^2(\mathbb{C}^2)$. If $v \in V$ we will say that v is nilpotent if $T(v)T(v)^{\#}$ is nilpotent (that is, some power of $T(v)T(v)^{\#}$ is 0). This is the same as saying that $f_{2j}(v) = 0$ for all $j = 1, 2, \ldots$. Hilbert's criterion for this condition is

Theorem 10. *v is nilpotent if and only if there is a rational homomorphism,* ϕ, *of the group* $\mathbb{C}^{\times} = \{z \in \mathbb{C} | z \neq 0\}$ *into* G *such that* $\lim_{z \to 0} \phi(z)v = 0$. *We note that the action of* G *stabilizes the set of nilpotent elements.*

If $v \in V$ set $G_v = \{g \in G | gv = v\}$. We can now state the basic result on the orbit structure of G on V. We will call an element of $G\mathfrak{a}$ semi-simple. Then the Jordan decomposition of [KR] implies

Theorem 11. *An element $v \in V$ is semi-simple if and only if Gv is closed. Let v be an element of V then $v = s + n$ with s semi-simple and n nilpotent such that $[s,n]_i = 0$ for $i = 1, 2, 3, 4$. If s, s' are semi-simple and n, n' are nilpotent and commute with s, s' respectively then $s + n = s' + n'$ if an only if $s = s'$ and $n = n'$. If $g \in G$, $v \in \mathfrak{a}$ and $gv \in \mathfrak{a}$ then there exists $w \in W$ such that $wv = gv$. If $s \in \mathfrak{a}$ and $n, n' \in V$ are nilpotent and commute with s then if there exists $g \in G$ such that $g(s + n) = s + n'$ then there exists $h \in G_s$ such that $hn = n'$. Finally, if $s \in \mathfrak{a}$ and if $\mathbb{N}_s = \{v \in V | v$ is nilpotent and commutes with $s\}$ then $\mathbb{N}_s$ consists of a finite number of G_s orbits.*

We will next give a quantitative version of this theorem. We will first establish a bit more terminology.

We will say that a nilpotent element, n, is regular if setting $U = T(n)T(n)^\#$, $R = T(n)^\#$ then $R, RU + UR, RU^2 + U^2R, RU^3 + U^3R$ are linearly independent operators. A family of such examples is

$$a \left| 0011 \right\rangle + b \left| 0100 \right\rangle + c \left| 1001 \right\rangle + d \left| 1010 \right\rangle$$

with $abcd \neq 0$. It is easily seen that all of the regular elements of the above form are in the G orbit of the element with a, b, c, d all equal to 1. Let us call this element n_o. It turns out that there are 4 distinct regular nilpotent orbits. There are 20 distinct nilpotent orbits. The general theory also allows us to determine the general orbits. The number of different "types" of orbits is 90. The term "type" will become clear in the course of the discussion below leading to an explanation of the quantitative statement.

For each $i = 1, 2, 3, 4$ we define $\varepsilon_i \in V^*$ by $\varepsilon_i(v_j) = \delta_{ij}$. Let $\Phi = \{\pm(\varepsilon_i + \varepsilon_j) | 1 \leq i < j \leq 4\} \cup \{\varepsilon_i - \varepsilon_j | 1 \leq i \neq j \leq 4\}$. Set $\Delta = \{\alpha_1 = \varepsilon_1 - \varepsilon_2, \alpha_2 = \varepsilon_2 - \varepsilon_3, \alpha_3 = \varepsilon_3 - \varepsilon_4, \alpha_4 = \varepsilon_3 + \varepsilon_4\}$. If $s \in \mathfrak{a}$ then we define $\Phi_s = \{\alpha \in \Phi | \alpha(s) = 0\}$. One can show that if $s \in \mathfrak{a}$ then there exists $w \in W$ such that $\Phi_{ws} = \Phi \cap span_{\mathbb{Z}}(\Delta \cap \Phi_{ws})$. The main theorem implies that we need only look at elements s satisfying

$$\Phi_s = \Phi \cap span_{\mathbb{Z}}(\Delta \cap \Phi_s).$$

Here are the possibilities with $|\Delta \cap \Phi_s| \leq 1$.

$\Delta \cap \Phi_s = \emptyset$, $s = x_1 v_1 + x_2 v_2 + x_3 v_3 + x_4 v_4$, $x_i \neq \pm x_j$ for all $i \neq j$,

$\Delta \cap \Phi_s = \{\alpha_1\}$, $s = x_1(v_1 + v_2) + x_3 v_3 + x_4 v_4$, $x_i \neq \pm x_j$ for all $i \neq j$,

$\Delta \cap \Phi_s = \{\alpha_2\}$, $s = x_1 v_1 + x_2(v_2 + v_3) + x_4 v_4$, $x_i \neq \pm x_j$ for all $i \neq j$,

$\Delta \cap \Phi_s = \{\alpha_3\}$, $s = x_1 v_1 + x_2 v_2 + x_3(v_3 + v_4)$, $x_i \neq \pm x_j$ for all $i \neq j$,

$\Delta \cap \Phi_s = \{\alpha_4\}$, $s = x_1 v_1 + x_2 v_2 + x_3(v_3 - v_4)$, $x_i \neq \pm x_j$ for all $i \neq j$.

We note that the permutation (123) maps the set $\{s | s = x_1(v_1 + v_2) + x_3 v_3 + x_4 v_4, x_i \neq \pm x_j\}$ for all $i \neq j$ bijectively onto the set $\{s | s = x_1 v_1 + x_2(v_2 + v_3) + x_4 v_4, x_i \neq \pm x_j$ for all $i \neq j\}$. Similarly, there is a permutation that maps the set indicated by $\Delta \cap \Phi_s = \{\alpha_2\}$ onto the set indicated by $\Delta \cap \Phi_s = \{\alpha_3\}$.

Finally, the sign change $v_1 \to v_1, v_2 \to -v_2, v_3 \to v_3, v_4 \to -v_4$ takes the set indicated by $\Delta \cap \Phi_s = \{\alpha_3\}$ onto the set indicated by $\Delta \cap \Phi_s = \{\alpha_4\}$. Thus by the basic theorem we need only consider the first two in our list. For $|\Delta \cap \Phi_s| \geq 2$ we will only list the cases up to the action of signed permutations involving an even number of sign changes. Here are all of the examples

1. $\Delta \cap \Phi_s = \emptyset$, $s = x_1 v_1 + x_2 v_2 + x_3 v_3 + x_4 v_4$, $x_i \neq \pm x_j$ for all $i \neq j$.
2. $\Delta \cap \Phi_s = \{\alpha_1\}$, $s = x_1(v_1 + v_2) + x_3 v_3 + x_4 v_4$, $x_i \neq \pm x_j$ for all $i \neq j$.
3. $\Delta \cap \Phi_s = \{\alpha_1, \alpha_2\}$, $s = x_1(v_1 + v_2 + v_3) + x_4 v_4$, $x_1 \neq \pm x_4$.
4. $\Delta \cap \Phi_s = \{\alpha_1, \alpha_3\}$, $s = x_1(v_1 + v_2) + x_3(v_3 + v_4)$, $x_1 \neq \pm x_3$.
5. $\Delta \cap \Phi_s = \{\alpha_1, \alpha_4\}$, $s = x_1(v_1 + v_2) + x_3(v_3 - v_4)$, $x_1 \neq \pm x_3$.
6. $\Delta \cap \Phi_s = \{\alpha_1, \alpha_2, \alpha_3\}$, $s = x_1(v_1 + v_2 + v_3 + v_4)$, $x_1 \neq 0$.
7. $\Delta \cap \Phi_s = \{\alpha_1, \alpha_2, \alpha_4\}$, $s = x_1(v_1 + v_2 + v_3 - v_4)$, $x_1 \neq 0$.
8. $\Delta \cap \Phi_s = \{\alpha_2, \alpha_3, \alpha_4\}$, $s = x_1 v_1$, $x_1 \neq 0$.
9. $\Delta \cap \Phi_s = \{\alpha_1, \alpha_3, \alpha_4\}$, $s = x_1(v_1 + v_2)$, $x_1 \neq 0$.
10. $\Delta \cap \Phi_s = \{\alpha_1, \alpha_2, \alpha_3, \alpha_4\}$, $s = 0$.

We now count the number of G_s orbits in $\mathcal{N}_s$ in each of the 10 cases above. Case 1 yields 1 since $\mathcal{N}_s = \{0\}$. Case 2 yields 2. Case 3 yields 3. Cases 4 and 5 yield 8. Cases 6, 7 and 8 yield 7. Case 9 yields 27. Case 10 yields 20. The total is our promised 90.

Here are some examples. The extremes in case 10 of the list involving the non-zero orbits are the 4 regular nilpotent orbits and the orbit of product states

$$\{u_1 \otimes u_2 \otimes u_3 \otimes u_4 | u_i \in \mathbb{C}^2 - \{0\}\}.$$

We now look at the so called WHZ state. This is (up to normalization) $s = |0000\rangle + |1111\rangle = v_1 + v_2$. It appears in case 9. Thus there are 26 additional orbits with s-component the WHZ state. Here is how you find them. We note that

$$G_s = \left\{ \left(\begin{bmatrix} a_1 & 0 \\ 0 & a_1^{-1} \end{bmatrix}, \begin{bmatrix} a_2 & 0 \\ 0 & a_2^{-1} \end{bmatrix}, \begin{bmatrix} a_3 & 0 \\ 0 & a_3^{-1} \end{bmatrix}, \begin{bmatrix} a_4 & 0 \\ 0 & a_4^{-1} \end{bmatrix} \right) | a_1 a_2 . a_3 a_4 = 1 \right\}$$

The space of all elements $v \in V$ such that $[s, v]_i = 0$ for all i is spanned by s and

$$\{|0, 0, 1, 1\rangle, |0, 1, 0, 1\rangle, |1, 0, 0, 1\rangle, |1, 0, 1, 0\rangle, |1, 1, 0, 0\rangle\}.$$

Let

$$S_1 = \{|0, 0, 1, 1\rangle, |1, 1, 0, 0\rangle\},$$
$$S_2 = \{|0, 1, 0, 1\rangle, |1, 0, 1, 0\rangle\},$$
$$S_3 = \{|1, 0, 0, 1\rangle, |0, 1, 1, 0\rangle\}.$$

Then the orbits corresponding to s are the orbits through $s + \sum_{j \in J} n_j$ where J is a subset of $\{1, 2, 3\}$ and $n_j \in S_j$. There are 27 such orbits. The orbits with minimal stability groups are the ones corresponding to $|J| = 3$. There are 8 of them.

We note that for 2 and 3 qubits the state $\frac{|00...\rangle + |11...\rangle}{\sqrt{2}}$ was arguably the most entangled. In the case of 4 qubits this state is just $\frac{v_1 + v_2}{\sqrt{2}}$ and so it is not even in $\mathfrak{a}'$.

This discussion indicates that the measures of entanglement for 4 qubits will form a complicated algebra. One useful invariant of such an algebra is the Hilbert series.

5.2 Some Hilbert Series of Measures of Entanglement

If V is a real vector space then we set $\mathcal{P}^j(V)$ equal to the complex vector space of all polynomials on V that are homogeneous of degree j. If W is a complex vector space, then $\mathcal{P}^j_{\mathbb{R}}(W) = \mathcal{P}^j(V)$ where V is W as a real vector space. We say that a subalgebra, A, of $\mathcal{P}^j_{\mathbb{R}}(W)$ is homogeneous if it is the direct sum of $A^j = \mathcal{P}^j_{\mathbb{R}}(W) \cap A$. If A is a homogeneous subalgebra of $\mathcal{P}_{\mathbb{R}}(W)$ then the formal power series

$$h_A(q) = \sum_{j \geq 0} q^j \dim A^j$$

is called the *Hilbert series* of A.

The results we have described for 2 and 3 qubits imply that

$$h_{\mathcal{P}_{\mathbb{R}}(\mathbb{C}^2 \otimes \mathbb{C}^2)^K} = \frac{1}{(1 - q^2)^3}$$

and

$$h_{\mathcal{P}_{\mathbb{R}}(\mathbb{C}^2 \otimes \mathbb{C}^2 \otimes \mathbb{C}^2)^K} = \frac{(1 + q^4)(1 + q^6)}{(1 - q^2)(1 - q^4)^6}.$$

As we predicted the case of 4 qubits is much more complicated. Here is the series (see [W])

Numerator: $1 + 3q^4 + 20q^6 + 76q^8 + 219q^{10} + 654q^{12} + 1539q^{14} + 3119q^{16} + 5660q^{18} + 9157q^{20} + 12876q^{22} + 16177q^{24} + 18275q^{26} + 18275q^{28} + 16177q^{30} + 12876q^{32} + 9157q^{34} + 5660q^{36} + 3119q^{38} + 1539q^{40} + 654q^{42} + 219q^{44} + 76q^{46} + 20q^{48} + 3q^{50} + q^{54}$

Denominator: $(1 - q^2)^3 (1 - q^4)^{11} (1 - q^6)^6$.

5.3 A Measure of Entanglement for n Qubits

In this subsection we will describe a specific measure of entanglement introduced in [M-W2] that has been used experimentally as test of entanglement. We will give a formula for it in terms of representation theory and show how it can be slightly modified to be an entanglement monotone.

Let $V = \mathbb{C}^2 \otimes \cdots \otimes \mathbb{C}^2$ n-fold product. We look upon $V \otimes V$ as $(\mathbb{C}^2 \otimes \mathbb{C}^2) \otimes \cdots \otimes (\mathbb{C}^2 \otimes \mathbb{C}^2)$. Let

$$S : \mathbb{C}^2 \otimes \mathbb{C}^2 \to S^2(\mathbb{C}^2)$$

and

$$A : \mathbb{C}^2 \otimes \mathbb{C}^2 \to \mathbb{C}^2 \wedge \mathbb{C}^2$$

be the canonical orthogonal projections. If $F \subset \{1, \ldots, n\}$ then we define p_F to be the product

$$R_1 \otimes \cdots \otimes R_n$$

with $R_i = A$ if $i \in F$ and $R_i = S$ otherwise. Then if $v \in V$ we have

$$v \otimes v = \sum_{|F| \text{ even}} p_F(v \otimes v).$$

The p_F are orthogonal projections so we have in particular

$$\|v\|^4 = \sum_{|F| \text{ even}} \|p_F(v \otimes v)\|^2.$$

We set

$$\Upsilon(v) = \|v\|^4 - \|p_\emptyset(v \otimes v)\|^2.$$

The following result is not completely obvious. We will sketch a reduction to the same assertion for another measure of entanglement.

Theorem 12. *A state $v \in V$ is a product state if and only if $\Upsilon(v) = 0$.*

This measure of entanglement is related to one denoted Q in [MW2] (they are the same for 2 and 3 qubits) and which was defined as follows. If $0 \leq j < N = 2^n$ and $j = \sum_{m=0}^{n-1} j_m 2^m$ then if $0 \leq i < n$ define $t_i(j) = \sum_{0 \leq m < i} j_m 2^m + \sum_{i < m < n} j_m 2^{m-1}$. If $v = \sum_{0 \leq j < N} v_j |j\rangle$ then we set $v_{i,0} = \sum_{j_i=0} v_j |t_i(j)\rangle$ and $v_{i,1} = \sum_{j_i=1} v_j |t_i(j)\rangle$. Thus if

$$v = \frac{|111\rangle + |001\rangle + |010\rangle + |100\rangle}{2}$$

then

$$v_{2,0} = \frac{|01\rangle + |10\rangle}{2}, v_{2,1} = \frac{|00\rangle + |11\rangle}{2}.$$

We set $Q(v) = \sum_{i=0}^{n-1} \|v_{i,0} \wedge v_{i,1}\|^2$. Here in $W \otimes W$, $u \wedge w = \frac{u \otimes w - w \otimes u}{2}$ and we use the tensor product inner product. We note that $Q\left(\frac{|000\rangle + |111\rangle}{\sqrt{2}}\right) = \frac{3}{8}$ and $Q\left(\frac{|001\rangle + |010\rangle + |100\rangle}{\sqrt{3}}\right) = \frac{1}{3}$. One can show that if $v \in \bigotimes^n \mathbb{C}^2$ then

$$Q(v) = \sum_{k=1}^{\frac{n}{2}} k \sum_{|F|=2k} \|p_F(v \otimes v)\|^2.$$

We note that in [MW2] we proved the (relatively easy) result that the Theorem above is true for Q replacing Υ. Since $Q(v) = 0$ if and only if $\|p_F(v \otimes v)\|^2 = 0$ for all $|F| > 0$ and Υ has the same property. Hence $\Upsilon(v) = 0$ if and only if $Q(v) = 0$.

In a forthcoming article we will prove that Υ is an *entanglement monotone*. This essentially means that quantum operations (such as measurements and local transformations) cannot increase its value. This condition is sometimes included in the definition of a measure of entanglement.

References

[KR] B. Kostant and S. Rallis, *Orbits and Lie group representations associated to symmetric spaces*, Amer. J. Math., **93** (1971), 753–809.

[MW2] David Meyer and Nolan Wallach, *Global entanglement in multiparticle systems*. Quantum information theory. J. Math. Phys. **43** (2002), no. 9, 4273–4278.

[R] Erik Rains, *Quantum codes of minimum distance two*. IEEE Trans. Inform. Theory **45** (1999), no. 1, 266–271.

[W] Nolan R. Wallach, *The Hilbert series of measures of entanglement for 4 qubits*, Acta Appl. Math. **86** (2005), no.1-2, 203–220.

List of Participants

(1) Ahmed Abouelaz
Hassan II Univ., Casablanca,
Morocco
a.abouelaz@fsac.ac.ma

(2) Andrea Altomani
Univ. della Basilicata, Italy
altomani@sns.it

(3) Federica Andreano
Univ. di Roma La Sapienza,
Italy
andreano@dmmm.uniroma1.it

(4) Laura Atanasi
Univ. di Roma Tor Vergata,
Italy
atanasi@mat.uniroma2.it

(5) Martina Balagovic
Univ. of Zagreb, Croatia
martinab@math.hr

(6) Abdelhamid Boussejra
Ibn Tofail Univ., Morocco
boussejra@lycos.com

(7) Paolo Bravi
Univ. di Roma La Sapienza,
Italy
bravi@mat.uniroma1.it

(8) Jarolim Bures
Charles Univ., Praha,
Czech Republic
jbures@karlin.mff.cuni.cz

(9) Matthieu Carette
Univ. di Padova, Italy
mcarette@math.unipd.it

(10) Giovanna Carnovale
Univ. di Padova, Italy
carnoval@math.unipd.it

(11) Enrico Casadio Tarabusi
(**editor**)
Univ. di Roma La Sapienza,
Italy
casadio@mat.uniroma1.it

(12) Paolo Ciatti
Univ. di Padova, Italy
ciatti@dmsa.unipd.it

(13) Michael Cowling (**lecturer**)
Univ. of New South Wales,
Australia
michaelc@maths.unsw.edu.au

(14) Andrea D'Agnolo (**editor**)
Univ. di Padova, Italy
dagnolo@math.unipd.it

(15) Alessandro D'Andrea
Univ. di Roma La Sapienza,
Italy
dandrea@mat.uniroma1.it

(16) Radouan Daher
Univ. of Hassan 2, Morocco
ra_daher@yahoo.fr

(17) Emilie David-Guillou
Univ. Paris 6, France
davidg@math.jussieu.fr

(18) Irina Denisova
San Petersburg, Russia
ira@wave.ipme.ru

(19) Omar El Fourchi
Faculty of Sciences Ain Chock,

Casablanca, Morocco
elfourchi_omar@yahoo.com

(20) Fouzia El Wassouli
Ibn Tofail Univ., Morocco
f_elwassouli@yahoo.fr

(21) Francesco Esposito
Univ. di Roma La Sapienza,
Italy
esposito@mat.uniroma1.it

(22) Chuying Fang
MIT, USA
cyfang@mit.edu

(23) Veronique Fischer
Univ. Paris-Sud, France
veronique.fischer@math.
u-psud.fr

(24) Peter Franek
Charles Univ., Praha,
Czech Republic
peto.franek@matfyz.cz

(25) Edward Frenkel (**lecturer**)
Univ. of California at Berkeley,
USA
frenkel@math.berkeley.edu

(26) Swiatoslaw Gal
Wrocklaw Univ., Poland
sgal@math.uni.wroc.pl

(27) Laura Geatti
Univ. di Roma Tor Vergata,
Italy
geatti@mat.uniroma2.it

(28) Allal Ghanmi
Univ. of Mohammed V - Agdal,
Rabat, Moroccco
aghanmi@math.net

(29) Anna Gori
Univ. di Firenze, Italy
gori@math.unifi.it

(30) Neven Grbac
Univ. of Zagreb, Croatia
neven.grbac@zpm.fer.hr

(31) Jane Gu
MIT, USA
zerin@mit.edu

(32) Stéphane Guillermou
Institut Fourier, France
Stephane.Guillermou@ujf-
grenoble.fr

(33) Valentina Guizzi
Univ. Roma 3, Italy
guizzi@uniroma3.it

(34) Marcela Hanzer
Univ. of Zagreb, Croatia
hanmar@math.hr

(35) Ahmed Intissar
Univ. of Mohammed V - Agdal,
Rabat, Moroccco
intissar@fsr.ac.ma

(36) Daniel Juteau
Univ. Paris 7 Denis Diderot,
France
juteau@math.jussieu.fr

(37) Galina Kamyshova
Saratov State Agrarian Univ.,
Russia
gkamichova@ssau.saratov.ru

(38) Masaki Kashiwara (**lecturer**)
RIMS, Kyoto Univ., Japan
masaki@kurims.kyoto-u.ac.jp

(39) Andja Kelava
Univ. of Zagreb, Croatia
akelava@math.hr

(40) Oleksandr Khomenko
Freiburg Univ., Germany
Oleksandr.Khomenko@math.
unifreiburg.de

(41) Rémi Lambert
Univ. de Liège, Belgium
R.Lambert@ulg.ac.be

(42) Dominique Luna
Grenoble I, France
d.luna@wanadoo.fr

(43) Andrea Maffei
Univ. di Roma La Sapienza,
Italy
amaffei@mat.uniroma1.it

(44) Corrado Marastoni
Univ. di Padova, Italy
maraston@math.unipd.it

(45) Raffaella Mascolo
Univ. di Padova, Italy
mascolo@math.unipd.it

(46) Luca Migliorini
Univ. di Bologna, Italy
migliori@dm.unibo.it

(47) Giovanni Morando
Univ. di Padova, Italy
gmorando@math.unipd.it

(48) Anne Moreau
Univ. Paris 7, France
moreau@math.jussieu.fr

(49) Christoph Mueller
Darmstadt Univ. of Technology,
Germany
cmueller@mathematik.
tu-darmstadt.de

(50) Kyo Nishiyama
Kyoto Univ., Japan
kyo@math.kyoto-u.ac.jp

(51) Hiroyuki Ochiai
Nagoya Univ., Japan
ochiai@math.nagoya-u.ac.jp

(52) Alessandro Ottazzi
Univ. di Genova, Italy
ottazzi@calvino.polito.it

(53) Pavle Pandzic
Univ. of Zagreb, Croatia
pandzic@math.hr

(54) Alessandra Pantano
Princeton Univ., USA
ale@math.mit.edu

(55) Nikolaos Papalexiou
Univ. of the Aegean, Greece
papalexi@aegean.gr

(56) Michael Pevzner
Reims Univ., France
pevzner@univ-reims.fr

(57) Guido Pezzini
Italy
pezziniguido@yahoo.it

(58) Massimo Picardello (**editor**)
Univ. di Roma Tor Vergata,
Italy
picard@mat.uniroma2.it

(59) Anke Pohl
Univ. of Paderborn, Germany
pohl@math.tu-clausthal.de

(60) Pietro Polesello
Univ. di Padova, Italy
pietro@math.unipd.it

(61) Luca Prelli
Univ. Padova Italy/ Univ. Paris
6, France
lprelli@math.unipd.it

(62) Nicolas Prudhon
Univ. Neuchatel, Switzerland
nicolas.prudhon@unine.ch

(63) David Renard
Ecole Polytechnique, France
renard@math.polytechnique.fr

(64) Elena Rubei
Univ. di Firenze, Italy
rubei@math.unifi.it

(65) Alessandro Ruzzi
Univ. di Roma La Sapienza,
Italy
ruzzi@mat.uniroma1.it

(66) Pierre Schapira
Univ. Paris 6, France
schapira@math.jussieu.fr

(67) Jean-Pierre Schneiders
Univ. de Liège, Belgium
jpschneiders@ulg.ac.be

(68) Dalibor Šmíd
Charles Univ., Praha,
Czech Republic
smid@karlin.mff.cuni.cz

(69) Petr Somberg
Charles Univ., Praha,
Czech Republic
somberg@karlin.mff.cuni.cz

(70) Vladimir Soucek
Charles Univ., Praha,
Czech Republic
soucek@karlin.mff.cuni.cz

(71) Giovanni Stegel
Univ. di Sassari, Italy
stegel@uniss.it

(72) Alexis Tchoudjem
Univ. de Lyon I, France
tchoudjem@igd.univ-lyon1.fr

(73) Alain Valette (**lecturer**)
Univ. Neuchatel, Switzerland
alain.valette@unine.ch

(74) David Vogan (**lecturer**)
MIT, USA
dav@math.mit.edu

(75) Nolan Wallach (**lecturer**)
Univ. of California,
San Diego, USA
nwallach@ucsd.edu

(76) Ingo Waschkies
Univ. de Nice, France
ingo@math.unice.fr

(77) Satoru Watanabe
Univ. Paris 6, France
watanabe@math.jussieu.fr

(78) Christoph Wockel
Darmstadt Univ. of Technology,
Germany
*wockel@mathematik.
tudarmstadt.de*

(79) Wai Ling Yee
MIT, USA
wlyee@math.mit.edu

LIST OF C.I.M.E. SEMINARS

Published by C.I.M.E

Published by Ed. Cremonese, Firenze

Published by Ed. Liguori, Napoli

Published by Ed. Liguori, Napoli & Birkhäuser

Published by Springer-Verlag

Lecture Notes in Mathematics

For information about earlier volumes
please contact your bookseller or Springer
LNM Online archive: springerlink.com

Vol. 1789: Y. Sommerhäuser, Yetter-Drinfel'd Hopf algebras over groups of prime order (2002)

Vol. 1790: X. Zhan, Matrix Inequalities (2002)

Vol. 1791: M. Knebusch, D. Zhang, Manis Valuations and Prüfer Extensions I: A new Chapter in Commutative Algebra (2002)

Vol. 1792: D. D. Ang, R. Gorenflo, V. K. Le, D. D. Trong, Moment Theory and Some Inverse Problems in Potential Theory and Heat Conduction (2002)

Vol. 1793: J. Cortés Monforte, Geometric, Control and Numerical Aspects of Nonholonomic Systems (2002)

Vol. 1794: N. Pytheas Fogg, Substitution in Dynamics, Arithmetics and Combinatorics. Editors: V. Berthé, S. Ferenczi, C. Mauduit, A. Siegel (2002)

Vol. 1795: H. Li, Filtered-Graded Transfer in Using Noncommutative Gröbner Bases (2002)

Vol. 1796: J.M. Melenk, hp-Finite Element Methods for Singular Perturbations (2002)

Vol. 1797: B. Schmidt, Characters and Cyclotomic Fields in Finite Geometry (2002)

Vol. 1798: W.M. Oliva, Geometric Mechanics (2002)

Vol. 1799: H. Pajot, Analytic Capacity, Rectifiability, Menger Curvature and the Cauchy Integral (2002)

Vol. 1800: O. Gabber, L. Ramero, Almost Ring Theory (2003)

Vol. 1801: J. Azéma, M. Émery, M. Ledoux, M. Yor (Eds.), Séminaire de Probabilités XXXVI (2003)

Vol. 1802: V. Capasso, E. Merzbach, B. G. Ivanoff, M. Dozzi, R. Dalang, T. Mountford, Topics in Spatial Stochastic Processes. Martina Franca, Italy 2001. Editor: E. Merzbach (2003)

Vol. 1803: G. Dolzmann, Variational Methods for Crystalline Microstructure – Analysis and Computation (2003)

Vol. 1804: I. Cherednik, Ya. Markov, R. Howe, G. Lusztig, Iwahori-Hecke Algebras and their Representation Theory. Martina Franca, Italy 1999. Editors: V. Baldoni, D. Barbasch (2003)

Vol. 1805: F. Cao, Geometric Curve Evolution and Image Processing (2003)

Vol. 1806: H. Broer, I. Hoveijn. G. Lunther, G. Vegter, Bifurcations in Hamiltonian Systems. Computing Singularities by Gröbner Bases (2003)

Vol. 1807: V. D. Milman, G. Schechtman (Eds.), Geometric Aspects of Functional Analysis. Israel Seminar 2000-2002 (2003)

Vol. 1808: W. Schindler, Measures with Symmetry Properties (2003)

Vol. 1809: O. Steinbach, Stability Estimates for Hybrid Coupled Domain Decomposition Methods (2003)

Vol. 1810: J. Wengenroth, Derived Functors in Functional Analysis (2003)

Vol. 1811: J. Stevens, Deformations of Singularities (2003)

Vol. 1812: L. Ambrosio, K. Deckelnick, G. Dziuk, M. Mimura, V. A. Solonnikov, H. M. Soner, Mathematical Aspects of Evolving Interfaces. Madeira, Funchal, Portugal 2000. Editors: P. Colli, J. F. Rodrigues (2003)

Vol. 1813: L. Ambrosio, L. A. Caffarelli, Y. Brenier, G. Buttazzo, C. Villani, Optimal Transportation and its Applications. Martina Franca, Italy 2001. Editors: L. A. Caffarelli, S. Salsa (2003)

Vol. 1814: P. Bank, F. Baudoin, H. Föllmer, L.C.G. Rogers, M. Soner, N. Touzi, Paris-Princeton Lectures on Mathematical Finance 2002 (2003)

Vol. 1815: A. M. Vershik (Ed.), Asymptotic Combinatorics with Applications to Mathematical Physics. St. Petersburg, Russia 2001 (2003)

Vol. 1816: S. Albeverio, W. Schachermayer, M. Talagrand, Lectures on Probability Theory and Statistics. Ecole d'Eté de Probabilités de Saint-Flour XXX-2000. Editor: P. Bernard (2003)

Vol. 1817: E. Koelink, W. Van Assche (Eds.), Orthogonal Polynomials and Special Functions. Leuven 2002 (2003)

Vol. 1818: M. Bildhauer, Convex Variational Problems with Linear, nearly Linear and/or Anisotropic Growth Conditions (2003)

Vol. 1819: D. Masser, Yu. V. Nesterenko, H. P. Schlickewei, W. M. Schmidt, M. Waldschmidt, Diophantine Approximation. Cetraro, Italy 2000. Editors: F. Amoroso, U. Zannier (2003)

Vol. 1820: F. Hiai, H. Kosaki, Means of Hilbert Space Operators (2003)

Vol. 1821: S. Teufel, Adiabatic Perturbation Theory in Quantum Dynamics (2003)

Vol. 1822: S.-N. Chow, R. Conti, R. Johnson, J. Mallet-Paret, R. Nussbaum, Dynamical Systems. Cetraro, Italy 2000. Editors: J. W. Macki, P. Zecca (2003)

Vol. 1823: A. M. Anile, W. Allegretto, C. Ringhofer, Mathematical Problems in Semiconductor Physics. Cetraro, Italy 1998. Editor: A. M. Anile (2003)

Vol. 1824: J. A. Navarro González, J. B. Sancho de Salas, $\mathscr{C}^{\infty}$ – Differentiable Spaces (2003)

Vol. 1825: J. H. Bramble, A. Cohen, W. Dahmen, Multiscale Problems and Methods in Numerical Simulations, Martina Franca, Italy 2001. Editor: C. Canuto (2003)

Vol. 1826: K. Dohmen, Improved Bonferroni Inequalities via Abstract Tubes. Inequalities and Identities of Inclusion-Exclusion Type. VIII, 113 p, 2003.

Vol. 1827: K. M. Pilgrim, Combinations of Complex Dynamical Systems. IX, 118 p, 2003.

Vol. 1828: D. J. Green, Gröbner Bases and the Computation of Group Cohomology. XII, 138 p, 2003.

Vol. 1829: E. Altman, B. Gaujal, A. Hordijk, Discrete-Event Control of Stochastic Networks: Multimodularity and Regularity. XIV, 313 p, 2003.

Vol. 1830: M. I. Gil', Operator Functions and Localization of Spectra. XIV, 256 p, 2003.

Vol. 1831: A. Connes, J. Cuntz, E. Guentner, N. Higson, J. E. Kaminker, Noncommutative Geometry, Martina Franca, Italy 2002. Editors: S. Doplicher, L. Longo (2004)

Vol. 1832: J. Azéma, M. Émery, M. Ledoux, M. Yor (Eds.), Séminaire de Probabilités XXXVII (2003)

Vol. 1833: D.-Q. Jiang, M. Qian, M.-P. Qian, Mathematical Theory of Nonequilibrium Steady States. On the Frontier of Probability and Dynamical Systems. IX, 280 p, 2004.

Vol. 1834: Yo. Yomdin, G. Comte, Tame Geometry with Application in Smooth Analysis. VIII, 186 p, 2004.

Vol. 1835: O.T. Izhboldin, B. Kahn, N.A. Karpenko, A. Vishik, Geometric Methods in the Algebraic Theory of Quadratic Forms. Summer School, Lens, 2000. Editor: J.-P. Tignol (2004)

Vol. 1836: C. Năstăsescu, F. Van Oystaeyen, Methods of Graded Rings. XIII, 304 p, 2004.

Vol. 1837: S. Tavaré, O. Zeitouni, Lectures on Probability Theory and Statistics. Ecole d'Eté de Probabilités de Saint-Flour XXXI-2001. Editor: J. Picard (2004)

Vol. 1838: A.J. Ganesh, N.W. O'Connell, D.J. Wischik, Big Queues. XII, 254 p, 2004.

Vol. 1839: R. Gohm, Noncommutative Stationary Processes. VIII, 170 p, 2004.

Vol. 1840: B. Tsirelson, W. Werner, Lectures on Probability Theory and Statistics. Ecole d'Eté de Probabilités de Saint-Flour XXXII-2002. Editor: J. Picard (2004)

Recent Reprints and New Editions